Supported by the National Fund for Academic Publication in Science and Technology

Mathematics Monograph Series 38

Index Theory in Nonlinear Analysis

(非线性分析的指标理论)

Chungen Liu (刘春根)

Responsible Editor: Yuzhuo Chen

Chungen Liu
School of Mathematical and Information
Science
Guangzhou University
Guangzhou, Guangdong, China

ISBN 978-7-03-059566-9/O-7528
Science Press, Beijing, China

Jointly published with Springer Nature Singapore Pte Ltd.

The print edition is only for sale in Mainland China. Customers from Mainland China please order the print book from: Science Press.

Published by Science Press
16 Donghuangchenggen North Street
Beijing 100717,P.R.China
Printed in Beijing

Preface to Mathematics Monograph Series

Science Press asked me to write a preface for their series of books called "Mathematics Monograph Series". They told me that the Press had published nearly 30 mathematical monographs in this series since 2006. This reminded me that, also in 2006, I received an email message from the Editor in Chief of "Sugaku Tushin" ("Mathematical Communications", the membership magazine of the Mathematical Society of Japan). The Editor in Chief told me that they were planning to have a special section on "Recent development in Chinese mathematical community", and invited me to write an article for the special section. As a result, I published in their magazine an article called "Some Aspects of Mathematical Community in China". Among other things, in the article I demonstrated that, with the favorable environment, the Chinese mathematical community had made great progress since the late 1970s (when China started to implement "Reform and Opening-up Policy"): A large number of publications (including articles and monographs) have been written by Chinese mathematicians, there have been always Chinese mathematicians presenting their speeches at various international academic conferences or workshops, many Chinese mathematicians have served as editors of international academic journals, or as members in various academic organizations. All these reveal that the Chinese mathematical community which has been growing rapidly has exerted more and more influence in the world.

Indeed, the series of mathematical monographs published by Science Press reflects partly more and more influence of Chinese mathematical community in the world. Chinese people are good at mathematics. In the past, Science Press published many high level mathematical monographs and textbooks. Among them some were written in Chinese and some were written in English. Some monographs which appeared originally in Chinese have been purchased by international publishers who then re-published them abroad in English, and this has gained influence in corresponding areas of the international mathematical community.

In recent years most Chinese mathematicians have mastered good English. In accordance with this situation, Science Press has decided to publish "Mathematics Monograph Series" — a series in which high level mathematical monographs and textbooks are written directly in English. The goal of this series is to provide further good service for Chinese mathematicians and to enhance further the influence of the mathematics study in China in the international mathematical community.

I would like to conclude this short preface with the following wish which I expressed also at the end of the afore mentioned article "Some Aspects of Mathematical Community in China":

The Chinese mathematical community will continuously make its effort to work hard, and to strengthen its international exchanges and collaborations, so as to make more contributions to the study and development of mathematics in the world.

Zhiming Ma
March 15, 2015

Foreword

This book contains three aspects of index theories with applications. From them, we choose one aspect to demonstrate how to use the L-index theory to study the multiplicity of brake orbits on a symmetric convex hypersurface. We begin from a famous Seifert conjecture about the number of brake orbits on a compact hypersurface in $\mathbb{R}^{2n}$.

1 Seifert Conjecture

Let us recall the famous conjecture proposed by H. Seifert in his pioneer work [268] in 1948 concerning the multiplicity of brake orbits of certain Hamiltonian systems in $\mathbb{R}^{2n}$.

We assume $H \in C^2(\mathbb{R}^{2n}, \mathbb{R})$ possesses the following form:

$$H(p,q) = \frac{1}{2}A(q)p \cdot p + V(q), \tag{1}$$

where $p, q \in \mathbb{R}^n$, $A(q)$ is a positive definite $n \times n$ symmetric matrix for any $q \in \mathbb{R}^n$, A is C^2, $V \in C^2(\mathbb{R}^n, \mathbb{R})$ is the potential energy. The solution of the following problem for Hamiltonian system

$$\dot{x} = JH'(x), \quad x = (p,q), \tag{2}$$

$$p(0) = p(\frac{\tau}{2}) = 0. \tag{3}$$

is called a brake orbit. Moreover, if h is the total energy of a brake orbit (q, p), i.e., $H(p(t), q(t)) = h$, then $V(q(0)) = V(q(\tau)) = h$ and $q(t) \in \bar{\Omega} \equiv \{q \in \mathbb{R}^n | V(q) \leq h\}$ for all $t \in \mathbb{R}$, where $J = \begin{pmatrix} 0 & -I \\ I & 0 \end{pmatrix}$ is the standard symplectic matrix and I is the $n \times n$ identity matrix.

In [268] of 1948, H. Seifert studied the existence of brake orbit for system (2)–(3) with the Hamiltonian function H in the form of (1) and proved that the set $\mathcal{J}_b(\Sigma)$ of brake orbits on the energy surface $\Sigma = H^{-1}(h)$ is not empty, i.e., $\mathcal{J}_b(\Sigma) \neq \emptyset$, provided $V' \neq 0$ on $\partial\Omega$, V is analytic, and $\bar{\Omega}$ is bounded and homeomorphic to the unit ball $B_1^n(0)$ in $\mathbb{R}^n$. The set of geometrically distinct brake orbits on the energy surface Σ is denoted by $\tilde{\mathcal{J}}_b(\Sigma)$. Then, in the same paper, he proposed the following conjecture which is still open for $n \geq 2$ now:

$$^{\#}\tilde{\mathcal{J}}_b(\Sigma) \geq n \textit{ under the same conditions.}$$

We note that for the Hamiltonian function,

$$H(p, q) = \frac{1}{2}|p|^2 + \sum_{j=1}^{n} a_j^2 q_j^2, \qquad q, p \in \mathbb{R}^n,$$

where $a_i/a_j \notin \mathbb{Q}$ for all $i \neq j$ and $q = (q_1, q_2, \ldots, q_n)$. There are exactly n geometrically distinct brake orbits on the energy hypersurface $\Sigma = H^{-1}(h)$.

2 The Generalized Seifert Conjecture in Linear Symplectic Space

For the standard symplectic space $(\mathbb{R}^{2n}, \omega_0)$ with $\omega_0(x, y) = \langle Jx, y\rangle$, an involution matrix defined by $N = \begin{pmatrix} -I & 0 \\ 0 & I \end{pmatrix}$ is clearly anti-symplectic, i.e., $NJ = -JN$. The fixed point set of N and $-N$ are the Lagrangian subspaces $L_0 = \{0\} \times \mathbb{R}^n$ and $L_1 = \mathbb{R}^n \times \{0\}$ of $(\mathbb{R}^{2n}, \omega_0)$, respectively.

In general, we suppose $H \in C^2(\mathbb{R}^{2n}\backslash\{0\}, \mathbb{R}) \cap C^1(\mathbb{R}^{2n}, \mathbb{R})$ satisfies the following reversible condition:

$$H(Nx) = H(x), \qquad \forall x \in \mathbb{R}^{2n}. \tag{4}$$

We consider the following fixed energy problem of nonlinear Hamiltonian system with Lagrangian boundary conditions

$$\dot{x}(t) = JH'(x(t)), \tag{5}$$

$$H(x(t)) = h, \tag{6}$$

$$x(0) \in L_0,\ x(\tau/2) \in L_0. \tag{7}$$

Here, we require that h is a regular value of the function H. It is clear that a solution (τ, x) of (5)–(7) is a characteristic chord on the contact submanifold $\Sigma := H^{-1}(h) = \{y \in \mathbb{R}^{2n} \mid H(y) = h\}$ of $(\mathbb{R}^{2n}, \omega_0)$ and satisfies

$$x(-t) = Nx(t), \tag{8}$$

$$x(\tau + t) = x(t). \tag{9}$$

In general, this kind of τ-periodic characteristic (τ, x) is called a *brake orbit* on the hypersurface Σ. We note that the problem (2)–(3) with the Hamiltonian function H defined in (1) is a special case of the problem (5)–(7). We denote by $\mathcal{J}_b(\Sigma, H)$ the set of all brake orbits on Σ. Two brake orbits $(\tau_i, x_i) \in \mathcal{J}_b(\Sigma, H),\ i = 1, 2$ are equivalent if the two brake orbits are geometrically the same, i.e., $x_1(\mathbb{R}) = x_2(\mathbb{R})$. We denote by $[(\tau, x)]$ the equivalence class of $(\tau, x) \in \mathcal{J}_b(\Sigma, H)$ in this equivalence relation and by $\tilde{\mathcal{J}}_b(\Sigma,\ H)$ the set of $[(\tau, x)]$ for all $(\tau, x) \in \mathcal{J}_b(\Sigma, H)$. In fact, $\tilde{\mathcal{J}}_b(\Sigma,\ H)$ is the set of geometrically distinct brake orbits on Σ, which is independent on the choice of H. So from now on, we simply denote it by $\tilde{\mathcal{J}}_b(\Sigma)$, and in the notation $[(\tau, x)]$, we always assume x has minimal period τ. We also denote by $\tilde{\mathcal{J}}(\Sigma)$ the set of all geometrically distinct closed characteristics on Σ. The number of elements in a set S is denoted by $^{\#}S$. It is well-known that $^{\#}\tilde{\mathcal{J}}_b(\Sigma)$ (and also $^{\#}\tilde{\mathcal{J}}(\Sigma)$) is only depending on Σ, that is to say, for simplicity, we take $h = 1$; if H and G are two C^2 functions satisfying (4) and $\Sigma_H := H^{-1}(1) = \Sigma_G := G^{-1}(1)$, then $^{\#}\mathcal{J}_b(\Sigma_H) =^{\#} \mathcal{J}_b(\Sigma_G)$. So we can consider the brake orbit problem in a more general setting. Let Σ be a C^2 compact hypersurface in $\mathbb{R}^{2n}$ bounding a compact set C with nonempty interior. Suppose Σ has nonvanishing Gaussian curvature and satisfies the reversible condition $N(\Sigma - x_0) = \Sigma - x_0 := \{x - x_0 | x \in \Sigma\}$ for some $x_0 \in C$. Without loss of generality, we may assume $x_0 = 0$. We denote the set of all such hypersurfaces in $\mathbb{R}^{2n}$ by $\mathcal{H}_b(2n)$. For $x \in \Sigma$, let $n_\Sigma(x)$ be the unit outward normal vector at $x \in \Sigma$. Note that, here, by the reversible condition, there holds $n_\Sigma(Nx) = Nn_\Sigma(x)$. We consider the dynamics problem of finding $\tau > 0$ and a C^1 smooth curve $x : [0, \tau] \to \mathbb{R}^{2n}$ such that

$$\dot{x}(t) = Jn_\Sigma(x(t)), \qquad x(t) \in \Sigma, \tag{10}$$

$$x(-t) = Nx(t), \qquad x(\tau + t) = x(t), \qquad \text{for all } t \in \mathbb{R}. \tag{11}$$

A solution (τ, x) of the problem (10)–(11) determines a brake orbit on Σ. Now, the generalized Seifert conjecture can be represented as
Generalized Seifert conjecture: For any $\Sigma \in \mathcal{H}_b(2n)$, there holds

$$^{\#}\tilde{\mathcal{J}}_b(\Sigma) \geq n.$$

We can take the above estimate as a result on the number of Lagrangian intersection (more precisely the Legendre intersection since $\Sigma \cap L_0$ is a Legendre submanifold of the contact manifold Σ) of "reversible" Hamiltonian map φ_b:

$$^{\#}\{\Sigma \cap L_0 \cap \varphi_b(L_0)\} \geq n.$$

The famous Arnold conjecture has relation with the Lagrangian boundary problem, which says that the number of Lagrangian intersection of a Hamiltonian map on a closed symplectic manifold M can be estimated from below by the Betti number of M in the nondegenerate case and by the cuplength of M (cf. [8]). In this direction, one can refer to [100, 137], and [191].

We denote by

$$\mathcal{H}_b^c(2n) = \{\Sigma \in \mathcal{H}_b(2n) |\ \Sigma \text{ is strictly convex } \},$$
$$\mathcal{H}_b^{s,c}(2n) = \{\Sigma \in \mathcal{H}_b^c(2n) |\ -\Sigma = \Sigma\}.$$

For the multiplicity of the brake orbits on a symmetrical compact convex hypersurface, we have the following result:

Theorem A *For any $\Sigma \in \mathcal{H}_b^{s,c}(2n)$, there holds*

$$^{\#}\tilde{\mathcal{J}}_b(\Sigma) \geq n.$$

3 L-Index Theory

Historically, as far as the author knows, the classification and an index theory for linear Hamiltonian systems with periodic boundary condition began with the work of H. Amann and E. Zehnder in their paper [5] of 1980. They established the corresponding index theory for linear Hamiltonian systems with constant coefficients. In their celebrated paper [57] of 1984, C. Conley and E. Zehnder defined an index theory for any nondegenerate path in symplectic group $\mathrm{Sp}(2n) = \{M \in \mathrm{GL}(\mathbb{R}^{2n}) |\ M^T J M = J\}$ with $n \geq 2$, where $J = \begin{pmatrix} 0 & -I_n \\ I_n & 0 \end{pmatrix}$. This index was further defined for the nondegenerate paths in Sp(2) by Y. Long and E. Zehnder in [231] of 1990. The index theory for degenerate linear Hamiltonian systems was defined by Y. Long in [224] and C. Viterbo in [280] of 1990 independently. Then, in the paper[225] of 1997, this index theory was further extended to any symplectic paths, and an axiomatic characterization of the index theory was given by Y. Long. We now call this index the Maslov-type index. For a symplectic path γ starting from the identify matrix, we denote its Maslov-type index by

$$(i_1(\gamma), \nu_1(\gamma)) \in \mathbb{Z} \times \{0, 1, \cdots, 2n\}.$$

In [226] of 1999, Y. Long established a generalized index function theory $(i_\omega(\gamma), \nu_\omega(\gamma))$ parameterized by ω in the unit circle $\mathbf{U}$ of the complex plane $\mathbb{C}$ for every symplectic path γ starting from the identity matrix. The Maslov-type index theory for symplectic paths is essentially a classification of the periodic linear Hamiltonian systems. So it is natural to use this index theory to study the periodic solutions of a Hamiltonian system.

We denote the set of symplectic paths defined in the interval $[0, \tau]$ by

$$\mathcal{P}_\tau(2n) = \{\gamma \in C([0, \tau], \mathrm{Sp}(2n)) | \gamma(0) = I_{2n}\}.$$

In [232] of 2006, Y. Long, D. Zhang, and C. Zhu studied the multiple solutions of the brake orbit problem on a convex hypersurface; in this paper, they introduced two indices $(\mu_1(\gamma), \nu_1(\gamma))$ and $(\mu_2(\gamma), \nu_2(\gamma))$ for symplectic path $\gamma \in \mathcal{P}_\tau(2n)$. After that, in [189] of 2007, C. Liu introduced an index theory associated with a Lagrangian subspace for symplectic paths. For a symplectic path $\gamma \in \mathcal{P}_\tau(2n)$, and a Lagrangian subspace L, by definition, the L-index is assigned to a pair of integers $(i_L(\gamma), \nu_L(\gamma)) \in \mathbb{Z} \times \{0, 1, \cdots, n\}$. This index theory is suitable for studying the Lagrangian boundary value problems (L-solution, for short) related to nonlinear Hamiltonian systems. In [192], C. Liu applied this index theory to study the L-solutions of some asymptotically linear Hamiltonian systems. The indices $\mu_1(\gamma)$ and $\mu_2(\gamma)$ are essentially special cases of the L-index $i_L(\gamma)$ for Lagrangian subspaces $L_0 = \{0\} \times \mathbb{R}^n$ and $L_1 = \mathbb{R}^n \times \{0\}$, respectively, up to a constant n.

The key ingredients in the study of the multiplicity of brake orbits on a symmetric convex hypersurface are the index iteration formulas (Bott-type formulas) and the (L_0, L_1)-estimate.

For simplicity, we suppose $\gamma \in \mathcal{P}(2n)$, i.e., we take $\tau = 1$. For $j \in \mathbb{N}$, we define the j-times iteration path $\gamma^j : [0, j] \to \mathrm{Sp}(2n)$ of γ by

$$\gamma^1(t) = \gamma(t), \ t \in [0, 1],$$

$$\gamma^2(t) = \begin{cases} \gamma(t), \ t \in [0, 1], \\ N\gamma(2-t)\gamma(1)^{-1}N\gamma(1), \ t \in [1, 2], \end{cases}$$

and in general, for $k \in \mathbb{N}$, we define $\gamma(2) = N\gamma(1)^{-1}N\gamma(1)$ and

$$\gamma^{2k-1}(t) = \begin{cases} \gamma(t), \ t \in [0, 1], \\ N\gamma(2-t)\gamma(1)^{-1}N\gamma(1), \ t \in [1, 2], \\ \cdots\cdots \\ N\gamma(2k-2-t)N\gamma(2)^{k-1}, \ t \in [2k-3, 2k-2], \\ \gamma(t-2k+2)\gamma(2)^{k-1}, \ t \in [2k-2, 2k-1], \end{cases} \tag{12}$$

$$\gamma^{2k}(t) = \begin{cases} \gamma(t), \ t \in [0, 1], \\ N\gamma(2-t)\gamma(1)^{-1}N\gamma(1), \ t \in [1, 2], \\ \cdots\cdots \\ \gamma(t-2k+2)\gamma(2)^{k-1}, \ t \in [2k-2, 2k-1], \\ N\gamma(2k-t)N\gamma(2)^k, \ t \in [2k-1, 2k]. \end{cases} \tag{13}$$

For $\gamma \in \mathcal{P}_\tau(2n)$, we define

$$\gamma^k(\tau t) = \tilde{\gamma}^k(t) \text{ with } \tilde{\gamma}(t) = \gamma(\tau t). \tag{14}$$

Proposition B (mixed (L_0, L_1)-concavity estimate) *For $\gamma \in \mathcal{P}_\tau(2n)$, let $P = \gamma(\tau)$. If $i_{L_0}(\gamma) \geq 0$, $i_{L_1}(\gamma) \geq 0$, $i_1(\gamma) \geq n$, $\gamma^2(t) = \gamma(t-\tau)\gamma(\tau)$ for all $t \in [\tau, 2\tau]$, then*

$$i_{L_1}(\gamma) + S^+_{P^2}(1) - \nu_{L_0}(\gamma) \geq 0. \tag{15}$$

where $S^+_M(1)$ is the splitting number of the symplectic matrix M (see [226] and [223]).

Proposition C (Bott-type formulas) *Suppose $\gamma \in \mathcal{P}_\tau(2n)$, for the iteration symplectic paths γ^k defined in* (12)–(14) *above, when k is odd, there hold*

$$\begin{aligned} i_{L_0}(\gamma^k) &= i_{L_0}(\gamma^1) + \textstyle\sum_{i=1}^{\frac{k-1}{2}} i_{\omega_k^{2i}}(\gamma^2), \\ \nu_{L_0}(\gamma^k) &= \nu_{L_0}(\gamma^1) + \textstyle\sum_{i=1}^{\frac{k-1}{2}} \nu_{\omega_k^{2i}}(\gamma^2), \end{aligned} \tag{16}$$

when k is even, there hold

$$\begin{aligned} i_{L_0}(\gamma^k) &= i_{L_0}(\gamma^1) + i^{L_0}_{\sqrt{-1}}(\gamma^1) + \textstyle\sum_{i=1}^{\frac{k}{2}-1} i_{\omega_k^{2i}}(\gamma^2), \\ \nu_{L_0}(\gamma^k) &= \nu_{L_0}(\gamma^1) + \nu^{L_0}_{\sqrt{-1}}(\gamma^1) + \textstyle\sum_{i=1}^{\frac{k}{2}-1} \nu_{\omega_k^{2i}}(\gamma^2), \end{aligned} \tag{17}$$

where $\omega_k = e^{\pi\sqrt{-1}/k}$ and $(i_\omega(\gamma), \nu_\omega(\gamma))$ is the ω index pair of the symplectic path γ introduced in [226] *and the index pair $(i^{L_0}_{\sqrt{-1}}(\gamma^1), \nu^{L_0}_{\sqrt{-1}}(\gamma^1))$ was defined in* [209].

For an L_0-solution (x, τ) of a nonlinear Hamiltonian system

$$\dot{x}(t) = JH'(x(t)), \quad x(0) \in L_0, \quad x(\tau) \in L_0,$$

the linearized system at x of the above nonlinear Hamiltonian system is a linear Hamiltonian system

$$\dot{z}(t) = JB(t)z(t), \quad B(t) = H''(x(t)).$$

Its fundamental solution $\gamma_x : [0, \tau] \to \mathrm{Sp}(2n)$ is a symplectic path starting from the identity matrix, so we have the L_0-index $(i_{L_0}(x), \nu_{L_0}(x))$ defined by

$$(i_{L_0}(x), \nu_{L_0}(x)) = (i_{L_0}(\gamma_x), \nu_{L_0}(\gamma_x)).$$

The k-times iteration ($k \in \mathbb{N}$) of x denoted by $(x^k, k\tau)$ is also an L_0-solution of the above nonlinear Hamiltonian system; we write $(i_{L_0}(x^k), \nu_{L_0}(x^k))$ as $(i_{L_0}(x,k), \nu_{L_0}(x,k))$. The formulas (16)–(17) help us to understand the indices of the iteration sequence x^k. We will, in this book, establish some important properties of $\{(i_{L_0}(x,k), \nu_{L_0}(x,k))\}_{k\in\mathbb{N}}$ and get a proof of Theorem A.

4 Index Information from Variational Method

For $\Sigma \in \mathcal{H}_b^{s,c}(2n)$, let $j_\Sigma : \Sigma \to [0, +\infty)$ be the gauge function of Σ defined by

$$j_\Sigma(0) = 0, \quad \text{and} \quad j_\Sigma(x) = \inf\{\lambda > 0 \mid \frac{x}{\lambda} \in C\}, \quad \forall x \in \mathbb{R}^{2n} \setminus \{0\},$$

where C is the domain enclosed by Σ.

Define

$$H_\Sigma(x) == (j_\Sigma(x))^2, \ \forall x \in \mathbb{R}^{2n}. \tag{18}$$

Then, $H_\Sigma \in C^2(\mathbb{R}^{2n}\setminus\{0\}, \mathbb{R}) \cap C^{1,1}(\mathbb{R}^{2n}, \mathbb{R})$.

We consider the following fixed energy problem:

$$\dot{x}(t) = JH'_\Sigma(x(t)), \tag{19}$$

$$H_\Sigma(x(t)) = 1, \tag{20}$$

$$x(-t) = Nx(t), \tag{21}$$

$$x(\tau + t) = x(t), \quad \forall t \in \mathbb{R}. \tag{22}$$

Denote by $\mathcal{J}_b(\Sigma)$ the set of all solutions (τ, x) of problem (19)–(22) and by $\tilde{\mathcal{J}}_b(\Sigma)$ the set of all geometrically distinct solutions of (19)–(22).

For $S^1 = \mathbb{R}/\mathbb{Z}$, as in [232], we define the Hilbert space E by

$$E = \left\{ x \in W^{1,2}(S^1, \mathbb{R}^{2n}) \,\middle|\, x(-t) = Nx(t), \quad \text{for all } t \in \mathbb{R} \text{ and } \int_0^1 x(t)dt = 0 \right\}.$$

The inner product on E is given by

$$(x, y) = \int_0^1 \langle \dot{x}(t), \dot{y}(t) \rangle dt. \tag{23}$$

The $C^{1,1}$ Hilbert manifold $M_\Sigma \subset E$ associated with Σ is defined by

$$M_\Sigma = \left\{ x \in E \,\middle|\, \int_0^1 H_\Sigma^*(-J\dot{x}(t))dt = 1 \text{ and } \int_0^1 \langle J\dot{x}(t), x(t)\rangle dt < 0 \right\}, \tag{24}$$

where H_Σ^* is the Fenchel conjugate function of the function H_Σ defined by

$$H_\Sigma^*(y) = \max\{(x \cdot y - H_\Sigma(x)) \,|\, x \in \mathbb{R}^{2n}\}. \tag{25}$$

Let $\mathbb{Z}_2 = \{-id, id\}$ be the usual $\mathbb{Z}_2$ group. We define the $\mathbb{Z}_2$-action on E by

$$-id(x) = -x, \quad id(x) = x, \qquad \forall x \in E.$$

Since H_Σ^* is even, M_Σ is symmetric to 0, i.e., $\mathbb{Z}_2$ invariant. M_Σ is a paracompact $\mathbb{Z}_2$-space. We define

$$\Phi(x) = \frac{1}{2}\int_0^1 \langle J\dot{x}(t), x(t)\rangle dt, \tag{26}$$

then Φ is a $\mathbb{Z}_2$ invariant function and $\Phi \in C^\infty(E, \mathbb{R})$. We denote by Φ_Σ the restriction of Φ to M_Σ.

Proposition D *If* $^{\#}\tilde{\mathcal{J}}_b(\Sigma) < +\infty$, *there is a sequence* $\{c_k\}_{k\in\mathbb{N}}$, *such that*

$$-\infty < c_1 < c_2 < \cdots < c_k < c_{k+1} < \cdots < 0, \tag{27}$$

$$c_k \to 0 \quad \text{as } k \to +\infty. \tag{28}$$

For any $k \in \mathbb{N}$, *there exists a brake orbit* $(\tau, x) \in \mathcal{J}_b(\Sigma)$ *with* τ *being the minimal period of* x *and* $m \in \mathbb{N}$ *satisfying* $m\tau = (-c_k)^{-1}$ *such that for*

$$z(x)(t) = (m\tau)^{-1}x(m\tau t) - \frac{1}{(m\tau)^2}\int_0^{m\tau} x(s)ds, \quad t \in S^1, \tag{29}$$

$z(x) \in M_\Sigma$ *is a critical point of* Φ_Σ *with* $\Phi_\Sigma(z(x)) = c_k$ *and*

$$i_{L_0}(x, m) \le k - 1 \le i_{L_0}(x, m) + \nu_{L_0}(x, m) - 1. \tag{30}$$

The above proposition says that for each critical value c_k, $k \in \mathbb{N}$, we have a brake orbit x_k on Σ and an integer number m_k such that its m_k-times iteration with the index interval $[i_{L_0}(x_k, m_k), i_{L_0}(x_k, m_k) + \nu_{L_0}(x_k, m_k) - 1]$ covers the positive integer number $k - 1$. This means that every positive integer number can be covered by this kind of index interval of an L_0-solution. So the indices of the L_0-solutions (x_k, m_k) are widespread like the set $\mathbb{N}$.

5 Index Jumping Formulas

In the following, we write $(i_{L_0}(\gamma, k), \nu_{L_0}(\gamma, k)) = (i_{L_0}(\gamma^k), \nu_{L_0}(\gamma^k))$ for any symplectic path $\gamma \in \mathcal{P}_\tau(2n)$ and $k \in \mathbb{N}$, where γ^k is defined by (12) and (13). We have the following jumping formulas for the L_0-index. For a symplectic path γ, its L_0-mean index $\bar{i}_{L_0}(\gamma)$ is defined by

$$\bar{i}_{L_0}(\gamma) = \lim_{k\to\infty} \frac{i_{L_0}(\gamma, k)}{k}.$$

Proposition E *Let $\gamma_j \in \mathcal{P}_{\tau_j}(2n)$ for $j = 1, \cdots, q$. Let $M_j = \gamma_j^2(2\tau_j) = N\gamma_j(\tau_j)^{-1}N\gamma_j(\tau_j)$, for $j = 1, \cdots, q$. Suppose*

$$\bar{i}_{L_0}(\gamma_j) > 0, \quad j = 1, \cdots, q.$$

Then there exist infinitely many $(R, m_1, m_2, \cdots, m_q) \in \mathbb{N}^{q+1}$ such that

(i) $\nu_{L_0}(\gamma_j, 2m_j \pm 1) = \nu_{L_0}(\gamma_j)$,
(ii) $i_{L_0}(\gamma_j, 2m_j - 1) + \nu_{L_0}(\gamma_j, 2m_j - 1) = R - (i_{L_1}(\gamma_j) + n + S^+_{M_j}(1) - \nu_{L_0}(\gamma_j))$,
(iii) $i_{L_0}(\gamma_j, 2m_j + 1) = R + i_{L_0}(\gamma_j)$,
(iv) $\nu_1(\gamma_j^2, 2m_j \pm 1) = \nu_1(\gamma_j^2)$,
(v) $i_1(\gamma_j^2, 2m_j - 1) + \nu_1(\gamma_j^2, 2m_j - 1) = 2R - (i(\gamma_j^2) + 2S^+_{M_j}(1) - \nu_1(\gamma_j^2))$,
(vi) $i_1(\gamma_j^2, 2m_j + 1) = 2R + i_1(\gamma_j^2)$,

where we have set $i_1(\gamma_j^2, n_j) = i_1(\gamma_j^{2n_j})$, $\nu_1(\gamma_j^2, n_j) = \nu_1(\gamma_j^{2n_j})$ for $n_j \in \mathbb{N}$.

Proposition F *For any $(\tau, x) \in \mathcal{J}_b(\Sigma, 2)$ and $m \in \mathbb{N}$, there hold*

$$i_{L_0}(x, m+1) - i_{L_0}(x, m) \geq 1, \tag{31}$$

$$i_{L_0}(x, m+1) + \nu_{L_0}(x, m+1) - 1 \geq i_{L_0}(x, m+1) > i_{L_0}(x, m) + \nu_{L_0}(x, m) - 1. \tag{32}$$

From Propositions D, E, and F, we can get that there are some L_0-solutions on Σ with their indices contributing densely as $\mathbb{N}$, but in some sense, the jumping steps of the iterated indices of a L_0-solution are wide. So there should be *enough* geometrically distinct brake orbits (at least n) on Σ. This is the main idea of the proof of Theorem A.

Preface

This monograph presents fundamental methods and topics in the relative Morse index theories and their efficient applications to some boundary value problems for Hamiltonian systems and partial differential equations. The Morse theory belongs to the categories of differential topology and nonlinear analysis. The finite-dimensional Morse theory deals with the topological properties of functions defined on finite-dimensional manifolds. The infinite-dimensional Morse theory concerns the topological properties of functionals defined on infinite-dimensional topological spaces such as Banach manifolds or, specially, Hilbert manifolds. One of the differences between the finite-dimensional Morse theory and the infinite-dimensional one is that in the latter case, the Morse index may be infinite. When the Morse index of a critical point of the functional is finite, one can fully understand the topological properties of the functional. But in many nonlinear variational problems, the Morse index at a critical point of the functional is infinite, namely, the functional is strongly indefinite. For instance, the functionals, with respect to first-order Hamiltonian systems, are strongly indefinite. When the Morse index of a critical point of a functional is infinite, one should define the so-called relative Morse index to understand the topological properties of the functional. The Conley-Zehnder index theory or, more generally, the Maslov-type index theory for periodic linear Hamiltonian systems is such a relative Morse index theory. For a linear Hamiltonian system, its fundamental solution is a symplectic path starting from the identity. In 1984, C. Conley and E. Zehnder in their celebrated paper [57] introduced the so-called Conley-Zehnder index theory for the nondegenerate symplectic paths in the real symplectic matrix group $\mathrm{Sp}(2n)$ with $n \geq 2$. In 1990, Y. Long and E. Zehnder in [231] generalized this index theory to the nondegenerate case with $n = 1$. Then, Y. Long in [224] and C. Viterbo in [280] further extended this index theory in 1990 independently to degenerate symplectic paths which are fundamental solutions of linear Hamiltonian systems. In [225], this index theory was further extended to continuous symplectic paths together with an axiomatic characterization for this index theory. In the literature after that, one usually calls this index the Maslov-type index. The book [223], written by Y. Long published in 2002, summarized the Maslov-type index theory for symplectic paths which are suitable for studying periodic solutions of nonlinear Hamiltonian systems. The Maslov-type index theory for symplectic paths is essentially a classification of the periodic linear Hamiltonian systems. Besides the problem of periodic solutions of nonlinear Hamiltonian systems, the important problems are the canonical boundary value problems for nonlinear Hamiltonian systems, such as the Lagrangian submanifold boundary value problems (L-boundary value problems for short), which solve the Hamiltonian systems for solutions x satisfying $x(0) \in L'$ and $x(\tau) \in L''$ for Lagrangian submanifolds L' and L'', and the boundary value problems related with symplectic matrix P(the P-boundary value problems for short) which solve the Hamiltonian systems for solutions x satisfying $x(\tau) = Px(0)$. These two kinds of boundary value problems were considered by A. Weinstein in [292–294]. As a sequel of the book [223], this book introduces some new index theories which are suitable for investigating some nonperiodic problems such as L-boundary value problems or P-boundary value problems of nonlinear Hamiltonian systems. The main point of the topics in this book is to systematically deal with the nonperiodic solution problems (open-string problems) or the symmetric periodic solution problems of Hamiltonian systems. The aims of this book are:

(1) to give an introduction to the index theory for symplectic paths with Lagrangian boundary conditions (L-index theory for short) and its iteration theory, which form a basis of Morse theoretical study about Hamiltonian systems with Lagrangian boundary conditions. The brake orbit problem is such a Lagrangian boundary problem.
(2) to give an introduction to the index theory for symplectic paths with P-boundary conditions (P-index theory for short) and its iteration theory, which form a basis of Morse theoretical study about Hamiltonian systems with P-boundary conditions. A periodic solution of some delay differential equations (systems) can be viewed as a P-boundary solution of Hamiltonian systems.
(3) to give an introduction to the index theory of abstract operator equations, where the essential spectrum of the linear self-adjoint operator possesses a gap on the real axis, and give applications of this theory to periodic and boundary value problems for some wave equations.
(4) to establish the relations among various index theories.
(5) to introduce some index theoretical methods in the study of nonlinear variational problems.

This book is divided into 11 chapters, with the first two chapters as its foundation parts covering some materials of linear symplectic spaces and the Maslov-type index theory published in [223]. The main body of this book contains two parts. The first part are Chaps. 3, 4, 5, 6, and 7, covering the theoretical material-relative Morse index theory, P-index theory, L-index theory, Maslov-type index theory for paths of Lagrangian pair, and relations among these index theories. The second part are Chaps. 8, 9, 10, and 11, which contain some applications of these index theories to the minimal periodic problems and subharmonic solution problems of Hamiltonian systems with symmetry, multiplicity of brake orbits, existence, and multiplicity for periodic-boundary value problems of some wave equations.

The first three chapters are foundation of this book. If one wants to skip some chapters, we suggest that from Chap. 4, one can skip directly to Chap. 8; from Chap. 5, one can skip to Chaps. 9 and 10; and from Chap. 3, one can skip directly to Chap. 11.

I am deeply indebted to Professor Yiming Long, who introduced me to this interesting area, for his constant encouragement and help. I have benefited a lot from his brilliant ideas and insights.

I would also like to thank the National Natural Science Foundation of China, the 973 Program of the Science and Technology Ministry of China, and the Research Fund for the Doctoral Program of Higher Education of the Education Ministry of China for their support. I am grateful to Professors Long Chen and Yifeng Yu of the Mathematical Department of the University of California, Irvine, for their hospitality and help when I visited UCI from June to September 2016. During that period, this book was prepared.

Nankai University, Tianjin, China
Guangzhou University, Guangzhou, China
July 2017

Chungen Liu

Contents

Chapter 1
Linear Algebraic Aspects

1.1 Linear Symplectic Spaces

In this book, we define by $\mathbb{N}$, $\mathbb{Z}$, $\mathbb{R}$ and $\mathbb{C}$ the sets of all natural, integral, real and complex numbers respectively. For a matrix M, we denote its transpose by M^T. For any $n \in \mathbb{N}$ and any field K, denote by K^n the linear space formed by all the column vectors of the form $x = (x_1, \cdots, x_n)^T$ with $x_i \in K$. We usually treat $x \in K^n$ as an $n \times 1$ matrix with no explain. Let $\mathcal{L}(K^n)$ denote the group of all $n \times n$ matrices with entries in the field K, and $\mathcal{L}_s(K^n)$ the subset of $\mathcal{L}(K^n)$ consists of symmetric matrices. Any linear map $T : K^n \to K^n$ corresponds to a matrix $T \in \mathcal{L}(K^n)$ in the usual way. We will not distinguish these two objectors.

Let V be an m-dimensional vector space over $\mathbb{R}$ with its dual space V^*, and let $\tilde{\omega} : V \times V \to \mathbb{R}$ be a bilinear map.

Definition 1.1 The map $\tilde{\omega}^* : V \to V^*$ is the linear map defined by $\tilde{\omega}^*(v)(u) = \tilde{\omega}(v, u)$.

The kernel of $\tilde{\omega}^*$ is the subspace U of V

Definition 1.2 A skew-symmetric bilinear map $\tilde{\omega}$ is symplectic (or nondegenerate) if $\tilde{\omega}^*$ is bijective, i.e., $U = \{0\}$. The map $\tilde{\omega}$ is then called a linear symplectic structure (or symplectic form) on V, and $(V, \tilde{\omega})$ is called a symplectic vector space.

The following are immediate properties of a linear symplectic structure $\tilde{\omega}$:

$\tilde{\omega}(u, v) = -\tilde{\omega}(v, u), \quad \forall u, v \in V.$

$\tilde{\omega}(u, v) = 0$ for some $u \in V$ and any $v \in V$ implies $u = 0$.

The map $\tilde{\omega}^* : V \to V^*$ is a bijection. By algebraic arguments, from the skew-symmetric of $\tilde{\omega}$ and $\dim U = 0$, we have that $\dim V = 2n$ is even.

Let $(V, \tilde{\omega})$ be a symplectic vector space. $u, v \in V$ are symplectic orthogonal and denoted by $u \bot_{\tilde{\omega}} v$, if $\tilde{\omega}(u, v) = 0$. For subspaces A, B of V, we define $A \bot_{\tilde{\omega}} B$ if

$u \bot_{\tilde{\omega}} v$ for any $u \in A$ and $v \in B$. We note that $u \bot_{\tilde{\omega}} u$ for any $u \in V$. For any linear subspace E of V, we define

$$E^{\bot_{\tilde{\omega}}} = \{u \in V \mid \tilde{\omega}(u, v) = 0, \quad \forall\, v \in E\}.$$

Then $E^{\bot_{\tilde{\omega}}}$ is a linear subspace of V, and there holds

$$(E^{\bot_{\tilde{\omega}}})^{\bot_{\tilde{\omega}}} = E, \quad \dim E + \dim E^{\bot_{\tilde{\omega}}} = 2n.$$

Definition 1.3 A linear subspace E of the symplectic vector space $(V, \tilde{\omega})$ is
isotropic, if $E \subset E^{\bot_{\tilde{\omega}}}$,
coisotropic, if $E \supset E^{\bot_{\tilde{\omega}}}$,
Lagrangian, if $E = E^{\bot_{\tilde{\omega}}}$,
symplectic, if $E \cap E^{\bot_{\tilde{\omega}}} = \{0\}$.

From the definition, we see that a linear subspace L of $(V, \tilde{\omega})$ is a Lagrangian subspace iff $\dim L = \frac{1}{2} \dim V = n$ and $\tilde{\omega}|_L = 0$ that is $\tilde{\omega}(u, v) = 0$ for any $u, v \in L$. A subspace $E \subset V$ is a symplectic subspace iff $(E, \tilde{\omega}|_E)$ is a symplectic vector space.

Lemma 1.4 *A dimension $2n$ symplectic vector space $(V, \tilde{\omega})$ has a basis $e_1, \cdots, e_n, f_1, \ldots, f_n$ satisfying $\tilde{\omega}(e_i, f_j) = \delta_{ij}$ and $\tilde{\omega}(e_i, e_j) = 0 = \tilde{\omega}(f_i, f_j)$.*

Proof Fix a non-zero vector $e_1 \in V$. By the non-degeneracy of $\tilde{\omega}$, there exists a vector $f_1 \in V$ satisfying $\tilde{\omega}(e_1, f_1) = 1$. So the subspace $E = \operatorname{span}\{e_1, f_1\}$ is a dimension 2 symplectic subspace of $(V, \tilde{\omega})$. If $\dim V = 2$, the lemma is proved. If $\dim V = 2n > 2$, then $E^{\bot_{\tilde{\omega}}}$ is a $2(n-1)$ dimensional symplectic subspace of $(V, \tilde{\omega})$, by an induction argument on the dimension the proof is complete. □

The basis $e_1, \cdots, e_n, f_1, \ldots, f_n$ in the above lemma is called a symplectic basis of $(V, \tilde{\omega})$. With a symplectic basis $e_1, \cdots, e_n, f_1, \ldots, f_n$, we can write $u = \sum_{i=1}^{n}(x_i e_i + y_i f_i)$, $x_i, \; y_i \in \mathbb{R}$. So the symplectic vector space $(V, \tilde{\omega})$ is symplectically isomorphic to the standard linear symplectic space $(\mathbb{R}^{2n}, \tilde{\omega}_0)$ with $\tilde{\omega}_0(z^1, z^2) = \sum_{i=1}^{n}(-x_i^2 y_i^1 + x_i^1 y_i^2) = (Jz^1, z^2) = (z^2)^T \cdot J \cdot z^1$ for $z^k = (x_1^k, \cdots, x_n^k, y_1^k, \cdots, y_n^k)^T$, $k = 1, 2$. Here the standard symplectic matrix J is defined by $J = \begin{pmatrix} 0 & -I_n \\ I_n & 0 \end{pmatrix}$. In other word, $\tilde{\omega}_0 = \sum_{i=1}^{n} dx_i \wedge dy_i$. For $z_i = (x_i, y_i) \in \mathbb{R}^n \times \mathbb{R}^n$, $i = 1, 2$, there holds

$$\tilde{\omega}_0(z_1, z_2) = \langle x_1, y_2 - \langle x_2, y_1 \rangle,$$

where $\langle \cdot, \cdot \rangle$ denotes the standard inner product in $\mathbb{R}^n$. A Lagrangian subspace $L \subset (\mathbb{R}^{2n}, \tilde{\omega}_0)$ is a dimensional n subspace with $\tilde{\omega}_0(z_1, z_2) = 0$ for all $z_1, \; z_2 \in L$.

Definition 1.5 Let $(V_k, \tilde{\omega}_k)$, $k = 1, 2$ be two dimension $2n$ symplectic vector spaces. A linear invertible map $T : V_1 \to V_2$ is called a linear symplectic map if

$$\tilde{\omega}_2(Tu, Tv) = \tilde{\omega}_1(u, v), \quad \forall u, v \in V_1, \tag{1.1}$$

that is to say $T^*\tilde{\omega}_2 = \tilde{\omega}_1$. In this case, the two symplectic vector spaces are called symplectic isomorphic. The linear map T is a symplectic isomorphism from $(V_1, \tilde{\omega}_1)$ to $(V_2, \tilde{\omega}_2)$.

So in this sense, the map $T : (V, \tilde{\omega}) \to (\mathbb{R}^{2n}, \tilde{\omega}_0)$, $T(u) = (x_1, \cdots, x_n, y_1, \cdots, y_n)$ as stated above is a symplectic map.

Corollary 1.6 *Any two symplectic vector spaces with the same dimension are symplectic isomorphic.*

Proof Since the isomorphic relation is an equivalent relation, and every dimension $2n$ symplectic vector space $(V, \tilde{\omega})$ is symplectic isomorphic to the standard symplectic space $(\mathbb{R}^{2n}, \tilde{\omega}_0)$, we prove the corollary. □

We note that the standard orthonormal basis of $\mathbb{R}^{2n}$ is a symplectic basis of the standard symplectic space $(\mathbb{R}^{2n}, \tilde{\omega}_0)$. In general, for linear symplectic space $(\mathbb{R}^{2n}, \tilde{\omega})$, there exists a skew-symmetric non-degenerate matrix $\mathcal{J}$ such that

$$\tilde{\omega}(u, v) = v^T \cdot \mathcal{J}^{-1} \cdot u, \tag{1.2}$$

namely, $\tilde{\omega} = \sum_{i,j=1}^{2n} a^{ij} dx_i \wedge dx_j$, here $\mathcal{J}^{-1} = (a^{ij})$. Then the $2n \times 2n$ matrix T as a symplectic map from $(\mathbb{R}^{2n}, \tilde{\omega})$ to $(\mathbb{R}^{2n}, \tilde{\omega}_0)$ should satisfy $T^T J T = \mathcal{J}^{-1}$.

1.2 Symplectic Matrices

Definition 2.1 A $2n \times 2n$ real matrix $M \in \mathcal{L}(\mathbb{R}^{2n})$ is called a symplectic matrix if it corresponds to a linear symplectic map

$$M : (\mathbb{R}^{2n}, \tilde{\omega}_0) \to (\mathbb{R}^{2n}, \tilde{\omega}_0)$$

in the standard orthonormal basis. So M is a symplectic matrix if and only if $M^T J M = J$. We denote by $\mathrm{Sp}(2n)$ the set of symplectic matrices, i.e.,

$$\mathrm{Sp}(2n) = \{M \in \mathcal{L}(\mathbb{R}^{2n}) | M^T J M = J\}.$$

In general for the linear symplectic space $(\mathbb{R}^{2n}, \tilde{\omega})$ with the corresponding skew-symmetric non-degenerate matrix $\mathcal{J}$ defined in (1.2), the matrix M is called $\mathcal{J}$-symplectic if $M^T \mathcal{J}^{-1} M = \mathcal{J}^{-1}$.

From definition, we know that any symplectic matrix is non-singular.

We remind that in the following proposition $T : (\mathbb{R}^{2n}, \tilde{\omega}) \to (\mathbb{R}^{2n}, \tilde{\omega}_0)$ is a symplectic map, and the corresponding matrix with respect to the standard basis is still denoted by T.

Proposition 2.2 *M is a $\mathcal{J}$-symplectic matrix if and only if $M' \equiv TMT^{-1}$ is a symplectic matrix.*

Proof By direct computation

$$\begin{aligned}(TMT^{-1})^T JTMT^{-1} &= (T^T)^{-1}M^T T^T JTMT^{-1}\\ &= (T^T)^{-1}M^T \mathcal{J}^{-1}MT^T = (T^T)^{-1}\mathcal{J}^{-1}T^{-1} = J.\end{aligned}$$

□

The above proposition says that the following diagram is symplectic.

$$\begin{array}{ccc}(\mathbb{R}^{2n}, \tilde{\omega}) & \xrightarrow{T} & (\mathbb{R}^{2n}, \tilde{\omega}_0)\\ M\downarrow & \curvearrowleft & \downarrow M'\\ (\mathbb{R}^{2n}, \tilde{\omega}) & \xrightarrow{T} & (\mathbb{R}^{2n}, \tilde{\omega}_0).\end{array}$$

Lemma 2.3 *Suppose $A, B \in \mathrm{Sp}(2n)$. Then $A^{-1}, A^T \in \mathrm{Sp}(2n)$ and $AB \in \mathrm{Sp}(2n)$. Moreover, I_{2n} and $J \in \mathrm{Sp}(2n)$.*

Proof From $A^T JA = J$, we get $(A^{-1})^T JA^{-1} = J$, so $A^{-1} \in \mathrm{Sp}(2n)$. Taking inverse in the second equality, we have $AJA^T = J$, so $A^T \in \mathrm{Sp}(2n)$. If further $B^T JB = J$, then $B^T A^T JAB = B^T JB = J$, so $AB \in \mathrm{Sp}(2n)$. That I_{2n} and $J \in \mathrm{Sp}(2n)$ are obvious. □

So $\mathrm{Sp}(2n)$ is a group, which is called symplectic group. From Proposition 2.2, we see that the set of all $2n \times 2n$ $\mathcal{J}$-symplectic matrices is also a group which is denoted by $\mathrm{Sp}_{\mathcal{J}}(2n)$. We call it the $\mathcal{J}$-symplectic group. We have

$$T\mathrm{Sp}_{\mathcal{J}}(2n)T^{-1} = \mathrm{Sp}(2n).$$

Lemma 2.4 *Any symplectic matrix $M \in \mathrm{Sp}(2n)$ has the square block form*

$$\begin{pmatrix} A & B\\ C & D\end{pmatrix}$$

with both $A^T C$ and $B^T D$ being symmetric, and $A^T D - C^T B = I_n$.

The proof of this lemma follows directly from definition.

Lemma 2.5 *For any symplectic matrix* $M = \begin{pmatrix} A & B \\ C & D \end{pmatrix}$, *we set* $(A\sqrt{-1} + B)(C\sqrt{-1} + D)^{-1} := P + Q\sqrt{-1}$ *with* P *and* Q *being real* $n \times n$ *matrices. Then we have that* P *and* Q *are symmetric matrices and* Q *is positively definite.*

Proof We have

$$\begin{aligned}(C^T\sqrt{-1} + D^T)(A\sqrt{-1} + B) = D^T B - C^T A + \sqrt{-1}(C^T B + D^T A),\\ (A^T\sqrt{-1} + B^T)(C\sqrt{-1} + D) = B^T D - A^T C + \sqrt{-1}(B^T C + A^T D).\end{aligned}$$

By the results in Lemma 2.4, we obtain

$$(P + Q\sqrt{-1})^T = P + Q\sqrt{-1}.$$

We now prove $Q > 0$. It is suffice to prove $\mathrm{Re}((A-iB)(D+iC)^{-1}) > 0. i = \sqrt{-1}$.

For any $x \in \mathbb{R}^n,\ x \neq 0$, set $y = (D + Ci)^{-1}x = a + bi,\ a, b \in \mathbb{R}^n$. Then there holds $x = (D + Ci)(a + bi) = Da - Cb + i(Ca + Db) \in \mathbb{R}^n$. It implies that

$$Ca + Db = 0. \tag{1.3}$$

By definition we have

$$\begin{aligned}&\mathrm{Re}\ (x^T(A - Bi)(D + Ci)^{-1}x)\\ &= \mathrm{Re}((a^T + b^T i)(D^T + C^T i)(A - Bi)(a + bi))\\ &= a^T(D^T A + C^T B)a - 2a^T(C^T A - D^T B)b - b^T(D^T A + C^T B)b.\end{aligned} \tag{1.4}$$

From (1.3), we have

$$a^T C^T Ba = -b^T D^T Ba,\ b^T D^T Ab = -a^T C^T Ab,$$

and by using Lemma 2.4 and (1.4), we obtain

$$\mathrm{Re}(x^T(A - Bi)(D + Ci)^{-1}x) = |a|^2 + |b|^2 + b^T(B^T C - C^T B)b. \tag{1.5}$$

We note that the matrix $B^T C - C^T B$ is anti-symmetric, so we have $b^T(B^T C - C^T B)b = 0$ and so we have

$$\mathrm{Re}(x^T(A - Bi)(D + Ci)^{-1}x) = |a|^2 + |b|^2 > 0.$$

□

The following lemma says that any symplectic matrix has its polar decomposition.

Lemma 2.6 ([227]) *Any symplectic matrix $M \in \mathrm{Sp}(2n)$ can be represented in the form*

$$M = PU, \tag{1.6}$$

where P is a symmetric symplectic positive definite matrix, U is a symplectic orthogonal matrix, and they are uniquely determined by M.

Proof In fact any non-singular square matrix has unique polar decomposition (1.6) with A being symmetric positive definite and U being orthogonal matrix. Let $P = \sqrt{MM^T}$. Then P is symmetric positive definite. Let $U = P^{-1}M$. From

$$UU^T = P^{-1}MM^TP^{-1} = P^{-1}P^2P^{-1} = I_{2n},$$

we obtain that U is orthogonal. Suppose we have two polar decompositions $M = P_1U_1 = P_2U_2$, then $U_1^TP_1 = U_2^TP_2$. Thus

$$P_1^2 = P_1U_1U_1^TP_1 = A_2U_2U_2^TP_2 = P_2^2.$$

Since P_1 and P_2 are symmetric positive definite, it implies that $P_1 = P_2$, so $U_1 = U_2$.

Next we need to prove that P and U are symplectic. From $M^TJM = J$, we have

$$M = J^{-1}[(PU)^T]^{-1}J = J^{-1}P^{-1}JJ^{-1}(U^T)^{-1}J.$$

It is easy to see that we have another polar decomposition $M = P'U'$ with $P' = J^{-1}P^{-1}J$ and $U' = J^{-1}(U^T)^{-1}J$. So we have $P = J^{-1}P^{-1}J$ and $U = J^{-1}(U^T)^{-1}J$. Thus both P and U are symplectic. □

It is well known that every positive definite symmetric matrix P can be uniquely represented by

$$P = \exp(Q),$$

where Q is a real symmetric matrix (the logarithm of P), and

$$\exp(Q) = \sum_{i=0}^{\infty}\frac{1}{i!}Q^i.$$

Lemma 2.7 ([223, 227]) *A symmetric positive definite matrix $P \in \mathcal{L}(\mathbb{R}^{2n})$ is symplectic, if and only if it has the form*

$$P = \exp(Q),$$

where $Q = \begin{pmatrix} A & B \\ B & -A \end{pmatrix}$*, with A and B being symmetric* $n \times n$ *matrices.*

Proof Since P is symmetric, Q is symmetric.

Necessity Since $P = \exp(Q)$ is symplectic,

$$\exp(Q) = J^{-1}\exp(-Q)J = \exp(-J^{-1}QJ).$$

Since $-J^{-1}QJ$ is real symmetric, by the uniqueness of the above representation we obtain

$$Q = -J^{-1}QJ. \tag{1.7}$$

Setting $Q = \begin{pmatrix} A & B \\ C & D \end{pmatrix}$ in (1.7) yields $D = -A$ and $C = B$. By $Q^T = Q$, we have $B^T = B$ and $A^T = A$.

Sufficiency If $Q = \begin{pmatrix} A & B \\ B & -A \end{pmatrix}$ with $A^T = A,\ B^T = B$, then Q solves the Eq. (1.7), and so P is symplectic. □

From the viewpoint of Lie group, Sp(2n) is a Lie group, its Lie algebra is $\mathrm{sp}(2n) = \{M \in \mathcal{L}(\mathbb{R}^{2n}) | M^T J + JM = 0\}$. It is valuable to note that the exponential map $\exp : \mathrm{sp}(2n) \to \mathrm{Sp}(2n)$ is not surjective, i.e., $\exp(\mathrm{sp}(2n))$ is only a proper subgroup of Sp(2n). The above result shows that the subset of symplectic symmetric positive definite matrices lies in the range $\exp(\mathrm{sp}(2n))$ of the exponential map.

Lemma 2.8 ([31], Theorem IV.2.2) *For any compact connected Lie group G with its Lie algebra* $\mathfrak{g}$*, the exponential map is surjective, i.e., there holds*

$$G = \exp(\mathfrak{g}).$$

Corollary 2.9 *We have*

$$\mathrm{Osp(2n)} \equiv \mathrm{Sp}(2n) \cap O(2n) = \exp(\mathrm{sp}(2n) \cap o(2n)),$$

where

$$\mathrm{sp}(2n) \cap o(2n) = \{N \in \mathcal{L}(\mathbb{R}^{2n}) \mid N^T J + JN = 0,\ N^T + N = 0\}. \tag{1.8}$$

Proof Since the Lie subgroup Sp(2n) ∩ O(2n) is connected and compact, it is a direct consequence of Lemma 2.8. □

As a consequence, we have the following result.

Proposition 2.10 *For any $M \in \mathrm{Sp}(2n)$, there exist matrices A and B satisfying $A^T = A$, $JA + AJ = 0$, $B^T = -B$, $JB - BJ = 0$ such that*

$$M = \exp(A)\exp(B). \tag{1.9}$$

Proof From Lemma 2.6, we have the polar decomposition

$$M = PU \tag{1.10}$$

with P being symplectic symmetric positive definite and U being symplectic orthogonal. So the result follows directly from Lemma 2.7 and Corollary 2.9 □

Example For $t \in \mathbb{R}$, the matrix $tJ \in \mathrm{sp}(2n) \cap o(2n)$, and

$$\begin{aligned} \exp(tJ) &= \sum_{k=0}^{\infty} \frac{t^k}{k!} J^k \\ &= \sum_{m=0}^{\infty} \frac{(-1)^m t^{2m}}{(2m)!} I_{2n} + \sum_{m=1}^{\infty} \frac{(-1)^{m-1} t^{2m-1}}{(2m-1)!} J \\ &= (\cos t) I_{2n} + (\sin t) J \end{aligned} \tag{1.11}$$

is an orthogonal symplectic matrix.

Lemma 2.11 ([223, 227]) *An orthogonal matrix $U \in \mathcal{L}(\mathbb{R}^{2n})$ is symplectic, if and only if it has the form*

$$U = \begin{pmatrix} A & B \\ -B & A \end{pmatrix} \tag{1.12}$$

where $A^T B$ is symmetric, and $A^T A + B^T B = I_n$, or equivalently $A \pm \sqrt{-1}B$ are unitary matrices.

Proof An orthogonal matrix $U \in \mathcal{L}(\mathbb{R}^{2n})$ is symplectic, if and only if

$$U^T (I_{2n} + \sqrt{-1}J) U = I_{2n} + \sqrt{-1}J.$$

We set $U = \begin{pmatrix} A & B \\ C & D \end{pmatrix}$. It implies that

$$\begin{aligned} &(A^T + \sqrt{-1}C^T)(A - \sqrt{-1}C) = I_n, \\ &(A^T + \sqrt{-1}C^T)(B - \sqrt{-1}D) = -\sqrt{-1}I_n, \\ &(B^T + \sqrt{-1}D^T)(B - \sqrt{-1}D) = I_n, \\ &(B^T + \sqrt{-1}D^T)(A - \sqrt{-1}C) = \sqrt{-1}I_n. \end{aligned}$$

From the first two equalities, we obtain $A - \sqrt{-1}C = D + \sqrt{-1}B$, so $D = A$ and $C = -B$. Together with the first equality, we see that $A + \sqrt{-1}B$ are unitary matrix. Thus $A - \sqrt{-1}B$ is also unitary matrix. That is $A \pm \sqrt{-1}B \in U(n)$. □

The above lemma establishes actually an isomorphism between the symplectic orthogonal group and the unitary group

$$\mathrm{Sp}(2n) \cap O(2n) \simeq U(n).$$

Note that in the sense of the topological groups, this is also a homeomorphism.

Corollary 2.12 ([227]) *For any* $M \in \mathrm{Sp}(2n)$, *there holds*

$$\det M = 1. \tag{1.13}$$

Proof By the polar decomposition (1.6), we have

$$\det M = \det P \det U.$$

From the $M^T JM = J$ and $\det J \neq 0$, we have $(\det M)^2 = 1$. Since $\det P > 0$, we have $\det P = 1$. Taking a complex matrix

$$T = \begin{pmatrix} I_n & \sqrt{-1}I_n \\ I_n & -\sqrt{-1}I_n \end{pmatrix},$$

we have

$$TUT^{-1} = \begin{pmatrix} A + \sqrt{-1}B & 0 \\ 0 & A - \sqrt{-1}B \end{pmatrix}.$$

So $\det U = \det(A + \sqrt{-1}B)\det(A - \sqrt{-1}B) = |\det(A + \sqrt{-1}B)|^2 = 1$. □

Lemma 2.13 *For any symplectic matrix* $M = \begin{pmatrix} A & B \\ C & D \end{pmatrix}$, *there hold*

$$(A - \sqrt{-1}C)(A + \sqrt{-1}C)^{-1} \in U(n) \tag{1.14}$$

and

$$(D - \sqrt{-1}B)(D + \sqrt{-1}B)^{-1} \in U(n). \tag{1.15}$$

Proof We only prove (1.14). Since M is non-singular, we get $A \pm \sqrt{-1}C$ is non-singular. By Lemma 2.4, we obtain

$$(A^T + \sqrt{-1}C^T)(A - \sqrt{-1}C) = (A^T - \sqrt{-1}C^T)(A + \sqrt{-1}C).$$

So there holds

$$\begin{aligned}&(A^T - \sqrt{-1}C^T)^{-1}(A^T + \sqrt{-1}C^T)(A - \sqrt{-1}C)(A + \sqrt{-1}C)^{-1}\\&= (A^T - \sqrt{-1}C^T)^{-1}(A^T - \sqrt{-1}C^T)(A + \sqrt{-1}C)(A + \sqrt{-1}C)^{-1}\\&= I_n.\end{aligned}$$

□

In the same way, we get

$$(A + \sqrt{-1}C)(A - \sqrt{-1}C)^{-1} \in U(n) \tag{1.16}$$

and

$$(D + \sqrt{-1}B)(D - \sqrt{-1}B)^{-1} \in U(n). \tag{1.17}$$

Lemma 2.14 *Let $\alpha_1, \cdots, \alpha_{2n}$ denote the columns of a symplectic matrix M respectively. Then $L_1 = span\{\alpha_1, \cdots, \alpha_n\}$ and $L_2 = span\{\alpha_{n+1}, \cdots, \alpha_{2n}\}$ are Lagrangian subspaces of $(\mathbb{R}^{2n}, \tilde{\omega}_0)$. For any Lagrangian subspace L of $(\mathbb{R}^{2n}, \tilde{\omega}_0)$, there is a symplectic orthogonal matrix U such that*

$$L = U(L_0)$$

with $L_0 = \{0\} \times \mathbb{R}^n \subset \mathbb{R}^{2n}$. The matrix U is determined uniquely up to an orthogonal matrix $O_n \in O(n)$ in the sense that for all such matrices U', there is a matrix $O_n \in O(n)$ such that

$$U' = U\begin{pmatrix} O_n & 0 \\ 0 & O_n \end{pmatrix}.$$

Proof It is clear that $M(L)$ is Lagrangian subspace of $(\mathbb{R}^{2n}, \tilde{\omega}_0)$ for any Lagrangian subspace L of $(\mathbb{R}^{2n}, \tilde{\omega}_0)$ and any $M \in \mathrm{Sp}(2n)$.

Let $M = \begin{pmatrix} A & B \\ C & D \end{pmatrix}$, then

$$\begin{pmatrix} B \\ D \end{pmatrix} = \begin{pmatrix} A & B \\ C & D \end{pmatrix} \cdot \begin{pmatrix} 0 \\ I_n \end{pmatrix}.$$

The Lagrangian subspace $L_0 = \{0\} \times \mathbb{R}^n$ is the subspace $span\{e_{n+1}, \cdots, e_{2n}\}$ of $\mathbb{R}^{2n}$, where $e_1, \cdots, e_{2n}$ is the standard basis of $\mathbb{R}^{2n}$). So $L_2 = M(L_0)$ is a Lagrangian subspace of $(\mathbb{R}^{2n}, \tilde{\omega}_0)$. We can prove that L_1 is a Lagrangian subspace of $(\mathbb{R}^{2n}, \tilde{\omega}_0)$ in a similar way.

Suppose L is a Lagrangian subspace of $(\mathbb{R}^{2n}, \tilde{\omega}_0)$ with orthonormal basis $\beta_{n+1}, \cdots, \beta_{2n}$. We extend it to an orthogonal basis of $\mathbb{R}^{2n}$ as $\beta_1, \cdots, \beta_{2n}$ such that

$$\tilde{\omega}_0(\beta_i, \beta_{n+j}) = \delta_{ij}, \ i, j = 1, \cdots, n.$$

Then the matrix U with the columns $\beta_1, \cdots, \beta_{2n}$ is an orthogonal matrix and satisfies

$$L = U(L_0).$$

In fact, let $U = \begin{pmatrix} A & C \\ B & D \end{pmatrix}$, then by the orthogonality and by the fact that L is Lagrangian subspace of $(\mathbb{R}^{2n}, \tilde{\omega}_0)$, there hold

$$\begin{pmatrix} C^T & D^T \end{pmatrix} \begin{pmatrix} C \\ D \end{pmatrix} = I_n, \ \begin{pmatrix} C^T & D^T \end{pmatrix} \begin{pmatrix} 0 & -I_n \\ I_n & 0 \end{pmatrix} \begin{pmatrix} C \\ D \end{pmatrix} = 0.$$

They are exactly the conditions in Lemma 2.11, i.e., $D \pm \sqrt{-1}C \in U(n)$. Since U is orthogonal symplectic, then $D = A$ and $C = -B$. In other word, $\beta_i = J^{-1}\beta_{n+i}$, $i = 1, \cdots, n$. The orthonormal basis $\beta_{n+1}, \cdots, \beta_{2n}$ of L determine the orthogonal symplectic matrix U. For another orthonormal basis $\alpha_{n+1}, \cdots, \alpha_{2n}$ of L, there is an orthogonal matrix $O_n \in O(n)$ such that $\alpha_{n+i} = O_n(\beta_{n+i})$, $i = 1, \cdots, n$. Suppose $\begin{pmatrix} -B' \\ A' \end{pmatrix}$ is the matrix with its ith-column being α_{n+i}, then

$$\begin{pmatrix} -B' \\ A' \end{pmatrix} = \begin{pmatrix} -BO_n \\ AO_n \end{pmatrix}.$$

i.e.,

$$U' = \begin{pmatrix} A' & -B' \\ B' & A' \end{pmatrix} = \begin{pmatrix} A & -B \\ B & A \end{pmatrix} \begin{pmatrix} O_n & 0 \\ 0 & O_n \end{pmatrix}.$$

□

A part of Lemma 2.14 was proved in [238].

We denote by $\Lambda(n)$ the set of all Lagrangian subspaces of $(\mathbb{R}^{2n}, \tilde{\omega}_0)$. From Lemmas 2.11 and 2.14, we obtain that

$$\Lambda(n) \simeq U(n)/O(n). \tag{1.18}$$

Given any two $2k_i \times 2k_i$ matrices of square block form, $M_i = \begin{pmatrix} A_i & B_i \\ C_i & D_i \end{pmatrix}$ with $i = 1, 2$, we define an operation $\diamond$-sum (*symplectic direct sum*) of M_1 and M_2 to be the $2(k_1 + k_2) \times 2(k_1 + k_2)$ symplectic matrix given by (cf. [223, 226])

$$M_1 \diamond M_2 = \begin{pmatrix} A_1 & 0 & B_1 & 0 \\ 0 & A_2 & 0 & B_2 \\ C_1 & 0 & D_1 & 0 \\ 0 & C_2 & 0 & D_2 \end{pmatrix}. \tag{1.19}$$

We denote by $M^{\diamond k}$ the k-fold $\diamond$-sum $M \diamond \cdots \diamond M$.

1.3 Lagrangian Subspaces

For a $2n$ dimensional real linear symplectic space $(V, \tilde{\omega})$, we have a complex structure $J : V \to V$ satisfying $\tilde{\omega}(Jv, Jw) = \tilde{\omega}(v, w)$ for any $v, w \in V$, $\tilde{\omega}(v, Jw) = \langle v, w\rangle$ is an inner product of V and $J^2 = -I_{2n}$. Associated to this situation, there is a Hermitian inner product $(\cdot, \cdot)$ on V defined by

$$(v, w) = -\tilde{\omega}(Jv, w) - \sqrt{-1}\tilde{\omega}(v, w).$$

We understand (V, J) as a complex vector space by identifying $(a + b\sqrt{-1})v$ with $(a + bJ)v$ for all $a, b \in \mathbb{R}$ and $v \in V$. Let $\mathrm{Lag}(V)$ denote the set of Lagrangian subspaces in $(V, \tilde{\omega})$. For the standard symplectic space $(\mathbb{R}^{2n}, \tilde{\omega}_0)$, we denote the set of its Lagrangian subspaces by $\Lambda(n)$, i.e., $\Lambda(n) = \mathrm{Lag}(\mathbb{R}^{2n})$. Let $L \in \mathrm{Lag}(V)$ be a Lagrangian subspace of $(V, \tilde{\omega})$. By definition $\dim L = n$ and $\tilde{\omega}(v, w) = 0$ for all $v, w \in L$. Suppose $\{e_j\}_{j=1,\cdots,n}$ is a basis of L with $\langle e_i, e_j\rangle = \delta_{ij}$, then by direct computation, we have $(e_i, e_j) = \delta_{ij}$. So $\{e_j\}_{j=1,\cdots,n}$ is also an orthogonal basis of the complex vector space (V, J) with the Hermitian inner product $(\cdot, \cdot)$. Thus any Lagrangian subspace L' of $(V, \tilde{\omega})$ is the real span

$$L' = \bigoplus_{j=1}^{n} \mathbb{R}\{e'_j\}$$

of some orthogonal basis $\{e'_j\}_{j=1,\cdots,n}$ of the complex vector space (V, J), and as a complex vector space we have

$$V = \bigoplus_{j=1}^{n} \mathbb{C}\{e'_j\} = \bigoplus_{j=1}^{n} \mathbb{R}\{e'_j\} \oplus \mathbb{R}\{Je'_j\}.$$

We denote the unitary group of the complex vector space $(V, (\cdot, \cdot))$ by $\tilde{U}(n)$, that is to say for every linear mapping $\tilde{U} \in \tilde{U}(n)$, there holds $(\tilde{U}v, \tilde{U}w) = (v, w)$ for all $v, w \in V$. Based on the Lagrangian subspace $L = \bigoplus_{j=1}^{n} \mathbb{R}\{e_j\}$, there is a natural mapping

$$\phi : \tilde{U}(n) \to \mathrm{Lag}(V) \tag{1.20}$$

sending $A \in \tilde{U}(n)$ to the real span $\phi(A) = \bigoplus_{j=1}^{n} \mathbb{R}\{Ae_j\}$. This mapping is obviously surjective and in fact it give rise to a bijection

$$\phi(L) : \tilde{U}(n)/\tilde{O}(n) \xrightarrow{\simeq} \mathrm{Lag}(V)$$

which depends only on L(not on the choice of basis $\{e_j\}_{j=1,\cdots,n}$ and the choice of the complex structure J), where $\tilde{O}(n) = \left\{\tilde{O} := \begin{pmatrix} O & 0 \\ 0 & O \end{pmatrix} : O \in O(n)\right\}$. In order to see this clearly, by Lemma 1.4, we choose a symplectic basis such that $(V, \tilde{\omega})$ can be viewed as the standard symplectic space $(\mathbb{R}^{2n}, \tilde{\omega}_0)$ with the complex structure $J = \begin{pmatrix} 0 & -I_n \\ I_n & 0 \end{pmatrix}$, which is the standard symplectic matrix. Suppose $\tilde{U} \in \tilde{U}(n)$. Since V is a real $2n$ dimensional space, we can assume $\tilde{U}$ in the form

$$\tilde{U} = \begin{pmatrix} A & B \\ C & D \end{pmatrix}.$$

Then by direct computation, from $(\tilde{U}v, \tilde{U}w) \equiv (v, w)$, we have

$$A^T A + C^T C = I_n, \ \ B^T B + D^T D = I_n, \ \ A^T B + C^T D = 0 \tag{1.21}$$

and

$$A^T D - C^T B = I_n, \ \ A^T C = C^T A, \ \ B^T D = D^T B. \tag{1.22}$$

The conditions in (1.21) means that $\tilde{U}$ is an orthogonal matrix. The conditions in (1.22) means that $\tilde{U}$ is a symplectic matrix. So $A = D$, $B = -C$ and $A \pm \sqrt{-1}B$ are unitary matrices from Lemma 2.11. Therefore, we have

$$\tilde{U}(n) = \mathrm{Sp}(2n) \cap O(2n).$$

If $\{e_j\}_{j=1,\cdots,n}$ is a orthogonal basis of L, then $e_1, \cdots, e_n, Je_1, \cdots, Je_n$ is a orthogonal basis of V. Taking Je_j as $\sqrt{-1}e_j$, we get $v = (z_1, \cdots, z_n)^T$ with $z_j = x_j + \sqrt{-1}y_j$ for a vector $\tilde{v} \in V$ with $\tilde{v} = (x_1, \cdots, x_n, y_1, \cdots, y_n)^T = \sum_{i=1}^{n}(x_i e_i + y_i Je_i)$. In this sense, we see that an unitary matrix $U = A - \sqrt{-1}B \in U(n)$ acts on a complex vector $v = (z_1, \cdots, z_n)^T$ is the same thing as that an orthogonal symplectic matrix $\tilde{U} = \begin{pmatrix} A & B \\ -B & A \end{pmatrix}$ acts on the real vector $\tilde{v} = (x_1, \cdots, x_n, y_1, \cdots, y_n)^T$. Thus we can identify U and $\tilde{U}$ in this sense. In fact, the mapping

$$\psi : U(n) \to \mathrm{Osp}(2n), \quad \psi(U) = \tilde{U}$$

is a group isomorphism, where we have denoted the orthogonal symplectic subgroup of $\mathrm{Sp}(2n)$ by $\mathrm{Osp}(2n)$. I.e. $\mathrm{Osp}(2n) = \mathrm{Sp}(2n) \cap O(2n) \cong U(n)$. It sends an orthogonal matrix $O(n) \ni C : L \to L$ to a $2n \times 2n$ orthogonal symplectic matrix $\psi(C) = \tilde{C} = \begin{pmatrix} C & 0 \\ 0 & C \end{pmatrix}$. It implies that

$$\tilde{U}(n)/\tilde{O}(n) \cong U(n)/O(n).$$

Lemma 3.1 ([32]) *Let L_1 and L_2 be two Lagrangian subspaces in V. Then*

(1) *$e^{J\theta}L_2$ is a Lagrangian subspace of V for all θ;*
(2) *There exists an $\epsilon \in (0, \pi)$ such that $L_1 \cap e^{J\theta}L_2 = \{0\}$ for all $0 < |\theta| < \epsilon$.*

Proof We only need to prove the statement (2). Let $a = \dim_{\mathbb{R}}(L_1 \cap L_2)$ and let $\{e_j\}_{j=1,\cdots,n}$ be an orthogonal basis of L_1 with the first a of it form a basis of $L_1 \cap L_2$.

Let U and W be the complex subspaces of V generated respectively by $\{e_j\}_{j=1,\cdots,a}$ and $\{e_j\}_{j=a+1,\cdots,n}$, i.e.

$$U = \bigoplus_{j=1}^{a} \mathbb{C}\{e_j\}, \quad W = \bigoplus_{j=a+1}^{n} \mathbb{C}\{e_j\}. \tag{1.23}$$

Then we have

$$V = U \oplus W \ \text{ with } L_1 \cap L_2 \subset U. \tag{1.24}$$

By dimension counting, the real span of $\{e_j\}_{j=1,\cdots,a}$ is $L_1 \cap L_2$ and the real span of $\{e_j\}_{j=a+1,\cdots,n}$ is $L_1 \cap W$. Thus we have $L_1 = (L_1 \cap L_2) \oplus (L_1 \cap W)$.

In a similar manner, we choose an orthogonal basis $\{e'_j\}_{j=1,\cdots,n}$ of L_2 such that $e'_j = e_j, \ j = 1, \cdots, a$. Let W' be the complex subspace of V generated by $\{e'_j\}_{j=a+1,\cdots,n}$, i.e., $W' = \bigoplus_{j=a+1}^{n} \mathbb{C}\{e'_j\}$. Then we have the following orthogonal sum decomposition:

$$V = U \oplus W', \quad L_2 = (L_1 \cap L_2) \oplus (L_2 \cap W').$$

Since the inner product $\langle \cdot, \cdot \rangle$ is nondegenerate. U has a unique orthogonal complement, so we have $W = W'$. Thus we have two orthogonal decompositions:

$$\begin{aligned} L_1 &= (L_1 \cap L_2) \oplus (L_1 \cap W), \\ L_2 &= (L_1 \cap L_2) \oplus (L_2 \cap W). \end{aligned} \tag{1.25}$$

Since $\{e_j\}_{j=1,\cdots,a}$ is a real basis for $L_1 \cap L_2$ and a complex basis for U, the multiplication by $e^{J\theta}$, $0 < |\theta| < \pi$, on $\{e_j\}_{j=1,\cdots,a}$ gives us an $\mathbb{R}$-linear independent one $\{e^{J\theta}e_j\}_{j=1,\cdots,a}$. In other words, we have

$$(L_1 \cap L_2) \cap e^{J\theta}(L_1 \cap L_2) = \{0\}. \tag{1.26}$$

However, we have

$$(L_1 \cap W) \cap (L_2 \cap W) = (L_1 \cap L_2) \cap W \subset U \cap W = \{0\},$$

so the subspaces $L_1 \cap W$ and $L_2 \cap W$ are transverse to each other. Since transversality is an open condition, there exists $\epsilon > 0$ such that

$$(L_1 \cap W) \cap e^{J\theta}(L_2 \cap W) = \{0\} \text{ for all } |\theta| < \epsilon. \tag{1.27}$$

Combining (1.25), (1.26) and (1.27), we have, for $0 < |\theta| < \epsilon$,

$$L_1 \cap e^{J\theta} L_2 = (L_1 \cap L_2) \cap e^{J\theta}(L_1 \cap L_2) \oplus (L_1 \cap W) \cap e^{J\theta}(L_2 \cap W) = \{0\}.$$

□

The following results are due to Arnold [9], Hörmander [143] (see also [76]).

Theorem 3.2 *The set of Lagrangian subspaces* $Lag(V)$ *is a connected regular algebraic subvariety of the Grassmann variety* $G_{V,n}$ *of* n*-dimensional linear subspaces of* V, $\dim Lag(V) = \frac{n(n+1)}{2}$. *For any* $L \in Lag(V)$, $\Lambda^0(V, L) = \{L' \in Lag(V) | \ L' \cap L = \{0\}\}$ *is (algebraically) diffeomorphic to the vector space* $Symm(L')$ *of all symmetric bilinear forms on* L', *for any* $L' \in \Lambda^0(V, L)$. *Moreover,* $\Lambda^0(V, L)$ *is open and dense in* $Lag(V)$. *Finally the tangent space* $T_{L'}Lag(V)$ *is canonically isomorphic to* $Symm(L')$ *for every* $L' \in Lag(V)$.

Proof Let $L, \ L' \in \mathrm{Lag}(V), \ L \cap L' = \{0\}$. Then each n-dimensional subspace M that is transversal to L is of the form $\{x + Ax | \ x \in L'\}$ for some linear map $A : L' \to L$. Then

$$Q(M)(x, y) = \tilde{\omega}(Ax, y), \quad x, y \in L' \tag{1.28}$$

defines a bilinear form on L', which is symmetric if and only if M is Lagrangian. In fact, if $M \in \mathrm{Lag}(V)$, there holds

$$\tilde{\omega}(x + Ax, y + Ay) = 0, \quad \forall \, x, y \in L'.$$

It implies $\tilde{\omega}(Ax, y) = \tilde{\omega}(Ay, x)$, so $Q(M)$ is symmetric. The reverse is obvious. So Q defines a bijective mapping: $\Lambda^0(V, L) \to \mathrm{Symm}(L')$, the inverse being an algebraic embedding: $\mathrm{Symm}(L') \to G_{V,n}$. So the Q's form a set of local coordinates of $\mathrm{Lag}(V)$; one can prove that 2^n of the $\Lambda^0(V, N)$ cover $\mathrm{Lag}(V)$. This proves the first part of the theorem.

Clearly $\Lambda^0(V, L)$ is open in Lag(V). For its density we choose, for any $L' \in$ Lag(V), an $\tilde{L} \in \Lambda^0(V, L) \cap \Lambda^0(V, L')$. Then $\Lambda^0(V, \tilde{L})$ is an open neighborhood of L', identified with Symm(L), in which $\Lambda^0(V, \tilde{L})$ corresponds to the nonsingular elements of Symm(L), which is dense.

Clearly Q depends on the choice of L', L, but for fixed L' its differential at L' does not depend on L. In fact, let $L'' \in \Lambda^0(V, L')$, then $L'' = \{z + Bz| \in L\}$ for some linear map $B : L \to L'$. If $\hat{L}$ is close enough to L', $\hat{L}$ still is transversal both to L and L'' and we can write $\hat{L} = \{x + Ax| \ x \in L'\} = \{y + \tilde{A}y| \ y \in L'\}$, for some linear maps $A : L' \to L$, $\tilde{A} : L' \to L''$. Then $x + Ax = y + \tilde{A}y$, $\tilde{A}y = z + Bz$ for some $z \in L$, so $z = Ax$, $y = x - Bz$, $x + Ax = (x - BAx) + \tilde{A}(x - BAx)$. Taking symplectic product with $u \in L'$:

$$\tilde{\omega}(Ax, u) = \tilde{\omega}(\tilde{A}x, u) - \tilde{\omega}(\tilde{A}BAx, u). \tag{1.29}$$

The second term in the right hand side vanishes of second order when $\hat{L} \to L'$, so Q and $\tilde{Q}$ have the same differential at $\hat{L} = L'$. □

Theorem 3.3 *For any $k \geq 1$, $L \in$ Lag(V), the set*

$$\Lambda^k(V, L) = \{L' \in \mathrm{Lag}(V)| \ \dim(L \cap L') = k\} \tag{1.30}$$

is a regular part of its closure $\overline{\Lambda^k(V, L)} = \bigcup_{l \geq k} \Lambda^l(V, L)$, which in turn is a connected algebraic subvariety of Lag(V) *of codimension $\frac{k(k+1)}{2}$ in* Lag(V). *If $L' \in \Lambda^k(V, L)$, then the tangent space $T_{L'}\Lambda^k(V, L)$ corresponds to the $Q \in$ Symm(L') such that $Q(x, y) = 0$ for all $x, y \in L \cap L'$.*

Proof In view of Theorem 3.2 also $\Lambda^0(V, L) \cap \Lambda^0(V, L')$ is dense in Lag(V) so it contains at leat some N. Let Q be the coordinization: $\Lambda^0(V, N) \to$ Symm(L') described in the proof of Theorem 3.2. Then, if $L' \in \Lambda^k(V, L)$, we have $\tilde{L} \in \Lambda^k(V, L) \cap \Lambda^0(V, N)$ if and only if $\dim\ker(Q(L) - Q(\tilde{L})) = k$. The reason is that for $\tilde{L} = \{x + Bx| \ x \in L'\}$ and $L = \{y + Ay| \ y \in L'\}$, where $B : L' \to N$ and $A : L' \to N$, $x + Bx \in L \cap \tilde{L} \Leftrightarrow (A - B)x = 0$.

On a suitable basis of L' we can write

$$Q(L) = \begin{pmatrix} 0 & 0 \\ 0 & S \end{pmatrix}, \quad Q(\tilde{L}) = \begin{pmatrix} T & U \\ U^T & W \end{pmatrix},$$

where S is a nonsingular $(n - k) \times (n - k)$ matrix.

If W is sufficiently small, then $S - W$ is still nonsingular and rank$(Q(L) - Q(\tilde{L})) = n - k$ if and only if $T = -UD$, $U^T = (S - W)D$ for some D. So the equation $\tilde{L} \in \Lambda^k(V, L)$ reads $T = -U(S - W)^{-1}U^T$ if $\tilde{L}$ is close to L'. This proves that $\Lambda^k(V, L)$ is a smooth algebraic manifold of codimension $\frac{k(k+1)}{2}$ and also that $T_{L'}\Lambda^k(V, L)$ corresponds to the symmetric matrices $\begin{pmatrix} T & U \\ U^T & W \end{pmatrix}$ such that $T = 0$.

To prove that $\Lambda^k(V, L)$ is connected we remark that $L' \to L \cap L'$ defines a fibration: $\Lambda^k(V, L) \to G_{L,k}$, with fiber over L_0 equal to $\mathrm{Lag}(L_0^{\perp_{\tilde{\omega}}}/L_0)$. To explain this, let $L_0 \subset L$, $\dim L_0 = k$. Then $L_0 \subset L = L^{\perp_{\tilde{\omega}}} \subset L_0^{\perp_{\tilde{\omega}}}$. Furthermore $L_0^{\perp_{\tilde{\omega}}}/L_0$ is a symplectic vector space with symplectic form $\tilde{\omega}_{mod L_0}$ defined by

$$\tilde{\omega}_{mod L_0}(e_1 + L_0, e_2 + L_0) = \tilde{\omega}(e_1, e_2), \quad e_1,\ e_2 \in L_0^{\perp_{\tilde{\omega}}}. \tag{1.31}$$

If $L' \in \mathrm{Lag}(V)$, $L \cap L' = L_0$, then also $L_0 \subset L' = (L')^{\perp_{\tilde{\omega}}} \subset L_0^{\perp_{\tilde{\omega}}}$ and it is easily verified that $L' \mapsto L'/L_0$ is an isomorphism between the fiber over L_0 and $\mathrm{Lag}(L_0^{\perp_{\tilde{\omega}}}/L_0)$. Because both $G_{L,k}$ and $\mathrm{Lag}(L_0^{\perp_{\tilde{\omega}}}/L_0)$ are connected it follows that $\Lambda^k(V, L)$ is connected. □

1.4 Linear Hamiltonian Systems

A linear Hamiltonian system can be written in the following way

$$\dot{z}(t) = JB(t)z(t), \tag{1.32}$$

where $B(t)$ is a continuous $2n \times 2n$ symmetric matrix function. Its fundamental solution $\gamma(t)$ is a differentiable $2n \times 2n$ matrix function satisfying

$$\begin{cases} \dot{\gamma}(t) = JB(t)\gamma(t), \\ \gamma(0) = I_{2n}. \end{cases}$$

For the reason of simplicity, we also call $\gamma(t)$ the fundamental solution of $B(t)$.

Example For the constant coefficient Hamiltonian system

$$\dot{z}(t) = JBz(t),$$

the fundamental solution is $\gamma(t) = \exp(tJB)$.

Lemma 4.1 *The linear Hamiltonian system* (1.32) *always possesses fundamental solution* $\gamma(t)$. *Furthermore, every solution* $x(t)$ *of* (1.32) *satisfying initial condition* $z(0) = \xi \in \mathbb{R}^{2n}$ *can be represented as*

$$z(t) = \gamma(t)\xi. \tag{1.33}$$

Proof It is a well known result. For reader's convenience, we give a proof here.

Suppose $\{e_1, \cdots, e_{2n}\}$ is the standard basis of $\mathbb{R}^{2n}$. Consider the initial value problem

$$\begin{cases} \dot{x}(t) = JB(t)x(t), \\ x(0) = e_i. \end{cases} \tag{1.34}$$

We know that the problem (1.34) has unique solution $\alpha_i(t)$. We write $\xi = (a_1, \cdots, a_{2n})^T$. Then the solution of the initial value problem

$$\{ \dot{z}(t) = JB(t)z(t), \ \mathbb{Z}(0) = \xi$$

can be written as

$$z(t) = \sum_{k=1}^{2n} a_k \alpha_k(t) = \gamma(t)\xi,$$

where the matrix $\gamma(t)$ is composed by taking $\alpha_i(t)$ as its ith-column. So $\gamma(t)$ is the fundamental solution of (1.32). □

Lemma 4.2 *The fundamental solution $\gamma(t)$ of (1.32) is a symplectic path, i.e., $\gamma(t) \in \mathrm{Sp}(2n)$ for any $t \in \mathbb{R}$. On the other hand, any C^1-symplectic path $\gamma(t)$ with $\gamma(0) = I_{2n}$ is the fundamental solution of some linear Hamiltonian system, i.e., $J\dot{\gamma}(t)\gamma(t)^{-1}$ is symmetric for any $t \in \mathbb{R}$.*

Proof By direct computation

$$\begin{aligned} &\tfrac{d}{dt}(\gamma(t)^T J\gamma(t)) = \dot{\gamma}(t)^T J\gamma(t) + \gamma(t)^T J\dot{\gamma}(t) \\ &= \gamma(t)^T B(t) J^T J\gamma(t) + \gamma(t)^T JJB(t)\gamma(t) = 0, \end{aligned}$$

we have $\gamma(t)^T J\gamma(t) = constant$. Since $\gamma(0)^T J\gamma(0) = J$, we prove that

$$\gamma(t)^T J\gamma(t) \equiv J, \ t \in \mathbb{R}. \tag{1.35}$$

If $\gamma(t)$ is a C^1 symplectic path satisfying $\gamma(0) = I_{2n}$, take $B(t) = -J\dot{\gamma}(t)\gamma(t)^{-1}$. We should prove $B^T(t) = B(t)$.

We have

$$B^T(t) = \gamma^T(t)^{-1}\dot{\gamma}(t)^T J.$$

From (1.35) there holds $\gamma^T(t)^{-1} = -J\gamma(t)J$. By differentiating (1.35), there holds

$$\dot{\gamma}(t)^T J = -\gamma^T(t)J\dot{\gamma}(t)\gamma^{-1}(t).$$

Since $\gamma(t)^T$ is symplectic

$$B^T(t) = J\gamma(t)J\gamma^T(t)J\dot{\gamma}(t)\gamma^{-1}(t) = -J\dot{\gamma}(t)\gamma^{-1}(t) = B(t).$$

□

Lemma 4.3 *For any matrix $M \in \mathrm{Sp}(2n)$, there is a C^0-symmetric positive definite matrix function $B(t)$ such that $M = \gamma(1)$, where $\gamma(t)$ is the fundamental solution of $B(t)$.*

We call the property stated in Lemma 4.3 the *positive definite path connectivity* of $\mathrm{Sp}(2n)$.

Proof From Proposition 2.10, for $M \in \mathrm{Sp}(2n)$, there exist matrices A and B satisfying $A^T = A,\ JA + AJ = 0$, $B^T = -B,\ JB - BJ = 0$ such that

$$M = \exp(A)\exp(B). \tag{1.36}$$

Set $\gamma_1(t) = \exp(tA)\exp(tB)$, then $\gamma_1(0) = I_{2n}$ and $\gamma_1(1) = M$. From Lemma 4.2, we have $B_1(t) := -J\dot{\gamma}_1(t)\gamma_1(t)^{-1}$ is symmetric for any $t \in \mathbb{R}$. So $\gamma_1(t)$ is the fundamental solution of $B_1(t)$. If $B_1(t) > 0$, the proof is done. In general, we set $\gamma(t) = \exp(2k\pi tJ)\gamma_1(t),\ k \in \mathbb{Z}$. Then $\gamma(0) = I_{2n}$ and $\gamma(1) = M$.

$$B(t) = -J\dot{\gamma}(t)\gamma(t)^{-1} = 2k\pi + \exp(2k\pi tJ)B_1(t)\exp(-2k\pi tJ).$$

is symmetric positive definite for large integer k. □

1.5 Eigenvalues of Symplectic Matrices

Denote by $\mathbf{U} = \{z \in \mathbb{C} \mid |z| = 1\}$ the unit circle on the complex plane $\mathbb{C}$. For any $m \times m$ matrix M, we denote by $\sigma(M)$ the spectrum of M, i.e.,

$$\sigma(M) = \{\lambda \in \mathbb{C} \mid \det(M - \lambda I_m) = 0\}.$$

Suppose $\lambda \in \sigma(M)$. The *geometric multiplicity* $\mu_g(\lambda)$ of λ is defined to be $\dim_{\mathbb{C}} \ker_{\mathbb{C}}(M - \lambda I_m)$. Denote the complex root vector space of M belonging to λ by

$$E_\lambda(M) = \cup_{k\ge 1} \ker_{\mathbb{C}}(M - \lambda I_m)^k \subset \mathbb{C}^m.$$

The *algebraic multiplicity* $\mu_a(\lambda)$ of $\lambda \in \sigma(M)$ is defined to be $\dim_{\mathbb{C}} E_\lambda(M)$.

It is clear that $\mu_g(\lambda) \le \mu_a(\lambda)$. If $\mu_g(\lambda) = \mu_a(\lambda)$, $\lambda \in \sigma(M)$ is called semi-simple, in this case, each irreducible invariant subspace of $E_\lambda(M)$ is one dimensional.

Fix $M \in \mathrm{Sp}(2n)$. Its characteristic polynomial is defined by

$$f_M(\lambda) = \det(M - \lambda I_{2n}),\ \lambda \in \mathbb{C}.$$

This is a polynomial of degree $2n$ in λ with real coefficients. Since $J^{-1}M^T JM = I_{2n}$ and $\det M = 1$, there holds

$$f_M(\lambda) = \det(M - \lambda J^{-1}M^T JM) = \det(I_{2n} - \lambda M^T).$$

Thus

$$f_M(\lambda) = \lambda^{2n} f_M(\lambda^{-1}).$$

Therefore $f_M(\lambda)$ can be written in the symmetric form

$$f_M(\lambda) = \lambda^n \sum_{k=0}^{n} a_k(\lambda^k + \lambda^{-k}),$$

where $a_k \in \mathbb{R}$, $k = 0, 1, \cdots, n$ and $a_n = 1$. The following result is very clear now.

Lemma 5.1 *Suppose $M \in \mathrm{Sp}(2n)$. If $\lambda \in \sigma(M)$, then also $\bar{\lambda}$, λ^{-1} and $\bar{\lambda}^{-1}$ belong to $\sigma(M)$, and each of them possesses the same geometric and algebraic multiplicities as λ if it is different from λ. Particularly, if 1 or $-1 \in \sigma(M)$, then its algebraic multiplicity is even.*

Definition 5.2 ([223, 226]) We define the homotopic set of $M \in \mathrm{Sp}(2n)$ by

$$\Omega(M) = \{N \in \mathrm{Sp}(2n) \,|\, \sigma(N) \cap \mathbf{U} = \sigma(M) \cap \mathbf{U} \text{ and}$$
$$\dim_{\mathbb{C}} \ker_{\mathbb{C}}(N - \lambda I) = \dim_{\mathbb{C}} \ker_{\mathbb{C}}(M - \lambda I),\ \forall \lambda \in \sigma(M) \cap \mathbf{U}\}.$$

The path connected component of $\Omega(M)$ which contains M is denoted by $\Omega_0(M)$, and is called the *homotopy component* of M in $\mathrm{Sp}(2n)$.

Definition 5.3 We say that a symplectic matrix $M \in \mathrm{Sp}(2n)$ is *elliptic* if $\sigma(M) \subset \mathbf{U}$, and M is *hyperbolic* if $\sigma(M) \cap \mathbf{U} = \emptyset$. M is *autonomous hyperbolic* if $\sigma(M) \cap \mathbf{U} = \{1\}$ and the algebraic multiplicity of $1 \in \sigma(M)$ is 2. The total algebraic multiplicities of eigenvalues $\lambda \in \sigma(M)$ belonging to $\mathbf{U}$ is called the *elliptic hight* of M and denote it by $e(M)$. The linear Hamiltonian system (1.32) in the interval $[0, \tau]$ is elliptic (resp. hyperbolic, or autonomous hyperbolic) if $M = \gamma(\tau)$ is elliptic (resp. hyperbolic, or autonomous hyperbolic), where $\gamma(t)$ the fundamental solution of (1.32).

A linear system $\dot{x} = M(t)x$ is called positively (resp. negatively) stable if all its real solutions remain bounded for all $t > 0$ (resp. $t < 0$). It is called stable if it is both positively and negatively stable, that is, its real solutions are bounded for all times $t \in \mathbb{R}$. If the coefficient matrix $B(t)$ in (1.32) is τ-periodic and the linear system is not elliptic in $[0, \tau]$, then the linear Hamiltonian system must be unstable. For this topic, the readers can refer the monograph [78] for further arguments.

Notes and Comments As far as our knowledge, there are many monographs and lectures taking linear symplectic spaces, symplectic group and linear Hamiltonian systems as preliminary material. For example, the monographs [78] of I. Ekeland, [223, 227] of Y. Long, [270] of A. C. da Silva and [238] of D. McDuff and D. Salamon. In Ekeland's book [78], there is an introduction to the Floquet theory of linear Hamiltonian systems and some arguments of stability of linear Hamiltonian systems. In Long's books [223, 227], some basic properties about the symplectic matrices were introduced, and the normal forms for eigenvalues on the unit circle of symplectic matrices were developed.

Chapter 2
A Brief Introduction to Index Functions

For $n \in \mathbb{N}$, we recall that the symplectic group is defined as

$$\mathrm{Sp}(2n) \equiv Sp(2n, \mathbb{R}) = \{M \in \mathcal{L}(\mathbb{R}^{2n}) \mid M^T J M = J\},$$

where $\begin{pmatrix} 0 & -I_n \\ I_n & 0 \end{pmatrix}$, I_n is the identity matrix on $\mathbb{R}^n$, and $\mathcal{L}(\mathbb{R}^{2n})$ is the space of $2n \times 2n$ real matrices. Without confusion, we shall omit the subindex of the identity matrices. For $\tau > 0$, suppose $H \in C^2(S_\tau \times \mathbb{R}^{2n}, \mathbb{R})$, where $S_\tau \equiv \mathbb{R}/(\tau\mathbb{Z})$. Let x be a τ-periodic solution of the nonlinear Hamiltonian system

$$\dot{x}(t) = JH'(t, x(t)), \quad x \in \mathbb{R}^{2n}. \tag{2.1}$$

Then the fundamental solution of the linearized Hamiltonian system of (2.1) at x

$$\dot{y}(t) = JB(t)y, \quad y \in \mathbb{R}^{2n}, \tag{2.2}$$

with $B(t) = H''(t, x(t))$, is a path $\gamma_x \in C([0, +\infty), \mathrm{Sp}(2n))$ with $\gamma_x(0) = I$.

In order to study the properties of periodic solutions of (2.1), C. Conley and E. Zehnder in 1984 established an index theory in their celebrated work [57] for nondegenerate paths in Sp(2n) started from the identity matrix with $n \geq 2$. This index theory was extended to the nondegenerate case with $n = 1$ by Y. Long and E. Zehnder in [231]. The index theory for the degenerate Hamiltonian systems was established by Y. Long in [224] and C. Viterbo in [280] via different methods. Then in [225], Y. Long further established the index theory for the general degenerate symplectic paths. In this book we call it the Maslov-type index for periodic boundary condition to distinguished it from the Maslov type index theories with some different boundary conditions introduced in the sequel of this book. In [226], Y. Long introduced the ω-index theory parametrized by all ω on the unit circle in the complex plane, and used it to establish the Bott-type formula and the iteration theory of the Maslov-type index for periodic boundary condition. We define the set of symplectic paths by

$$\mathcal{P}(2n) \equiv \mathcal{P}_\tau(2n) = \{\gamma \in C([0,\tau], \mathrm{Sp}(2n)) \mid \gamma(0) = I\}.$$

For any $\gamma \in \mathcal{P}_\tau(2n)$, $\omega \in \mathbf{U}$, as in [226] the ω-nullity of γ is defined by

$$\nu_\omega(\gamma) = \dim_{\mathbb{C}} \ker_{\mathbb{C}}(\gamma(\tau) - \omega I).$$

If $\nu_\omega(\gamma) > 0$, we say that the symplectic path γ is ω-degenerate, otherwise it is called ω-non-degenerate. The ω-index theory assigns a pair of integers $(i_\omega(\gamma), \nu_\omega(\gamma)) \in \mathbb{Z} \times \{0, 1, \ldots 2n\}$ to each $\gamma \in \mathcal{P}_\tau(2n)$. When $\omega = 1$, it coincides with the Maslov-type index theory for periodic boundary condition. If $\gamma(t)$ is the fundamental solution of the linear system (2.2), we denote the index pair of γ also by $(i_\omega(B), \nu_\omega(B))$. It is a classification of the linear Hamiltonian systems. If $x(t)$ is a τ-periodic solution of the nonlinear Hamiltonian system (2.1), we denote $(i_\omega(x), \nu_\omega(x)) = (i_\omega(B), \nu_\omega(B))$ in this case. We refer to the monographs [223] and [227] for more details and applications of this index theory. In Chaps. 4 and 5 below, we will introduce some new development of the Maslov type index theories which are suitable to be used in the studies of Hamiltonian systems with non-periodic boundary conditions such as P-boundary condition (P-index) and Lagrangian boundary condition (L-index).

2.1 Maslov Type Index $i_1(\gamma)$

We now recall the definition of the Maslov-type index for periodic boundary condition. For simplicity we take $\tau = 1$ now. For a symplectic path $\gamma \in \mathcal{P}_1(2n)$, suppose its polar decomposition is $\gamma(t) = U(t)P(t)$ with $U(t) \in O(2n) \cap \mathrm{Sp}(2n)$ and $P(t)$ the symmetrical positive definite symplectic matrix for $t \in [0, 1]$. We recall that a symplectic matrix M is non-degenerate if $\det(M - I) \neq 0$. γ is non-degenerate if $\det(\gamma(1) - I) \neq 0$. We denote the set of nondegenerate symplectic matrices by $\mathrm{Sp}(2n)^*$ and the nondegenerate symplectic paths by $\mathcal{P}_1(2n)^*$. $\mathrm{Sp}(2n)^0 = \mathrm{Sp}(2n) \setminus \mathrm{Sp}(2n)^*$, $\mathcal{P}_1(2n)^0 = \mathcal{P}_1(2n) \setminus \mathcal{P}_1(2n)^*$ the degenerate set and the degenerate path set. $\mathrm{Sp}(2n)^*$ is 2-path connected. M^+ and M^- belong to different path connected components of $\mathrm{Sp}(2n)^*$, where

$$M^+ = D(2)^{\diamond n}, \quad M^- = D(-2) \diamond D(2)^{\diamond(n-1)}, \quad D(a) = \mathrm{diag}\{a, a^{-1}\},$$

and $M^{\diamond k}$ is the k-folds symplectic sum defined in (1.19). We choose a symplectic path $\beta : [0, 1] \to \mathrm{Sp}(2n)^*$ with $\beta(0) = \gamma(1)$ and $\beta(1) \in \{M^+, M^-\}$. By joining γ with β, we get a path $\tilde{\gamma} = \beta * \gamma \in \mathcal{P}_1(2n)^*$. Its polar decomposition is $\tilde{\gamma}(t) =$

$\tilde{U}(t)\tilde{P}(t)$. Writing $\tilde{U}(t) = \begin{pmatrix} \tilde{A}(t) & \tilde{B}(t) \\ -\tilde{B}(t) & \tilde{A}(t) \end{pmatrix}$, we have $\tilde{A}(t) - \sqrt{-1}\tilde{B}(t) \in U(n)$. Suppose $\det(\tilde{A}(t) - \sqrt{-1}\tilde{B}(t)) = e^{\sqrt{-1}\tilde{\theta}(t)}$, then $(\tilde{\theta}(1) - \tilde{\theta}(0))/\pi := k \in \mathbb{Z}$, and there holds

$$\begin{cases} k \text{ is odd}, & \text{if } \beta(1) = M^-, \\ k \text{ is even}, & \text{if } \beta(1) = M^+. \end{cases} \tag{2.3}$$

Definition 1.1 ([57, 231]) *For a non-degenerate path* $\gamma \in \mathcal{P}_1(2n)^*$, we define

$$i_1(\gamma) = k, \quad \nu_1(\gamma) = 0. \tag{2.4}$$

If $\gamma \in \mathcal{P}_1(2n)^0$ is a degenerate path, then one can perturb the path near the end point $\gamma(\tau)$ slightly to get a non-degenerate path γ_s for $s \in [-1, 1]$ such that

$$\begin{cases} \gamma_0 = \gamma, \\ \gamma_s(t) = \gamma(t), \quad \forall 0 \le t \le 1-\epsilon, \quad 0 < \epsilon << 1, \\ \nu_1(\gamma_s) = 0, \quad \forall s \ne 0, \\ i_1(\gamma_s) = i_1(\gamma_{s'}), \quad \forall ss' > 0, \\ i_1(\gamma_s) - i_1(\gamma_{s'}) = \nu_1(\gamma), \quad s > 0, \quad s' < 0. \end{cases}$$

The perturbed path γ_s can be chosen as $\gamma_s(t) = \gamma(t)e^{s\rho_\epsilon(t)J}$ with the continuous function ρ_ϵ defined as

$$\rho_\epsilon(t) = \begin{cases} 0, & 0 \le t \le 1-\epsilon, \\ (2-\epsilon)(t-1+\epsilon), & 1-\epsilon < t < 1-\frac{\epsilon}{2}, \\ \epsilon t, & 1-\frac{\epsilon}{2} \le t \le 1. \end{cases}$$

Definition 1.2 ([223, 224]) For a degenerate path $\gamma \in \mathcal{P}_1(2n)^0$, we define

$$i_1(\gamma) = i_1(\gamma_{-s}), \quad s \in (0, 1], \quad \nu_1(\gamma) = \dim_{\mathbb{C}} \ker_{\mathbb{C}}(\gamma(1) - I).$$

For the properties of the index pair $(i_1(\gamma), \nu_1(\gamma))$, we refer the monographs [223] and [227] for detail arguments. We now prove some new results about this index theory.

Theorem 1.3 *For any*

$$\gamma = \begin{pmatrix} A & B \\ -B & A \end{pmatrix} \in \mathcal{P}_1(2n) \cap O(2n) \ (orthogonal\ path),$$

suppose $\lambda_j(t) = e^{\sqrt{-1}\theta_j(t)}$, $j = 1, \cdots, n$ *are eigenvalues of the unitary matrix function* $A(t) - \sqrt{-1}B(t)$ *with* $\theta_j \in C([0, 1], \mathbb{R})$. *There holds*

$$i_1(\gamma) = 2\sum_{j=1}^{n} E\left(\frac{\theta_j(1)-\theta_j(0)}{2\pi}\right) + n, \tag{2.5}$$

where $E(a) = \max\{k \in \mathbb{Z}|\ k < a\}$.

Proof It is easy to see that $1 \in \sigma(A(t) - \sqrt{-1}B(t))$ if and only if $1 \in \sigma(\gamma(1))$.

Step 1 Suppose $\gamma \in \mathcal{P}_1(2n)^*$, then $\theta_j(1)-\theta_j(0) \neq 2k\pi,\ \ k \in \mathbb{Z}$. So for the integer part we have

$$\left[\frac{\tilde{\theta}_j(1)-\tilde{\theta}_j(0)}{2\pi}\right] = \left[\frac{\theta_j(1)-\theta_j(0)}{2\pi}\right] = E\left(\frac{\theta_j(1)-\theta_j(0)}{2\pi}\right), \tag{2.6}$$

and

$$\frac{\tilde{\theta}_j(1)-\tilde{\theta}_j(0)}{\pi} \in 2\mathbb{Z}+1,$$

where $[a] = \max\{k \in \mathbb{Z}|k \leq a\}$ is the Gaussian function. So we have

$$\frac{\tilde{\theta}_j(1)-\tilde{\theta}_j(0)}{2\pi} = \left[\frac{\tilde{\theta}_j(1)-\tilde{\theta}_j(0)}{2\pi}\right] + \frac{1}{2}.$$

Thus we get

$$\frac{\tilde{\theta}_j(1)-\tilde{\theta}_j(0)}{\pi} = 2\left[\frac{\tilde{\theta}_j(1)-\tilde{\theta}_j(0)}{2\pi}\right] + 1 = 2\left[\frac{\theta_j(1)-\theta_j(0)}{2\pi}\right] + 1. \tag{2.7}$$

Taking sum from $j = 1$ to $j = n$, we have

$$i_1(\gamma) = \sum_{j=1}^{n} \frac{\tilde{\theta}_j(1)-\tilde{\theta}_j(0)}{\pi} = \sum_{j=1}^{n} \left\{2\left[\frac{\theta_j(1)-\theta_j(0)}{2\pi}\right] + 1\right\}. \tag{2.8}$$

And (2.8) implies (2.5).

Step 2 Suppose $\gamma \in \mathcal{P}_1(2n)^0$, there are some θ_j with

$$\frac{\theta_j(1)-\theta_j(0)}{2\pi} = \left[\frac{\theta_j(1)-\theta_j(0)}{2\pi}\right] \tag{2.9}$$

and others with

$$\frac{\theta_j(1)-\theta_j(0)}{2\pi} > \left[\frac{\theta_j(1)-\theta_j(0)}{2\pi}\right]. \tag{2.10}$$

In the two cases (2.9) and (2.10), by some suitable rotations and a similar argument as in the above step, we get

$$\frac{\tilde{\theta}_{-s,j}(1)-\tilde{\theta}_{-s,j}(0)}{2\pi} = E\left(\frac{\theta_j(1)-\theta_j(0)}{2\pi}\right)+\frac{1}{2}. \tag{2.11}$$

Thus we have

$$i_1(\gamma) = i_1(\gamma_{-s}) = 2\sum_{j=1}^{n} E\left(\frac{\theta_j(1)-\theta_j(0)}{2\pi}\right)+n.$$

□

Remark 1.4 By the same reason as in Step 2 above, we have

$$-n \le i_1(\gamma) - \frac{\theta_\gamma(1)-\theta_\gamma(0)}{\pi} < n,$$

where $\theta_\gamma \in C([0,1],\mathbb{R})$ is defined by $\det(A(t)-\sqrt{-1}B(t)) = e^{\sqrt{-1}\theta_\gamma(t)}$, i.e., $\theta_\gamma(t) = \sum_{j=1}^{n}\theta_j(t)$. The left hand side equality holds only if $\gamma(1) = I_{2n}$.

Theorem 1.5 *For a symmetric positive definite path $\gamma \in \mathcal{P}_\tau(2n)$, there holds*

$$i_1(\gamma) = -\frac{1}{2}\nu_1(\gamma) \in [-n, 0]. \tag{2.12}$$

Proof We note that if $\gamma \in \mathcal{P}_\tau(2n)^*$, by definition, we can connect $\gamma(1)$ with M^+ by symmetric positive definite path. So there holds $i_1(\gamma) = 0$. If $\gamma \in \mathcal{P}_\tau(2n)^0$, the multiplicity of the eigenvalue $1 \in \sigma(\gamma(1))$ is even. I.e., $\nu_1(\gamma) = 2k \in 2\mathbb{Z}$. By k-times rotations, we get

$$\gamma_s(t) = R_{h_1}(st\theta_0)\cdots R_{h_k}(st\theta_0)\gamma(t).$$

Thus by definition again, we get

$$i_1(\gamma) = -k = -\frac{1}{2}\nu_1(\gamma) \ge -n.$$

□

2.2 ω-Index $i_\omega(\gamma)$

The ω-index theory for continuous symplectic paths starting from the identity matrix I_{2n} was first established in [226]. In this subsection we give a brief introduction of this ω-index theory without proofs and refer to [223] and [226] for the details. For any $\omega \in \mathbf{U}$, where $\mathbf{U}$ is the unite circle in complex plane, and $M \in \mathrm{Sp}(2n)$, as in [226] we define

$$D_\omega(M) = (-1)^{n-1}\omega^{-n}\det(M - \omega I).$$

One can easily see that $D_\omega(M) = D_{\bar{\omega}}(M)$ for all $\omega \in \mathbf{U}$, $M \in \mathrm{Sp}(2n)$ and $D \in C^\infty(\mathbf{U} \times \mathrm{Sp}(2n), \mathbb{R})$. For $\omega \in \mathbf{U}$, we set

$$\begin{aligned}
\mathrm{Sp}(2n)_\omega^\pm &= \{M \in \mathrm{Sp}(2n) | \pm D_\omega(M) < 0\},\\
\mathrm{Sp}(2n)_\omega^* &= \mathrm{Sp}(2n)_\omega^+ \cup \mathrm{Sp}(2n)_\omega^-, \quad \mathrm{Sp}(2n)_\omega^0 = \mathrm{Sp}(2n) \setminus \mathrm{Sp}(2n)_\omega^*,
\end{aligned}$$

and

$$\mathcal{P}_{\tau,\omega}^*(2n) = \{\gamma \in \mathcal{P}_\tau(2n) | \gamma(\tau) \in \mathrm{Sp}(2n)_\omega^*\}.$$

Definition 2.1 For any $\tau > 0$ and $\gamma \in \mathcal{P}_\tau(2n)$, we define

$$\nu_\omega(\gamma) = \dim_{\mathbb{C}} \ker_{\mathbb{C}}(\gamma(\tau) - \omega I), \quad \forall \omega \in \mathbf{U}.$$

Definition 2.2 For $\tau > 0$ and $\omega \in \mathbf{U}$, given two paths γ_0 and $\gamma_1 \in \mathcal{P}_\tau(2n)$, if there exists a map $\delta \in C([0,1] \times [0,\tau], \mathrm{Sp}(2n))$ such that $\delta(0,\cdot) = \gamma_0(\cdot)$, $\delta(1,\cdot) = \gamma_1(\cdot)$, $\delta(s,0) = I$ and $\nu_\omega(\delta(s,\cdot))$ is constant for $0 \le s \le 1$, the paths γ_0 and γ_1 are *ω-homotopic* on $[0,\tau]$ along $\delta(\cdot,\tau)$ and we write $\gamma_0 \sim_\omega \gamma_1$. If $\gamma_0 \sim_\omega \gamma_1$ for all $\omega \in \mathbf{U}$, then γ_0 and γ_1 are *homotopic* on $[0,\tau]$ along $\delta(\cdot,\tau)$ and we write $\gamma_0 \sim \gamma_1$.

As well known, every $M \in \mathrm{Sp}(2n)$ has its unique polar decomposition $M = AU$, where $A = (MM^T)^{1/2}$, $U = \begin{pmatrix} u_1 & -u_2 \\ u_2 & u_1 \end{pmatrix}$ with $u = u_1 + \sqrt{-1}u_2 \in \mathcal{L}(\mathbb{C}^n)$ being a unitary matrix. For $\gamma \in \mathcal{P}_\tau(2n)$, we have the corresponding $u(t) = u_1(t) + \sqrt{-1}u_2(t) \in U(n)$. So there exists a continuous real function $\Delta(t)$ satisfying $\det u(t) = \exp(\sqrt{-1}\Delta(t))$. We define $\Delta_\tau(\gamma) = \Delta(\tau) - \Delta(0) \in \mathbb{R}$ which depends only on γ.

For any $\gamma \in \mathcal{P}_{\tau,\omega}^*(2n)$, we can connect $\gamma(\tau)$ to M_n^- or M_n^+ by a path β within $\mathrm{Sp}(2n)_\omega^*$ and get a product path $\beta * \gamma$ defined by $\beta * \gamma(t) = \gamma(2t)$ if $0 \le t \le \tau/2$, $\beta * \gamma(t) = \beta(2t - \tau)$ if $\tau/2 \le t \le \tau$. Then

$$k \equiv \frac{1}{\pi}\Delta_\tau(\beta * \gamma) \in \mathbb{Z}.$$

This integer k is independent of the special choice of the path β. As in [226], we define

$$i_\omega(\gamma) = k \in \mathbb{Z}.$$

For $\gamma \in \mathcal{P}^0_{\tau,\omega}(2n) = \mathcal{P}_\tau(2n) \setminus \mathcal{P}^*_{\tau,\omega}(2n)$, we define

$$i_\omega(\gamma) = \inf\{\, i_\omega(\beta) |\ \beta \in \mathcal{P}^*_\tau(2n) \text{ and } \beta \text{ is sufficiently } C^0\text{-close to } \gamma\}.$$

Theorem 2.3 ([226]) *For any $\gamma \in \mathcal{P}_\tau(2n)$ and $\omega \in \mathbf{U}$, the above definition yields*

$$(i_\omega(\gamma), \nu_\omega(\gamma)) \in \mathbb{Z} \times \{0, 1, \cdots, 2n\},$$

*which is called the ω-**index** of γ.*

Note that the Maslov-type index coincides with the 1-index for any $\gamma \in \mathcal{P}_\tau(2n)$:

$$i_\omega(\gamma) = i_1(\gamma), \quad \nu_\omega(\gamma) = \nu_1(\gamma), \quad \omega = 1.$$

This ω-index theory generalizes also corresponding Bott functions $\Lambda(\omega)$ and $N(\omega)$ for closed geodesics defined in [28] and Ekeland index functions for convex Hamiltonian systems defined in [78].

For any $\gamma \in \mathcal{P}_\tau(2n)$, we define the iteration path $\tilde{\gamma} \in C([0, +\infty), \mathrm{Sp}(2n))$ of γ by

$$\tilde{\gamma}(t) = \gamma(t - j\tau)\gamma(\tau)^j, \qquad \text{for } j\tau \le t \le (j+1)\tau \text{ and } j \in \{0\} \cup \mathbb{N},$$

and denote by $\gamma^m = \tilde{\gamma}|_{[0,m\tau]}$ for $m \in \mathbb{N}$.

Theorem 2.4 ([226]) *For any $\gamma \in \mathcal{P}_\tau(2n)$ and $k \in \mathbb{N}$, there hold*

$$i_1(\gamma^k) = \sum_{\omega^k=1} i_\omega(\gamma), \quad \nu_1(\gamma^k) = \sum_{\omega^k=1} \nu_\omega(\gamma), \tag{2.13}$$

$$i_{\omega_0}(\gamma^k) = \sum_{\omega^k=\omega_0} i_\omega(\gamma), \quad \nu_{\omega_0}(\gamma^k) = \sum_{\omega^k=\omega_0} \nu_\omega(\gamma), \tag{2.14}$$

$$\bar{i}_1(\gamma) \equiv \lim_{k\to\infty} \frac{i_1(\gamma^k)}{k} = \frac{1}{2\pi}\int_0^{2\pi} i_{\exp(\sqrt{-1}\theta)}(\gamma)\, d\theta \in \mathbb{R}. \tag{2.15}$$

$\bar{i}_1(\gamma)$ is called the Maslov-type mean index per period τ of $\gamma \in \mathcal{P}_\tau(2n)$.

Equation (2.13) is called the Bott-type formula for the index pair (i_1, ν_1).

Note that there holds

$$\bar{i}_1(\gamma^k) = k\bar{i}_1(\gamma), \qquad \forall \gamma \in \mathcal{P}_\tau(2n), k \in \mathbb{N}. \tag{2.16}$$

Theorem 2.5 (Homotopy invariance, [226]) *For any two paths γ_0 and $\gamma_1 \in \mathcal{P}_\tau(2n)$, if $\gamma_0 \sim_\omega \gamma_1$ on $[0, \tau]$, there hold*

$$i_\omega(\gamma_0) = i_\omega(\gamma_1), \quad \nu_\omega(\gamma_0) = \nu_\omega(\gamma_1).$$

Theorem 2.6 (Symplectic additivity, [226]) *For any $\gamma_j \in \mathcal{P}_\tau(2n_j)$ with $n_j \in \mathbb{N}$, $j = 0,\ 1$, there holds*

$$i_\omega(\gamma_0 \diamond \gamma_1) = i_\omega(\gamma_0) + i_\omega(\gamma_1).$$

Lemma 2.7 ([223]) *For a symplectic path $\gamma \in \mathcal{P}(2n)$ with $\gamma(1) = M_1 \diamond M_2$, $M_j \in \mathrm{Sp}(2n_j)$, $j = 1, 2$, $n_1 + n_2 = n$, there exists two symplectic paths $\gamma_j \in \mathcal{P}(2n_j)$ such that $\gamma \sim \gamma_1 \diamond \gamma_2$ and $\gamma_j(1) = M_j$.*

As proved in [226], the homotopic invariance, symplectic additivity, and the values on elements in $\mathcal{P}_\tau(2) \cup \mathcal{P}_\tau(4)$ uniquely determine the ω-index theory.

For any $\tau > 0$, $\gamma \in \mathcal{P}_\tau(2n)$, the ω-index pair $(i_\omega(\gamma), \nu_\omega(\gamma))$ is determined by the homotopic class of γ in $\mathcal{P}_\tau(2n)$. In particular, $i_\omega(\gamma)$ is completely determined by the homotopic component $\Omega_0(\gamma(\tau))$ up to an additive constant, and $\nu_\omega(\gamma)$ is completely determined by $\Omega_0(\gamma(\tau))$.

We define the basic normal forms of eigenvalues in $\mathbf{U}$ as:

$$N_1(\lambda, b) = \begin{pmatrix} \lambda & b \\ 0 & \lambda \end{pmatrix}, \quad \lambda = \pm 1, \quad b = \pm 1,\ 0,$$

$$R(\theta) = \begin{pmatrix} \cos\theta & -\sin\theta \\ \sin\theta & \cos\theta \end{pmatrix}, \theta \in (0, \pi) \cup (\pi, 2\pi),$$

$$N_2(\omega, B) = \begin{pmatrix} R(\theta) & B \\ 0 & R(\theta) \end{pmatrix}, \quad \theta \in (0, \pi) \cup (\pi, 2\pi),$$
$$B = \begin{pmatrix} b_{11} & b_{12} \\ b_{21} & b_{22} \end{pmatrix}, \quad b_{ij} \in \mathbb{R}, \ b_{12} \neq b_{21}.$$

A basic normal form M is trivial, if for sufficiently small $\alpha > 0$, $MR((t-1)\alpha)^{\diamond n}$ possesses no eigenvalue on $\mathbf{U}$ for $t \in [0, 1)$, and is non-trivial otherwise.

Note that the matrix $N_1(\lambda, b)$ is non-trivial if $\lambda b \neq -1$ and trivial if $\lambda b = -1$.

For the Maslov-type index function $i_\omega(\gamma)$, we have the following result which was proved in [197] (see also [196] and [223]).

Proposition 2.8 ([197])

1° *For any $\gamma \in \mathcal{P}(2n)$ and $\omega \in \mathbf{U} \setminus \{1\}$, there always holds*

$$i_1(\gamma) + \nu_1(\gamma) - n \leq i_\omega(\gamma) \leq i_1(\gamma) + n - \nu_\omega(\gamma). \tag{2.17}$$

2° *The left equality in (2.17) holds for some* $\omega \in \mathbf{U}^+\setminus\{1\}$ *(or* $\mathbf{U}^-\setminus\{1\}$*) if and only if there holds* $I_{2p}\diamond N_1(1,-1)^{\diamond q}\diamond K \in \Omega_0(\gamma(\tau))$ *for some non-negative integers* p *and* q *satisfying* $0\le p+q\le n$ *and* $K\in \mathrm{Sp}(2(n-p-q))$ *with* $\sigma(K)\subset \mathbf{U}\setminus\{1\}$ *satisfying that all eigenvalues of* K *located within the arc between* 1 *and* ω *including* ω *in* $\mathbf{U}^+$ *(or* $\mathbf{U}^-$*) possess total multiplicity* $n-p-q$*. If* $\omega\neq -1$*, all eigenvalues of* K *are in* $\mathbf{U}\setminus\mathbb{R}$ *and those in* $\mathbf{U}^+\setminus\mathbb{R}$ *(or* $\mathbf{U}^-\setminus\mathbb{R}$*) are all Krein negative (or positive) definite. If* $\omega=-1$*, it holds that* $-I_{2s}\diamond N_1(-1,1)^{\diamond t}\diamond H\in \Omega_0(K)$ *for some non-negative integers* s *and* t *satisfying* $0\le s+t\le n-p-q$*, and some* $H\in \mathrm{Sp}(2(n-p-q-s-t))$ *satisfying* $\sigma(H)\subset \mathbf{U}\setminus\mathbb{R}$ *and that all elements in* $\sigma(H)\cap\mathbf{U}^+$ *(or* $\sigma(H)\cap\mathbf{U}^-$*) are all Krein-negative (or Krein-positive) definite.*

3° *The left equality in (2.17) holds for all* $\omega\in\mathbf{U}\setminus\{1\}$ *if and only if* $I_{2p}\diamond N_1(1,-1)^{\diamond(n-p)}\in\Omega_0(\gamma(\tau))$ *for some integer* $p\in[0,n]$*. Specifically in this case, all the eigenvalues of* $\gamma(\tau)$ *equal to* 1 *and* $\nu_\tau(\gamma)=n+p\ge n$*.*

4° *The right equality in (2.17) holds for some* $\omega \in \mathbf{U}^+\setminus\{1\}$ *(or* $\mathbf{U}^-\setminus\{1\}$*) if and only if there holds* $I_{2p}\diamond N_1(1,1)^{\diamond r}\diamond K\in\Omega_0(\gamma(\tau))$ *for some non-negative integers* p *and* r *satisfying* $0\le p+r\le n$ *and* $K\in\mathrm{Sp}(2(n-p-r))$ *with* $\sigma(K)\subset\mathbf{U}\setminus\{1\}$ *satisfying the condition that all eigenvalues of* K *located within the closed arc between* 1 *and* ω *in* $\mathbf{U}^+\setminus\{1\}$ *(or* $\mathbf{U}^-\setminus\{1\}$*) possess total multiplicity* $n-p-r$*. If* $\omega\neq-1$*, all eigenvalues in* $\sigma(K)\cap\mathbf{U}^+$ *(or* $\sigma(K)\cap\mathbf{U}^-$*) are all Krein positive (or negative) definite; if* $\omega=-1$*, there holds* $(-I_{2s})\diamond N_1(-1,1)^{\diamond t}\diamond H\in\Omega_0(K)$ *for some non-negative integers* s *and* t *satisfying* $0\le s+t\le n-p-r$*, and some* $H\in\mathrm{Sp}(2(n-p-r-s-t))$ *satisfying* $\sigma(H)\subset\mathbf{U}\setminus\mathbb{R}$ *and that all elements in* $\sigma(H)\cap\mathbf{U}^+$ *(or* $\sigma(H)\cap\mathbf{U}^-$*) are all Krein positive (or negative) definite.*

5° *The right equality in (2.17) holds for all* $\omega\in\mathbf{U}\setminus\{1\}$ *if and only if* $I_{2p}\diamond N_1(1,1)^{\diamond(n-p)}\in\Omega_0(\gamma(\tau))$ *for some integer* $p\in[0,n]$*. Specifically in this case, all the eigenvalues of* $\gamma(\tau)$ *must be* 1*, and there holds* $\nu_\tau(\gamma)=n+p\ge n$*.*

6° *Both equalities in (2.17) hold for all* $\omega\in\mathbf{U}\setminus\{1\}$ *if and only if* $\gamma(\tau)=I_{2n}$*.*

The following results are useful iteration inequalities for the Maslov-type index $i_1(\gamma)$.

Proposition 2.9 ([197])

1° *For any* $\gamma\in\mathcal{P}(2n)$ *and* $m\in\mathbb{N}$*, there always holds*

$$m\bar{i}_1(\gamma)-n\le i_1(\gamma^m)\le m\bar{i}_1(\gamma)+n-\nu_1(\gamma^m). \tag{2.18}$$

2° *The right equality in (2.18) holds for all* $m\in\mathbb{N}$ *if and only if* $I_{2p}\diamond N_1(1,-1)^{\diamond(n-p)}\in\Omega_0(\gamma(\tau))$ *for some integer number* $p\in[0,n]$*. Especially in this case, all the eigenvalues of* $\gamma(\tau)$ *are equal to* 1 *and* $\nu_1(\gamma)=n+p\ge n$*.*

3° *The left equality in (2.18) holds for all* $m\in\mathbb{N}$ *if and only if* $I_{2p}\diamond N_1(1,1)^{\diamond(n-p)}\in\Omega_0(\gamma(\tau))$ *for some integer number* $p\in[0,n]$*. Especially in this case, all the eigenvalues of* $\gamma(\tau)$ *equal to* 1 *and* $\nu_\tau(\gamma)=n+p\ge n$*.*

4° *Both equalities in (2.18) hold for all* $m\in\mathbb{N}$ *if and only if* $\gamma(\tau)=I_{2n}$*.*

Proposition 2.10 ([197])

1° *For any $\gamma \in \mathcal{P}(2n)$ and $m \in \mathbb{N}$, there always holds*

$$m(i_1(\gamma)+\nu_1(\gamma)-n)+n-\nu_1(\gamma) \le i_1(\gamma^m) \le m(i_1(\gamma)+n)-n-(\nu_1(\gamma^m)-\nu_1(\gamma)). \tag{2.19}$$

2° *The left equality of* (2.19) *holds for some $m > 1$ if and only if $I_{2p}\diamond N_1(1,-1)^{\diamond q}\diamond K \in \Omega_0(\gamma(\tau))$ for some non-negative integers p and q satisfying $p+q \le n$ and some $K \in \mathrm{Sp}(2(n-p-q))$ satisfying $\sigma(K) \subset \mathbf{U}\setminus\{1\}$. If $m = 2$ and $r = n-p-q > 0$, then $N_1(-1,-1)^{\diamond r} \in \Omega_0(K)$. If $m \ge 3$ and $r = n-p-q > 0$, then $R(\theta_1)\diamond\cdots\diamond R(\theta_r) \in \Omega_0(K)$ for some $\theta_j \in (0,\pi)$ satisfying the condition that $0 < \frac{m\theta_j}{2\pi} \le 1$ with $1 \le j \le r$. In this case, all eigenvalues of K on $\mathbf{U}^+\setminus\{1\}$ (on $\mathbf{U}^-\setminus\{1\}$) are located on the open arc between 1 and $\exp(2\pi\sqrt{-1}/m)$ (and $\exp(-2\pi\sqrt{-1}/m)$) in $\mathbf{U}^+$ (in $\mathbf{U}^-$) and are all Krein negative (or Krein positive) definite.*

3° *The right equality of* (2.19) *holds for some $m > 1$ if and only if $I_{2p}\diamond N_1(1,1)^{\diamond r}\diamond K \in \Omega_0(\gamma(\tau))$ for some non-negative integers p and r satisfying $p+r \le n$ and some $K \in \mathrm{Sp}(2(n-p-r))$ with $\sigma(K) = \{-1\}$ satisfying the following conditions:*

If $m > 2$, we must have $n = p+r$.
If $m = 2$ and $n-p-r > 0$, there holds $N_1(-1,1)^{\diamond t}\diamond N_1(-1,-1)^{\diamond s} \in \Omega_0(K)$ for some non-negative integers s and t satisfying $s+t = n-p-r$.

4° *The two equalities of* (2.19) *hold for some $m = m_1$ and $m = m_2 \ge 2$ respectively if and only if $\gamma(\tau) = I_{2p}\diamond N_1(-1,-1)^{\diamond(n-p)}$ for some non-negative integer $p \le n$. Here $p < n$ happens only when $m_1 = m_2 = 2$.*

Proposition 2.11 ([233]) *For any $\gamma \in \mathcal{P}(2n)$, and any $m_1,\ m_2 \in \mathbb{N}$, there hold*

$$\begin{aligned}\nu_1(\gamma^{m_1}) + \nu_1(\gamma^{m_2}) - \nu_1(\gamma^{(m_1,m_2)}) - \frac{e(\gamma(\tau))}{2} \\ \le i_1(\gamma^{m_1+m_2}) - i_1(\gamma^{m_1}) - i_1(\gamma^{m_2}) \\ \le \nu_1(\gamma^{(m_1,m_2)}) - \nu_1(\gamma^{m_1+m_2}) + \frac{e(\gamma(\tau))}{2},\end{aligned}$$

where (m_1, m_2) is the greatest common divisor of m_1 and m_2.

Proposition 2.12 ([233]) *For any $\gamma \in \mathcal{P}(2n)$, and any $m \in \mathbb{N}$, there hold*

$$\nu_1(\gamma^m) - \frac{e(\gamma(\tau))}{2} \le i_1(\gamma^{m+1}) - i_1(\gamma^m) - i_1(\gamma) \le \nu_1(\gamma) - \nu_1(\gamma^{m+1}) + \frac{e(\gamma(\tau))}{2}. \tag{2.20}$$

We remind that the upper and lower semi-circle $\mathbf{U}^{\pm}$ in $\mathbf{U}$ is defined by $\mathbf{U}^+ = \{e^{\sqrt{-1}\theta} \mid \theta \in [0,\pi]\}$ and $\mathbf{U}^- = \{e^{\sqrt{-1}\theta} \mid \theta \in [\pi, 2\pi]\}$.

Definition 2.13 ([226]) For any $M \in \mathrm{Sp}(2n)$ and $\omega \in \mathbf{U}$, choose $\tau > 0$ and $\gamma \in \mathcal{P}_\tau(2n)$ with $\gamma(\tau) = M$, and define

$$S_M^\pm(\omega) = \lim_{\epsilon \to 0^+} i_{\exp \pm \epsilon \sqrt{-1}\omega}(\gamma) - i_\omega(\gamma).$$

These two integers are independent of the choice of $\tau > 0$ and the path γ. They are called the splitting numbers of M at ω.

It is easy to see that $S_M^+(1) = S_M^-(1)$ and $S_M^+(-1) = S_M^-(-1)$. When $\omega \notin \sigma(M)$, we have $S_M^\pm(\omega) = 0$. So from definition, for $\omega_0 = e^{\sqrt{-1}\theta_0}$, $\theta_0 \in (0, \pi]$, there holds

$$i_{\omega_0}(\gamma) = i_1(\gamma) + S_M^+(1) + \sum_{\theta \in (0,\theta_0)} (S^+(e^{\sqrt{-1}\theta}) - S^-(e^{\sqrt{-1}\theta})) - S_M^-(e^{\sqrt{-1}\theta_0}). \tag{2.21}$$

From the definition and the Bott type formula (2.14), we know that

$$S_{M^k}^\pm(\omega_0) = \sum_{\omega^k = \omega_0} S_M^\pm(\omega). \tag{2.22}$$

As a special case, we have

$$S_{M^2}^\pm(1) = S^\pm(1) + S^\pm(-1). \tag{2.23}$$

We recall that the Krein form is defined by

$$\langle Gx, y\rangle \quad \forall\, x, y \in \mathbb{C}^{2n},$$

where $G = \sqrt{-1}J$ and $\langle\cdot,\cdot\rangle$ is the standard Hermitian inner product in $\mathbb{C}^{2n}$. This quadratic form is symmetric in $\mathbb{C}^{2n}$.

We suppose $M \in \mathrm{Sp}(2n)$. It is well known that if $\lambda, \mu \in \sigma(M)$, and $\lambda\bar{\mu} \neq 1$, then the generalized eigen-subspaces $E_\lambda(M)$ and $E_\mu(M)$ are G-orthogonal. When $|\lambda| = 1$, then the restriction of G to $E_\lambda(M)$ is non-degenerate. I.e., for $x \in E_\lambda(M)$, $\langle Gx, y\rangle = 0$ for any $y \in E_\lambda(M)$ implies $x = 0$.

Definition 2.14 For $M \in \mathrm{Sp}(2n)$, if $\lambda \in \sigma(M)$ with $|\lambda| = 1$. Denote the total multiplicities of positive and negative eigenvalues of $G|_{E_\lambda(M)}$ by p and q respectively. The integer pair (p, q) is called the *Krein type number* of λ. If $q = 0$, λ is *Krein positive*. If $p = 0$, λ is *Krein negative*.

Lemma 2.15 ([226]) *For any $M \in \mathrm{Sp}(2n)$ and $\omega \in \mathbf{U}$, denote by (p, q) the Krein type number of $\omega \in \sigma(M)$. The splitting numbers $S_M^\pm(\omega)$ are constant on $\Omega^0(M)$, and satisfy*

$$0 \le S_M^{\pm}(\omega) \le \dim_{\mathbb{C}} \ker_{\mathbb{C}}(M - \omega I),$$
$$0 \le S_M^{+}(\omega) \le p, \quad 0 \le S_M^{-}(\omega) \le q,$$
$$S_M^{+}(\omega) = S_M^{-}(\bar{\omega}).$$

We note that (2.17) can be deduced from formula (2.21) and Lemma 2.15.

Lemma 2.16 ([223]) *For the splitting numbers, we have the following list.*

(1) $(S_M^+(1), S_M^-(1)) = (1, 1)$, *for* $M = N_1(1, b)$ *with* $b = 0$ *or* 1.
(2) $(S_M^+(1), S_M^-(1)) = (0, 0)$, *for* $M = N_1(1, -1)$.
(3) $(S_M^+(-1), S_M^-(-1)) = (1, 1)$, *for* $M = N_1(-1, b)$ *with* $b = 0$ *or* -1.
(4) $(S_M^+(-1), S_M^-(-1)) = (0, 0)$, *for* $M = N_1(-1, 1)$.
(5) $(S_M^+(e^{\sqrt{-1}\theta}), S_M^-(e^{\sqrt{-1}\theta})) = (0, 1)$, *for* $M = R(\theta)$ *with* $\theta \in (0, \pi) \cup (\pi, 2\pi)$.
(6) $(S_M^+(e^{\sqrt{-1}\theta}), S_M^-(e^{\sqrt{-1}\theta})) = (1, 1)$, *for* $M = N_2(\omega, b)$ *being non-trivial with* $\omega = e^{\sqrt{-1}\theta}$, $\theta \in \mathbb{R}$.
(7) $(S_M^+(e^{\sqrt{-1}\theta}), S_M^-(e^{\sqrt{-1}\theta})) = (0, 0)$, *for* $M = N_2(\omega, B)$ *being trivial with* $\omega = e^{\sqrt{-1}\theta}$, $\theta \in \mathbb{R}$.
(8) $(S_M^+(e^{\sqrt{-1}\theta}), S_M^-(e^{\sqrt{-1}\theta})) = (0, 0)$, *for any* $\omega \in \mathbf{U}$ *and* $M \in \mathrm{Sp}(2n)$ *satisfying* $\omega \in \sigma(M)$.

Chapter 3
Relative Morse Index

3.1 Relative Index via Galerkin Approximation Sequences

Let E be a separable Hilbert space, and $Q = A - B : E \to E$ be a bounded self-adjoint linear operators with $B : E \to E$ being a compact self-adjoint operator. Suppose that $N = \ker Q$ and $\dim N < +\infty$. $Q|_{N^\perp}$ is invertible. $P : E \to N$ is the orthogonal projection. We denote $0 < d \leq \frac{1}{4}\|(Q|_{N^\perp})^{-1}\|^{-1}$. Suppose $\Gamma = \{P_k | k = 1, 2, \cdots\}$ is the Galerkin approximation sequence of A with

(1) $E_k := P_k E$ is finite dimensional for all $k \in \mathbb{N}$,
(2) $P_k \to I$ strongly as $k \to +\infty$
(3) $P_k A = A P_k$.

For a self-adjoint operator T, we denote by $M^*(T)$ the eigenspaces of T with eigenvalues belonging to $(0, +\infty)$, $\{0\}$ and $(-\infty, 0)$ with $* = +, 0$ and $* = -$, respectively. We denote by $m^*(T) = \dim M^*(T)$. Similarly, we denote by $M_d^*(T)$ the d-eigenspaces of T with eigenvalues belonging to $(d, +\infty)$, $(-d, d)$ and $(-\infty, -d)$ with $* = +, 0$ and $* = -$, respectively. We denote by $m_d^*(T) = \dim M_d^*(T)$. For any adjoint operator L, we denote $L^\sharp = (L|_{ImL})^{-1}$.

Lemma 1.1 *There exists $m_0 \in \mathbb{N}$ such that for all $m \geq m_0$, there hold*

$$m^-(P_m(Q + P)P_m) = m_d^-(P_m(Q + P)P_m) \tag{3.1}$$

and

$$m^-(P_m(Q + P)P_m) = m_d^-(P_m Q P_m). \tag{3.2}$$

Proof The proof of (3.1) is essential the same as that of Theorem 2.1 of [95], we note that $\dim\ker(Q + P) = 0$.

By considering the operators $Q + sP$ and $Q - sP$ for small $s > 0$, for example $s < \min\{1, d/2\}$, there exists $m_1 \in \mathbb{N}$ such that

$$m_d^-(P_m Q P_m) \le m^-(P_m(Q + sP)P_m), \ \forall m \ge m_1 \tag{3.3}$$

and

$$m_d^-(P_m Q P_m) \ge m^-(P_m(Q - sP)P_m) - m_d^0(P_m Q P_m), \ \forall m \ge m_1. \tag{3.4}$$

In fact, the claim (3.3) follows from

$$P_m(Q + sP)P_m = P_m Q P_m + s P_m P P_m$$

and for $x \in M_d^-(P_m Q P_m)$,

$$(P_m(Q + sP)P_m x, x) \le -d\|x\|^2 + s\|x\|^2 \le -\frac{d}{2}\|x\|^2.$$

The claim (3.4) follows from that for $x \in M^-(P_m(Q - sP)P_m)$,

$$(P_m Q P_m x, x) \le s(P_m P P_m x, x) < d\|x\|^2.$$

By the Floquet theory, for $m \ge m_1$ we have $m_d^0(P_m Q P_m) = \dim N = \dim Im(P_m P P_m)$, and by $Im(P_m P P_m) \subseteq M_d^0(P_m Q P_m)$ we have $Im(P_m P P_m) = M_d^0(P_m Q P_m)$. It is easy to see that $M_d^0(P_m Q P_m) \subseteq M_d^+(P_m(Q + sP)P_m)$. By using

$$P_m(Q - sP)P_m = P_m(Q + sP)P_m - 2s P_m P P_m$$

we have

$$m^-(P_m(Q - sP)P_m) \ge m^-(P_m(Q + sP)P_m) + m_d^0(P_m Q P_m), \ \forall m \ge m_1. \tag{3.5}$$

Now (3.2) follows from (3.3), (3.4), and (3.5). □

Since $M^-(Q + P) = M^-(Q)$ and the two operators $Q + P$ and Q have the same negative spectrum, moreover, $P_m(Q + P)P_m \to Q + P$ and $P_m Q P_m \to Q$ strongly, one can prove (3.2) by the spectrum decomposition theory.

Lemma 1.2 *Let B be a linear self-adjoint compact operator as above. Then $m_d^0(P_m(A - B)P_m)$ eventually becomes a constant independent of m and for large m, there holds*

$$m_d^0(P_m(A - B)P_m) = m^0(A - B). \tag{3.6}$$

Proof It is easy to show that there is a constant $m_1 > 0$ such that for $m \ge m_1$

$$\dim P_m \ker(A-B) = \dim \ker(A-B).$$

Since B is compact, there is $m_2 \ge m_1$ such that for $m \ge m_2$

$$\|(I-P_m)B\| \le 2d.$$

Take $m \ge m_2$, let $E_m = P_m \ker(A-B) \bigoplus Y_m$, then $Y_m \subseteq \mathrm{Im}(A-B)$. For $y \in Y_m$ we have

$$y = (A-B)^\sharp(A-B)y = (A-B)^\sharp(P_m(A-B)P_m y + (I-P_m)By).$$

It implies

$$\|P_m(A-B)P_m y\| \ge 2d\|y\|, \ \forall y \in Y_m.$$

Thus we have

$$m_d^0(P_m(A-B)P_m) \le m^0(A-B). \tag{3.7}$$

On the other hand, for $x \in P_m \ker(A-B)$, there exists $y \in \ker(A-B)$, such that $x = P_m y$. Since $P_m \to I$ strongly, there exists $m_3 \ge m_2$ such that for $m \ge m_3$

$$\|I-P_m\| < \frac{1}{2}, \quad \|P_m(A-B)(I-P_m)\| \le \frac{d}{2}.$$

So we have

$$\|P_m(A-B)P_m x\| = \|P_m(A-B)(I-P_m)y\| \le \frac{d}{2}\|y\| < d\|x\|.$$

It implies that

$$m_d^0(P_m(A-B)P_m) \ge m^0(A-B). \tag{3.8}$$

Equation (3.6) holds from (3.7) and (3.8). □

In [95], the following result was proved.

$$m_d^*(P_m(A-B)P_m) = m^*(A-B) \text{ with } * = 0, \ + \text{and } -. \tag{3.9}$$

From the above proof, we see that for any two operators B_1 and $B_2 \in \mathcal{L}_s(E)$, the exist $d > 0$ and $m^* > 0$ such that

$$m_d^0(P_m(A - B(s))P_m) = m^0(A - B(s)), \quad m > m^*,$$

where $B(s) = (1 - s)B_1 + sB_2$, $s \in [0, 1]$.

Theorem 1.3 *For any two operators B_1 and $B_2 \in \mathcal{L}_s(E)$ with $B_1 < B_2$, there is $m^* > 0$ such that*

$$m_d^-(P_m(A - B_2)P_m) - m_d^-(P_m(A - B_1)P_m) = \sum_{s \in [0,1)} m^0(A - B(s)), \quad m > m^*. \tag{3.10}$$

Proof Denote by $m_d^-(s) = m_d^-(P_m(A - B(s))P_m)$, $m_d^0(s) = m_d^0(P_m(A - B(s))P_m) = m^0(A - B(s))$. If $m^0(A - B(s_0)) = 0$, then there is a neighborhood $B(s_0, \delta)$ of s_0 such that for $s \in B(s_0, \delta)$, there hold $m_d^0(P_m(A - B(s))P_m) = m^0(A - B(s)) = 0$. Thus $m_d^-(s)$ is constant in $B(s_0, \delta)$. If $m^0(A - B(s_0)) \neq 0$, we claim that $m_d^-(s_0 + 0) - m_d^-(s_0) = \nu(s_0)$. In fact, on one side hand, by the continuity of the eigenvalue of continuous operator function, we have $m^-(s_0+0) - m^-(s_0) \le \nu(s_0)$. On the other side hand, since $(A - B(s_0)) > (A - B(s))$, for $s_0 < s$, we see that $m^0(A - B(s)) = 0$ for $s > s_0$ but $s - s_0$ small enough. So $m_d^0(s) = m_d^0(P_m(A - B(s))P_m) = 0$. Since $P_m(A - B(s_0))P_m \ge P_m(A - B(s))P_m$, so we have $m_d^-(s_0 + 0) + m_d^0(s_0 + 0) \ge m_d^-(s_0) + m_d^0(s_0)$. Thus the claim is true by using Lemma 1.1. Therefore we have the equality (3.10). □

Lemma 1.4 *Let B be a linear self-adjoint compact operator. Then the difference of the d-Morse indices*

$$m_d^-(P_m(A - B)P_m) - m_d^-(P_mAP_m) \tag{3.11}$$

eventually becomes a constant independent of m, where $d > 0$ is determined by the operators A and $A - B$.

A similar result was proved in [38].

Proof We can choose $B_0 < 0$ and $B_0 < B$, so we have

$$\begin{aligned} & m_d^-(P_m(A - B)P_m) - m_d^-(P_mAP_m) \\ & = (m_d^-(P_m(A - B)P_m) - m_d^-(P_m(A - B_0)P_m)) \\ & \quad -(m_d^-(P_mAP_m) - m_d^-(P_m(A - B_0)P_m)). \end{aligned}$$

□

Definition 1.5 For the self-adjoint bounded Fredholm operator A with a Galerkin approximation sequence Γ and the self-adjoint compact operator B on Hilbert space E, we define the relative Morse index by

$$I(A, A - B) = m_d^-(P_m(A - B)P_m) - m_d^-(P_mAP_m), \quad m \ge m^*, \tag{3.12}$$

where $m^* > 0$ is a constant large enough such that the difference in (3.12) becomes a constant independent of $m \geq m^*$.

By Lemma 1.4 we have the following

Remark 1.6 Let $\tilde{E}$ be another separable Hilbert space, $\tilde{A}$ be a linear self-adjoint Fredholm operator on $\tilde{E}$ and B be a compact linear self-adjoint operator on $\tilde{E}$. There holds

$$I(A \oplus \tilde{A}, (A \oplus \tilde{A}) - (B \oplus \tilde{B})) = I(A, A - B) + I(\tilde{A}, \tilde{A} - \tilde{B}),$$

where $(A \oplus \tilde{A})(x \oplus y) = Ax \oplus \tilde{A}y$ and $(B \oplus \tilde{B})(x \oplus y) = Bx \oplus \tilde{B}y$ for $x \oplus y \in E \oplus \tilde{E}$.

The spectral flow for a parameter family of linear self-adjoint Fredholm operators was introduced by Atiyah, Patodi and Singer in [11]. The following result shows that the relative index in Definition 1.5 is a spectral flow.

Lemma 1.7 *For the operators A and B in Definition* 1.5*, there holds*

$$I(A, A - B) = -\mathrm{sf}\{A - sB,\ 0 \leq s \leq 1\}, \tag{3.13}$$

where $\mathrm{sf}(A - sB,\ 0 \leq s \leq 1)$ *is the spectral flow of the operator family* $A - sB$, $s \in [0, 1]$ *(cf.* [316]*).*

Proof For simplicity, we set $I_{\mathrm{sf}}(A, A - B) = -\mathrm{sf}\{A - sB,\ 0 \leq s \leq 1\}$ which is exact the relative Morse index defined in [316]. By the Galerkin approximation formula in Theorem 3.1 of [316],

$$I_{\mathrm{sf}}(A, A - B) = I_{\mathrm{sf}}(P_m A P_m,\ P_m(A - B)P_m) \tag{3.14}$$

if $\ker(A) = \ker(A - B) = 0$.

By (2.17) of [316], we have

$$\begin{aligned} I_{\mathrm{sf}}(P_m A P_m,\ P_m(A - B)P_m) &= m^-(P_m(A - B)P_m) - m^-(P_m A P_m) \\ &= m_d^-(P_m(A - B)P_m) - m_d^-(P_m A P_m) \\ &= I(A, A - B) \end{aligned} \tag{3.15}$$

for $d > 0$ small enough. Hence (3.13) holds in the nondegenerate case. In general, if $\ker(A) \neq 0$ or $\ker(A - B) \neq 0$, we can choose $d > 0$ small enough such that $\ker(A + d\mathrm{Id}) = \ker(A - B + d\mathrm{Id}) = 0$, here $\mathrm{Id} :\ E \to E$ is the identity operator. By (2.14) of [316] we have

$$\begin{aligned} I_{\mathrm{sf}}(A, A - B) &= I_{\mathrm{sf}}(A, A + d\mathrm{Id}) + I_{\mathrm{sf}}(A + d\mathrm{Id},\ A - B + d\mathrm{Id}) \\ &\quad + I_{\mathrm{sf}}(A - B + d\mathrm{Id},\ A - B) \\ &= I_{\mathrm{sf}}(A + d\mathrm{Id},\ A - B + d\mathrm{Id}) = I(A + d \cdot \mathrm{Id},\ A - B + d \cdot \mathrm{Id}) \end{aligned}$$

$$= m^-(P_m(A - B + d\mathrm{Id})P_m) - m^-(P_m(A + d\mathrm{Id})P_m)$$

$$= m_d^-(P_m(A - B)P_m) - m_d^-(P_m A P_m) = I(A, A - B). \quad (3.16)$$

In the second equality of (3.16) we note that $I_{\mathrm{sf}}(A, A + d\mathrm{Id}) = I_{\mathrm{sf}}(A - B + d\mathrm{Id}, A - B) = 0$ for $d > 0$ small enough since the spectrum of A is discrete and B is a compact operator, in the third and the forth equalities of (3.16) we have applied (3.15). □

A similar way to define the relative index of two operators was shown in [38]. A different way to study the relative index theory was presented in [91].

3.2 Relative Morse Index via Orthogonal Projections

In general, let $\mathbb{E}$ be a separable Hilbert space, for any self-adjoint operator A on $\mathbb{E}$, there is a unique A-invariant orthogonal splitting

$$\mathbb{E} = \mathbb{E}^+(A) \oplus \mathbb{E}^-(A) \oplus \mathbb{E}^0(A), \quad (3.17)$$

where $\mathbb{E}^0(A)$ is the null space of A, A is positive definite on $\mathbb{E}^+(A)$ and negative definite on $\mathbb{E}^-(A)$. We denote by P_A the orthogonal projection from $\mathbb{E}$ to $\mathbb{E}^-(A)$. For any bounded self-adjoint Fredholm operator $\mathcal{F}$ and a compact self-adjoint operator $\mathcal{T}$ on $\mathbb{E}$, $P_{\mathcal{F}} - P_{\mathcal{F}-\mathcal{T}}$ is compact (cf. Lemma 2.7 of [316]), where $P_{\mathcal{F}} : \mathbb{E} \to \mathbb{E}^-(\mathcal{F})$ and $P_{\mathcal{F}-\mathcal{T}} : \mathbb{E} \to \mathbb{E}^-(\mathcal{F} - \mathcal{T})$ are the respective projections. Then by Fredholm operator theory, $P_{\mathcal{F}}|_{\mathbb{E}^-(\mathcal{F}-\mathcal{T})} : \mathbb{E}^-(\mathcal{F} - \mathcal{T}) \to \mathbb{E}^-(\mathcal{F})$ is a Fredholm operator. Here and in the sequel, we denote by ind $(\cdot)$ the Fredholm index of a Fredholm operator.

Definition 2.1 For any bounded self-adjoint Fredholm operator $\mathcal{F}$ and a compact self-adjoint operator $\mathcal{T}$ on $\mathbb{E}$, the relative Morse index pair $(\mu_{\mathcal{F}}(\mathcal{T}), \upsilon_{\mathcal{F}}(\mathcal{T}))$ is defined by

$$\mu_{\mathcal{F}}(\mathcal{T}) = \mathrm{ind}(P_{\mathcal{F}}|_{\mathbb{E}^-(\mathcal{F}-\mathcal{T})}). \quad (3.18)$$

and

$$\upsilon_{\mathcal{F}}(\mathcal{T}) = \dim \mathbb{E}^0(\mathcal{F} - \mathcal{T}). \quad (3.19)$$

Let $\{\mathcal{F}_\theta | \theta \in [0, 1]\}$ be a continuous path of self-adjoint Fredholm operators on the Hilbert space $\mathbb{E}$. It is well known that the concept of spectral flow $Sf(\mathcal{F}_\theta)$ was first introduced by Atiyah, Patodi and Singer in [11], and then extensively studied in [32, 98, 259, 260, 316]. The following proposition displays the relationship between spectral flow and the relative Morse index defined above.

Proposition 2.2 (cf. [40]) *Suppose that, for each $\theta \in [0, 1]$, $\mathcal{F}_\theta - \mathcal{F}_0$ is a compact operator on $\mathbb{E}$, then*

$$\mathrm{ind}(P_{\mathcal{F}_0}|_{\mathbb{E}^-(\mathcal{F}_1)}) = -sf(\mathcal{F}_\theta).$$

Thus, from Definition 2.1,

$$\mu_{\mathcal{F}_0}(\mathcal{T}) = -sf(\mathcal{F}_\theta,\ 0 \le \theta \le 1),$$

where $\mathcal{F}_\theta = \mathcal{F} - \theta\mathcal{T}$, $\mathcal{T}$ is a compact operator. Moreover, if $\sigma(\mathcal{T}) \subset [0, \infty)$ and $0 \notin \sigma_P(\mathcal{T})$, from the definition of Spectral flow, we have

$$\begin{aligned} \mu_{\mathcal{F}_0}(\mathcal{T}) &= -sf(\mathcal{F}_\theta,\ 0 \le \theta \le 1) \\ &= \sum_{\theta\in[0,1)} \upsilon_{\mathcal{F}}(\theta\mathcal{T}) \\ &= \sum_{\theta\in[0,1)} \dim \mathbb{E}^0(\mathcal{F} - \theta\mathcal{T}). \end{aligned} \tag{3.20}$$

3.3 Morse Index via Dual Methods

Let H be an infinitely dimensional separable Hilbert space with inner product $(\cdot, \cdot)_H$ and norm $\|\cdot\|_H$.

Denote by $\mathcal{O}(H)$ the set of all linear self-adjoint operators on H. For $A \in \mathcal{O}(H)$, we denote by $\sigma(A)$ the spectrum of A and $\sigma_e(A)$ the essential spectrum of A. We define three subsets of $\mathcal{O}(H)$ as follows

$$\mathcal{O}_e^-(\mu) = \{A \in \mathcal{O}(H)|\ \sigma_e(A) \cap (-\infty, \mu) = \emptyset \text{ and } \sigma(A) \cap (-\infty, \mu) \ne \emptyset\},$$

$$\mathcal{O}_e^+(\mu) = \{A \in \mathcal{O}(H)|\ \sigma_e(A) \cap (\mu, +\infty) = \emptyset \text{ and } \sigma(A) \cap (\mu, +\infty) \ne \emptyset\},$$

$$\mathcal{O}_e^0(a, b) = \{A \in \mathcal{O}(H)|\ \sigma_e(A) \cap (a, b) = \emptyset \text{ and } \sigma(A) \cap (a, b) \ne \emptyset\}.$$

We note that if $\mu = +\infty$ and $A \in \mathcal{O}_e^-(\mu)$, then $\sigma_e(A) = \emptyset$. If $\sigma_e(A) \ne \emptyset$ and $A \in \mathcal{O}_e^-(\mu)$ for some μ, then $-\infty < \mu < +\infty$ is a real number. Setting $\lambda^- = \inf(\sigma_e(A))$, we have $-\infty < \lambda^- < +\infty$ is real number and $A \in \mathcal{O}_e^-(\lambda^-)$. Similarly, if $\mu = -\infty$ and $A \in \mathcal{O}_e^+(\mu)$, then $\sigma_e(A) = \emptyset$. If $\sigma_e(A) \ne \emptyset$ and $A \in \mathcal{O}_e^+(\mu)$ for some μ, then $-\infty < \mu < +\infty$ is a real number. Setting $\lambda^+ = \sup(\sigma_e(A))$, we have $-\infty < \lambda^+ < +\infty$ is real number and $A \in \mathcal{O}_e^+(\lambda^+)$. If the operator A is fixed and $A \in \mathcal{O}_e^-(\mu)$ or $A \in \mathcal{O}_e^+(\mu)$, we always write it in $A \in \mathcal{O}_e^-(\lambda^-)$ or $A \in \mathcal{O}_e^+(\lambda^+)$ with $\lambda^\mp$ in the above sense. We remind that $\inf\emptyset = +\infty$ and $\sup\emptyset = -\infty$.

We have a remark that when $A \in \mathcal{O}(H)$ with $\sigma_e(A) \cap (-\infty, \mu) = \emptyset$ and $\sigma(A) \cap (-\infty, \mu) = \emptyset$, the index $i_A^-(B)$ in *case 1* below is still well defined for

$B \in \mathcal{L}_s^-(H, \lambda^-)$, but all the indices are the same, i.e., $i_A^-(B)$ is a constant function on $\mathcal{L}_s^-(H, \lambda^-)$ and so the index $i_A^-(B)$ take no further information. We also have a similar remark for other two cases.

Let $A \in \mathcal{O}(H)$ satisfying $\sigma(A) \setminus \sigma_e(A) \neq \emptyset$. Now, we consider the following cases:

Case 1. $A \in \mathcal{O}_e^-(\lambda^-)$.
Case 2. $A \in \mathcal{O}_e^+(\lambda^+)$.
Case 3. $A \in \mathcal{O}_e^0(\lambda_a, \lambda_b)$, $-\infty < \lambda_a < \lambda_b < +\infty$ and $\lambda_a, \lambda_b \notin \sigma(A)$.

Denote $\mathcal{L}_s(H)$ the set of all linear bounded self-adjoint operators on H. Corresponding to *Case 1*, *Case 2* and *Case 3*, we define $\mathcal{L}_s^-(H, \lambda^-)$, $\mathcal{L}_s^+(H, \lambda^+)$ and $\mathcal{L}_s^0(H, \lambda_a, \lambda_b)$ three subsets of $\mathcal{L}_s(H)$ respectively by

$$\mathcal{L}_s^-(H, \lambda^-) = \{B \in \mathcal{L}_s(H),\ B < \lambda^- \cdot I\}, \tag{3.21}$$

$$\mathcal{L}_s^+(H, \lambda^+) = \{B \in \mathcal{L}_s(H),\ B > \lambda^+ \cdot I\}, \tag{3.22}$$

and

$$\mathcal{L}_s^0(H, \lambda_a, \lambda_b) = \{B \in \mathcal{L}_s(H),\ \lambda_a \cdot I < B < \lambda_b \cdot I \tag{3.23}$$

where I is the identity map on H, the inequality $B < \lambda^- \cdot I$ means that there exists $\delta > 0$ such that $(\lambda^- - \delta) \cdot I - B$ is positive define, $B > \lambda^+ \cdot I$ and $\lambda_a \cdot I < B < \lambda_b \cdot I$ have similar meanings. It is easy to see $\mathcal{L}_s^-(H, \lambda^-)$, $\mathcal{L}_s^+(H, \lambda^+)$ and $\mathcal{L}_s^0(H, \lambda_a, \lambda_b)$ are open and convex subsets of $\mathcal{L}_s(H)$. We will define the index pair $(i_A^\mp(B), \nu_A^\mp(B))$ and $(i_A^0(B), \nu_A^0(B))$ in three cases.

3.3.1 The Definition of Index Pair in Case 1 and 2

In this subsection, we will give the definition of our index pair in *Case 1* and *Case 2*. Let's begin with some useful lemmas.

Lemma 3.1 ([284]) *If* $A \in \mathcal{O}_e^-(\lambda^-)$ *(or* $A \in \mathcal{O}_e^+(\lambda^+)$ *), for any* $B \in \mathcal{L}_s^-(H, \lambda^-)$ *(or* $B \in \mathcal{L}_s^+(H, \lambda^+)$*),* $\dim\ker(A - B) < \infty$*. Further more, if* $0 \in \sigma(A - B)$*, then* 0 *is an isolated point spectrum.*

Proof We only consider the situation of $A \in \mathcal{O}_e^-(\lambda^-)$ and $B \in \mathcal{L}_s^-(H, \lambda^-)$, the proof for other situations is similar. Since B is a bounded self-adjoint operator on H, we can choose a number $k \in \mathbb{R}$ satisfying

$$k \notin \sigma(A),\ k \cdot I < B. \tag{3.24}$$

Then we have $A - k \cdot I$ is invertible and

$$0 < B - k \cdot I < (\lambda^- - k) \cdot I. \tag{3.25}$$

That is to say $B - k \cdot I$ is a positive definite self-adjoint operator on H and

$$\|B - k \cdot I\| < \lambda^- - k. \tag{3.26}$$

If $\sigma(A) \cap (-\lambda^- + 2k, \lambda^-) = \emptyset$, that is $\sigma(A - k \cdot I) \cap (-\lambda^- + k, \lambda^- - k) = \emptyset$. From (3.26) it is easy to see $0 \notin \sigma(A - B)$. Now, we assume

$$\sigma(A) \cap (-\lambda^- + 2k, \lambda^-) \neq \emptyset. \tag{3.27}$$

For simplicity, we denote $A_k := A - k \cdot I$ and $B_k := B - k \cdot I$, then $A - B = A_k - B_k$, specially $\ker(A - B) = \ker(A_k - B_k)$ and $\sigma(A - B) = \sigma(A_k - B_k)$. We need the idea of Lyapunov-Schmidt reduction which is also the idea of saddle point reduction. Let $E(z)$ be the spectral measure of A_k. From (3.26) and (3.27), there exists $\delta > 0$ small enough such that

$$\|B_k\| < \lambda^- - k - \delta, \tag{3.28}$$

and

$$\sigma(A_k) \cap (-\lambda^- + k + \delta, \lambda^- - k - \delta) \neq \emptyset.$$

Denote

$$P_0 := \int_{-(\lambda^- - k - \delta)}^{\lambda^- - k - \delta} 1 dE(z),$$

and

$$P_1 := I - P_0.$$

Then H has the decomposition $H = H_0 \oplus H_1$ with $H_0 = P_0 H$ and $H_1 = P_1 H$. From our assumption there are only finite eigenvalues of A_k in set $(-\lambda^- + k + \delta, \lambda^- - k - \delta)$, each of them has finite dimensional eigenspace. So H_0 is a finite dimensional space. Denote $E = D(|A_k|^{1/2})$, since $k \notin \sigma(A)$ and $0 \notin \sigma(A_k)$, thus E is a Hilbert space with the inner product $(\cdot, \cdot)_E$ and corresponding norm $\| \cdot \|_E$ defined by

$$(x, y)_E := (|A_k|^{1/2} x, |A_k|^{1/2} y)_H, \ \forall x, y \in E,$$

$$\|x\|_E^2 := (x, x)_E, \ \forall x \in E.$$

We also have the following decomposition

$$E = E_0 \oplus E_1, \tag{3.29}$$

with $E_0 = E \cap H_0$ and $E_1 = E \cap H_1$. The operator A_k and B_k will define two bounded self-adjoint operators $\tilde{A}_k$ and $\tilde{B}_k$ on E by

$$(\tilde{A}_k x, y)_E := (A_k x, y)_H, \ \forall x, y \in E,$$

and

$$(\tilde{B}_k x, y)_E := (B_k x, y)_H, \ \forall x, y \in E.$$

It is easy to see $\tilde{A}_k = |A_k|^{-1} A_k$ and $\tilde{B}_k = |A_k|^{-1} B_k$. Thus

$$\ker(A_k - B_k) = \ker(\tilde{A}_k - \tilde{B}_k).$$

Further more, we can write $\tilde{A}_k$ and $\tilde{B}_k$ in the block form

$$\tilde{A}_k = \begin{pmatrix} \tilde{A}_{k,1} & 0 \\ 0 & \tilde{A}_{k,2} \end{pmatrix}, \quad \tilde{B}_k = \begin{pmatrix} \tilde{B}_{k,11} & \tilde{B}_{k,12} \\ \tilde{B}_{k,21} & \tilde{B}_{k,22} \end{pmatrix}, \tag{3.30}$$

with respect to the decomposition (3.29). For any $u \in E$, $u = x + y$ with $x \in E_0$ and $y \in E_1$, the equation

$$\tilde{A}_k u = \tilde{B}_k u$$

can be rewritten as

$$\begin{pmatrix} \tilde{A}_{k,1} & 0 \\ 0 & \tilde{A}_{k,2} \end{pmatrix} \begin{pmatrix} x \\ y \end{pmatrix} = \begin{pmatrix} \tilde{B}_{k,11} & \tilde{B}_{k,12} \\ \tilde{B}_{k,21} & \tilde{B}_{k,22} \end{pmatrix} \begin{pmatrix} x \\ y \end{pmatrix}.$$

That is

$$\begin{cases} \tilde{A}_{k,1} x = \tilde{B}_{k,11} x + \tilde{B}_{k,12} y, \\ \tilde{A}_{k,2} y = \tilde{B}_{k,21} x + \tilde{B}_{k,22} y. \end{cases}$$

From (3.28), the definitions of P_0, P_1, $\tilde{A}_{k,2}$ and $\tilde{B}_{k,22}$, we have $\tilde{A}_{k,2}$ is invertible on E_1 and $\|\tilde{A}_{k,2}^{-1} \tilde{B}_{k,22}\|_E < 1$. Thus we have

$$y = (\tilde{A}_{k,2} - \tilde{B}_{k,22})^{-1} \tilde{B}_{k,21} x,$$

and

$$\tilde{A}_k u = \tilde{B}_k u \Longleftrightarrow \tilde{A}_{k,1} x = [\tilde{B}_{k,11} + \tilde{B}_{k,12} (\tilde{A}_{k,2} - \tilde{B}_{k,22})^{-1} \tilde{B}_{k,21}] x.$$

It's easy to see $\dim \ker(A - B) \leq \dim(E_0) < \infty$ and we have proved the first part. In order to prove the rest part of the lemma, by arguing indirectly, assume 0 is not an isolated point spectrum of $A - B$ for some fixed $B \in \mathcal{L}_s^-(H, \lambda^-)$. That is to say $H_\varepsilon := Q_\varepsilon H$ is an infinite dimensional space for any small $\varepsilon > 0$, where Q_ε is a projection map corresponding to the spectral measure $\hat{E}(z)$ of $A - B$ defined by

$$Q_\varepsilon = \int_{-\varepsilon/2}^{\varepsilon/2} 1 d\hat{E}(z).$$

Now we can choose $\varepsilon > 0$ small enough, such that $B < (\lambda^- - \varepsilon) \cdot I$ and define a self-adjoint operator

$$C_\varepsilon = \int_{-\varepsilon/2}^{\varepsilon/2} z d\hat{E}(z).$$

Then we have $B + C_\varepsilon < (\lambda^- - \frac{1}{2}\varepsilon) \cdot I$, that is to say $B + C_\varepsilon \in \mathcal{L}_s^-(H, \lambda^-)$. On the other hand we have $\ker(A - B - C_\varepsilon) = H_\varepsilon$, which is contradict to the fact $\dim \ker(A - B - C_\varepsilon) < \infty$. Thus we have proved the lemma. □

Now for any $B \in \mathcal{L}_s^-(H, \lambda^-)$ and $k \in \mathbb{R}$ satisfying (3.24), consider the bounded self-adjoint operator $T_{B,k}^-$ on H defined by

$$T_{B,k}^- := B_k^{-1} - A_k^{-1}, \ \forall B \in \mathcal{L}_s^-(H, \lambda^-). \tag{3.31}$$

Firstly, the invertible map B_k^{-1} establishes the one-to-one correspondence between $\ker(T_{B,k}^-)$ and $\ker(A - B)$, so we have $\dim \ker(T_{B,k}^-) = \dim \ker(A - B)$. Secondly, from (3.25) and the definition of H_1, we have

$$(T_{B,k}^- y, y)_H > c(y, y)_H, \ \forall y \in H_1,$$

for some fixed $c > 0$. Since H has the decomposition $H = H_0 \oplus H_1$ and $\dim H_0 < \infty$, $T_{B,k}^-$ has only finite dimensional negative definite subspace, that is to say $(-\infty, 0) \cap \sigma(T_{B,k}^-)$ has only finite points with finite dimensional eigenvalue space. Summed up, we have the following lemma.

Lemma 3.2 ([284]) *Suppose $A \in \mathcal{O}_e^-(\lambda^-)$. For any $B \in \mathcal{L}_s^-(H, \lambda^-)$ and $k \in \mathbb{R}$ satisfying* (3.24), *there is an orthogonal decomposition of H*

$$H = H_{T_{B,k}^-}^- \oplus H_{T_{B,k}^-}^0 \oplus H_{T_{B,k}^-}^+,$$

such that $T_{B,k}^-$ is negative definite, zero and positive definite on $H_{T_{B,k}^-}^-$, $H_{T_{B,k}^-}^0$ and $H_{T_{B,k}^-}^+$ respectively. Further more

$$\dim H_{T_{B,k}^-}^- < \infty, \ \dim H_{T_{B,k}^-}^0 = \dim \ker(A - B).$$

Thus if $A \in \mathcal{O}_e^-(\lambda^-)$, for any $B \in \mathcal{L}_s^-(H, \lambda^-)$ and $k \in \mathbb{R}$ satisfying (3.24) we denote the Morse index pair of $T_{B,k}^-$ by $(i_{A,k}^-(B), \nu_A^-(B))$, that is

$$i_{A,k}^-(B) := \dim H_{T_{B,k}^-}^-, \quad \nu_A^-(B) := \dim H_{T_{B,k}^-}^0. \tag{3.32}$$

Of cause the index $i_{A,k}^-(B)$ depends on the choose of k. But we will show that $i_{A,k}^-(B_1) - i_{A,k}^-(B_2)$ will not depend on k for any fixed $B_1, B_2 \in \mathcal{L}_s^-(H, \lambda^-)$, it only depends on B_1, B_2 and A. For this purpose, we need the following lemma.

Lemma 3.3 ([284]) *Suppose* $A \in \mathcal{O}_e^-(\lambda^-)$. *For any* $B_1, B_2 \in \mathcal{L}_s^-(H, \lambda^-)$ *satisfying* $B_1 < B_2$, *we have*

$$i_{A,k}^-(B_2) - i_{A,k}^-(B_1) = \sum_{s \in [0,1)} \nu(A - (1-s)B_1 - sB_2)$$

for any $k \in \mathbb{R} \setminus \sigma(A)$ *and* $k \cdot I < B_1, B_2$. *Where* $\nu(P) = \dim \ker P$ *is the nullity of the linear operator* P.

Proof Denote $i(s) := i_{A,k}^-(B(s))$ and $\nu(s) = \nu(A - B(s))$, where $B(s) := (1-s)B_1 + sB_2$. Since $B_1 < B_2$, we have $B(s_1) < B(s_2)$, for any $0 \le s_1 < s_2 \le 1$, so we have

$$B_k^{-1}(s_1) > B_k^{-1}(s_2) > 0, \ 0 \le s_1 < s_2 \le 1,$$

and

$$T(s_1) > T(s_2), \ 0 \le s_1 < s_2 \le 1,$$

where $T(s) := B_k^{-1}(s) - A_k^{-1}$ and the map $T(s) : [0,1] \to \mathcal{L}_s(H)$ is continuous. Firstly, from the definition of $i_{A,k}^-(\cdot)$, it's easy to see $i(s)$ is left continuous and

$$0 \le i(s_1) \le i(s_2) \le i_{A,k}^-(B_2), \ \forall\, 0 \le s_1 < s_2 \le 1.$$

Further more, for $s_0 \in [0,1]$, if $\nu(s_0) = 0$ then $i(s)$ is continuous at s_0. If $\nu(s) \ne 0$, we have $i(s+0) - i(s) = \nu(s)$. In fact, by the continuity of the eigenvalue of continuous operator function, we have $i(s+0) - i(s) \le \nu(s)$. On the other side, since $T(s_1) > T(s_2)$, for $s_1 < s_2$, we see that $i(s+0) - i(s) \ge \nu(s)$. From the above properties of $i(s)$ and the fact that $i(s) \in [0, i_{A,k}^-(B_2)] \cap \mathbb{Z}$, thus there are only finite number of $s \in [0,1]$ such that $\nu(s) \ne 0$ and

$$i_{A,k}^-(B_2) - i_{A,k}^-(B_1) = \sum_{s \in [0,1)} \nu(A - (1-s)B_1 - sB_2).$$

Thus we have proved the lemma. □

Remark 3.4 With the same methods, if $A \in \mathcal{O}_e^+(\lambda^+)$, we can define the index pair for $B \in \mathcal{L}_s^+(H, \lambda^+)$, with any $k \in \mathbb{R}$ satisfying

$$k \notin \sigma(A),\ k \cdot I > B. \tag{3.33}$$

In this case, we redefine the operators $A_k := k \cdot I - A$ and $B_k := k \cdot I - B$. Consider the bounded self-adjoint operator

$$T_{B,k}^+ := B_k^{-1} - A_k^{-1},\ \forall B \in \mathcal{L}_s^+(H, \lambda^+).$$

It is easy to see $T_{B,k}^+$ has finite dimensional negative definite space and $\dim\ker T_{B,k}^+ = \dim\ker(A - B)$. Similarly, define $i_{A,k}^+(B)$ the dimension of the negative definite space of $T_{B,k}^+$ and $\nu_A^+(B) = \dim\ker(A - B)$.

And we also have the following lemma.

Lemma 3.5 ([284]) *If $A \in \mathcal{O}_e^+(\lambda^+)$, for any $B_1, B_2 \in \mathcal{L}_s^+(H, \lambda^+)$ satisfying $B_2 < B_1$, we have*

$$i_{A,k}^+(B_2) - i_{A,k}^+(B_1) = \sum_{s \in [0,1)} \nu(A - (1-s)B_1 - sB_2)$$

for any $k \in \mathbb{R} \setminus \sigma(A)$ and $k \cdot I > B_1, B_2$.

The proof is similar to the proof of Lemma 3.3, so we omit it here. Let $B^- \in \mathcal{L}_s^-(H, \lambda^-)$ and $B^+ \in \mathcal{L}_s^+(H, \lambda^+)$ be fixed.

Definition 3.6 ([284]) If $A \in \mathcal{O}_e^\mp(\lambda^\mp)$, for any $B \in \mathcal{L}_s^\mp(H, \lambda^\mp)$, define the index pair $(i_A^\mp(B), \nu_A^\mp(B))$ by

$$\begin{aligned} i_A^\mp(B) &:= i_{A,k}^\mp(B) - i_{A,k}^\mp(B^\mp), \\ \nu_A^\mp(B) &:= \dim\ker(A - B), \end{aligned}$$

with some $k \in \mathbb{R} \setminus \sigma(A)$ satisfying $\mp k \cdot I > \mp B$, and $\mp k \cdot I > \mp B^\mp$. We note that $\nu_A^-(B) = \nu_A^+(B)$ when they all make sense, so in sequel, we write it in $\nu_A(B)$ for simplicity.

The definition is well defined, we will prove that it only depends on the choice of $B^\mp$. We only consider the case of $B \in \mathcal{L}_s^-(H, \lambda^-)$, by Lemma 3.3, for any $\tilde{k} \in \mathbb{R}$ satisfying $B, B^- < \tilde{k} \cdot I$ and $\tilde{k} < \lambda^-$,

$$\begin{aligned} i_{A,k}^-(B) - i_{A,k}^-(B^-) &= (i_{A,k}^-(\tilde{k} \cdot I) - i_{A,k}^-(B^-)) - (i_{A,k}^-(\tilde{k} \cdot I) - i_{A,k}^-(B)) \\ &= \sum_{s \in [0,1)} \nu(A - (1-s)B^- - s\tilde{k} \cdot I) - \sum_{s \in [0,1)} \nu(A - (1-s)B - s\tilde{k} \cdot I), \end{aligned}$$

where the right hand side does not depend on the choice of k and we have proved that the definition of $i_A^{\mp}(B)$ is well defined. In this definition, for the fixed operators $B^{\mp}$, we have $i_A^{\mp}(B^{\mp}) = 0$. For any other choice of the operators $B^{\mp}$, the corresponding index is different up to a constant.

Remark 3.7

A. It is worth to note that we don't care about the choice of $B^{\mp}$ though they have affects on the definition of $i_A^{\mp}(B)$. What we care about is the difference between $i_A^{\mp}(B_1)$ and $i_A^{\mp}(B_2)$ with $B_1, B_2 \in \mathcal{L}_s^{\mp}(H, \lambda^{\mp})$. From the definition, we have $i_A^{\mp}(B_2) - i_A^{\mp}(B_1) = i_{A,k}^{\mp}(B_2) - i_{A,k}^{\mp}(B_1)$, that is to say $i_A^{\mp}(B_2) - i_A^{\mp}(B_1)$ is the difference between the Morse indices of $T_{B_2,k}^{\mp}$ and $T_{B_1,k}^{\mp}$. Thus for any $B \in \mathcal{L}_s^{\mp}(H, \lambda^{\mp})$, we call $i_A^{\mp}(B)$ the relative Morse index of B.
B. In the definition of our index $i_A^{\mp}(B)$, the number k, in fact the operator $k \cdot I$, can be replaced by bounded self-adjoint operator $\hat{B}$. For example, in Definition 3.6, for any $B \in \mathcal{L}_s^{-}(H, \lambda^{-})$, we can choose a bounded self-adjoint operator $\hat{B}$ satisfying $0 \notin \sigma(A - \hat{B})$ and $\hat{B} < B$.
C. If the self-adjoint operator A on H has no essential spectrum that is to say A has compact resolvent (i.e., A is Fredholm operator), in this case we define $\lambda^{-} = +\infty$ then $\mathcal{L}_s^{-}(H, \lambda^{-}) = \mathcal{L}_s(H)$ and for any $B \in \mathcal{L}_s(H)$ we can also define the index pair $(i_A^{-}(B), \nu_A^{-}(B))$.

Definition 3.8 Suppose A is a Fredholm operator on H. For any bounded self-adjoint operator $B \in \mathcal{L}_s(H)$, we define the index pair $(i_A(B), \nu_A(B))$ by

$$(i_A(B), \nu_A(B)) = (i_A^{-}(B), \nu_A^{-}(B)).$$

By Lemma 3.3 it's easy to see the index pair $(i_A(B), \nu_A(B))$ will coincide with the indexes defined in [69, 282] and [283] up to a constant, so the index theory introduced in this subsection can be regarded as a generalization of the ones in [69, 282] and [283].

3.3.2 *The Definition of Index Pair in Case 3*

Now, as a supplement to Case 1 and Case 2, we are in the position of defining the index pair in Case 3. Let $A \in \mathcal{O}_e^0(\lambda_a, \lambda_b)$ for some $\lambda_a, \lambda_b \in \mathbb{R}$, from the definition of $\mathcal{O}_e^0(\lambda_a, \lambda_b)$, we can assume that $\lambda_a, \lambda_b \notin \sigma(A)$. Firstly, we have the same result of Lemma 3.1 for any $B \in \mathcal{L}_s^0(H, \lambda_a, \lambda_b)$. We give a brief proof here. Choose the number $k = \frac{\lambda_a+\lambda_b}{2}$ in Lemma 3.1. If $k \notin \sigma(A)$, the rest part of the proof will be same as done before. Otherwise, $k \in \sigma(A)$, since there is no essential spectrum in (λ_a, λ_b), k is an eigenvalue of A with finitely dimensional eigenspace. Let P_k the projection map on the eigenspace of k, so P_k is a self-adjoint operator with finite rank. Then $k \notin \sigma(A+P_k)$, the results in Lemma 3.1 will keep valid for $(A+P_k)-B$.

Since P_k is a finite rank self-adjoint operator, the results in Lemma 3.1 will keep valid for $(A + P_k) - (B + P_k)$.

Secondly, compared to the *Case 1* and *Case 2*, instead of using dual variational method to define the index here, we define the index pair for $B \in \mathcal{L}_s^0(H, \lambda_a, \lambda_b)$ by the Morse index of saddle point reduction. Recall $k = \frac{\lambda_a+\lambda_b}{2}$, and P_k defined above, let

$$A_k := \begin{cases} A - k \cdot I, & k \notin \sigma(A), \\ A - k \cdot I + P_k, & k \in \sigma(A), \end{cases}$$

and

$$B_k := \begin{cases} B - k \cdot I, & k \notin \sigma(A), \\ B - k \cdot I + P_k, & k \in \sigma(A). \end{cases}$$

Thus, we can also define the Hilbert space E, the bounded self-adjoint operator $\tilde{A}_k$ and $\tilde{B}_k$ on E, and we also have the decomposition as (3.29) and the block form (3.30) of $\tilde{A}_k$ and $\tilde{B}_k$, the only difference is the definition of P_0. Let $E(z)$ be the spectral measure of A_k, redefine the projection map P_0 by

$$P_0 = \int_{-(\lambda_b-\lambda_a)/2}^{(\lambda_b-\lambda_a)/2} 1dE(z).$$

With the above discussion, we have the following definition.

Definition 3.9 ([284]) Assume $A \in \mathcal{O}_e^0(\lambda_a, \lambda_b)$. For any $B \in \mathcal{L}_s^0(H, \lambda_a, \lambda_b)$, define the index pair $(i_A^0(B), \nu_A^0(B))$ by

$$\begin{aligned} i_A^0(B) &:= m^-(\tilde{A}_{k,1} - [\tilde{B}_{k,11} + \tilde{B}_{k,12}(\tilde{A}_{k,2} - \tilde{B}_{k,22})^{-1}\tilde{B}_{k,21}]), \\ \nu_A^0(B) &:= \dim\ker(A - B), \end{aligned}$$

where $\tilde{A}_{k,1} - [\tilde{B}_{k,11} + \tilde{B}_{k,12}(\tilde{A}_{k,2} - \tilde{B}_{k,22})^{-1}\tilde{B}_{k,21}]$ is a self-adjoint operator on finite dimensional space E_0, and $m^-(P)$ is the Morse index of the quadratic form $(Pz, z)_H$ defined by the self-adjoint operator P.

3.4 Saddle Point Reduction for the General Cases

Let H be a Hilbert space with inner product $(\cdot, \cdot)_H$ and norm $\|\cdot\|_H$, $A \in \mathcal{O}(H)$ be a self-adjoint linear operator with compact resolvent and dense domain $D(A)$. $B \in \mathcal{L}_s(H)$ is a bounded self-adjoint linear operator on H with its operator norm $\|B\|_H < c$, $\pm c \notin \sigma(A)$. Denote by $N = \ker(A)$ the kernel of A and $P_0 = H \to N$ the projection. We set $\tilde{A} = A + P_0$. Denote by E_λ the spectral resolution of the self-adjoint operator $\tilde{A}$, we define the projections on H by

$$\mathcal{P} = \int_{-c}^{c} dE_\lambda, \quad \mathcal{P}^+ = \int_{c}^{+\infty} dE_\lambda, \quad \mathcal{P}^- = \int_{-\infty}^{-c} dE_\lambda.$$

The Hilbert space H possesses an orthogonal decomposition

$$H = H^+ \oplus H^- \oplus X,$$

where $H^\pm = \mathcal{P}^\pm H$, and $X = \mathcal{P}H$ is a finite dimensional space. We consider the quadratic functional

$$f(z) = \frac{1}{2}((A - B)z, z)_H, \quad z \in D(A) \subset H.$$

Theorem 4.1 ([284]) *There exist a function $a \in C^2(X, \mathbb{R})$ and a linear map $u : X \to H$ satisfying the following conditions:*

1° *The map u has the form $u(x) = w(x) + x$ with $\mathcal{P}w(x) = 0$.*

2° *The function a satisfies*

$$a(x) = f(u(x)) = \frac{1}{2}((A - B)u(x), u(x))_H = \frac{1}{2}((A - B')x, x)_H,$$

where $B' : X \to X$ is defined in (3.37) *below.*

3° *$x \in X$ is a critical point of a, if and only if $z = u(x)$ is a critical point of f, i.e. $z = u(x) \in \ker(A - B)$.*

Proof Denote $E = D(|\tilde{A}|^{1/2})$, since $0 \notin \sigma(\tilde{A})$, thus E is a Hilbert space with the inner product $(\cdot, \cdot)_E$ and corresponding norm $\|\cdot\|_E$ defined by

$$(x, y)_E := (|\tilde{A}|^{1/2}x, |\tilde{A}|^{1/2}y)_H, \quad \forall x, y \in E,$$

$$\|x\|_E^2 := (x, x)_E, \quad \forall x \in E.$$

We also have the following decomposition

$$E = E_0 \oplus E_1, \tag{3.34}$$

with $E_0 = E \cap X$ and $E_1 = E \cap (H^+ \cup H^-)$. The operators A and B will define two bounded self-adjoint operator $\bar{A}$ and $\bar{B}$ on E by

$$(\bar{A}x, y)_E := (Ax, y)_H, \quad \forall x, y \in E,$$

and

$$(\bar{B}x, y)_E := (Bx, y)_H, \quad \forall x, y \in E.$$

It is easy to see $\bar{A} = |\tilde{A}|^{-1}A$ and $\bar{B} = |\tilde{A}|^{-1}B$. Thus

$$\ker(\bar{A} - \bar{B}) = \ker(A - B).$$

Further more, we can write $\bar{A}$ and $\bar{B}$ in the block form

$$\bar{A} = \begin{pmatrix} \bar{A}_1 & 0 \\ 0 & \bar{A}_2 \end{pmatrix}, \quad \bar{B} = \begin{pmatrix} \bar{B}_{11} & \bar{B}_{12} \\ \bar{B}_{21} & \bar{B}_{22} \end{pmatrix}, \tag{3.35}$$

with respect to the decomposition (3.34). For any $u \in E$, $u = x + y$ with $x \in E_0$ and $y \in E_1$, the equation

$$\bar{A}u = \bar{B}u$$

can be rewritten as

$$\begin{pmatrix} \bar{A}_1 & 0 \\ 0 & \bar{A}_2 \end{pmatrix} \begin{pmatrix} x \\ y \end{pmatrix} = \begin{pmatrix} \bar{B}_{11} & \bar{B}_{12} \\ \bar{B}_{21} & \bar{B}_{22} \end{pmatrix} \begin{pmatrix} x \\ y \end{pmatrix}.$$

That is

$$\begin{cases} \bar{A}_1 x = \bar{B}_{11}x + \bar{B}_{12}y \\ \bar{A}_2 y = \bar{B}_{21}x + \bar{B}_{22}y. \end{cases}$$

From (3.28), the definitions of $\mathcal{P}$, $\mathcal{P}^{\pm}$, $\bar{A}_2$ and $\bar{B}_{22}$, we have $\bar{A}_2$ is invertible on E_1 and $\|\bar{A}_2^{-1}\bar{B}_{22}\|_E < 1$. Thus we have

$$y = w(x) = (\bar{A}_2 - \bar{B}_{22})^{-1}\bar{B}_{21}x,$$

and

$$\bar{A}u - \bar{B}u = 0 \Longleftrightarrow \bar{A}_1 x - [\bar{B}_{11} + \bar{B}_{12}(\bar{A}_2 - \bar{B}_{22})^{-1}\bar{B}_{21}]x = 0. \tag{3.36}$$

So $u(x) = x + w(x) = x + (\bar{A}_2 - \bar{B}_{22})^{-1}\bar{B}_{21}x$ satisfies all the required properties with B' defined as

$$B' = \bar{B}_{11} + \bar{B}_{12}(\bar{A}_2 - \bar{B}_{22})^{-1}\bar{B}_{21}. \tag{3.37}$$

□

We note that in the nonlinear case, the same result is true. But in the proof one should use the contraction mapping principle and the implicit function theorem. We refer the papers [4, 5, 33, 34] and [223] for general settings.

Definition 4.2 ([284]) For any $B \in \mathcal{L}_s(H)$ with $\|B\|_H < c$, we define

$$\mu_A^c(B) = m^-(\bar{A}|_{E_0} - B'), \quad \nu_A(B) = \dim\ker(A - B).$$

Theorem 4.3 ([284]) *For any two operators $B_1, B_2 \in \mathcal{L}_s(H)$ with $\|B_i\|_H < c$, $i = 1, 2$ and $B_1 < B_2$, there holds*

$$\mu_A^c(B_2) - \mu_A^c(B_1) = \sum_{s\in[0,1)} \nu_A((1-s)B_1 + sB_2). \tag{3.38}$$

Proof We set $B_s = (1-s)B_1 + sB_2$, $i(s) = \mu_A^c((1-s)B_1 + sB_2)$, $\nu(s) = \nu_A^c((1-s)B_1 + sB_2)$ and

$$a_s(x) = \frac{1}{2}((\bar{A}_1 - B_s')x, x)_E,$$

where $B_s' = ((1-s)\bar{B}_1 + s\bar{B}_2)_{11} + ((1-s)\bar{B}_1 + s\bar{B}_2)_{12}(\bar{A}_2 - ((1-s)\bar{B}_1 + s\bar{B}_2)_{22})^{-1}((1-s)\bar{B}_1 + s\bar{B}_2)_{21}$. We denote $b(s) = \bar{A}_1 - B_s'$.

For any $s_0 \in [0, 1]$, if $\nu(s_0) = 0$, that is to say $b(s_0)$ has zero nullity subspace of E_0, so from the continuous dependent of the quadratic function a_s on s, there exists a neighbourhood $U(s_0)$ of s_0 in $[0, 1]$, such that

$$i(s) = i(s_0) \text{ and } \nu(s) = \nu(s_0) = 0, \ \forall s \in U(s_0). \tag{3.39}$$

If $\nu(s_0) \neq 0$, we have the following decomposition

$$E_0 = E_0^- \oplus E_0^0 \oplus E_0^+,$$

such that $b(s_0)$ is negative definite, zero and positive definite on E_0^-, E_0^0 and E_0^+ respectively. For any $x_0 \in \ker b(s_0)$ with $\|x_0\| = 1$, that is $b(s_0)x_0 = 0$. Define a smooth function $a(s) : [0, 1] \to \mathbb{R}$ by

$$a(s) := (b(s)x_0, x_0)_E,$$

and we have $a(s_0) = 0$. From the definition of $b(s)$ and denote $\xi(s) := (\bar{A}_2 - \bar{B}_{22}(s))^{-1}\bar{B}_{21}(s)$ for simplicity, we have

$$a(s) = ((\bar{A} - \bar{B}(s))(x_0 + \xi(s)x_0), (x_0 + \xi(s)x_0))_E,$$

and

$$(\bar{A} - \bar{B}(s_0))(x_0 + \xi(s_0)x_0) = 0.$$

So, we have

$$a'(s_0) = -(\bar{B}'(s_0)(x_0 + \xi(s_0)x_0), (x_0 + \xi(s_0)x_0)_E$$
$$= -((B_2 - B_1)(x_0 + \xi(s_0)x_0), (x_0 + \xi(s_0)x_0)_H.$$

Since $B_1 < B_2$ and $x_0 \neq 0$, we have $a'(s_0) < 0$. Summing up, there exists $\delta > 0$, such that $a(s) < 0$ for any $s \in (s_0, s_0 + \delta)$. So from the continuous of $b(s)$, there exists $\bar{\delta} \leq \delta$, such that

$$(b(s)x, x)_E < 0, \ \forall x \in E_0^- \oplus E_0^0, \ s \in (s_0, s_0 + \bar{\delta}),$$

$$(b(s)x, x)_E > 0, \ \forall x \in E_0^0, \ s \in (s_0 - \bar{\delta}, s_0)$$

and

$$(b(s)x, x)_E > 0, \ \forall x \in E_0^+, \ s \in (s_0, s_0 + \bar{\delta}).$$

That is to say

$$i(s) = i(s_0) + \nu(s_0) \text{ and } \nu(s) = 0, \ \ \forall s \in (s_0, s_0 + \bar{\delta}) \tag{3.40}$$

and

$$i(s) = i(s_0) \text{ and } \nu(s) = 0, \ \ \forall s \in (s_0 - \bar{\delta}, s_0). \tag{3.41}$$

So from (3.39), (3.40) and (3.41), we have

$$\mu_A^c(B_2) - \mu_A^c(B_1) = \sum_{s\in[0,1)} \nu_A((1-s)B_1 + sB_2).$$

Thus we have proved the lemma. □

From Theorem 4.3, we know that $\mu_A^c(B_2) - \mu_A^c(B_1)$ is independent of c for any B_1 and $B_2 \in \mathcal{L}_s(H)$ with $c > \max\{\|B_1\|_H, \|B_2\|_H\}$. In fact, for any such two operators, we can choose an operator $B_0 \in \mathcal{L}_s(H)$ such that $B_0 < B_i$, $i = 1, 2$ and $\|B_0\|_H < c$. Then we have

$$\mu_A^c(B_2) - \mu_A^c(B_1) = (\mu_A^c(B_2) - \mu_A^c(B_0)) - (\mu_A^c(B_1) - \mu_A^c(B_0)),$$

which is independent of c.

Definition 4.4 ([284]) For any $B \in \mathcal{L}_s(H)$, we define

$$\mu_A(B) = \mu_A^c(B) - \mu_A^c(0), \ \ c > \|B\|_H. \tag{3.42}$$

So the index pair $(\mu_A(B), \nu_A(B))$ is well defined.

From Theorems 1.3 and 4.3, we have the following result.

Theorem 4.5 ([284]) *Suppose both the indices* $I(A, A - B)$ *and* $\mu_A(B)$ *are well defined for the operator pair* (A, B)*. Then we have*

$$I(A, A - B) = \mu_A(B). \tag{3.43}$$

Proof Firstly, we claim that for the positively definite operator $B > 0$, (3.43) is true. In fact, since $I(A, A - 0) = \mu_A(0) = 0$, there holds

$$I(A, A - B) = \sum_{s \in [0,1)} m^0(A - sB) = \sum_{s \in [0,1)} \nu_A(sB) = \mu_A(B).$$

In general, we choose a positively definite operator B_0 such that $B < B_0$, so we have

$$I(A, A - B_0) - I(A, A - B) = \sum_{s \in [0,1)} m^0(A - (1 - s)B - sB_0) = \mu_A(B_0) - \mu_A(B).$$

Therefore from $I(A, A - B_0) = \mu_A(B_0)$, we have the desired equality (3.43). □

Chapter 4
The P-Index Theory

4.1 P-Index Theory

If the Hamiltonian function $H \in C^2(\mathbb{R}\times\mathbb{R}^{2n}, \mathbb{R})$ satisfying $H(t+\tau, Px) = H(t, x)$ with $P \in \mathrm{Sp}(2n)$, it is natural to consider the following nonlinear Hamiltonian system with P-boundary condition

$$\begin{cases} \dot{x}(t) = JH'(t, x(t)), \\ x(\tau) = Px(0). \end{cases} \tag{4.1}$$

So if x is a solution of the problem Eq. (4.1), the matrix $B(t) = H''(t, x(t))$ should satisfy the following condition

$$B(t+\tau) = (P^{-1})^T B(t) P^{-1}.$$

The fundamental solution of the linear system

$$\dot{z}(t) = JB(t)z(t) \tag{4.2}$$

in the interval $[0, \tau]$ is a symplectic path $\gamma(t)$, i.e., $\gamma \in \mathcal{P}_\tau(2n)$.

Definition 1.1 ([202]) For any $\gamma \in \mathcal{P}_\tau(2n)$, $P \in \mathrm{Sp}(2n)$, $\omega \in \mathbf{U}$, the (P, ω)-nullity $\nu_\omega^P(\gamma)$ is defined by

$$\nu_\omega^P(\gamma) = \dim_{\mathbb{C}} \ker_{\mathbb{C}}(\gamma(\tau) - \omega P). \tag{4.3}$$

We note that for the fundamental solution $\gamma \in \mathcal{P}_\tau(2n)$ of the linear Hamiltonian system (4.2), from Lemma 4.1 of Chap. 1, the $(P, 1)$-nullity $\nu_1^P(\gamma)$ is just the dimensional of the real solution space of the linear Hamiltonian system (4.2) satisfying the boundary condition $x(\tau) = Px(0)$. And the (P, ω)-nullity $\nu_\omega^P(\gamma)$

is just the complex dimensional of the complex solution space of the linear Hamiltonian system (4.2) satisfying the boundary condition $x(\tau) = \omega Px(0)$.

In the following, we define the (P, ω)-index part $i_\omega^P(\gamma)$ parameterized by all ω on the unit circle for any symplectic path γ starting from the identity, and study some of its basic properties. We fix $n \in \mathbb{N}$, $\tau > 0$ and firstly introduce some notations and definitions.

For $P, M \in \mathrm{Sp}(2n)$, $\omega \in \mathbf{U}$, we define

$$D_{\omega,P}(M) = (-1)^{n-1}\omega^{-n}det(M - \omega P),$$

$$\mathrm{Sp}(2n)^{\pm}_{\omega,P} = \{M \in \mathrm{Sp}(2n) \mid \pm D_{\omega,P}(M) < 0\},$$

$$\mathrm{Sp}(2n)^{*}_{\omega,P} = \mathrm{Sp}(2n)^{+}_{\omega,P} \cup \mathrm{Sp}(2n)^{-}_{\omega,P},$$

$$\mathrm{Sp}(2n)^{0}_{\omega,P} = \mathrm{Sp}(2n) \setminus \mathrm{Sp}(2n)^{*}_{\omega,P}.$$

For $\tau > 0$, we define four subsets of $\mathcal{P}_\tau(2n)$ by

$${}^{0}_{P}\mathcal{P}^{*}_{\tau,\omega}(2n) = \{\gamma \in \mathcal{P}_\tau(2n) \mid \gamma(0) \in \mathrm{Sp}(2n)^{0}_{\omega,P}, \gamma(\tau) \in \mathrm{Sp}(2n)^{*}_{\omega,P}\},$$

$${}^{0}_{P}\mathcal{P}^{0}_{\tau,\omega}(2n) = \{\gamma \in \mathcal{P}_\tau(2n) \mid \gamma(0) \in \mathrm{Sp}(2n)^{0}_{\omega,P}, \gamma(\tau) \in \mathrm{Sp}(2n)^{0}_{\omega,P}\},$$

$${}^{*}_{P}\mathcal{P}^{*}_{\tau,\omega}(2n) = \{\gamma \in \mathcal{P}_\tau(2n) \mid \gamma(0) \in \mathrm{Sp}(2n)^{*}_{\omega,P}, \gamma(\tau) \in \mathrm{Sp}(2n)^{*}_{\omega,P}\},$$

$${}^{*}_{P}\mathcal{P}^{0}_{\tau,\omega}(2n) = \{\gamma \in \mathcal{P}_\tau(2n) \mid \gamma(0) \in \mathrm{Sp}(2n)^{*}_{\omega,P}, \gamma(\tau) \in \mathrm{Sp}(2n)^{0}_{\omega,P}\}.$$

$${}_{P}\mathcal{P}^{0}_{\tau,\omega}(2n) = {}^{0}_{P}\mathcal{P}^{0}_{\tau,\omega}(2n) \cup {}^{*}_{P}\mathcal{P}^{0}_{\tau,\omega}(2n),$$

$${}_{P}\mathcal{P}^{*}_{\tau,\omega}(2n) = {}^{0}_{P}\mathcal{P}^{*}_{\tau,\omega}(2n) \cup {}^{*}_{P}\mathcal{P}^{*}_{\tau,\omega}(2n),$$

$${}^{0}_{P}\mathcal{P}_{\tau,\omega}(2n) = {}^{0}_{P}\mathcal{P}^{*}_{\tau,\omega}(2n) \cup {}^{0}_{P}\mathcal{P}^{0}_{\tau,\omega}(2n),$$

$${}^{*}_{P}\mathcal{P}_{\tau,\omega}(2n) = {}^{*}_{P}\mathcal{P}^{*}_{\tau,\omega}(2n) \cup {}^{*}_{P}\mathcal{P}^{0}_{\tau,\omega}(2n).$$

We denote $\mathrm{Sp}(2n)^{*}_{\omega} = \mathrm{Sp}(2n)^{*}_{\omega,I}$, $\mathrm{Sp}(2n)^{0}_{\omega} = \mathrm{Sp}(2n)^{0}_{\omega,I}$.

For any two paths $\gamma_1 : [0, \tau] \to \mathrm{Sp}(2n)$ and $\gamma_2 : [0, \tau] \to \mathrm{Sp}(2n)$ with $\gamma_1(\tau) = \gamma_2(0)$, we define their joint path by

$$\gamma_2 * \gamma_1(t) = \begin{cases} \gamma_1(2t), & 0 \le t \le \tau/2, \\ \gamma_2(2t - \tau), & \tau/2 \le t \le \tau. \end{cases}$$

For $\theta \in \mathbb{R}$ define $R(\theta) = \begin{pmatrix} \cos\theta & -\sin\theta \\ \sin\theta & \cos\theta \end{pmatrix}$.

Let $D(a) = \mathrm{diag}(a, a^{-1})$ for $a \in \mathbb{R} \setminus \{0\}$, and define

$$M_n^+ = D(2)^{\diamond n}, \quad M_n^- = D(-2) \diamond D(2)^{\diamond(n-1)}.$$

We have $M_n^+ \in \mathrm{Sp}(2n)_\omega^+$ and $M_n^- \in \mathrm{Sp}(2n)_\omega^-$.

From Lemma 2.6 of Chap. 1, every $M \in \mathrm{Sp}(2n)$ has its unique polar decomposition $M = AU$, where $A = (MM^T)^{1/2}$ is a symmetric symplectic positive definite matrix, U is a symplectic orthogonal matrix. Therefore U can be written as $U = \begin{pmatrix} u_1 & -u_2 \\ u_2 & u_1 \end{pmatrix}$, where $u = u_1 + \sqrt{-1}u_2$ is a unitary matrix. So for every path γ in $C([0, \tau], \mathrm{Sp}(2n))$ we can associate uniquely a path $u(t)$ in the unitary group on $\mathbb{C}^n$ to it. Let $\Delta : [0, \tau] \to \mathbb{R}$ be any continuous real function satisfying $\det u(t) = \exp(\sqrt{-1}\Delta(t))$. We define the rotation number of γ on $[0, \tau]$ by $\Delta_\tau(\gamma) = \Delta(\tau) - \Delta(0)$ which depends only on the symplectic path γ but not on the choice of the function Δ.

Definition 1.2 For $\tau > 0$ and $\omega \in \mathbf{U}$, given two paths $\gamma_0, \gamma_1 \in C([0, \tau], \mathrm{Sp}(2n))$, if there exists a map $\delta \in C([0, 1] \times [0, \tau], \mathrm{Sp}(2n))$ such that $\delta(0, \cdot) = \gamma_0(\cdot)$, $\delta(1, \cdot) = \gamma_1(\cdot)$, and both $\dim_{\mathbb{C}} \ker_{\mathbb{C}}(\delta(s, 0) - \omega I)$ and $\dim_{\mathbb{C}} \ker_{\mathbb{C}}(\delta(s, 1) - \omega I)$ are constant for $s \in [0, 1]$, then γ_0 and γ_1 are ω-homotopic on $[0, \tau]$ and we write $\gamma_0 \sim_\omega \gamma_1$. This homotopy possesses fixed end points if $\delta(s, 0) = \gamma_0(0)$, $\delta(s, \tau) = \gamma_0(\tau)$ for all $s \in [0, 1]$.

The following lemma studies the relation between the above homotopy and the homotopy which fixes the end points. It generalizes the Lemma 5.2.2 in [223] but with almost the same proof.

Lemma 1.3 *If $\gamma_0, \gamma_1 \in C([0, \tau], \mathrm{Sp}(2n))$ possess common end points $\gamma_0(0) = \gamma_1(0)$, $\gamma_0(\tau) = \gamma_1(\tau)$. Suppose $\gamma_0 \sim_\omega \gamma_1$ on $[0, \tau]$ via a homotopy $\delta \in C([0, 1] \times [0, \tau], \mathrm{Sp}(2n))$ such that $\delta(\cdot, 0)$, $\delta(\cdot, \tau)$ are contractible in* Sp(2n). *Then the homotopy δ can be modified to fix the end points all the time, i.e. $\delta(s, 0) = \gamma_0(0)$, $\delta(s, \tau) = \gamma_0(\tau)$ for all $s \in [0, 1]$.*

Remark 1.4 It is known that $Sp_\omega^k(2n) = \{M \in \mathrm{Sp}(2n) | \dim_{\mathbb{C}} \ker_{\mathbb{C}}(M - \omega I) = k\}$ is contractible in $\mathrm{Sp}(2n)$ for $0 \le k \le 2n$(see for example p14 of [227] for a proof for the case $\omega = 1$ and $k = 0$), so the closed curves $\delta(\cdot, 0) \subset Sp_\omega^k(2n)$ and $\delta(\cdot, \tau) \subset Sp_\omega^l(2n)$ for some integers k and l must be contractible in $\mathrm{Sp}(2n)$.

Lemma 1.5 ([223]) *If $\gamma_0, \gamma_1 \in C([0, \tau], \mathrm{Sp}(2n))$ possess common end points $\gamma_0(0) = \gamma_1(0)$, $\gamma_0(\tau) = \gamma_1(\tau)$, then $\Delta_\tau(\gamma_0) = \Delta_\tau(\gamma_1)$ if and only if $\gamma_0 \sim \gamma_1$ on $[0, \tau]$ with fixed points.*

By [57] and [231], $\mathrm{Sp}(2n)^*_\omega$ contains precisely two connected components $\mathrm{Sp}(2n)^\pm_\omega$ which contains $M_n^\pm$ respectively. Thus for any $\gamma \in {}^*_P\mathcal{P}^*_{\tau,\omega}(2n)$, we can connect $P^{-1}\gamma(0)$ to M_n^+ or M_n^- by β_0 in $\mathrm{Sp}(2n)^*_\omega$, connect $P^{-1}\gamma(\tau)$ to M_n^+ or M_n^- by β_1 in $\mathrm{Sp}(2n)^*_\omega$, so we get a joint path $\beta_1 * P^{-1}\gamma * \beta_0^-$. Here we remind that the path β_0^- is the backward path of β_0, i.e., $\beta_0^-(t) = \beta_0(\tau - t)$. By the definition of $M_n^\pm$, $k \equiv \Delta_\tau(\beta_1 * P^{-1}\gamma * \beta_0^-)/\pi$ is an integer, in this case we say $\gamma \in {}^*_P\mathcal{P}^*_{\tau,\omega,k}(2n)$. Since $\mathrm{Sp}(2n)^*_\omega$ is simply connected in $\mathrm{Sp}(2n)$, this integer k is independent of the choice of the paths β_0, β_1. This integer also satisfies

$$k \in \begin{cases} 2\mathbb{Z}, & \text{if } \beta_0(\tau) = \beta_1(\tau), \\ 2\mathbb{Z}+1, & \text{if } \beta_0(\tau) \neq \beta_1(\tau). \end{cases}$$

Definition 1.6 ([186, 202]) The (P, ω)-index $i^P_\omega(\gamma)$ of a path $\gamma \in \mathcal{P}_\tau(2n)$ is defined as

(1) We define $i^P_\omega(\gamma) = k$, if $\gamma \in {}^*_P\mathcal{P}^*_{\tau,\omega,k}(2n)$.
(2) If $\gamma \in {}^*_P\mathcal{P}^0_{\tau,\omega}(2n)$, then

$$i^P_\omega(\gamma) = \inf\{i^P_\omega(\beta) \mid \beta \in {}_P\mathcal{P}^*_{\tau,\omega}(2n) \text{ and } \beta \text{ is sufficiently } C^0 \text{ close to } \gamma \text{ in } \mathcal{P}_\tau(2n)\}.$$

(3) If $\gamma \in {}^0_P\mathcal{P}^*_{\tau,\omega}(2n)$, then

$$i^P_\omega(\gamma) = \max\{\, i^P_\omega(\beta) \mid \beta \in C([0, \tau], \mathrm{Sp}(2n)),\ \beta(0) \in \mathrm{Sp}(2n)^*_{\omega,P} \text{ and } \beta \text{ is sufficiently } C^0 \text{ close to } \gamma\}.$$

Where by saying that β is sufficiently C^0-close to γ in $\mathcal{P}_\tau(2n)$ we mean that β is close enough to γ in the topology of $\mathcal{P}_\tau(2n)$ so that there exists a homotopy $\delta : [0, 1] \to \mathcal{P}_\tau(2n)$ with the properties $\delta(0) = \beta$, $\delta(1) = \gamma$, and $\delta(s) \in {}^*_P\mathcal{P}^*_{\tau,\omega}(2n)$ for all $0 \le s < 1$.

In another way, we define the Maslov type (P, ω)-index in the following way by using the ω-index function.

Definition 1.7 ([202]) For any $\tau > 0$, $\omega \in \mathbf{U}$, $P \in \mathrm{Sp}(2n)$ and $\gamma \in \mathcal{P}_\tau(2n)$, we define the Maslov type (P, ω)-index

$$i^P_\omega(\gamma) = i_\omega(P^{-1}\gamma * \xi) - i_\omega(\xi), \tag{4.4}$$

where $\xi \in \mathcal{P}_\tau(2n)$ such that $\xi(\tau) = P^{-1}\gamma(0) = P^{-1}$.

Remark 1.8 The above definition depends only on γ but not on ξ, and therefore it is well defined. Definitions 1.1, 1.6, and 1.7 give a pair of integers $(i_\omega^P(\gamma), \nu_\omega^P(\gamma)) \in \mathbb{Z} \times \{0, 1, \dots 2n\}$ for any $\gamma \in \mathcal{P}_\tau(2n)$ and $P \in \mathrm{Sp}(2n)$. We call $i_\omega^P(\gamma)$ the (P, ω)-index of γ and $\nu_\omega^P(\gamma)$ the (P, ω)-nullity of γ. When $P \neq I$, the Maslov type P-index in [186] is the $(P, 1)$-index. We show it in Theorem 1.9 below.

When $\omega = 1$, we denote $(i^P(\gamma), \nu^P(\gamma)) = (i_\omega^P(\gamma), \nu_\omega^P(\gamma))$. We also denote $i^P(\gamma) = i^P(B)$ when γ is the fundamental solution of (4.2).

From the definition, $i_\omega^P(I) = 0$. When $P = I, \omega = 1$, if we use Definition 1.7 to define $i_\omega^P(\gamma)$, now we take $\xi \equiv I$, then

$$\begin{aligned} i^I(\gamma) &= i_1(P^{-1}\gamma * \xi) - i_1(\xi) \\ &= i_1(\gamma) - (-n) \\ &= i_1(\gamma) + n. \end{aligned}$$

Thus $i^I(\gamma)$ and $i_1(\gamma)$ are different up to n, where $i_1(\gamma)$ is the Maslov-type index for periodic boundary condition introduced in Chap. 2 (cf. [223–225]).

Theorem 1.9 ([202]) *The two definitions of index $i_\omega^P(\gamma)$ in Definitions* 1.6 *and* 1.7 *are consistent.*

Proof

(1) If $\gamma \in {}_P^*\mathcal{P}_{\tau,\omega}^*(2n)$, we choose $\xi \in \mathcal{P}_\tau(2n)$ such that $\xi(\tau) = P^{-1}$, and we can connect $P^{-1}\gamma(0)$ to M_n^+ or M_n^- by a path $\beta_0 : [0, \tau] \to \mathrm{Sp}(2n)_\omega^*$, connect $P^{-1}\gamma(\tau)$ to M_n^+ or M_n^- by a path $\beta_1 : [0, \tau] \to \mathrm{Sp}(2n)_\omega^*$. Then we have

$$\begin{aligned} & i_\omega(P^{-1}\gamma * \xi) - i_\omega(\xi) \\ =& \Delta_\tau(\beta_1 * P^{-1}\gamma * \xi)/\pi - \Delta_\tau(\beta_0 * \xi)/\pi \\ =& \Delta_\tau(\beta_1 * P^{-1}\gamma)/\pi + \Delta_\tau(\xi)/\pi - (\Delta_\tau(\beta_0) + \Delta_\tau(\xi))/\pi \\ =& \Delta_\tau(\beta_1 * P^{-1}\gamma)/\pi - \Delta_\tau(\beta_0)/\pi \\ =& \Delta_\tau(\beta_1 * P^{-1}\gamma * \beta_0^-)/\pi. \end{aligned}$$

(2) If $\gamma \in {}_P^*\mathcal{P}_{\tau,\omega}^0(2n)$, by the Definition 6.2.13 of the Maslov-type index in [223], we have

$$\begin{aligned} & i_\omega(P^{-1}\gamma * \xi) \\ &\quad = \inf\{i_\omega(P^{-1}\beta * \xi) \mid \beta \in {}_P\mathcal{P}_{\tau,\omega}^*(2n), \\ &\qquad\quad \beta \text{ is sufficiently } C^0\text{-close to } \gamma \text{ in } \mathcal{P}_\tau(2n)\} \\ &\quad = \inf\{i_\omega^P(\beta) \mid \beta \in {}_P\mathcal{P}_{\tau,\omega}^*(2n), \\ &\qquad\quad \beta \text{ is sufficiently } C^0\text{-close to } \gamma \text{ in } \mathcal{P}_\tau(2n)\} + i_\omega(\xi). \end{aligned}$$

We get the conclusion by putting the second term on the left.

(3) If $\gamma \in {}^{0}_{P}\mathcal{P}_{\tau,\omega}(2n)$, first we choose ξ in $\mathcal{P}_\tau(2n)$ such that $\xi(\tau) = P^{-1}$. Then we choose $\xi' \in \mathcal{P}_\tau(2n)$ is sufficiently C^0-close to ξ in $\mathcal{P}_\tau(2n)$ such that $\xi'(\tau) \in \mathrm{Sp}(2n)^*_{\omega,P}$; further, we choose β is sufficiently C^0-close to γ such that $P^{-1}\beta(0) = \xi'(\tau)$, $\beta(\tau) = \gamma(\tau)$.
So we have $P^{-1}\gamma * \xi \sim P^{-1}\beta * \xi'$ on $[0, \tau]$ with fixed points and obtain

$$\begin{aligned} i_\omega^P(\gamma) &= i_\omega(P^{-1}\gamma * \xi) - i_\omega(\xi) \\ &= i_\omega(P^{-1}\beta * \xi') - i_\omega(\xi) \\ &= i_\omega^P(\beta) + i_\omega(\xi') - i_\omega(\xi). \end{aligned}$$

By the definition of Maslov-type index, $i_\omega(\xi') - i_\omega(\xi) \geq 0$, the equality can be achieved when $\xi' = \xi_{-s}$, $\beta = {}_s\gamma$, $\forall s \in (0, 1]$. We will explain it in sequel.

□

Definition 1.10 For $\omega \in \mathbf{U}$, given two paths γ_0 and γ_1 in $\mathcal{P}_\tau(2n)$, if there exists a map $\delta \in C([0, 1] \times [0, \tau], \mathrm{Sp}(2n))$ such that $\delta(0, \cdot) = \gamma_0(\cdot)$, $\delta(1, \cdot) = \gamma_1(\cdot)$, $\delta(s, 0) = I$ and $\dim\ker(\delta(s, \tau) - \omega P)$ are constant for $s \in [0, 1]$, then we say they are (P, ω)-homotopic, and write $\gamma_0 \sim {}^P_\omega \gamma_1$.

Proposition 1.11 *For $\omega \in \mathbf{U}$, $P \in \mathrm{Sp}(2n)$, suppose $\gamma_j \in {}_P\mathcal{P}^*_{\tau,\omega}(2n)$ for $j = 0, 1$. Then $i_\omega^P(\gamma_0) = i_\omega^P(\gamma_1)$ if and only if $\gamma_0 \sim {}^P_\omega \gamma_1$.*

Proof Sufficiency. It is the same as the following Theorem 1.14.

Necessity. We choose $\xi \in \mathcal{P}_\tau(2n)$ such that $\xi(\tau) = P^{-1}$, $\gamma_j \in {}_P\mathcal{P}^*_{\tau,\omega}(2n)$ together with $i_\omega^P(\gamma_0) = i_\omega^P(\gamma_1)$ imply that there exists a map $\delta \in C([0, 1] \times [0, \tau], \mathrm{Sp}(2n))$ such that $P^{-1}\gamma_0 * \xi \sim {}_\omega P^{-1}\gamma_1 * \xi$. We can modify δ such that $\delta(s, t) = \xi(2t)$, $0 \leq s \leq 1$, $0 \leq t \leq \frac{\tau}{2}$. So we have $P^{-1}\gamma_0 \sim_\omega P^{-1}\gamma_1$ with the initial point fixed. It implies $\gamma_0 \sim {}^P_\omega \gamma_1$. □

Modifying the result in Corollary 6.2.10 of [223], we get the following result.

Proposition 1.12 *If γ_0, $\gamma_1 \in \mathcal{P}_\tau(2n)$ with the same end points possess the same Maslov (P, ω)-index if and only if they can be continuously deformed to each other with the end points fixed.*

Proposition 1.13 ([186, 202]) *For $\omega \in \mathbf{U}$, suppose $P_j \in Sp(2n_j)$, $\gamma_j \in \mathcal{P}_\tau(2n_j)$ for $j = 0, 1$. Then*

$$i_\omega^P(\gamma_0 \diamond \gamma_1) = i_\omega^{P_0}(\gamma_0) + i_\omega^{P_1}(\gamma_1), \tag{4.5}$$

where $P = P_0 \diamond P_1$ and $\gamma_0 \diamond \gamma_1(t) = \gamma_0(t) \diamond \gamma_1(t)$ for all $t \in [0, \tau]$.

Proof We choose $\xi_j \in \mathcal{P}_\tau(2n_j)$ such that $\xi_j(\tau) = P_j^{-1}$ and set $\xi(t) = \xi_0 \diamond \xi_1(t) \in \mathcal{P}_\tau(2n_0 + 2n_1)$.

Suppose W is the product matrices of the elementary transformations of changing the rows satisfying $\gamma_0(t) \diamond \gamma_1(t) = W \begin{pmatrix} \gamma_0(t) & 0 \\ 0 & \gamma_1(t) \end{pmatrix} W^{-1}$. Note that $W = W^{-1}$, $W^2 = I$.

Then we obtain

$$\begin{aligned} i_\omega^P(\gamma_0 \diamond \gamma_1) &= i_\omega(P^{-1}(\gamma_0 \diamond \gamma_1) * \xi) - i_\omega(\xi) \\ &= i_\omega\left(W \begin{pmatrix} P_0^{-1}\gamma_0 & 0 \\ 0 & P_1^{-1}\gamma_1 \end{pmatrix} W^{-1} * \xi\right) - i_\omega(\xi) \\ &= i_\omega((P_0^{-1}\gamma_0 * \xi_0) \diamond (P_1^{-1}\gamma_1 * \xi_1)) - i_\omega(\xi_0 \diamond \xi_1) \\ &= i_\omega(P_0^{-1}\gamma_0 * \xi_0) + i_\omega(P_1^{-1}\gamma_1 * \xi_1) - i_\omega(\xi_0) - i_\omega(\xi_1) \\ &= i_\omega^{P_0}(\gamma_0) + i_\omega^{P_1}(\gamma_1). \end{aligned}$$

□

Theorem 1.14 ([186, 202]) *Let $\omega \in \mathbf{U}$, $P \in \mathrm{Sp}(2n)$. For any two paths γ_0 and $\gamma_1 \in \mathcal{P}_\tau(2n)$, if $\gamma_0 \sim_\omega^P \gamma_1$ on $[0, \tau]$, then*

$$i_\omega^P(\gamma_0) = i_\omega^P(\gamma_1), \quad \nu_\omega^P(\gamma_0) = \nu_\omega^P(\gamma_1). \tag{4.6}$$

Proof By Definition 1.10 and $\gamma_0 \sim_\omega^P \gamma_1$, we obtain $\nu_\omega^P(\gamma_0) = \nu_\omega^P(\gamma_1)$.

To prove the first equality in (4.6), we choose a path $\xi \in \mathcal{P}_\tau(2n)$ so that $\xi(\tau) = P^{-1}$. By $\gamma_0 \sim_\omega^P \gamma_1$, we obtain $P^{-1}\gamma_0 * \xi \sim_\omega P^{-1}\gamma_1 * \xi$. Then according to Theorem 6.2.3 in [223], we have $i_\omega(P^{-1}\gamma_0 * \xi) = i_\omega(P^{-1}\gamma_1 * \xi)$. This proves the first equality of (4.6). □

Next we study the degenerate paths, and get the most important property of rotational perturbation paths. Fix $\omega \in \mathbf{U}$, $P \in \mathrm{Sp}(2n)$ and $P \neq I$, $\gamma \in {}_P^0\mathcal{P}_{\tau,\omega}(2n)$. Using Theorem 7.5 in [226], there exists $W_1 \in \mathrm{Sp}(2n)$ such that (7.8) in [226] holds for $M = \xi(\tau)$. As discussed in [226], we define

$$Q(M, s_1, \ldots, s_{p+2q}) \equiv M W_1^{-1} R_{m_1}(s_1\theta_0) \cdots R_{m_{p+2q}}(s_{p+2q}\theta_0) W_1. \tag{4.7}$$

For any $(s, t) \in [-1, 1] \times [0, \tau]$, we define the rotational perturbation paths of ξ

$$\xi_s(t) = \xi(t) W_1^{-1} R_{m_1}(s\rho(t)\theta_0) \cdots R_{m_{p+2q}}(s\rho(t)\theta_0) W_1.$$

Where s_i, m_i, θ_0, $\rho(t)$ are all defined as in [226].

For $t_0' \in (0, \tau)$, Let $\kappa \in C^2([0, \tau], [0, 1])$ such that $\kappa(t) = 0$ for $t_0' \leq t \leq \tau$, $\dot{\kappa}(t) \leq 0$ for $0 \leq t \leq \tau$, $\kappa(0) = 1$ and $\dot{\kappa}(0) = 0$. Whenever $t_0' \in (0, \tau)$ is sufficiently close to 0, there holds $P^{-1}\gamma([0, t_0']) \subset B_\epsilon(P^{-1})$. For any $(s, t) \in [-1, 1] \times [0, \tau]$, we define the left rotational perturbation paths of γ

$$_s\gamma(t) = \gamma(t)W_1^{-1}R_{m_1}(s\kappa(t)\theta_0)\cdots R_{m_{p+2q}}(s\kappa(t)\theta_0)W_1. \tag{4.8}$$

Corresponding, we define $_s(P^{-1}\gamma) = P^{-1}{}_s\gamma$. Note that $\xi_s(\tau) = {}_s(P^{-1}\gamma)(0)$, $_s\gamma(0) \in \mathrm{Sp}(2n)^*_{\omega,P}$.

Similarly, when $\gamma \in {}_P\mathcal{P}^0_{\tau,\omega}(2n)$, we using Theorem 7.5 in [226], there exists $W_2 \in \mathrm{Sp}(2n)$ such that (7.8) in [226] holds for $M' = \gamma(\tau)$. As discussed in [226], we define

$$Q'(M', s_1, \ldots, s_{p'+2q'}) \equiv M'W_2^{-1}R_{m_1'}(s_1\theta_0)\cdots R_{m'_{p'+2q'}}(s_{p'+2q'}\theta_0)W_2. \tag{4.9}$$

We define the right rotational perturbation paths of γ

$$\gamma_s(t) = \gamma(t)W_2^{-1}R_{m_1'}(s\rho(t)\theta_0)\cdots R_{m'_{p'+2q'}}(s\rho(t)\theta_0)W_2. \tag{4.10}$$

Corresponding, we define $(P^{-1}\gamma)_s = P^{-1}\gamma_s$. Note that $\gamma_s \in {}_P\mathcal{P}^*_{\tau,\omega}(2n)$.

Theorem 1.15 ([202]) *For $\omega \in \mathbf{U}$, $P \in \mathrm{Sp}(2n)$ and $0 < s \le 1$.*
If $\gamma \in {}^0_P\mathcal{P}_{\tau,\omega}(2n)$, the left perturbation paths of γ defined by (4.8) *satisfy*

$$i^P_\omega({}_s\gamma) - i^P_\omega({}_{-s}\gamma) = -\dim\ker(P^{-1} - \omega I). \tag{4.11}$$

If $\gamma \in {}_P\mathcal{P}^0_{\tau,\omega}(2n)$, the right rotational perturbation paths of γ defined by (4.10) *satisfy*

$$i^P_\omega(\gamma_s) - i^P_\omega(\gamma_{-s}) = \nu^P_\omega(\gamma). \tag{4.12}$$

Proof If $\gamma \in {}^0_P\mathcal{P}_{\tau,\omega}(2n)$, for any $(s,t) \in [-1,1] \times [0,\tau]$, we define $\zeta_s(t) = Q(P^{-1}, st/\tau, \ldots, st/\tau)$ for $s \in [-1,1]$. Note that $\xi_s(\tau) = \zeta_s(\tau)$. Fix $s \in (0,1]$. Then by Lemma 1.5 we obtain:

- $\zeta_{-s} * \xi \sim \xi_{-s}$ on $[0,\tau]$ with fixed points;
- $\zeta_s * \xi \sim \xi_s$ on $[0,\tau]$ with fixed points;
- $P^{-1}\gamma * \zeta^-_{-s} \sim {}_{-s}(P^{-1}\gamma)$ on $[0,\tau]$ with fixed points;
- $P^{-1}\gamma * \zeta^-_s \sim {}_s(P^{-1}\gamma)$ on $[0,\tau]$ with fixed points;

where $\zeta^-_s(t) = \zeta_s(\tau - t)$, $0 \le t \le \tau$ is different from ζ_s only with two opposite orientations.

Naturally, we get ${}_{-s}(P^{-1}\gamma) * \xi_{-s} \sim {}_s(P^{-1}\gamma) * \xi_s$ on $[0,\tau]$ with fixed points, then

$$i_\omega({}_{-s}(P^{-1}\gamma) * \xi_{-s}) = i_\omega({}_s(P^{-1}\gamma) * \xi_s).$$

Therefore (4.11) follows from Definition 1.7 and Theorem 5.4.1 in [223].

Equation (4.12) is directly obtained by Definition 1.7 and Theorem 5.4.1 in [223].

□

Theorem 1.16 ([202]) *For $\omega \in \mathbf{U}$ and $\tau > 0$. Let $P \in \mathrm{Sp}(2n)$, $\gamma \in {}_P\mathcal{P}^0_{\tau,\omega}(2n)$. Then for any paths α and $\beta \in {}_P\mathcal{P}^*_{\tau,\omega}(2n)$ which are sufficiently C^0-close to γ, there holds*

$$| i^P_\omega(\beta) - i^P_\omega(\alpha) | \leq \nu^P_\omega(\gamma). \tag{4.13}$$

Proof We choose $\xi \in \mathcal{P}_\tau(2n)$ such that $\xi(\tau) = P^{-1}$. By Theorem 6.1.8 in [223], we obtain

$$i_\omega((P^{-1}\gamma)_{-s} * \xi) \leq i_\omega((P^{-1}\alpha) * \xi) \leq i_\omega((P^{-1}\gamma)_s * \xi).$$

By Definition 2.6, we get

$$i^P_\omega(\gamma_{-s}) \leq i^P_\omega(\alpha) \leq i^P_\omega(\gamma_s) = i^P_\omega(\gamma_{-s}) + \nu^P_\omega(\gamma). \tag{4.14}$$

The proof is complete by (4.14). □

As in Sect. 1.2, the general symplectic space $(\mathbb{R}^{2n}, \tilde{\omega})$ has its symplectic transform group as the $\mathcal{J}$-symplectic matrix group $Sp_{\mathcal{J}}(2n)$. We denote by $\mathcal{P}_{\tau,\mathcal{J}}(2n) = \{\gamma \in C^0([0,\tau], Sp_{\mathcal{J}}(2n)) | \gamma(0) = I\}$ the set of $\mathcal{J}$-symplectic paths. From Proposition 2.2 of Chap. 1, we have

$$T\mathcal{P}_{\tau,\mathcal{J}}(2n)T^{-1} = \mathcal{P}_\tau(2n),$$

where the symplectic transform $T : (\mathbb{R}^{2n}, \tilde{\omega}) \to (\mathbb{R}^{2n}, \tilde{\omega}_0)$ defined in Sect. 1.1 corresponds to a matrix T satisfying $T^T JT = \mathcal{J}^{-1}$.

Definition 1.17 For $\omega \in \mathbf{U}$, $P \in Sp_{\mathcal{J}}(2n)$, $\gamma \in \mathcal{P}_{\tau,\mathcal{J}}(2n)$, the (P,ω)-index of γ is defined as

$$i^{\mathcal{J},P}_\omega(\gamma) = i^{P_0}_\omega(\gamma_0), \quad \nu^{\mathcal{J},P}_\omega(\gamma) = \nu^{P_0}_\omega(\gamma_0), \tag{4.15}$$

where $P_0 = TPT^{-1}$, and $\gamma_0(t) = T\gamma(t)T^{-1}$.

Remark 1.18

I. We note that there holds

$$\nu^{\mathcal{J},P}_\omega(\gamma) = \dim_{\mathbb{C}} \ker_{\mathbb{C}}(\gamma(\tau) - \omega P). \tag{4.16}$$

II. The Definition 1.17 is natural in the following sense. Consider the linear Hamiltonian system

$$\begin{cases} \dot{x}(t) &= \mathcal{J}B(t)x(t), \\ x(\tau) &= Px(0). \end{cases} \tag{4.17}$$

If we take $x(t) = T^{-1}x_0(t)$, the linear Hamiltonian system (4.17) becomes

$$\begin{cases} \dot{x}_0(t) &= JB_0(t)x_0(t), \\ x_0(\tau) &= P_0x_0(0), \end{cases} \tag{4.18}$$

where $B_0(t) = -(T^{-1})^T B(t)T^{-1}$. Suppose $\gamma(t)$ and $\gamma_0(t)$ be the fundamental solutions of the linear Hamiltonian systems (4.17) and (4.18), respectively. Then there holds

$$\gamma_0(t) = T\gamma(t)T^{-1}. \tag{4.19}$$

For a general continuous symplectic path $\rho : [a, b] \to \mathrm{Sp}(2n)$, we define its Maslov type P-index as follow.

Definition 1.19

$$\hat{i}^P(\rho) = i^P(\gamma_b) - i^P(\gamma_a), \tag{4.20}$$

where $\gamma_a \in \mathcal{P}(2n)$ is a symplectic path ended at $\rho(a)$ and $\gamma_b = \rho * \gamma_a \in \mathcal{P}(2n)$ is a symplectic path ended at $\rho(b)$ which is the composite of γ_a with ρ.

We remind that for the constant path $\gamma = I$ there holds $i^P(I) = 0$, so $\hat{i}^P(\gamma) = i^P(\gamma)$ for $\gamma \in \mathcal{P}(2n)$.

Lemma 1.20 *The index $\hat{i}^P(\rho)$ is well defined, i.e., it is independent of the choice of γ_a.*

Proof For two symplectic paths $\gamma_a,\ \gamma_a' \in \mathcal{P}(2n)$, we denote $\gamma_b,\ \gamma_b' \in \mathcal{P}(2n)$ correspondingly as in Definition 1.19. We should prove

$$i^P(\gamma_b) - i^P(\gamma_b') = i^P(\gamma_a) - i^P(\gamma_a'). \tag{4.21}$$

Taking ξ as in Definition 1.17, then it should be proved that

$$i_1(P^{-1}\gamma_b * \xi) - i_1(P^{-1}\gamma_b' * \xi) = i_1(P^{-1}\gamma_a * \xi) - i_1(P^{-1}\gamma_a' * \xi). \tag{4.22}$$

Using the pole decomposition, we suppose

$$P^{-1}\gamma_b * \xi = P_bU_b,\ P^{-1}\gamma_b' * \xi = P_b'U_b',\ P^{-1}\gamma_a * \xi = P_aU_a,\ P^{-1}\gamma_a' * \xi = P_a'U_a',$$

and

$$\det U_b = e^{\sqrt{-1}\Delta_b},\ \det U_b' = e^{\sqrt{-1}\Delta_b'},\ \det U_a = e^{\sqrt{-1}\Delta_a},\ \det U_a' = e^{\sqrt{-1}\Delta_a'}.$$

By definition, the functions Δ's have the path additivity, we prove the equality (4.22) in the non-degenerate cases. For the degenerate cases, since the two paths have the same end points, we can prove the equality (4.22) by taking non-degenerate paths close to the degenerate ones and taking the infimum in every terms. □

Theorem 1.21 *The index $\hat{i}^P$ has the following properties*

(1) *(Affine Scale Invariance). For $k > 0, l \geq 0$, we have the affine map $\varphi : [a, b] \to [ka + l, kb + l]$ defined by $\varphi(t) = kt + l$. For a given continuous path $\rho : [ka + l, kb + l] \to \mathrm{Sp}(2n)$, there holds*

$$\hat{i}^P(\rho) = \hat{i}^P(\rho \circ \varphi). \tag{4.23}$$

(2) *(Homotopy Invariance rel. End Points). If $\delta : [0, 1] \times [a, b] \to \mathrm{Sp}(2n)$ is a continuous map with $\delta(0, t) = \rho_1(t)$, $\delta(1, t) = \rho_2(t)$, $\delta(s, a) = \rho_1(a) = \rho_2(a)$ and $\delta(s, b) = \rho_1(b) = \rho_2(b)$ for $s \in [0, 1]$, then*

$$\hat{i}^P(\rho_1) = \hat{i}^P(\rho_2). \tag{4.24}$$

(3) *(Path Additivity). If $a < b < c$, and $\rho_{[a,c]} : [a, c] \to \mathrm{Sp}(2n)$ is concatenate path of $\rho_{[a,b]}$ and $\rho_{[b,c]}$, the, there holds*

$$\hat{i}^P(\rho_{[a,c]}) = \hat{i}^P(\rho_{[a,b]}) + \hat{i}^P(\rho_{[b,c]}). \tag{4.25}$$

(4) *(Symplectic Additivity). Let $P_k \in \mathrm{Sp}(2n_k)$, $\rho_k : [a, b] \to \mathrm{Sp}(2n_k)$, $k = 1, 2$, $P = P_1 \diamond P_2$, $\rho = \rho_1 \diamond \rho_2$. Then we have*

$$\hat{i}^P(\rho) = \hat{i}_{P_1}(\rho_1) + \hat{i}_{P_2}(\rho_2). \tag{4.26}$$

(5) *(Symplectic Invariance). For any $M \in \mathrm{Sp}(2n)$, there holds*

$$\hat{i}^{MP}(M\rho) = \hat{i}^P(\rho), \quad \hat{i}^{PM}(\rho M) = \hat{i}^P(\rho). \tag{4.27}$$

(6) *(Normalization). For $P = I$ and $\rho : [-\varepsilon, \varepsilon] \to \mathrm{Sp}(2)$ with $\varepsilon > 0$ small and $\rho(t) = e^{Jt}\begin{pmatrix} 1 & 1 \\ 0 & 1 \end{pmatrix}$, we have*

$$\begin{aligned} &(i) \ \ \hat{i}_I(\rho) = 1; \\ &(ii) \ \ \hat{i}_I(\rho_{[-\varepsilon,0]}) = 0; \\ &(iii) \ \hat{i}_I(\rho_{[0,\varepsilon]}) = 1. \end{aligned} \tag{4.28}$$

Proof We prove the statement (6) only. The remainders are direct consequence of the definition. From definition,

$$\hat{i}_I(\rho) = i_1(\gamma_\varepsilon) - i_1(\gamma_{-\varepsilon}).$$

But it is easy to see $i_1(\gamma_{-\varepsilon}) = 0$ and $i_1(\gamma_\varepsilon) = 1$. This proves statement (i). The other two statements are similar. □

We will see that the six properties in Theorem 1.21 character the index $\hat{i}^P$ uniquely. In the next chapter, we will define an index $\mu_P(\gamma)$ as the Maslov type index of a pair of Lagrangian paths and prove that this index also satisfies the six properties of Theorem 1.21. So we have

$$\hat{i}^P(\gamma) = i^P(\gamma) = \mu_P(\gamma)$$

for every $\gamma \in \mathcal{P}(2n)$.

Notes and comments The Maslov P-index theory for a symplectic path was first studied in [68] and [186] independently for any symplectic matrix P with different treatment. In fact, in the case of P is a special orthogonal matrix, the (P, ω)-index theory and its iteration theory were studied in [71].

4.2 Relative Index via Saddle Point Reduction Method

Let $B \in C([0, \tau], \mathcal{L}_s(2n))$ be a continuous symmetric $2n \times 2n$ matrix value function, we denote by $(i_\omega(B), \nu_\omega(B)) = (i_\omega(\gamma_B), \nu_\omega(\gamma_B))$ the ω-index of γ_B, where γ_B is the fundamental solution of the linear Hamiltonian system $\dot{x}(t) = JB(t)x(t)$.

Lemma 2.1 *Suppose $B_0, B_1 \in C(\mathbb{R}, \mathcal{L}_s(2n))$ such that $B_0 < B_1$ and $B_i(t + \tau) = B_i(t)$, $i = 0, 1$, there holds*

$$i_\omega(B_1) - i_\omega(B_0) = \sum_{s \in [0,1)} \nu_\omega((1 - s)B_0 + sB_1). \tag{4.29}$$

Proof By the saddle point reduction formula of ω-index (cf. [223]), there holds

$$m^-(B_i) = d_\omega + i_\omega(B_i), i = 0, 1.$$

Therefore, by using the boundary condition to define the operators A, B_i, there holds

$$i_\omega(B_1) - i_\omega(B_0) = m^-(B_1) - m^-(B_0) = \mu_A^c(B_1) - \mu_A^c(B_0).$$

The remainder is the same as the proof of Theorem III.4.3, we remind in this case, the nullity in the formula (3.38) should be ω-nullity. □

We now denoted by $\gamma_P(t)$ the symplectic path with $\gamma_P(0) = I$ and $\gamma_P(\tau) = P$. The specific choice is as follows.

For $P \in Sp(2n)$, we know that there exists a unique polar decomposition $P = AU$, where $A = \exp(M_1)$, M_1 satisfying

$$M_1^T J + J M_1 = 0 \text{ and } M_1^T = M_1; \tag{4.30}$$

U is a symplectic orthogonal matrix. $Sp(2n) \cap O(2n)$ is a compact connected Lie group and its Lie algebra is $sp(2n) \cap o(2n)$ constituted by the matrices M_2 satisfying

$$M_2^T J + J M_2 = 0 \text{ and } M_2^T + M_2 = 0. \tag{4.31}$$

Then there exists a matrix $M_2 \in sp(2n) \cap o(2n)$ such that $U = \exp(M_2)$. So P takes the form $P = \exp(M_1)\exp(M_2)$ (cf. Proposition I.2.10).

We set $P(t) = \exp(tM_1/\tau)\exp(tM_2/\tau)$. Note that $P(t) \in \mathcal{P}_\tau(2n)$ and $P(\tau) = P$.

In order to get the relationship between the Morse index and the Maslov (P, ω)-index, we should demand that

$$\gamma_P(0)^T J \dot{\gamma}_P(0) = \gamma_P(\tau)^T J \dot{\gamma}_P(\tau), \tag{4.32}$$

then the corresponding $B(t)$ satisfies $B(0) = B(\tau)$ while using Theorem 3.2 (one needs the matrix function $\tilde{B}_{\gamma_P}(t)$ in (4.125) below be τ-periodic). So when $P = \exp(M)$ with $M^T J + JM = 0$, we take $\gamma_P(t) = \exp(tM/\tau)$. It can be easily verified that $\gamma_P(t)$ satisfies (4.32). When P has the general form $P = \exp(M_1)\exp(M_2)$, M_1 and M_2 satisfy (4.30) and (4.31) respectively, (4.32) does not always hold, we should do some modification on $P(t)$. We let

$$\widetilde{P}(t) = \begin{cases} I, & 0 \le t \le \tau/3, \\ P(3t - \tau), & \tau/3 \le t \le 2\tau/3, \\ P, & 2\tau/3 \le t \le \tau. \end{cases}$$

Note that $\widetilde{P}(t) \in \mathcal{P}_\tau(2n)$ and $\widetilde{P}(\tau) = P$. We get $\widetilde{P}_\epsilon(t) \in C^1([0, \tau], Sp(2n))$ by taking small perturbation of $\widetilde{P}(t)$ near $t = \tau/3$ and $t = 2\tau/3$ in the space of symplectic paths, the end points will be fixed in the process of the perturbation. At the moment, we have $\widetilde{P}_\epsilon(t) \sim \widetilde{P}(t)$ on $[0, \tau]$ with fixed end points and $\dot{\widetilde{P}}_\epsilon(0) = \dot{\widetilde{P}}_\epsilon(\tau) = 0$. We set $\gamma_P(t) = \widetilde{P}_\epsilon(t)$, then $\gamma_P(t)$ satisfies (4.32). Obviously, $P(t) \sim \widetilde{P}(t)$ on $[0, \tau]$ with fixed end points, thus $\gamma_P(t) \sim P(t)$ on $[0, \tau]$ with fixed end points and $i_\omega^P(\gamma_P(t)) = i_\omega^P(P(t))$ by Proposition 1.12.

Lemma 2.2 ([194]) *Suppose $B \in C(\mathbb{R}, \mathcal{L}_s(2n))$ satisfying $B(t + \tau) = (P^{-1})^T B(t) P^{-1}$, there holds*

$$\nu_\omega^P(B) = \nu_\omega(\tilde{B}), \tag{4.33}$$

where $\tilde{B}(t) = \gamma_P(t)^T J \dot{\gamma}_P(t) + \gamma_P(t)^T B(t) \gamma_P(t)$.

Proof It is easy to check that the fundamental solution of the linear Hamiltonian system $\dot{x}(t) = J\tilde{B}(t)x(t)$ is the following symplectic path $\gamma_2(t) = \gamma_P(t)^{-1}\gamma(t)$. But $\gamma_2(\tau) = \gamma_P(\tau)^{-1}\gamma(\tau) = P^{-1}\gamma(\tau)$. Thus by definition, there holds

$$\nu_\omega^P(B) = \dim\ker(\gamma(\tau) - P) = \dim\ker(P^{-1}\gamma(\tau) - I) = \nu_\omega(\tilde{B}).$$

□

Theorem 2.3 ([186, 202]) *Let γ, γ_P, and $\gamma_2 \in \mathcal{P}_\tau(2n)$ be defined as above, there holds*

$$i_\omega^P(\gamma) - i_\omega^P(\gamma_P) = \begin{cases} i_\omega(\gamma_2), & \omega \neq 1, \\ i_\omega(\gamma_2) + n, & \omega = 1. \end{cases} \tag{4.34}$$

Thus the number $i_\omega(\gamma_2) + i_\omega^P(\gamma_P)$ depends only on P but not on the choice of M_1 and M_2, where M_1 and M_2 are defined by $P = \exp(M_1)\exp(M_2)$ as above.

Proof If $P = I$, then $\gamma_P \equiv I$ and $\gamma_2 = \gamma$, by the definition of Maslov ω-index in [223] and Definition 1.7, we have

$$i_\omega^I(\gamma) - i_\omega^I(\gamma_P) = i_\omega(\gamma_2) - i_\omega(I) = \begin{cases} i_\omega(\gamma_2), & \omega \neq 1, \\ i_\omega(\gamma_2) + n, & \omega = 1. \end{cases}$$

If $P \neq I$, when $\omega = 1$, it is just the result of Theorem 4.2 in [186].

Next we prove the case of $\omega \neq 1$.

We define three symplectic paths $\gamma_3, \gamma_5 \in \mathcal{P}_\tau(2n)$ and $\gamma_4 \in C([0, \tau], Sp(2n))$ by

$$\gamma_3(t) = \gamma_P(t)^{-1}, \quad \gamma_4(t) = P^{-1}\gamma(t), \quad \gamma_5 = \gamma_4 * \gamma_3.$$

Then we have $\gamma_5 \sim_\omega \gamma_2$ and the homotopy is defined by

$$\delta(s, t) = \begin{cases} \gamma_3(2st), & 0 \leq t \leq \tau/2, \\ \gamma_P((2 - 2s)t + (2s - 1)\tau)^{-1}\gamma(2t - \tau), & \tau/2 \leq t \leq \tau. \end{cases}$$

Hence we have

$$i_\omega(\gamma_2) = i_\omega(\gamma_5). \tag{4.35}$$

By the definition of (P, ω)-index, we get

$$i_\omega(\gamma_5) = i_\omega^P(\gamma) + i_\omega(\gamma_3). \tag{4.36}$$

We note that

$$P^{-1}P(t) \sim_\omega P(\tau - t)^{-1} \text{ on } [0, \tau] \text{ with fixed points}$$

by the homotopy map $\widetilde{\delta}(s,t) = e^{-M_2}e^{stM_2/\tau}e^{(t-\tau)M_1/\tau}e^{(1-s)tM_2/\tau}$. By the choice of $\gamma_P(t)$, we know that $\gamma_P(t) \sim P(t)$ on $[0, \tau]$ with fixed end points. Thus

$$P^{-1}\gamma_P(t) \sim_\omega \gamma_3(\tau - t) \text{ on } [0, \tau] \text{ with fixed points.} \tag{4.37}$$

Therefore by definition we have

$$\begin{aligned} i_\omega(\gamma_3) &= i_\omega(P^{-1}\gamma_P(\tau - t)) \\ &= -i_\omega^P(\gamma_P(t)) + i_\omega(P^{-1}\gamma_P(t) * P^{-1}\gamma_P(\tau - t)) \\ &= -i_\omega^P(\gamma_P(t)). \end{aligned} \tag{4.38}$$

We complete the proof from (4.35), (4.36), and (4.38). □

Lemma 2.4 ([194]) *Suppose $B_0, B_1 \in C(\mathbb{R}, \mathcal{L}_s(2n))$ satisfying $B(t+\tau) = (P^{-1})^T B(t) P^{-1}$ and $B_0 < B_1$, there holds*

$$i^P(B_1) - i^P(B_0) = \sum_{s\in[0,1)} \nu^P((1-s)B_0 + sB_1). \tag{4.39}$$

Proof From Theorem 2.3, there holds

$$i^P(B_1) - i^P(B_0) = i_1(\tilde{B}_1) - i_1(\tilde{B}_0) = \sum_{s\in[0,1)} \nu_1((1-s)\tilde{B}_0 + s\tilde{B}_1).$$

We see that $(1-s)\tilde{B}_0(t) + s\tilde{B}_1(t) = \gamma_P(t)^T J\dot{\gamma}_P(t) + \gamma_P(t)^T[(1-s)B_0(t) + sB_1(t)]\gamma_P(t) = \tilde{B}_s(t)$ with $B_s(t) = (1-s)B_0(t) + sB_1(t)$. Now from Lemma 2.2, we get the result (4.39). □

Theorem 2.5 ([194]) *Suppose $B \in C(\mathbb{R}, \mathcal{L}_s(2n))$ satisfying $B(t+\tau) = (P^{-1})^T B(t) P^{-1}$, there holds*

$$I(A, A-B) = \mu_A(B) = i^P(B). \tag{4.40}$$

Proof We only need to prove the case $P \neq I$. From definition, we have $i^P(0) = 0$. So by using Lemma 2.4 and the same computations as in the proof of Theorem 4.5, we have the formulas (4.40). □

4.3 Galerkin Approximation for the (P, ω)-Boundary Problem of Hamiltonian Systems

Let $\tau > 0, \omega = e^{\theta\sqrt{-1}} \in \mathbf{U}$. Define $W^\omega = W_\tau^\omega$ to be the subspace of $L^2([0, \tau], \mathbb{C}^{2n})$ composed by all y with the form

$$y(t)=\sum_{k\in\mathbb{Z}}e^{\sqrt{-1}(\theta+2k\pi)t/\tau}a_k,\quad a_k\in\mathbb{C}^{2n},$$

satisfying

$$\|y\|^2=\|y\|_{1/2,2}^2=\sum_{k\in\mathbb{Z}}((\theta+2k\pi)^2+1)\tau|a_k|^2<\infty.$$

This space is a Hilbert space with the norm $\|\cdot\|$ and the inner product $\langle\cdot,\cdot\rangle$. Define $W_k^\omega=W_k^{\omega,+}\bigoplus W_k^{\omega,-}$ with

$$W_k^{\omega,\pm}=e^{\sqrt{-1}(\theta+2k\pi)t/\tau}(J\pm\sqrt{-1}I)\mathbb{R}^{2n}. \tag{4.41}$$

Then

$$W^\omega=\bigoplus_{k\in\mathbb{Z}}W_k^\omega. \tag{4.42}$$

For $P\in Sp(2n)$, define the (P,ω)-space by

$$W_{\gamma_P}^\omega=W_{\tau,\gamma_P}^\omega=\{x\in L^2([0,\tau],\mathbb{C}^{2n})\mid x(t)=\gamma_P(t)\xi(t),\xi(t)\in W^\omega\}. \tag{4.43}$$

By (4.42) and (4.43), correspondingly we have the space splitting

$$W_{\gamma_P}^\omega=\bigoplus_{k\in\mathbb{Z}}W_{\gamma_P,k}^\omega,$$

where $W_{\gamma_P,k}^\omega=\{x\in L^2([0,\tau],\mathbb{C}^{2n})\mid x(t)=\gamma_P(t)\xi(t),\xi(t)\in W_k^\omega\}$. And we similarly define $W_{\gamma_P,k}^{\omega,\pm}$ the subspaces of $W_{\gamma_P,k}^\omega$ with $W_{\gamma_P,k}^\omega=W_{\gamma_P,k}^{\omega,+}\oplus W_{\gamma_P,k}^{\omega,-}$.

We define the norm $\|\cdot\|_1$ and the inner product $\langle\cdot,\cdot\rangle_1$ in the space $W_{\gamma_P}^\omega$ such that it is a Hilbert space. For $z_i(t)=\gamma_P(t)\xi_i(t)$, $i=1,2$, define

$$\|z_1\|_1=\|\xi\|,\quad\langle z_1,z_2\rangle_1=\langle\xi_1,\xi_2\rangle.$$

We define the operator $A_P^\omega:W_{\gamma_P}^\omega\to W_{\gamma_P}^\omega$ by

$$\langle A_P^\omega x,y\rangle_1=\int_0^\tau(-J\dot{x}(t),y(t))dt,\quad\forall x,y\in W_{\gamma_P}^\omega, \tag{4.44}$$

where $(\cdot,\cdot)$ is the standard Hermitian product in $\mathbb{C}^{2n}$. Then A is bounded linear and self-adjoint with finite dimensional kernel $N=\ker A_P^\omega$. The range of A_P^ω is closed and the restriction $A_P^\omega|_{N^\perp}$ is invertible. Define the functional on $W_{\gamma_P}^\omega$ by

$$f(x)=\frac{1}{2}\langle A_P^\omega x,x\rangle_1-\int_0^\tau H(t,x)dt,\ \ \forall x\in W_{\gamma_P}^\omega. \tag{4.45}$$

Define the compact self-adjoint operator $B_P^\omega: W_{\gamma_P}^\omega\to W_{\gamma_P}^\omega$ by

$$\langle B_P^\omega z_1,z_2\rangle_1=\int_0^\tau (B(t)z_1(t),z_2(t))dt,\ \ \forall z_1,z_2\in W_{\gamma_P}^\omega,\ \ B(t)=H''(t,x(t)). \tag{4.46}$$

We define an operator $\widetilde{M}_P^\omega: W_{\gamma_P}^\omega\to W_{\gamma_P}^\omega$ by

$$\langle \widetilde{M}_P^\omega x,y\rangle_1=\int_0^\tau (-J\dot{\gamma}_P(t)\gamma_P(t)^{-1}x(t),y(t))dt,\ \ \forall x,y\in W_{\gamma_P}^\omega. \tag{4.47}$$

For simplicity, we set $A=A_P^\omega$, $B=B_P^\omega$ and $\widetilde{M}=\widetilde{M}_P^\omega$. Then we have

$$\langle (A-\widetilde{M})x,y\rangle_1=\int_0^\tau (-J\dot{\xi}(t),\eta(t))dt, \tag{4.48}$$

where $x(t)=\gamma_P(t)\xi(t)$, $y(t)=\gamma_P(t)\eta(t)$, $\xi,\eta\in W^\omega$. By direct computation, we get

$$\sigma(A-\widetilde{M})=\{\lambda_k^\pm\},\ \ \lambda_k^\pm=\pm\frac{\theta+2k\pi}{((\theta+2k\pi)^2+1)\tau}, \tag{4.49}$$

each eigenvalue $\lambda_k^\pm$ of $A-\widetilde{M}$ has multiplicity $2n$ and the corresponding eigenspace is $W_{\gamma_P,k}^{\omega,\pm}$.

Let $Z_{m,\gamma_P}=\bigoplus_{|k|\le m}W_{\gamma_P,k}^\omega$ and $P_m: W_{\gamma_P}^\omega\to Z_{m,\gamma_P}$ the projection map. It is readily seen that $\Gamma=\{P_m\mid m=1,2,\cdots\}$ be the Galerkin approximation sequence with respect to $A-\widetilde{M}$:

(1) $Z_{m,\gamma_P}=P_mW_{\gamma_P}^\omega$ is finite dimensional for all $m\in\mathbb{N}$,
(2) $P_m\to I$ strongly as $m\to+\infty$,
(3) $[P_m,A-\widetilde{M}]:=P_m(A-\widetilde{M})-(A-\widetilde{M})P_m=0$.

Define the operator $A^\omega: W^\omega\to W^\omega$ by

$$\langle A^\omega\xi,\eta\rangle=\int_0^\tau (-J\dot{\xi}(t),\eta(t))dt,\ \ \forall\xi,\eta\in W^\omega. \tag{4.50}$$

and the compact self-adjoint operator $B^\omega: W^\omega\to W^\omega$ by

$$\langle B^\omega\xi,\eta\rangle=\int_0^\tau (B(t)\xi(t),\eta(t))dt \tag{4.51}$$

for a symmetric matrix function $B(t)\in C(S_\tau,\mathcal{L}(\mathbb{R}^{2n}))$. By the Floquet theory we have

$$\nu_\omega(B) = \dim\ker(A^\omega - B^\omega). \tag{4.52}$$

Let $Z_m^\omega = \bigoplus_{|k|\le m} W_k^\omega$ and $P_m^\omega : W^\omega \to Z_m^\omega$ the projection map. Then the sequence $\Gamma^\omega = \{P_m^\omega \mid m = 1, 2, \cdots\}$ be the Galerkin approximation sequence of A^ω:

(1) $Z_m^\omega = P_m^\omega W^\omega$ is finite dimensional for all $m \in \mathbb{N}$,
(2) $P_m^\omega \to I$ strongly as $m \to +\infty$,
(3) $P_m^\omega A^\omega = A^\omega P_m^\omega, \forall m \ge 1$.

For a self-adjoint operator S, we denote by $M^*(S)$ the eigenspaces of S with eigenvalues belonging to $(0, +\infty)$, $\{0\}$ and $(-\infty, 0)$ with $* = +, 0$ and $* = -$, respectively. We denote by $m^*(S) = \dim M^*(S)$. Similarly, for any $d > 0$, we denote by $M_d^*(S)$ the d-eigenspaces of S with eigenvalues belonging to $[d, +\infty)$, $(-d, d)$ and $(-\infty, -d]$ with $* = +, 0$ and $* = -$, respectively. We also denote by $m_d^*(S) = \dim M_d^*(S)$. We denote $S^\sharp = (S|_{ImS})^{-1}$, $P_m S P_m = (P_m S P_m)|_{Z_{m,\gamma_P}} : Z_{m,\gamma_P} \to Z_{m,\gamma_P}$ and $P_m^\omega S P_m^\omega = (P_m^\omega S P_m^\omega)|_{Z_m^\omega} : Z_m^\omega \to Z_m^\omega$.

Lemma 3.1 *For any symmetric matrix function $B(t) \in C(S_\tau, \mathcal{L}(\mathbb{R}^{2n}))$, there exists an $m_1 > 0$ such that for $m \ge m_1$, there holds*

$$\dim\ker(P_m^\omega(A^\omega - B^\omega)P_m^\omega) \le \dim\ker(A^\omega - B^\omega), \tag{4.53}$$

where B^ω is defined by (4.112).

Proof We follow the idea of [95].
There is an $m_0 > 0$ such that for $m \ge m_0$,

$$\dim P_m^\omega \ker(A^\omega - B^\omega) = \dim\ker(A^\omega - B^\omega). \tag{4.54}$$

For otherwise, there exists $\xi_j \in \ker(A^\omega - B^\omega) \cap (I - P_{m_j}^\omega)W^\omega$ such that $\|\xi_j\| = 1$. Note that $A^\omega \xi_j = (I - P_{m_j}^\omega)B^\omega \xi_j$. Then we have

$$\|A^\omega \xi_j\| \ge \|(A^\omega)^\sharp\|^{-1} > 0,$$

and

$$\|(I - P_{m_j}^\omega)B^\omega \xi_j\| \le \|(I - P_{m_j}^\omega)B^\omega\| \to 0$$

as $j \to +\infty$, a contradiction. Thus (4.54) holds.

Take $m \ge m_0$, let $X_m = P_m^\omega \ker(A^\omega - B^\omega)$ and $Z_m^\omega = X_m \oplus Y_m$. Then we have

$$Y_m \subset Im(A^\omega - B^\omega).$$

Let $d = \frac{1}{4}\|(A^\omega - B^\omega)^\sharp\|^{-1}$. Since B^ω is compact, we have

$$\|(I - P_m^{\omega})B^{\omega}\| \to 0 \text{ as } m \to +\infty.$$

Hence there is an $m_1 \geq m_0$ such that for $m \geq m_1$,

$$\|(I - P_m^{\omega})B^{\omega}\| \leq 2d. \tag{4.55}$$

For $m \geq m_1$, $\forall \eta \in Y_m$, we have

$$\eta = (A^{\omega} - B^{\omega})^{\sharp}(A^{\omega} - B^{\omega})\eta = (A^{\omega} - B^{\omega})^{\sharp}(P_m^{\omega}(A^{\omega} - B^{\omega})P_m^{\omega}\eta + (P_m^{\omega} - I)B^{\omega}\eta).$$

This implies that

$$\|\eta\| \leq \frac{1}{2d}\|P_m^{\omega}(A^{\omega} - B^{\omega})P_m^{\omega}\eta\|. \tag{4.56}$$

Thus by (4.54) and (4.56) we have

$$\dim\ker(P_m^{\omega}(A^{\omega} - B^{\omega})P_m^{\omega}) \leq \dim X_m = \dim\ker(A^{\omega} - B^{\omega}).$$

□

Theorem 3.2 ([202]) *For any symmetric matrix function $B(t) \in C(S_\tau, \mathcal{L}(\mathbb{R}^{2n}))$ with the Maslov ω-index $(i_\omega(B), \nu_\omega(B))$ and any constant $0 < d \leq \frac{1}{4}\|(A^{\omega} - B^{\omega})^{\sharp}\|^{-1}$, there exists an $m^* > 0$ such that for $m \geq m^*$ we have*

$$\begin{aligned}
m_d^+(P_m^{\omega}(A^{\omega} - B^{\omega})P_m^{\omega}) &= \frac{1}{2}\dim Z_m^{\omega} - i_\omega(B) - \nu_\omega(B), \\
m_d^-(P_m^{\omega}(A^{\omega} - B^{\omega})P_m^{\omega}) &= \frac{1}{2}\dim Z_m^{\omega} + i_\omega(B), \\
m_d^0(P_m^{\omega}(A^{\omega} - B^{\omega})P_m^{\omega}) &= \nu_\omega(B).
\end{aligned} \tag{4.57}$$

Proof We follow the idea of [94].
We distinguish two cases.

Case 1. $\nu_\omega(B) = 0$. By (4.52) and Lemma 6.5, for $m \geq m_1$ we obtain that

$$\dim M^0(P_m^{\omega}(A^{\omega} - B^{\omega})P_m^{\omega}) = \dim\ker(A^{\omega} - B^{\omega}) = 0.$$

Since B^{ω} is compact, there exists $m_2 \geq m_1$ such that for $m \geq m_2$,

$$\|(I - P_m^{\omega})B^{\omega}\| \leq \frac{1}{2}\|(A^{\omega} - B^{\omega})^{-1}\|^{-1}.$$

Then $P_m^{\omega}(A^{\omega} - B^{\omega})P_m^{\omega} = (A^{\omega} - B^{\omega})P_m^{\omega} + (I - P_m^{\omega})B^{\omega}P_m^{\omega}$ implies that

$$\|P_m^{\omega}(A^{\omega}-B^{\omega})P_m^{\omega}\xi\| \ge \frac{1}{2}\|(A^{\omega}-B^{\omega})^{\sharp}\|^{-1}\|\xi\|. \ \forall \xi \in Z_m^{\omega}.$$

Thus

$$M_d^*(P_m^{\omega}(A^{\omega}-B^{\omega})P_m^{\omega}) = M^*(P_m^{\omega}(A^{\omega}-B^{\omega})P_m^{\omega}) \text{ for } * = +, -, 0.$$

As proved in [226], there is $m^* > 0$ such that for $m \ge m^*$ the relation (4.57) holds.

Case 2. $\nu_{\omega}(B) > 0$. By Lemma 3.1 there exists $m_3 > 0$ such that for $m \ge m_3$,

$$m_d^0(P_m^{\omega}(A^{\omega}-B^{\omega})P_m^{\omega}) \le \nu_{\omega}(B). \tag{4.58}$$

For otherwise, there exists $\eta \in M_d^0(P_m^{\omega}(A^{\omega}-B^{\omega})P_m^{\omega}) \cap Y_m$, $\| \eta \| = 1$, where

$$Z_m^{\omega} = P_m^{\omega}\ker(A^{\omega}-B^{\omega}) \oplus Y_m, \quad \dim P_m^{\omega}\ker(A^{\omega}-B^{\omega}) = \nu_{\omega}(B).$$

Then $\|(P_m^{\omega}(A^{\omega}-B^{\omega})P_m^{\omega})\eta\| \le d\|\eta\|$ contradicts to $\|\eta\| \le \frac{1}{2d}\|P_m^{\omega}(A^{\omega}-B^{\omega})P_m^{\omega}\|$.

Let γ be the fundamental solution of (1.2) and γ_s be as described in [226]. For $s \in [-1, 1]$, we define

$$B_s(t) = -J\dot{\gamma}_s(t)\gamma_s^{-1}(t), \quad t \in [0, \tau].$$

Let B_s^{ω} be the compact operator defined by (4.112) corresponding to $B_s(t)$. By the results in [226] we have

$$\begin{aligned} &M^0(A^{\omega}-B_s^{\omega}) = 0 \text{ for } s \ne 0, \quad \|B_s^{\omega}-B^{\omega}\| \to 0 \text{ as } s \to 0. \\ &i_{\omega}(\gamma_s) - i_{\omega}(\gamma_{-s}) = \nu_{\omega}(B), \quad i_{\omega}(\gamma_{-s}) = i_{\omega}(B), \ \forall s \in (0, 1]. \end{aligned} \tag{4.59}$$

Choose $0 < s_0 \le 1$ such that $\|B^{\omega}-B_{\pm s_0}^{\omega}\| \le \frac{1}{2}d$. By case 1, (4.58), (4.59) and the fact that

$$P_m^{\omega}(A^{\omega}-B_{\pm s_0}^{\omega})P_m^{\omega} = P_m^{\omega}(A^{\omega}-B^{\omega})P_m^{\omega} + P_m^{\omega}(B^{\omega}-B_{\pm s_0}^{\omega})P_m^{\omega},$$

there exists $m^* \ge m_3$ such that for $m \ge m^*$,

$$\begin{aligned} m_d^+(P_m^{\omega}(A^{\omega}-B^{\omega})P_m^{\omega}) &\le m^+(P_m^{\omega}(A^{\omega}-B_{s_0}^{\omega})P_m^{\omega}) = \frac{1}{2}\dim Z_m^{\omega} - i_{\omega}(B) - \nu_{\omega}(B), \\ m_d^+(P_m^{\omega}(A^{\omega}-B^{\omega})P_m^{\omega}) &\ge m^+(P_m^{\omega}(A^{\omega}-B_{-s_0}^{\omega})P_m^{\omega}) - m_d^0(P_m^{\omega}(A^{\omega}-B^{\omega})P_m^{\omega}) \\ &\ge \frac{1}{2}\dim Z_m^{\omega} - i_{\omega}(B) - \nu_{\omega}(B). \end{aligned}$$

Hence, $m_d^0(P_m^{\omega}(A^{\omega}-B^{\omega})P_m^{\omega})=\nu_{\omega}(B)$ and

$$m_d^+(P_m^{\omega}(A^{\omega}-B^{\omega})P_m^{\omega})=\frac{1}{2}\dim Z_m^{\omega}-i_{\omega}(B)-\nu_{\omega}(B).$$

Similarly, we have

$$\begin{aligned}
m_d^-(P_m^{\omega}(A^{\omega}-B^{\omega})P_m^{\omega}) &\le m^-(P_m^{\omega}(A^{\omega}-B^{\omega}_{-s_0})P_m^{\omega})=\frac{1}{2}\dim Z_m^{\omega}+i_{\omega}(B),\\
m_d^-(P_m^{\omega}(A^{\omega}-B^{\omega})P_m^{\omega}) &\ge m^-(P_m^{\omega}(A^{\omega}-B^{\omega}_{s_0})P_m^{\omega})-m_d^0(P_m^{\omega}(A^{\omega}-B^{\omega})P_m^{\omega})\\
&\ge \frac{1}{2}\dim Z_m^{\omega}+i_{\omega}(B),
\end{aligned}$$

Hence $m_d^-(P_m^{\omega}(A^{\omega}-B^{\omega})P_m^{\omega})=\frac{1}{2}\dim Z_m^{\omega}+i_{\omega}(B)$ and the proof is complete. □

Theorem 3.3 ([202]) *For* $0<d\le\frac{1}{4}\|(A_P^{\omega}-B_P^{\omega})^{\sharp}\|^{-1}$, *there exists an* $m^*>0$ *such that for* $m\ge m^*$ *we have*

$$\begin{aligned}
m_d^+(P_m(A_P^{\omega}-B_P^{\omega})P_m) &= \begin{cases}\frac{1}{2}\dim Z_{m,\gamma_P}+i_{\omega}^P(\gamma_P)-i_{\omega}^P(\gamma)-\nu_{\omega}^P(\gamma), & \omega\ne 1,\\ \frac{1}{2}\dim Z_{m,\gamma_P}+i_{\omega}^P(\gamma_P)-i_{\omega}^P(\gamma)-\nu_{\omega}^P(\gamma)+n, & \omega=1.\end{cases}\\
m_d^-(P_m(A_P^{\omega}-B_P^{\omega})P_m) &= \begin{cases}\frac{1}{2}\dim Z_{m,\gamma_P}-i_{\omega}^P(\gamma_P)+i_{\omega}^P(\gamma), & \omega\ne 1,\\ \frac{1}{2}\dim Z_{m,\gamma_P}-i_{\omega}^P(\gamma_P)+i_{\omega}^P(\gamma)-n, & \omega=1.\end{cases}\\
m_d^0(P_m(A_P^{\omega}-B_P^{\omega})P_m) &= \nu_{\omega}^P(\gamma).
\end{aligned} \tag{4.60}$$

Proof Let $x(t)=\gamma_P(t)\xi(t)$, $\xi\in W^{\omega}$. Then we have

$$\begin{aligned}
&\langle (A-B)x,x\rangle_1\\
&=\int_0^{\tau}[(-J\dot{x}(t),x(t))-(B(t)x(t),x(t))]dt\\
&=\int_0^{\tau}[(-J\dot{\xi}(t),\xi(t))-(\gamma_P(t)^TJ\dot{\gamma}_P(t)\xi(t),\xi(t))-(\gamma_P(t)^TB(t)\gamma_P(t)\xi(t),\xi(t))]dt\\
&=\int_0^{\tau}[(-J\dot{\xi}(t),\xi(t))-(\widetilde{B}_{\gamma_P}(t)\xi(t),\xi(t))]dt,
\end{aligned}$$

we remind that we have set $A=A_P^{\omega}$, $B=B_P^{\omega}$ and $\widetilde{B}_{\gamma_P}(t)=\gamma_P(t)^TJ\dot{\gamma}_P(t)+\gamma_P(t)^TB(t)\gamma_P(t)$. By the definition of $\gamma_P(t)$ and $B(t)$, we know that $\widetilde{B}_{\gamma_P}(t)$ is symmetric and $\widetilde{B}_{\gamma_P}(0)=\widetilde{B}_{\gamma_P}(\tau)$.

Consider the following linear Hamiltonian systems

$$\dot{z}(t)=J\widetilde{B}_{\gamma_P}(t)z(t),\quad z(t)\in\mathbb{R}^{2n}. \tag{4.61}$$

Suppose $\widetilde{\gamma}(t)$ is the fundamental solution of (4.125). Then by direct computation, we obtain

$$\widetilde{\gamma}(t) = \gamma_P(t)^{-1}\gamma(t) = \gamma_2(t).$$

By Theorem 3.2, there exists an $m^* > 0$ such that for $m \geq m^*$ such that

$$\begin{aligned}
m_d^+(P_m(A-B)P_m) &= m_d^+(P_m^\omega(A^\omega - \widetilde{B}_{\gamma_P}^\omega)P_m^\omega) = \frac{1}{2}\dim Z_m^\omega - i_\omega(\widetilde{B}_{\gamma_P}) - \nu_\omega(\widetilde{B}_{\gamma_P}),\\
m_d^-(P_m(A-B)P_m) &= m_d^-(P_m^\omega(A^\omega - \widetilde{B}_{\gamma_P}^\omega)P_m^\omega) = \frac{1}{2}\dim Z_m^\omega + i_\omega(\widetilde{B}_{\gamma_P}),\\
m_d^0(P_m(A-B)P_m) &= m_d^0(P_m^\omega(A^\omega - \widetilde{B}_{\gamma_P}^\omega)P_m^\omega) = \nu_\omega(\widetilde{B}_{\gamma_P}),
\end{aligned} \tag{4.62}$$

where $\widetilde{B}_{\gamma_P}^\omega$ be the compact operator defined by (4.112) corresponding to $\widetilde{B}_{\gamma_P}(t)$. Note that $\dim Z_{m,\gamma_P} = \dim Z_m^\omega$, $\|(A^\omega - \widetilde{B}_{\gamma_P}^\omega)^\sharp\|^{-1} = \|(A-B)^\sharp\|^{-1}$. Hence (4.60) follows from (4.34), (4.130) and $\nu_\omega(\widetilde{B}_{\gamma_P}) = \nu_\omega(\gamma_2) = \nu_\omega^P(\gamma)$. □

4.4 (P, ω)-Index Theory from Analytical Point of View

Let $Q^\omega = A^\omega - B^\omega : W^\omega \to W^\omega$. We note that A^ω is a bounded self-adjoint operator, B^ω is a compact self-adjoint operator. Suppose $N^\omega := \ker Q^\omega$, then N^ω is a finite dimensional subspace of W^ω and $Q^\omega|_{N^{\omega\perp}}$ is invertible. We denote by $P^\omega : W^\omega \to N^\omega$ the orthogonal projection. Set $d = \frac{1}{4}\|(Q^\omega|_{N^{\omega\perp}})^{-1}\|^{-1}$. Recall that $\Gamma^\omega = \{P_m^\omega \mid m = 1, 2, \cdots\}$ is the Galerkin approximation sequence of A^ω.

From Lemma III.1.4, we have the following result.

Lemma 4.1 *The difference between the d-Morse indices*

$$m_d^-(P_m^\omega(A^\omega - B^\omega)P_m^\omega) - m_d^-(P_m^\omega A^\omega P_m^\omega) \tag{4.63}$$

and $m_d^0(P_m^\omega(A^\omega - B^\omega)P_m^\omega)$ eventually becomes a constant independent of m, and for large m there holds

$$m_d^0(P_m^\omega(A^\omega - B^\omega)P_m^\omega) = m^0(A^\omega - B^\omega). \tag{4.64}$$

Following Definition III.1.5, we have the following definition for relative ω-index.

Definition 4.2 We define the relative index by

$$I(A^\omega, A^\omega - B^\omega) = m_d^-(P_m^\omega(A^\omega - B^\omega)P_m^\omega) - m_d^-(P_m^\omega A^\omega P_m^\omega),\ m \geq m^*, \tag{4.65}$$

where $m^* > 0$ is a constant large enough such that the difference in (4.63) becomes a constant independent of $m \geq m^*$.

Remark 4.3 Following Lemma III.1.7, up to a sign, the relative index $I(A^\omega, A^\omega - B^\omega)$ is exactly the spectral flow of the operator family $A^\omega - sB^\omega$. Precisely, there holds

$$I(A^\omega, A^\omega - B^\omega) = -sf(A^\omega - sB^\omega).$$

Let $\widetilde{B}^\omega_{\gamma_P} : W^\omega \to W^\omega$ the compact self-adjoint operator. The operator $\widetilde{B}^\omega_{\gamma_P}$ is defined by

$$\langle \widetilde{B}^\omega_{\gamma_P}\xi, \eta\rangle = \int_0^\tau (\widetilde{B}_{\gamma_P}(t)\xi(t), \eta(t))dt, \tag{4.66}$$

where

$$\widetilde{B}_{\gamma_P}(t) = \gamma_P(t)^T J\dot{\gamma}_P(t) + \gamma_P(t)^T B(t)\gamma_P(t) \tag{4.67}$$

and $B(t) \in C([0, \tau], \mathcal{L}(\mathbb{R}^{2n}))$ is a symmetric matrix function.

Define the operator $\bar{B}^\omega_{\gamma_P} : W^\omega \to W^\omega$ by

$$\langle \bar{B}^\omega_{\gamma_P}\xi, \eta\rangle = \int_0^\tau (\bar{B}_{\gamma_P}(t)\xi(t), \eta(t))dt, \tag{4.68}$$

where

$$\bar{B}_{\gamma_P}(t) = \gamma_P(t)^T J\dot{\gamma}_P(t). \tag{4.69}$$

Recall that $\gamma_P \in \mathcal{P}_\tau(2n)$ satisfies $\gamma_P(\tau) = P$. So we have that $\widetilde{B}_{\gamma_P}(t)$ and $\bar{B}_{\gamma_P}(t)$ are both symmetric matrix functions and $\widetilde{B}_{\gamma_P}(0) = \widetilde{B}_{\gamma_P}(\tau)$, $\bar{B}_{\gamma_P}(0) = \bar{B}_{\gamma_P}(\tau)$. Therefore we can give the following definition

Definition 4.4 ([202]) We defined the index pair $(j^{\gamma_P}_\omega(B), n^{\gamma_P}_\omega(B))$ of B by

$$\begin{aligned} j^{\gamma_P}_\omega(B) &= I(A^\omega, A^\omega - \widetilde{B}^\omega_{\gamma_P}) - I(A^\omega, A^\omega - \bar{B}^\omega_{\gamma_P}), \\ n^{\gamma_P}_\omega(B) &= m^0(A - B) = m^0(A^\omega - \widetilde{B}^\omega_{\gamma_P}). \end{aligned} \tag{4.70}$$

Theorem 4.5 *For $\omega \in \mathbf{U}$ and $P \in Sp(2n)$, we have*

$$j^{\gamma_P}_\omega(B) = i^P_\omega(\gamma), \tag{4.71}$$

$$n^{\gamma_P}_\omega(B) = \nu^P_\omega(\gamma), \tag{4.72}$$

where $\gamma(t)$ is the fundamental solution of $\dot{z}(t) = JB(t)z(t)$. Thus the two definitions of the (P, ω)-index are consistent.

Proof Recalling the notations in Theorem 2.3, by (4.70) and (4.65), we have

$$\begin{aligned} j_\omega^{\gamma_P}(B) &= I(A^\omega, A^\omega - \widetilde{B}_{\gamma_P}^\omega) - I(A^\omega, A^\omega - \bar{B}_{\gamma_P}^\omega) \\ &= m_d^-(P_m^\omega(A^\omega - \widetilde{B}_{\gamma_P}^\omega)P_m^\omega) - m_d^-(P_m^\omega A^\omega P_m^\omega) \\ &\quad - \left(m_d^-(P_m^\omega(A^\omega - \bar{B}_{\gamma_P}^\omega)P_m^\omega) - m_d^-(P_m^\omega A^\omega P_m^\omega)\right) \\ &= m_d^-(P_m^\omega(A^\omega - \widetilde{B}_{\gamma_P}^\omega)P_m^\omega) - m_d^-(P_m^\omega(A^\omega - \bar{B}_{\gamma_P}^\omega)P_m^\omega) \\ &= i_\omega(\gamma_2) - i_\omega(\gamma_3), \end{aligned}$$

where $\gamma_2(t) = \gamma_P(t)^{-1}\gamma(t)$ is the fundamental solution of $\dot{z}(t) = J\widetilde{B}_{\gamma_P}(t)z(t)$ and $\gamma_3(t) = \gamma_P(t)^{-1}$ is the fundamental solution of $\dot{z}(t) = J\bar{B}_{\gamma_P}(t)z(t)$. The last equality is achieved by Theorem 3.2.

When $P = I$, then $\gamma_2 = \gamma$ and $\gamma_3 \equiv I$, so

$$i_\omega(\gamma_2) - i_\omega(\gamma_3) = i_\omega(\gamma) - i_\omega(I) = i_\omega^I(\gamma). \tag{4.73}$$

When $P \neq I$, in the Definition 1.7, we take $\xi(t) = \gamma_3(t) = \gamma_P(t)^{-1}$, then by (4.35) and (4.36), we obtain

$$i_\omega(\gamma_2) - i_\omega(\gamma_3) = i_\omega^P(\gamma). \tag{4.74}$$

Note that $\nu_\omega(\widetilde{B}_{\gamma_P}) = \nu_\omega^P(\gamma)$, and by Floquet theory, (4.57) and (4.64) we get the second equality. □

Definition 4.6 If $z(t)$ is a solution of the nonlinear system

$$\dot{z}(t) = JH'(t, z(t)),\ x(t) \in \mathbb{R}^{2n}$$

and $B_z(t) = H''(t, z(t))$, we define

$$j_\omega^{\gamma_P}(z) = j_\omega^{\gamma_P}(B_z),\quad \nu_\omega^{\gamma_P}(z) = \nu_\omega^{\gamma_P}(B_z).$$

Remark 4.7 We note that the relation (4.32) and the homotopy in (4.37) are the only key ingredients in the arguments of Sects. 4.3 and 4.4, so $\forall\, \gamma_0 \in \mathcal{P}_\tau(2n)$ satisfies $\gamma_0(\tau) = P$ and

$$\gamma_0(0)^T J\dot{\gamma}_0(0) = \gamma_0(\tau)^T J\dot{\gamma}_0(\tau), \tag{4.75}$$

$$P^{-1}\gamma_0(t) \sim_\omega \gamma_0(\tau - t)^{-1} \text{ on } [0, \tau] \text{ with fixed points}, \tag{4.76}$$

we can obtain all results in these two sections if replacing γ_P by γ_0. Thus the index pair $(j_\omega^{\gamma_0}(B), n_\omega^{\gamma_0}(B))$ is not dependent on the choice of γ_0 in this sense and we denote $(j_\omega^P(B), n_\omega^P(B)) := (j_\omega^{\gamma_0}(B), n_\omega^{\gamma_0}(B))$.

4.5 Bott-Type Formula for the Maslov Type P-Index

In this section, we establish the Bott-type iteration formula for the Maslov type P-index theory. For $P \in Sp(2n)$, we consider the following P-boundary value problem

$$\begin{cases} \dot{x}(t) = JH'(t, x(t)), \\ x(\tau) = Px(0), \end{cases} \tag{4.77}$$

where $H \in C^2(\mathbb{R} \times \mathbb{R}^{2n}, \mathbb{R})$ satisfying $H(t+\tau, Px) = H(t, x)$, $H'(t, x)$ is the gradient of H with respect to the variable x. Clearly we have $P^T H''(t+\tau, Px)P = H''(t, x)$. Let $x(t)$ be a solution of (4.77). Linearizing the Hamiltonian system (4.77) at the solution x we get a linear Hamiltonian system

$$\dot{y} = JB(t)y, \tag{4.78}$$

where $B(t) = H''(t, x(t))$ satisfies

$$B(t+\tau) = H''(t+\tau, x(t+\tau)) = H''(t+\tau, Px(t)) = (P^{-1})^T H''(t, x(t))P^{-1}. \tag{4.79}$$

It implies $P^T B(t+\tau)P = B(t)$ (we say B satisfying (P, τ) periodic condition). We can extend the definition of $B(t)$ from $[0, \tau]$ to $[0, +\infty)$ in the obvious way.

Recall that

$$\hat{\mathcal{P}}_\tau(2n) = \{\gamma \in C^1([0, \tau], Sp(2n)) \mid \gamma(0) = I, \dot{\gamma}(\tau) = P\dot{\gamma}(0)P^{-1}\gamma(\tau)\}.$$

Note that the set $\hat{\mathcal{P}}_\tau(2n)$ formed by the fundamental solutions of all the linear Hamiltonian systems (4.78) with continuous symmetric and (P, τ)-periodic coefficients and $\hat{\mathcal{P}}_\tau(2n)$ is dense in $\mathcal{P}_\tau(2n)$.

Suppose that the continuous symplectic path $\gamma : [0, \tau] \to Sp(2n)$ is the fundamental solution of (4.78). For $k \in \mathbb{N}$, we define

$$B_k(t) = B(t)|_{[0, k\tau]} \tag{4.80}$$

and define the k-times iteration path $\gamma^k : [0, k\tau] \to Sp(2n)$ of γ by

$$\gamma^k(t) = \begin{cases} \gamma(t), & t \in [0, \tau], \\ P\gamma(t-\tau)P^{-1}\gamma(\tau), & t \in [\tau, 2\tau], \\ P^2\gamma(t-2\tau)(P^{-1}\gamma(\tau))^2, & t \in [2\tau, 3\tau], \\ P^3\gamma(t-3\tau)(P^{-1}\gamma(\tau))^3, & t \in [3\tau, 4\tau], \\ \cdots\cdots \\ P^{k-1}\gamma(t-(k-1)\tau)(P^{-1}\gamma(\tau))^{k-1}, & t \in [(k-1)\tau, k\tau]. \end{cases} \tag{4.81}$$

Note that $\gamma^k(t)$ is the fundamental solution of $\dot{y}(t) = JB_k(t)y(t)$.

We suppose $P^k = \exp(\bar{M}_1)\exp(\bar{M}_2)$, where $\bar{M}_1$, $\bar{M}_2$ satisfy (4.30) and (4.31) respectively. When the interval get changed from $[0, \tau]$ to $[0, k\tau]$, we have $P_k(t) = \exp(t\bar{M}_1/k\tau)\exp(t\bar{M}_2/k\tau)$ naturally and $P_k(t) \in \mathcal{P}_\tau(2n)$. Note that $P_k(k\tau) = P^k$. We obtain $\gamma_{P^k}(t)$ by the same argument of the choice of $\gamma_P(t)$.

We extend $\gamma_P(t)$ from $[0, \tau]$ to $[0, +\infty)$ by $\gamma_P(t + \tau) = P \cdot \gamma_P(t)$. And we define $\bar{\gamma}_{P^k}(t) = \gamma_P(t)|_{[0,k\tau]}$. Note that $\bar{\gamma}_P(t) = \gamma_P(t)$ and $\bar{\gamma}_{P^k}(k\tau) = P^k$. In order to simplify the calculation of the iteration of Malsov P-index, we want to replace $\gamma_{P^k}(t)$ by $\bar{\gamma}_{P^k}(t)$. In view of Remark 4.7, (4.75) holds with $\bar{\gamma}_{P^k}(t)$ on $[0, k\tau]$, so we only need the following lemma.

Lemma 5.1 *For the iterated path $\bar{\gamma}_{P^k}(t)$, we have the following ω-homotopy*

$$P^{-k}\bar{\gamma}_{P^k}(t) \sim_\omega \bar{\gamma}_{P^k}(k\tau - t)^{-1} \text{ on } [0, k\tau] \text{ with fixed points.} \tag{4.82}$$

Proof As consequences of definitions, we get

$$P^{-k}\bar{\gamma}_{P^k}(t) = \begin{cases} P^{-k}\gamma_P(t), & t \in [0, \tau], \\ P^{-k+1}\gamma_P(t-\tau), & t \in [\tau, 2\tau], \\ P^{-k+2}\gamma_P(t-2\tau), & t \in [2\tau, 3\tau], \\ \cdots\cdots \\ P^{-2}\gamma_P(t-(k-2)\tau), & t \in [(k-2)\tau, (k-1)\tau], \\ P^{-1}\gamma_P(t-(k-1)\tau), & t \in [(k-1)\tau, k\tau]. \end{cases} \tag{4.83}$$

$$\bar{\gamma}_{P^k}(k\tau - t)^{-1} = \begin{cases} \gamma_P(\tau - t)^{-1}P^{-(k-1)}, & t \in [0, \tau], \\ \gamma_P(2\tau - t)^{-1}P^{-(k-2)}, & t \in [\tau, 2\tau], \\ \gamma_P(3\tau - t)^{-1}P^{-(k-3)}, & t \in [2\tau, 3\tau], \\ \cdots\cdots \\ \gamma_P((k-1)\tau - t)^{-1}P^{-1}, & t \in [(k-2)\tau, (k-1)\tau], \\ \gamma_P(k\tau - t)^{-1}, & t \in [(k-1)\tau, k\tau]. \end{cases} \tag{4.84}$$

Note that $P^{-k}\bar{\gamma}_{P^k}(t)$ and $\bar{\gamma}_{P^k}(k\tau - t)^{-1}$ have the same values at $t = 0, \tau, 2\tau, \cdots, (k-1)\tau, k\tau$, which are $P^{-k}, P^{-k+1}, \cdots, P^{-1}, I$, respectively. We only prove here that $P^{-k}\bar{\gamma}_{P^k}(t)$ and $\bar{\gamma}_{P^k}(k\tau - t)^{-1}$ are homotopic on $[0, \tau]$ with fixed end points given by $\delta_1(s, t)$, the rest is similar, suppose $P^{-k}\bar{\gamma}_{P^k}(t)$ and $\bar{\gamma}_{P^k}(k\tau - t)^{-1}$ are homotopic on $[j\tau, (j+1)\tau]$ with fixed end points given by $\delta_j(s, t)$, $j = 1, 2, k-1$, so we complete the proof by the homotopy map $\delta_{k-1} * \delta_{k-2} * \cdots * \delta_1$.

If we have constructed a homotopic map $\delta(s, t)$ such that $P^{-k}\bar{\gamma}_{P^k}(t) \sim_\omega \bar{\gamma}_{P^k}(k\tau - t)^{-1}$ on $[0, \tau]$ along $\delta(\cdot, \tau)$, according to Lemma 1.3 and Remark 1.4, the homotopy δ can be modified to fix the end points all the time.

In the following, we use Corollary 6.2.2, Corollary 6.2.5 and Corollary 6.2.10 in [223] repeatedly (or apply Theorem II.2.5 and Theorem II.2.6) to give a direct proof of $P^{-k}\bar{\gamma}_{P^k}(t) \sim_\omega \bar{\gamma}_{P^k}(k\tau - t)^{-1}$ on $[0, \tau]$ with fixed points.

$$\begin{aligned}
&P^{-k}\gamma_P(t) \sim_\omega \gamma_P(\tau - t)^{-1}P^{-(k-1)} \text{ on } [0, \tau] \text{ with fixed end points}\\
&\iff \gamma_P(t)P^{k-1} \sim_\omega P^k\gamma_P(\tau - t)^{-1} \text{ on } [0, \tau] \text{ with fixed end points}\\
&\iff i_\omega(\gamma_P(t)P^{k-1}) = i_\omega(P^k\gamma_P(\tau - t)^{-1}) = i_\omega(\gamma_P(\tau - t)^{-1}P^k)\\
&\iff \gamma_P(t)P^{k-1} \sim_\omega \gamma_P(\tau - t)^{-1}P^k \text{ on } [0, \tau] \text{ with fixed end points}\\
&\iff \gamma_P(t) \sim_\omega \gamma_P(\tau - t)^{-1}P \text{ on } [0, \tau] \text{ with fixed end points}\\
&\iff i_\omega(\gamma_P(t)) = i_\omega(\gamma_P(\tau - t)^{-1}P).
\end{aligned}$$

The first and the forth implications " $\iff$ " follow from the definition of homotopy of two paths, the second and third " $\iff$ " follow from the results of Corollary 6.2.5 and Corollary 6.2.10 in [223], and the last " $\iff$ " follows from Corollary 6.2.2 in [223].

Recall that $P(t) = \exp(tM_1/\tau)\exp(tM_2/\tau)$, $\forall t \in [0, \tau]$. Then we obtain

$$\begin{aligned}
i_\omega(P(\tau - t)^{-1}P) &= i_\omega\big(\exp(M_2)P(\tau - t)^{-1}P\exp(-M_2)\big)\\
&= i_\omega\big(\exp(tM_2/\tau)\exp(tM_1/\tau)\big)\\
&= i_\omega\big(\exp(-tM_2/\tau)\exp(tM_2/\tau)\exp(tM_1/\tau)\exp(tM_2/\tau)\big)\\
&= i_\omega(P(t)).
\end{aligned}$$

Thus $P^{-k}P(t) \sim_\omega P(\tau - t)^{-1}P^{-(k-1)}$ on $[0, \tau]$ with fixed end points. Note that $\gamma_P(t) \sim P(t)$ on $[0, \tau]$ with fixed end points, thus $P^{-k}\gamma_P(t) \sim_\omega \gamma_P(\tau - t)^{-1}P^{-(k-1)}$ on $[0, \tau]$ with fixed end points. □

For any $\tau > 0$, $\gamma \in \hat{\mathcal{P}}_\tau(2n)$ and $\omega \in \mathbf{U}$, we define

$$j_\omega^{P^k}(\gamma^k) = j_\omega^{\bar{\gamma}_{P^k}}(\gamma^k), \quad n_\omega^{P^k}(\gamma^k) = n_\omega^{\bar{\gamma}_{P^k}}(\gamma^k). \tag{4.85}$$

When $\omega = 1$, we denote $j_\omega^{P^k}(\gamma^k)$ and $n_\omega^{P^k}(\gamma^k)$ by $j_{P^k}(\gamma^k)$, $n_{P^k}(\gamma^k)$ respectively.

We define the Hilbert space

$$W_k^1 = W^{1/2,2}(S_{k\tau}, \mathbb{C}^{2n}), \tag{4.86}$$

where $S_{k\tau} = \mathbb{R}/(k\tau\mathbb{Z})$, and the Hilbert space

$$W_k^{P^k} = \{x \in L^2([0, k\tau], \mathbb{C}^{2n}) \mid x(t) = \bar{\gamma}_{P^k}(t)\xi(t), \xi(t) \in W_k^1\}.$$

Recall that $W^{\omega}_{\gamma_P} = \{x \in L^2([0,\tau], \mathbb{C}^{2n}) \mid x(t) = \gamma_P(t)\xi(t), \xi(t) \in W^{\omega}\}$. If $\omega^k = 1$, we can identify $W^{\omega}_{\gamma_P}$ with the subspace $\{x \in W^{P^k}_k \mid x(t+\tau) = \omega P x(t)\}$ of $W^{P^k}_k$.

We define two self-adjoint operators and a quadratic form on $W^{P^k}_k$ by

$$\langle A_k x, y\rangle_1 = \int_0^{k\tau} (-J\dot{x}(t), y(t))dt, \quad \langle B_k x, y\rangle_1 = \int_0^{k\tau} (B_k(t)x(t), y(t))dt, \tag{4.87}$$

$$Q_k(x,y) = \langle (A_k - B_k)x, y\rangle_1. \tag{4.88}$$

We also define a quadratic form on $W^{\omega}_{\gamma_P}$ by

$$Q(x,y) = \langle (A-B)x, y\rangle_1, \tag{4.89}$$

where the operators A and B are defined by (4.44) and (4.46).

Lemma 5.2 ([202]) *Let $k \geq 1$, $P \in Sp(2n)$ be given. For $\omega^k = 1$, then $W^{\omega}_{\gamma_P}$ are orthogonal subspace of $W^1_{\bar{\gamma}_{P^k}}$ for Q_k, and $W^1_{\bar{\gamma}_{P^k}}$ splits into a direct sum*

$$W^1_{\bar{\gamma}_{P^k}} = \bigoplus_{\omega^k=1} W^{\omega}_{\gamma_P}. \tag{4.90}$$

Proof Any $x(t) \in W^1_{\bar{\gamma}_{P^k}}$ can be written as:

$$x(t) = \bar{\gamma}_{P^k}(t) \sum_{p\in\mathbb{Z}} \exp\left(\frac{2p\pi t\sqrt{-1}}{k\tau}\right) a_p, \quad a_p \in \mathbb{C}^{2n}. \tag{4.91}$$

For $q = 0, 1, \ldots, k-1$, we denote by $C(q)$ the set of all p such that $p - q \in k\mathbb{Z}$. We may write

$$x(t) = \sum_{q=0}^{k-1} x_q(t), \quad with \ x_q(t) = \bar{\gamma}_{P^k}(t) \sum_{p\in C(q)} \exp\left(\frac{2p\pi t\sqrt{-1}}{k\tau}\right) a_p. \tag{4.92}$$

We then check that

$$\begin{aligned} x_q(t+\tau) &= P \cdot \bar{\gamma}_{P^k}(t) \sum_{p\in C(q)} \exp\left(\frac{2p\pi t\sqrt{-1} + 2p\pi\tau\sqrt{-1}}{k\tau}\right) a_p \\ &= \omega_q P x_q(t), \end{aligned} \tag{4.93}$$

where $\omega_q = \exp\left(\frac{2\pi q\sqrt{-1}}{k}\right)$. So we have $x_q \in W^{\omega_q}_{\gamma_P}$. When q runs from 0 to $k-1$, then ω runs through the k-th roots of unity. Hence we get the splitting (4.90).

Letting $x \in W^{\omega}_{\gamma_P}$ and $y \in W^{\lambda}_{\gamma_P}$, where ω and λ are k-th roots of unity, there holds

$$\begin{aligned}&\int_j^{j+1}[(-J\dot{x}(t), y(t)) - (B(t)x(t), y(t))]\,dt \\ &= \omega\bar{\lambda}\int_{j-1}^{j}[(-J\dot{x}(t), y(t)) - (B(t)x(t), y(t))]\,dt.\end{aligned} \tag{4.94}$$

Then we get

$$\begin{aligned}Q_k(x, y) &= \int_0^{k\tau}[(-J\dot{x}(t), y(t)) - (B_k(t)x(t), y(t))]\,dt \\ &= \sum_{j=0}^{k-1}(\omega\bar{\lambda})^j\int_0^{\tau}[(-J\dot{x}(t), y(t)) - (B(t)x(t), y(t))]\,dt \\ &= \sum_{j=0}^{k-1}(\omega\bar{\lambda})^j Q(x, y).\end{aligned}$$

When $\omega = \lambda$,

$$\sum_{j=0}^{k-1}(\omega\bar{\lambda})^j = \sum_{j=0}^{k-1}(\omega\bar{\omega})^j = k.$$

When $\omega \neq \lambda$,

$$\sum_{j=0}^{k-1}(\omega\bar{\lambda})^j = \frac{1-(\omega\bar{\lambda})^k}{1-\omega\bar{\lambda}} = 0.$$

It implies $W^{\omega}_{\gamma_P}$ and $W^{\lambda}_{\gamma_P}$ are orthogonal for Q_k, $\forall\omega \neq \lambda$ and

$$Q_k(x, y) = kQ(x, y), \quad x, y \in W^{\omega}_{\gamma_P}. \tag{4.95}$$

□

Based on the results of the previous sections, following the idea of [28], we now prove the following result which is called the Bott-type formula for the (P, ω)-index.

Theorem 5.3 ([202]) *For any $\tau > 0$, $\omega_0 \in \mathbf{U}$, $\gamma \in \mathcal{P}_\tau(2n)$ and $k \in \mathbb{N}$, we have*

$$i^{P^k}_{\omega_0}(\gamma^k) = \sum_{\omega^k=\omega_0} i^P_{\omega}(\gamma), \quad \nu^{P^k}_{\omega_0}(\gamma^k) = \sum_{\omega^k=\omega_0} \nu^P_{\omega}(\gamma). \tag{4.96}$$

Proof We first assume that $\omega_0 = 1$. By definition, there hold

$$n_{P^k}(\gamma^k) = m^0(A_k - B_k), \;\; n_\omega^P(\gamma) = m^0(A - B).$$

By (4.95) and Lemma 5.2, we get

$$n_{P^k}(\gamma^k) = \sum_{\omega^k=1} n_\omega^P(\gamma).$$

Let $x(t), y(t) \in W^1_{\bar{\gamma}_{P^k}}$, we set $x(t) = \bar{\gamma}_{P^k}(t)\xi(t)$, $y(t) = \bar{\gamma}_{P^k}(t)\eta(t)$, $\xi, \eta \in W^1_k$. Then we have

$$\begin{aligned}\langle A_k x, y\rangle_1 &= \int_0^{k\tau} [(-J\dot{x}(t), y(t)) \\ &= \int_0^{k\tau} [(-J\dot{\xi}(t), \eta(t)) - (\bar{\gamma}_{P^k}(t)^T J\dot{\bar{\gamma}}_{P^k}(t)\xi(t), \eta(t)) \\ &= \int_0^{k\tau} [(-J\dot{\xi}(t), \eta(t)) - (\bar{B}_k(t)\xi(t), \eta(t))]dt,\end{aligned}$$

where

$$\begin{aligned}\bar{B}_k(t) &= \bar{\gamma}_{P^k}(t)^T J\dot{\bar{\gamma}}_{P^k}(t) \\ &= \begin{cases} \gamma_P(t)^T J\dot{\gamma}_P(t), & t \in [0, \tau], \\ \gamma_P(t-\tau)^T J\dot{\gamma}_P(t-\tau), & t \in [\tau, 2\tau], \\ \cdots\cdots \\ \gamma_P(t-(k-1)\tau)^T J\dot{\gamma}_P(t-(k-1)\tau), & t \in [(k-1)\tau, k\tau]. \end{cases}\end{aligned} \tag{4.97}$$

We can easily see that $\bar{B}_k(t)$ is symmetric by the definition of $\bar{\gamma}_{P^k}(t)$ and τ-periodic.

$$\begin{aligned}&\langle (A_k - B_k)x, y\rangle_1 \\ &= \int_0^{k\tau} [(-J\dot{x}(t), y(t)) - (B_k(t)x(t), y(t))]dt \\ &= \int_0^{k\tau} [(-J\dot{\xi}(t), \eta(t)) - (\bar{\gamma}_{P^k}(t)^T J\dot{\bar{\gamma}}_{P^k}(t)\xi(t), \eta(t)) \\ &\quad - (\bar{\gamma}_{P^k}(t)^T B_k(t)\bar{\gamma}_{P^k}(t)\xi(t), \eta(t))]dt \\ &= \int_0^{k\tau} [(-J\dot{\xi}(t), \eta(t)) - (\widetilde{B_k}(t)\xi(t), \eta(t))]dt,\end{aligned}$$

where

$$
\begin{aligned}
\widetilde{B}_k(t) &= \bar{\gamma}_{P^k}(t)^T J\dot{\bar{\gamma}}_{P^k}(t) + \bar{\gamma}_{P^k}(t)^T B_k(t)\bar{\gamma}_{P^k}(t) \\
&= \begin{cases}
\gamma_P(t)^T J\dot{\gamma}_P(t) + \gamma_P(t)^T B(t)\gamma_P(t), & t\in[0,\tau], \\
\gamma_P(t-\tau)^T J\dot{\gamma}_P(t-\tau) + \gamma_P(t-\tau)^T B(t-\tau)\gamma_P(t-\tau), & t\in[\tau,2\tau], \\
\cdots\cdots & \\
\gamma_P(t-(k-1)\tau)^T J\dot{\gamma}_P(t-(k-1)\tau)+ & \\
\gamma_P(t-(k-1)\tau)^T B(t-(k-1)\tau)\gamma_P(t-(k-1)\tau), & t\in[(k-1)\tau,k\tau].
\end{cases}
\end{aligned}
\tag{4.98}
$$

It is also symmetric and τ-periodic.

We define $A_k^1 : W_k^1 \to W_k^1$ by

$$\langle A_k^1\xi,\eta\rangle = \int_0^{k\tau} (-J\dot{\xi}(t),\eta(t))dt. \tag{4.99}$$

Define $\widetilde{B}_k^1 : W_k^1 \to W_k^1$ by

$$\langle \widetilde{B}_k^1\xi,\eta\rangle = \int_0^{k\tau} (\widetilde{B}_k(t)\xi(t),\eta(t))dt, \tag{4.100}$$

where $\widetilde{B}_k(t)$ is defined by (4.98).

Define $\bar{B}_k^1 : W_k^1 \to W_k^1$ by

$$\langle \bar{B}_k^1\xi,\eta\rangle = \int_0^{k\tau} (\bar{B}_k(t)\xi(t),\eta(t))dt,$$

where $\bar{B}_k(t)$ is defined by (4.97).

By definition, there hold

$$
\begin{aligned}
j_{P^k}(\gamma^k) &= I(A_k^1, A_k^1-\widetilde{B}_k^1) - I(A_k^1, A_k^1-\bar{B}_k^1), \\
j_\omega^P(\gamma) &= I(A^\omega, A^\omega-\widetilde{B}_{\gamma_P}^\omega) - I(A^\omega, A^\omega-\bar{B}_{\gamma_P}^\omega),
\end{aligned}
\tag{4.101}
$$

where the operator $\widetilde{B}_{\gamma_P}^\omega$ and $\bar{B}_{\gamma_P}^\omega$ is defined by (4.66) and (4.68) respectively.

As a special case of Lemma 5.2, we have

$$W_k^1 = \bigoplus_{\omega^k=1} W^\omega. \tag{4.102}$$

We can treat W^ω as $W^\omega = \{y \in W_k^1 \mid y(\tau) = \omega y(0)\}$, which is the subspace of W_k^1.

Taking $\xi \in W^\omega$ and $\eta \in W^\lambda$, where ω and λ are k-th roots of unity, by (4.97) and (4.98) there hold

$$\begin{aligned}\langle (A_k^1 - \widetilde{B}_k^1)\xi, \eta\rangle &= \int_0^{k\tau} (-J\dot{\xi}(t), \eta(t))dt - \int_0^{k\tau} (\widetilde{B}_k(t)\xi(t), \eta(t))dt \\ &= \sum_{j=0}^{k-1} (\omega\bar{\lambda})^j \int_0^{\tau} [(-J\dot{\xi}(t), \eta(t)) - (\widetilde{B}_{\gamma_P}(t)\xi(t), \eta(t))]\, dt \\ &= \sum_{j=0}^{k-1} (\omega\bar{\lambda})^j \langle (A^\omega - \widetilde{B}_{\gamma_P}^\omega)\xi, \eta\rangle,\end{aligned} \tag{4.103}$$

$$\begin{aligned}\langle (A_k^1 - \bar{B}_k^1)\xi, \eta\rangle &= \int_0^{k\tau} (-J\dot{\xi}(t), \eta(t))dt - \int_0^{k\tau} (\bar{B}_k(t)\xi(t), \eta(t))dt \\ &= \sum_{j=0}^{k-1} (\omega\bar{\lambda})^j \int_0^{\tau} [(-J\dot{\xi}(t), \eta(t)) - (\bar{B}_{\gamma_P}(t)\xi(t), \eta(t))]\, dt \\ &= \sum_{j=0}^{k-1} (\omega\bar{\lambda})^j \langle (A^\omega - \bar{B}_{\gamma_P}^\omega)\xi, \eta\rangle,\end{aligned} \tag{4.104}$$

where $\widetilde{B}_{\gamma_P}^\omega$, $\widetilde{B}_{\gamma_P}(t)$, $\bar{B}_{\gamma_P}^\omega$ and $\bar{B}_{\gamma_P}(t)$ satisfy (4.66), (4.67), (4.68), and (4.69) respectively.

By (4.103) and (4.104), we have

$$I(A_k^1, A_k^1 - \widetilde{B}_k^1) = \sum_{\omega^k=1} I(A^\omega, A^\omega - \widetilde{B}_{\gamma_P}^\omega), \quad I(A_k^1, A_k^1 - \bar{B}_k^1) = \sum_{\omega^k=1} I(A^\omega, A^\omega - \bar{B}_{\gamma_P}^\omega). \tag{4.105}$$

So by (4.101), we get

$$j_{P^k}(\gamma^k) = \sum_{\omega^k=1} j_\omega^P(\gamma), \quad n_{P^k}(\gamma^k) = \sum_{\omega^k=1} n_\omega^P(\gamma). \tag{4.106}$$

We complete the proof by Theorem 4.5 and (4.106).

For the general case, we should replace the space $W_k^{P^k}$ by $W_{k\tau,\gamma_{pk}}^{\omega_0}$ defined in (4.43) and ω_q by $\omega_{q,\omega_0} = \exp\left(\frac{(2\pi q+\theta_0)\sqrt{-1}}{k}\right)$ with $\omega_0 = \exp(\theta_0\sqrt{-1})$. It is easy to see $\omega_{q,\omega_0}^k = \omega_0$.

For the general case of $\gamma \in \mathcal{P}_\tau(2n)$. We can choose $\beta \in \hat{\mathcal{P}}_\tau(2n)$ such that $\beta(\tau) = \gamma(\tau)$ and $\beta \sim \gamma$ with fixed end points. This homotopy can be extended to $[0, 1] \times [0, k\tau]$ for any $k \in \mathbb{N}$. Thus we obtain $\beta^k \sim \gamma^k$, $\forall k \in \mathbb{N}$. By Proposition 1.12, for all $k \in \mathbb{N}$ and $\omega \in \mathbf{U}$ we obtain

$$i_\omega^{P^k}(\gamma^k) = i_\omega^{P^k}(\beta^k), \quad \nu_\omega^{P^k}(\gamma^k) = \nu_\omega^{P^k}(\beta^k). \tag{4.107}$$

Together with (4.96) for β, we obtain it for γ. This completes the proof of Theorem 5.3. □

4.6 Iteration Theory for P-Index

4.6.1 *Splitting Numbers*

Definition 6.1 For any $\tau > 0$, $P \in Sp(2n)$, $\gamma \in \mathcal{P}_\tau(2n)$, $m \in \mathbb{N}$, and $\omega \in \mathbf{U}$, we define

$$i_\omega^{P^m}(\gamma, m) = i_\omega^{P^m}(\gamma_P^m), \quad \nu_\omega^{P^m}(\gamma, m) = \nu_\omega^{P^m}(\gamma_P^m). \tag{4.108}$$

Where the iteration path $\gamma_P^m \in \mathcal{P}_{m\tau}(2n)$ is defined by (4.81).

If the subindex $\omega = 1$, we simply write $(i(\gamma, m), \nu(\gamma, m))$, $(i^{P^m}(\gamma, m), \nu^{P^m}(\gamma, m))$ etc, and omit the subindex 1 when there is no confusion.

Fix $\tau > 0$, $P \in Sp(2n)$ and a path $\gamma \in \mathcal{P}_\tau(2n)$, next we study the properties of the index function $i_\omega^P(\gamma)$ of γ at ω as a function $\omega \in \mathbf{U}$, and use the short notations when γ is a fixed path

$$i^P(\omega) = i_\omega^P(\gamma), \quad \nu^P(\omega) = \nu_\omega^P(\gamma). \tag{4.109}$$

By definition we have

$$i^P(\omega) = i^P(\bar{\omega}), \quad \nu^P(\omega) = \nu^P(\bar{\omega}). \tag{4.110}$$

Lemma 6.2 ([203]) *$i^P(\cdot)$ is locally constant on $\mathbf{U} \setminus (\sigma(P^{-1}\gamma(\tau)) \cup \sigma(P^{-1}))$, and thus is constant on each connected component of $\mathbf{U} \setminus (\sigma(P^{-1}\gamma(\tau)) \cup \sigma(P^{-1}))$. It holds that*

$$\nu^P(\omega) = 0, \quad \forall \omega \in \mathbf{U} \setminus (\sigma(P^{-1}\gamma(\tau)) \cup \sigma(P^{-1})). \tag{4.111}$$

Proof Note that $\mathbf{U} \cap (\sigma(P^{-1}\gamma(\tau)) \cup \sigma(P^{-1}))$ contains at most $4n$ points. For any $\omega_0 \in \mathbf{U} \setminus (\sigma(P^{-1}\gamma(\tau)) \cup \sigma(P^{-1}))$, let $\mathcal{N}(\omega_0)$ be an open connected neighborhood of ω_0 in $\mathbf{U} \setminus (\sigma(P^{-1}\gamma(\tau)) \cup \sigma(P^{-1}))$. By Definition 1.1 we get $\nu^P(\omega) = 0$ for all $\omega \in \mathcal{N}(\omega_0)$. Then (4.111) holds and $\gamma \in {}^*_P\mathcal{P}^*_{\tau,\omega}(2n)$ for all $\omega \in \mathcal{N}(\omega_0)$. As in Definition 1.7, we can connect P^{-1} to M_n^+ or M_n^- by β_0 in $Sp(2n)^*_{\omega_0}$, connect $P^{-1}\gamma(\tau)$ to M_n^+ or M_n^- by β_1 in $Sp(2n)^*_{\omega_0}$. By the compactness of the image of β_0, β_1 and the openness of $Sp(2n)^*_{\omega_0}$ in $Sp(2n)$, we can further require $\mathcal{N}(\omega_0)$ to be so small that both β_0 and β_1 are completely located within $Sp(2n)^*_{\omega}$ for all $\omega \in \mathcal{N}(\omega_0)$. We choose a path $\xi \in \mathcal{P}_\tau(2n)$ so that $\xi(\tau) = P^{-1}$. Then by definition, this implies $i_\omega(P^{-1}\gamma * \xi) = i_{\omega_0}(P^{-1}\gamma * \xi)$, $i_\omega(\xi) = i_{\omega_0}(\xi)$ for all $\omega \in \mathcal{N}(\omega_0)$. Thus by Definition 1.7, we have $i^P(\omega) = i^P(\omega_0)$ for all $\omega \in \mathcal{N}(\omega_0)$. □

Corollary 6.3 ([203]) *The discontinuity points of $i^P(\cdot)$ and $\nu^P(\cdot)$ are contained in $\mathbf{U} \cap \sigma(P^{-1}\gamma(\tau)) \cap \sigma(P^{-1})$.*

Analogy to the definition of splitting numbers about Maslov-type index, we give the following definition.

Definition 6.4 ([203]) For any $P \in Sp(2n)$, $M \in Sp(2n)$ and $\omega \in \mathbf{U}$, choosing $\tau > 0$ and $\gamma \in \mathcal{P}_\tau(2n)$ with $\gamma(\tau) = M$, we define

$$ {}_PS^{\pm}_M(\omega) = \lim_{\varepsilon \to 0+} i^P_{\exp(\pm\varepsilon\sqrt{-1})\omega}(\gamma) - i^P_\omega(\gamma). \tag{4.112}$$

Lemma 6.5 ([203]) *We have ${}_PS^{\pm}_M(\omega) = S^{\pm}_{P^{-1}M}(\omega) - S^{\pm}_{P^{-1}}(\omega)$. Thus ${}_PS^{\pm}_M(\omega)$ is well defined in the sense that it is independent of the path $\gamma \in \mathcal{P}_\tau(2n)$ with $\gamma(\tau) = M$ appearing in* (4.112).

Proof Choose $\xi \in \mathcal{P}_\tau(2n)$ such that $\xi(\tau) = P^{-1}$, then by definition we obtain

$$\begin{aligned}
{}_PS^{\pm}_M(\omega) &= \lim_{\varepsilon \to 0+} i^P_{\exp(\pm\varepsilon\sqrt{-1})\omega}(\gamma) - i^P_\omega(\gamma) \\
&= \lim_{\varepsilon \to 0+} (i_{\exp(\pm\varepsilon\sqrt{-1})\omega}(P^{-1}\gamma * \xi) \\
&\quad - i_{\exp(\pm\varepsilon\sqrt{-1})\omega}(\xi)) - i_\omega(P^{-1}\gamma * \xi) + i_\omega(\xi) \\
&= \lim_{\varepsilon \to 0+} (i_{\exp(\pm\varepsilon\sqrt{-1})\omega}(P^{-1}\gamma * \xi) - i_\omega(P^{-1}\gamma * \xi)) \\
&\quad - \lim_{\varepsilon \to 0+} (i_{\exp(\pm\varepsilon\sqrt{-1})\omega}(\xi) - i_\omega(\xi)) \\
&= S^{\pm}_{P^{-1}M}(\omega) - S^{\pm}_{P^{-1}}(\omega).
\end{aligned}$$

Since $S^{\pm}_{P^{-1}M}(\omega)$ and $S^{\pm}_{P^{-1}}(\omega)$ are independent of the choice of the path γ, so we complete the proof. □

Lemma 6.6 ([203]) *For $P, M \in Sp(2n)$ and $\omega \in \mathbf{U}$, there hold*

$$ {}_P S_M^{\pm}(\omega) = 0, \qquad \text{if } \omega \in \mathbf{U} \setminus (\sigma(P^{-1}\gamma(\tau)) \cup \sigma(P^{-1})), \tag{4.113}$$

$$ {}_P S_M^{+}(\omega) = {}_P S_M^{-}(\bar{\omega}). \tag{4.114}$$

For $P_i, M_i \in Sp(2n_i)$ with $i = 0$ and 1, $P = P_1 \diamond P_2$, there holds

$$ {}_P S_{M_0 \diamond M_1}^{\pm}(\omega) = {}_{P_1} S_{M_0}^{\pm}(\omega) + {}_{P_2} S_{M_1}^{\pm}(\omega), \quad \forall \omega \in \mathbf{U}. \tag{4.115}$$

Proof This is a direct consequence of Definition 6.4, Lemma 6.5 and the symplectic additivity of Maslov P-index. □

4.6.2 Abstract Precise Iteration Formulas

Definition 6.7 ([203]) We define the P-mean index of $\gamma \in \mathcal{P}_\tau(2n)$ by

$$\bar{i}^P(\gamma) \equiv \lim_{m \to +\infty} \frac{i^{P^m}(\gamma, m)}{m}. \tag{4.116}$$

The following result tall us that the P-mean index $\bar{i}^P(\gamma)$ is well defined and always finite.

Proposition 6.8 ([203]) *For any $\tau > 0$ and $\gamma \in \mathcal{P}_\tau(2n)$ there hold*

$$\bar{i}^P(\gamma) = \bar{i}(P^{-1}\gamma * \xi) - \bar{i}(\xi) = \frac{1}{2\pi}\int_0^{2\pi} i^P_{e^{\sqrt{-1}\theta}}(\gamma)\, d\theta, \tag{4.117}$$

where $\bar{i}(\rho)$ is the mean index of ρ for the Maslov-type index in the periodic case,

$$\bar{\nu}^P(\gamma) \equiv \lim_{m \to +\infty} \frac{\nu^{P^m}(\gamma, m)}{m} = \frac{1}{2\pi}\int_0^{2\pi} \nu_{e^{\sqrt{-1}\theta}}(P^{-1}\gamma * \xi)\, d\theta = 0. \tag{4.118}$$

Especially, $\bar{i}^P(\gamma)$ is always a finite real number.

Proof This is a direct consequence of Theorem 5.3 and (2.15). □

For $P, M \in Sp(2n)$, we define

$$C_P(M) = \sum_{0<\theta<2\pi} {}_P S_M^{-}(e^{\sqrt{-1}\theta}). \tag{4.119}$$

Proposition 6.9 ([203]) *For any path $\gamma \in \mathcal{P}_\tau(2n)$ with $M = \gamma(\tau)$, $P \in Sp(2n)$, we extend γ to $[0, \infty)$ by (4.81). Then for any $m \in \mathbb{N}$ we have*

$$\begin{aligned} i^{P^m}(\gamma, m) &= m(i^P(\gamma, 1) + {}_PS^{\pm}_M(1) - C_P(M)) \\ &\quad + 2\sum_{\theta \in (0,2\pi)} E(\frac{m\theta}{2\pi})\, {}_PS^-_M(e^{\sqrt{-1}\theta}) - ({}_PS^+_M(1) + C_P(M)). \end{aligned} \tag{4.120}$$

Where $E(a) = \min\{k \in \mathbb{Z} \mid k \geq a\}$, for any $a \in \mathbb{R}$.

Proof By Theorem 2.1 in [233]. We have

$$\begin{aligned} i(P^{-1}\gamma * \xi, m) &= m(i(P^{-1}\gamma * \xi, 1) + S^+_{P^{-1}M}(1) - C(P^{-1}M)) \\ &\quad + 2\sum_{\theta \in (0,2\pi)} E(\frac{m\theta}{2\pi}) S^-_{P^{-1}M}(e^{\sqrt{-1}\theta}) - (S^+_{P^{-1}M}(1) + C(P^{-1}M)) \end{aligned} \tag{4.121}$$

and

$$\begin{aligned} i(\xi, m) &= m(i(\xi, 1) + S^+_M(1) - C(M)) \\ &\quad + 2\sum_{\theta \in (0,2\pi)} E(\frac{m\theta}{2\pi}) S^-_M(e^{\sqrt{-1}\theta}) - (S^+_M(1) + C(M)). \end{aligned} \tag{4.122}$$

Thus by Theorem 5.3, for any $m \in \mathbb{N}$ we have

$$\begin{aligned} i^{P^m}(\gamma, m) &= \sum_{\omega^m = 1} i^P_\omega(\gamma) \\ &= \sum_{\omega^m = 1} i_\omega(P^{-1}\gamma * \xi) - \sum_{\omega^m = 1} i_\omega(\xi) \\ &= i(P^{-1}\gamma * \xi, m) - i(\xi, m). \end{aligned} \tag{4.123}$$

This concludes the proof. □

Corollary 6.10 ([203]) *For any path $\gamma \in \mathcal{P}_\tau(2n)$ with $M = \gamma(\tau)$, $P \in Sp(2n)$, there holds*

$$\bar{i}^P(\gamma, 1) = i^P(\gamma, 1) + {}_PS^+_M(1) - C_P(M) + \sum_{\theta \in (0,2\pi)} \frac{\theta}{\pi}\, {}_PS^-_M(e^{\sqrt{-1}\theta}). \tag{4.124}$$

4.6.3 Iteration Inequalities

Recall that for $M \in Sp(2n)$ and $\lambda \in \mathbf{U} \cap \sigma(M)$ being a l-fold eigenvalue, the Hermitian form $(\sqrt{-1}J\cdot, \cdot)_{\mathbb{C}^{2n}}$, which is called the Krein form, is always nondegenerate on the generalized eigenspace $E_\lambda(M) = \ker_{\mathbb{C}}(M - \lambda I)^l$, where $(\cdot, \cdot)_{\mathbb{C}^{2n}}$ denotes the inner product in $\mathbb{C}^{2n}$. The λ is of Krein type (p, q) with $p + q = l$ if the restriction of the Krein type on $E_\lambda(M)$ has signature (p, q). λ is Krein positive if it has Krein type $(p, 0)$ and is Krein negative if it has Krein type $(0, q)$. If $\lambda \in \mathbf{U} \setminus \sigma(M)$, we define the Krein type of λ to be $(0, 0)$. Recall that the elliptic height $e(M)$ of M is the total algebraic multiplicity of all eigenvalues of M on $\mathbf{U}$. If $e(M) = 2n$, then M is elliptic. We denote by $Sp^e(2n)$ the subset of all elliptic matrices in $Sp(2n)$ and $\mathbf{U}^{\pm} = \{z = e^{\sqrt{-1}\theta} | \pm\theta \in [0, \pi]\}$ the closed semi-circle in $\mathbf{U}$.

Proposition 6.11 ([203]) *We have the following statements.*

(1) *For any path $\gamma \in \mathcal{P}_\tau(2n)$, $P \in Sp(2n)$ and $\omega \in \mathbf{U} \setminus \{1\}$, it always hold that*

$$i^P(\gamma, 1) + \nu^P(\gamma, 1) - n + i_1(\xi) - i_\omega(\xi) \leq i^P_\omega(\gamma) \leq i^P(\gamma, 1) + n - \nu^P_\omega(\gamma) + i_1(\xi) - i_\omega(\xi). \tag{4.125}$$

ξ be any symplectic path in $\mathcal{P}_\tau(2n)$ such that $\xi(\tau) = P^{-1}$.

(2) *The left equality in (4.125) holds for some $\omega \in \mathbf{U}^+ \setminus \{1\}$ (or $\mathbf{U}^- \setminus \{1\}$) if and only if $I_{2p} \diamond N_1(1, -1)^{\diamond q} \diamond K \in \Omega_0(P^{-1}\gamma(\tau))$ for some non-negative integers p and q satisfying $0 \leq p + q \leq n$ and $K \in Sp(2(n - p - q))$ with $\sigma(K) \subset \mathbf{U} \setminus \{1\}$ satisfying the condition all eigenvalues of K located within the arc between 1 and ω including ω in $\mathbf{U}^+ \setminus \{1\}$ (or $\mathbf{U}^- \setminus \{1\}$) possess total multiplicity $n - p - q$. If $\omega \neq -1$, all eigenvalues of K are in $\mathbf{U} \setminus \mathbb{R}$ and those in $\mathbf{U}^+ \setminus \mathbb{R}$ (or $\mathbf{U}^- \setminus \mathbb{R}$) are all Krein-negative (or Krein-positive) definite. If $\omega = -1$, it holds that $-I_{2s} \diamond N_1(-1, -1)^{\diamond u} \diamond H \in \Omega_0(K)$ for some non-negative integers s and u satisfying $0 \leq s + u \leq n - p - q$, and some $H \in Sp(2(n - p - q - s - u))$ satisfying $\sigma(H) \subset \mathbf{U} \setminus \mathbb{R}$ and that all elements in $\sigma(H) \cap \mathbf{U}^+$ (or $\sigma(H) \cap \mathbf{U}^-$) are all Krein-negative (or Krein-positive) definite.*

(3) *The left equality in (4.125) holds for some $\omega \in \mathbf{U} \setminus \{1\}$ if and only if $I_{2p} \diamond N_1(1, -1)^{\diamond n-p} \in \Omega_0(P^{-1}\gamma(\tau))$ for some integer $p \in [0, n]$. Especially in this case, all the eigenvalues of $P^{-1}\gamma(\tau)$ are equal to 1 and $\nu^P(\gamma) = n + p \geq n$.*

(4) *The right equality in (4.125) holds for some $\omega \in \mathbf{U}^+ \setminus \{1\}$ (or $\mathbf{U}^- \setminus \{1\}$) if and only if $I_{2p} \diamond N_1(1, 1)^{\diamond r} \diamond K \in \Omega_0(P^{-1}\gamma(\tau))$ for some non-negative integers p and r satisfying $0 \leq p + r \leq n$ and $K \in Sp(2(n - p - r))$ with $\sigma(K) \subset \mathbf{U} \setminus \{1\}$ satisfying the condition all eigenvalues of K located within the closed arc between 1 and ω in $\mathbf{U}^+ \setminus \{1\}$ (or $\mathbf{U}^- \setminus \{1\}$) possess total multiplicity $n - p - r$. If $\omega \neq -1$, all eigenvalues in $\sigma(K) \cap \mathbf{U}^+$ (or $\sigma(K) \cap \mathbf{U}^-$) are all Kerin positive(or*

negative) definite. If $\omega = -1$, *it holds that* $-I_{2s} \diamond N_1(-1, 1)^{\diamond t} \diamond H \in \Omega_0(K)$ *for some non-negative integers* s *and* t *satisfying* $0 \le s + t \le n - p - r$, *and some* $H \in Sp(2(n - p - r - s - t))$ *satisfying* $\sigma(H) \subset \mathbf{U} \setminus \mathbb{R}$ *and all elements in* $\sigma(H) \cap \mathbf{U}^+$ *(or* $\sigma(H) \cap \mathbf{U}^-$*) are all Krein positive (or negative) definite.*

(5) *The right equality in* (4.125) *holds for some* $\omega \in \mathbf{U} \setminus \{1\}$ *if and only if* $I_{2p} \diamond N_1(1, 1)^{\diamond n-p} \in \Omega_0(P^{-1}\gamma(\tau))$ *for some integer* $p \in [0, n]$. *Especially in this case, all the eigenvalues of* $P^{-1}\gamma(\tau)$ *are equal to 1 and* $\nu^P(\gamma) = n + p \ge n$.

(6) *Both equalities in* (4.125) *hold for all* $\omega \in \mathbf{U}\setminus\{1\}$ *if and only if* $\gamma(\tau) = P = I_{2n}$.

Proof From Proposition II.2.8, we know that

$$i_1(\gamma) + \nu_1(\gamma) - n \le i_\omega(\gamma) \le i_1(\gamma) + n - \nu_\omega(\gamma) \tag{4.126}$$

and the equality conditions (see also Proposition 3.1 in [197]). $i_\omega^P(\gamma) = i_\omega(P^{-1}\gamma * \xi) - i_\omega(\xi)$ is independent of the choice of the path ξ. So we have

$$i_1(P^{-1}\gamma * \xi) + \nu_1(P^{-1}\gamma * \xi) - n \le i_\omega(P^{-1}\gamma * \xi) \le i_1(\gamma) + n - \nu_\omega(P^{-1}\gamma * \xi). \tag{4.127}$$

Then we get (4.125) by (4.127) and the definition of $i_\omega^P(\gamma)$. The equalities hold in (4.125) are equivalent of the equalities holding in (4.127). □

Proposition 6.12 ([203]) *We have the following statements.*

(1) *For any path* $\gamma \in \mathcal{P}_\tau(2n)$, $P \in Sp(2n)$ *and* $m \in \mathbb{N}$,

$$\begin{aligned} m\bar{i}^P(\gamma, 1) - n + m\bar{i}(\xi, 1) - i(\xi, m) &\le i^{P^m}(\gamma, m) \\ &\le m\bar{i}^P(\gamma, 1) + n - \nu^{P^m}(\gamma, m) + m\bar{i}(\xi, 1) - i(\xi, m). \end{aligned} \tag{4.128}$$

(2) *The right equality in* (4.128) *holds for all* $m \in \mathbb{N}$ *if and only if* $I_{2p} \diamond N_1(1, -1)^{\diamond(n-p)} \in \Omega_0(P^{-1}\gamma(\tau))$ *for some integer* $p \in [0, n]$. *Especially in this case, all the eigenvalues of* $P^{-1}\gamma(\tau)$ *are equal to* 1 *and* $\nu^P(\gamma) = n+p \ge n$.

(3) *The left equality in* (4.128) *holds for all* $m \in \mathbb{N}$ *if and only if* $I_{2p} \diamond N_1(1, 1)^{\diamond(n-p)} \in \Omega_0(P^{-1}\gamma(\tau))$ *for some integer* $p \in [0, n]$. *Especially in this case, all the eigenvalues of* $P^{-1}\gamma(\tau)$ *are equal to 1 and* $\nu^P(\gamma) = n + p \ge n$.

(4) *Both equalities in* (4.128) *hold for all* $m \in \mathbb{N}$ *if and only if* $\gamma(\tau) = I_{2n}$.

Proof From Proposition II.2.9, we know that

$$m\bar{i}(\gamma, 1) - n \le i(\gamma, m) \le m\bar{i}(\gamma, 1) + n - \nu(\gamma, m) \tag{4.129}$$

and the equality conditions. The results follow from (4.117) and (4.123) and replacing γ by $P^{-1}\gamma * \xi$ in (4.129). □

Proposition 6.13 ([203]) *We have the following statements.*

(1) *For any path $\gamma \in \mathcal{P}_\tau(2n)$, $P \in Sp(2n)$ and $m \in \mathbb{N}$,*

$$\begin{aligned} &m(i^P(\gamma,1)+\nu^P(\gamma,1)-n))+n-\nu^P(\gamma,1)+mi_1(\xi)-i(\xi,m) \\ &\leq i^{P^m}(\gamma,m) \\ &\leq m(i^P(\gamma,1)+n)-n-(\nu^{P^m}(\gamma,m)-\nu^P(\gamma,1))+mi_1(\xi)-i(\xi,m). \end{aligned} \tag{4.130}$$

(2) *The left equality of* (4.130) *holds for some $m \geq 2$ if and only if $I_{2p} \diamond N_1(1,-1)^{\diamond q} \diamond K \in \Omega_0(P^{-1}\gamma(\tau))$ for some non-negative integers p and q satisfying $0 \leq p+q \leq n$ and some $K \in Sp(2(n-p-q))$ with $\sigma(K) \subset \mathbf{U}\backslash\{1\}$ satisfying the following conditions:*
If $m = 2$ and $u = n-p-q > 0$, it holds that $-I_{2s} \diamond N_1(-1,-1)^{\diamond u} \in \Omega_0(K)$ for some non-negative integers s and t satisfying $s+t = n-p-q$.
If $m \geq 3$ and $u = n-p-q > 0$, then $R(\theta_1) \diamond \cdot \diamond R(\theta_u) \in \Omega_0(K)$ for some $\theta_j \in (0,\pi)$ satisfying the condition that $0 < m\theta_j/2\pi \leq 1$ with $1 \leq j \leq u$. In this case, all eigenvalues of K on $\mathbf{U}^+\backslash\{1\}$ (or $\mathbf{U}^-\backslash\{1\}$) are located on the open arc between 1 and $\exp(2\pi\sqrt{-1}/m)$ (or $\exp(-2\pi\sqrt{-1}/m)$) in $\mathbf{U}^+$ (in $\mathbf{U}^-$) and are all Krein-negative (or Krein-positive) definite.

(3) *The right equality of* (4.130) *holds for some $m \geq 2$ if and only if $I_{2p} \diamond N_1(1,1)^{\diamond r} \diamond K \in \Omega_0(P^{-1}\gamma(\tau))$ for some non-negative integers p and r satisfying $0 \leq p+r \leq n$ and some $K \in Sp(2(n-p-r))$ with $\sigma(K) \subset \mathbf{U}\backslash\{1\}$ satisfying the following conditions:*
If $m = 2$ and $n-p-r > 0$, it holds that $-I_{2s} \diamond N_1(-1,1)^{\diamond t} \in \Omega_0(K)$ for some non-negative integers s_1 and t satisfying $s+t = n-p-r$.
If $m \geq 3$, we must have $n = p+r$.

(4) *The two equalities of* (4.130) *hold for some m_1 and $m_2 \geq 2$ respectively if and only if $P^{-1}\gamma(\tau) = I_{2p} \diamond N_1(-1,-1)^{\diamond n-p}$ for some non-negative integers $p \leq n$. Here $p < n$ happens only when $m_1 = m_2 = 2$.*

Proof By Theorem 5.3 with $\omega_0 = 1$, summing (4.125) over all m-th roots of unity, we obtain

$$\begin{aligned} &(m-1)(i^P(\gamma,1)+\nu^P(\gamma,1)-n)+(m-1)i_1(\xi) \\ &\quad -\sum_{\omega^m=1,\omega\neq1} i_\omega(\xi)+i^P(\gamma,1) \leq i^{P^m}(\gamma,m) \\ &\leq (m-1)(i^P(\gamma,1)+n)+(m-1)i_1(\xi) \\ &\quad -\sum_{\omega^m=1,\omega\neq1} \nu_\omega^P(\gamma)-\sum_{\omega^m=1,\omega\neq1} i_\omega(\xi)+i^P(\gamma,1). \end{aligned}$$

This yields (4.130). The equality conditions follow from Proposition II.2.10. □

Proposition 6.14 ([203]) *For any path $\gamma \in \mathcal{P}_\tau(2n)$, $P \in Sp(2n)$, set $M = \gamma(\tau)$ and extend γ to $[0, \infty)$ by (4.81). Then for any m_1 and $m_2 \in \mathbb{N}$ we have*

$$\begin{aligned}
&\nu^{P^{m_1}}(\gamma, m_1) + \nu^{P^{m_2}}(\gamma, m_2) - \nu^{P^{(m_1,m_2)}}(\gamma, (m_1, m_2)) - \nu(\xi, (m_1, m_2))\\
&\quad + \nu(\xi, m_1 + m_2) - \frac{e(P^{-1}M)}{2} - \frac{e(P^{-1})}{2}\\
&\le i^{P^{(m_1+m_2)}}(\gamma, m_1 + m_2) - i^{P^{m_1}}(\gamma, m_1) - i^{P^{m_2}}(\gamma, m_2)\\
&\le \nu^{P^{(m_1,m_2)}}(\gamma, (m_1, m_2)) - \nu^{P^{(m_1+m_2)}}(\gamma, m_1 + m_2) - \nu(\xi, m_1) - \nu(\xi, m_2)\\
&\quad + \nu(\xi, (m_1, m_2)) + \frac{e(P^{-1}M)}{2} + \frac{e(P^{-1})}{2}.
\end{aligned} \tag{4.131}$$

where (m_1, m_2) is the greatest common divisor of m_1 and m_2.

Proof Proposition 6.14 is a direct result of (4.123) and Proposition II.2.11. □

The following result is a direct corollary of Proposition 6.14.

Corollary 6.15 ([203]) *For any path $\gamma \in \mathcal{P}_\tau(2n)$, $P \in Sp(2n)$, set $M = \gamma(\tau)$ and extend γ to $[0, \infty)$ by (4.81). Then for any $m \in \mathbb{N}$ we have*

$$\begin{aligned}
&\nu^{P^m}(\gamma, m) - \nu(\xi, 1) + \nu(\xi, m+1) - \frac{e(P^{-1}M)}{2} - \frac{e(P^{-1})}{2}\\
&\le i^{P^{(m+1)}}(\gamma, m+1) - i^{P^m}(\gamma, m) - i^P(\gamma, 1) \qquad (4.132)\\
&\le \nu^P(\gamma, 1) - \nu^{P^{(m+1)}}(\gamma, m+1) - \nu(\xi, m) + \frac{e(P^{-1}M)}{2} + \frac{e(P^{-1})}{2}.
\end{aligned}$$

Proposition 6.16 ([203]) *For any path $\gamma \in \mathcal{P}_\tau(2n)$, $P \in Sp(2n)$, set $M = \gamma(\tau)$ and extend γ to $[0, \infty)$ by (4.81). Suppose that there exist A_1, $A_2 \in Sp(2n)$ and B_1, $B_2 \in Sp(2n-2)$ such that*

$$P^{-1}M = A_1^{-1}(N_1(1, 1) \diamond B_1)A_1, \quad P^{-1} = A_2^{-1}(N_1(1, 1) \diamond B_2)A_2. \tag{4.133}$$

Then for any $m \in \mathbb{N}$ we have

$$\begin{aligned}
&\nu^{P^m}(\gamma, m) - \nu(\xi, 1) + \nu(\xi, m+1) - \frac{e(P^{-1}M)}{2} - \frac{e(P^{-1})}{2} + 1\\
&\le i^{P^{(m+1)}}(\gamma, m+1) - i^{P^m}(\gamma, m) - i^P(\gamma, 1)\\
&\le \nu^P(\gamma, 1) - \nu^{P^{(m+1)}}(\gamma, m+1) - \nu(\xi, m) + \frac{e(P^{-1}M)}{2} + \frac{e(P^{-1})}{2} - 1.
\end{aligned} \tag{4.134}$$

Proof It is a direct result of Theorem 5.3 and Proposition II.2.12. □

Chapter 5
The L-Index Theory

5.1 Definition of L-Index

In this subsection for $L \in \Lambda(n)$ we will define an index pair for any symplectic path $\gamma \in \mathcal{P}_\tau(2n)$ with L-boundary condition(L-index for short). Comparing with the Maslov-type index theory for periodic boundary condition which is suitable to be used in the study of periodic solution of a Hamiltonian systems, the L-index theory is suitable to be used in the study of Hamiltonian systems with L-boundary condition. Firstly, we consider a special case. Suppose $L = L_0 = \{0\} \oplus \mathbb{R}^n \subset \mathbb{R}^{2n}$ which is a Lagrangian subspace of the linear symplectic space $(\mathbb{R}^{2n}, \omega_0)$.

For a symplectic path $\gamma(t)$, we write it in the following form:

$$\gamma(t) = \begin{pmatrix} S(t) & V(t) \\ T(t) & U(t) \end{pmatrix}, \tag{5.1}$$

where $S(t), T(t), V(t), U(t)$ are $n \times n$ matrices. The n vectors coming from the column of the matrix $\begin{pmatrix} V(t) \\ U(t) \end{pmatrix}$ are linear independent and they span a Lagrangian subspace of $(\mathbb{R}^{2n}, \omega_0)$. Particularly, at $t = 0$, this Lagrangian subspace is $L_0 = \{0\} \oplus \mathbb{R}^n$.

For $L_0 = \{0\} \oplus \mathbb{R}^n$, we define the following two subspaces of $\mathrm{Sp}(2n)$ by

$$\mathrm{Sp}(2n)^*_{L_0} = \{M \in \mathrm{Sp}(2n) |\ \det V_M \neq 0\},$$

$$\mathrm{Sp}(2n)^0_{L_0} = \{M \in \mathrm{Sp}(2n) |\ \det V_M = 0\},$$

for $M = \begin{pmatrix} S_M & V_M \\ T_M & U_M \end{pmatrix}$. We note that in [232] the singular set $\mathrm{Sp}(2n)^0_{L_0}$ and an index theory with this singular set was defined via an analytic method.

Since the space $\mathrm{Sp}(2n)$ is path connected, and the $n \times n$ non-degenerate matrix space has two path connected components, one with $\det V_M > 0$, and another with $\det V_M < 0$, the space $\mathrm{Sp}(2n)^*_{L_0}$ has two path connected components as well. We denote by

$$\mathrm{Sp}(2n)^{\pm}_{L_0} = \{M \in \mathrm{Sp}(2n) | \pm \det V_M > 0\},$$

then we have $\mathrm{Sp}(2n)^*_{L_0} = \mathrm{Sp}(2n)^+_{L_0} \cup \mathrm{Sp}(2n)^-_{L_0}$. We call $\mathrm{Sp}(2n)^0_{L_0}$ the L_0-degenerate subspace of $\mathrm{Sp}(2n)$ and $\mathrm{Sp}(2n)^*_{L_0}$ the L_0-non-degenerate subspace of $\mathrm{Sp}(2n)$. We denote the corresponding symplectic path spaces by

$$\mathcal{P}_\tau(2n)^*_{L_0} = \{\gamma \in \mathcal{P}_\tau(2n) | \gamma(\tau) \in \mathrm{Sp}(2n)^*_{L_0}\}$$

and

$$\mathcal{P}_\tau(2n)^0_{L_0} = \{\gamma \in \mathcal{P}_\tau(2n) | \gamma(\tau) \in \mathrm{Sp}(2n)^0_{L_0}\}.$$

If $\gamma \in \mathcal{P}_\tau(2n)^0_{L_0}$, we call it the L_0-degenerate symplectic path, otherwise, if $\gamma \in \mathcal{P}_\tau(2n)^*_{L_0}$, we call it the L_0-nondegenerate symplectic path. For simplicity, we fix $\tau = 1$ now.

Definition 1.1 We define the L_0-*nullity of any symplectic path* $\gamma \in \mathcal{P}_1(2n)$ by

$$\nu_{L_0}(\gamma) \equiv \dim \ker_{L_0}(\gamma(1)) := \dim \ker V(1) = n - \operatorname{rank} V(1) \tag{5.2}$$

with the $n \times n$ matrix function $V(t)$ defined in (5.1).

We note that $\operatorname{rank}\begin{pmatrix} V(t) \\ U(t) \end{pmatrix} = n$, so the complex matrix $U(t) \pm \sqrt{-1}V(t)$ is invertible. We define a complex matrix function by

$$\mathcal{Q}(t) = \mathcal{Q}_\gamma(t) = [U(t) - \sqrt{-1}V(t)][U(t) + \sqrt{-1}V(t)]^{-1}. \tag{5.3}$$

From Lemma I.2.13 we know that the matrix $\mathcal{Q}(t) \in U(n)$ is unitary matrix for any $t \in [0, 1]$. We denote by

$$M_+ = \begin{pmatrix} 0 & I_n \\ -I_n & 0 \end{pmatrix}, \quad M_- = \begin{pmatrix} 0 & D_n \\ -D_n & 0 \end{pmatrix}, \quad D_n = \mathrm{diag}(-1, 1, \cdots, 1).$$

It is clear that $M_\pm \in \mathrm{Sp}(2n)^{\pm}_{L_0}$.

For a path $\gamma \in \mathcal{P}_1(2n)^*_{L_0}$, we first adjoin it with a simple symplectic path starting from $J = -M_+$, i.e., we define a symplectic path by

$$\tilde{\gamma}(t) = \begin{cases} I \cos \frac{(1-2t)\pi}{2} + J \sin \frac{(1-2t)\pi}{2}, & t \in [0, 1/2]; \\ \gamma(2t-1), & t \in [1/2, 1], \end{cases} \tag{5.4}$$

then we choose a symplectic path $\beta(t)$ in $\mathrm{Sp}(2n)^*_{L_0}$ starting from $\gamma(1)$ and ending at M_+ or M_- according to $\gamma(1) \in \mathrm{Sp}(2n)^+_{L_0}$ or $\gamma(1) \in \mathrm{Sp}(2n)^-_{L_0}$, respectively. We now define a joint path by

$$\bar{\gamma}(t) = \beta * \tilde{\gamma} := \begin{cases} \tilde{\gamma}(2t), & t \in [0, 1/2], \\ \beta(2t-1), & t \in [1/2, 1]. \end{cases} \tag{5.5}$$

By the definition, we see that the symplectic path $\bar{\gamma}$ starting from $-M_+$ and ending at either M_+ or M_-. As above, we define

$$\bar{\mathcal{Q}}(t) = [\bar{U}(t) - \sqrt{-1}\bar{V}(t)][\bar{U}(t) + \sqrt{-1}\bar{V}(t)]^{-1}. \tag{5.6}$$

for $\bar{\gamma}(t) = \begin{pmatrix} \bar{S}(t) & \bar{V}(t) \\ \bar{T}(t) & \bar{U}(t) \end{pmatrix}$. We can choose a continuous function $\bar{\Delta} : [0, 1] \to \mathbb{R}$ such that

$$\det\bar{\mathcal{Q}}(t) = e^{2\sqrt{-1}\bar{\Delta}(t)}. \tag{5.7}$$

By the above arguments, we see that the number $\frac{1}{\pi}(\bar{\Delta}(1) - \bar{\Delta}(0)) \in \mathbb{Z}$ and it does not depend on the choice of the function $\bar{\Delta}(t)$. We note that there is a positive continuous function $\rho : [0, 1] \to (0, +\infty)$ such that

$$\det(\bar{U}(t) - \sqrt{-1}\bar{V}(t)) = \rho(t)e^{\sqrt{-1}\bar{\Delta}(t)}.$$

Definition 1.2 ([189]) For a symplectic path $\gamma \in \mathcal{P}_1(2n)^*_{L_0}$, we define the L_0-*index of* γ by

$$i_{L_0}(\gamma) = \frac{1}{\pi}(\bar{\Delta}(1) - \bar{\Delta}(0)). \tag{5.8}$$

We now describe the index of L_0-nondegenerate symplectic path $\gamma \in \mathcal{P}_1(2n)^*_{L_0}$ by another way. In (5.3) we know that $\mathcal{Q}_\gamma(t) \in U(n)$ for any $t \in [0, 1]$ (here the subscript γ in $\mathcal{Q}_\gamma$ is to indicate the dependence of γ). By the non-degenerate condition, we have $\det V(1) \neq 0$. Suppose $\lambda_j(t) = e^{2\sqrt{-1}\theta_j(t)}$ are the eigenvalues of $\mathcal{Q}_\gamma(t)$ for $j = 1, \cdots, n$. The existence of the continuous functions $\theta_j(t)$ comes from that by the continuity of the matrix value function $\mathcal{Q}_\gamma(t)$, we know that the eigenvalues $\lambda_j(t)$ are all continuous functions on the interval $[0, 1]$, so by take the logarithm we get the continuity $\theta_j : [0, 1] \to \mathbb{R}/(2\pi\mathbb{Z})$, moreover, by a locally continuous extension method, one can get continuous functions $\theta_j : [0, 1] \to \mathbb{R}$ with $\theta_j(0) = 2k\pi$ for some fix $k \in \mathbb{Z}$, for example $k = 0$. More precisely, for any $t_0 \in [0, 1]$, there is a small $\delta_0 > 0$ such that in $N(t_0) := (t_0 - \delta_0, t_0 + \delta_0) \cap [0, 1]$ the functions $\theta_j^0 : N(t_0) \to \mathbb{R}$ are well defined(not unique, the difference of any two components may be $2k\pi$ for some $k \in \mathbb{Z}$). For another choice $t_1 \in [0, 1]$ with $N(t_0) \cap N(t_1) \neq \emptyset$ then by a suitable choice of the component of the functions, one can get well defined continuous functions $\theta_j^1 : N(t_1) \to \mathbb{R}$ such that $\theta_j^0(t) = \theta_j^1(t)$ for $t \in N(t_0) \cap N(t_1)$. Now one can start from $t = 0$ and extend step by step to define the continuous functions $\theta_j : [0, 1] \to \mathbb{R}$ by using the compactness of $[0, 1]$.

Proposition 1.3 ([189]) *For $\gamma \in \mathcal{P}_1(2n)^*_{L_0}$, with the above notations, there holds*

$$i_{L_0}(\gamma) = \sum_{j=1}^{n} \left[\frac{\theta_j(1) - \theta_j(0)}{\pi}\right], \tag{5.9}$$

where $[a] = \max\{k \in \mathbb{Z}|\ k \le a\}$.

Proof We take a symplectic path $\beta : [a, b] \to \mathrm{Sp}(2n)^*_{L_0}$. Suppose $\beta(t) = \begin{pmatrix} A_\beta(t) & C_\beta(t) \\ B_\beta(t) & D_\beta(t) \end{pmatrix}$ with $\det C_\beta(t) \ne 0$. Consider the determinant of the unitary matrix

$$\begin{aligned}\det \mathcal{Q}_\beta(t) &:= \det(D_\beta(t) - \sqrt{-1}C_\beta(t))(D_\beta(t) + \sqrt{-1}C_\beta(t))^{-1} \\ &= \det(D_\beta(t)C_\beta^{-1}(t) - \sqrt{-1}I)(D_\beta(t)C_\beta^{-1}(t) + \sqrt{-1}I)^{-1}.\end{aligned}$$

It is easy to see that $D_\beta(t)C_\beta^{-1}(t)$ is a symmetric matrix function. So the eigenvalue of the matrix $\mathcal{Q}_\beta(t)$ can be written as

$$\zeta_{\beta,j}(t) = \frac{a_j(t) - \sqrt{-1}}{a_j(t) + \sqrt{-1}} = \frac{a_j^2(t) - 1}{a_j^2(t) + 1} - \frac{2a_j(t)\sqrt{-1}}{a_j^2(t) + 1} \in S^1,\ a_j(t) \in \mathbb{R}.$$

It is clear that if we write $\zeta_{\beta,j}(t) = e^{2\sqrt{-1}\vartheta_{\beta,j}(t)}$, there holds $\vartheta_{\beta,j}(t) \notin \pi\mathbb{Z}$. By the continuity, we have $\left|\frac{\vartheta_{\beta,j}(b) - \vartheta_{\beta,j}(a)}{\pi}\right| < 1$. We now just take β as in (5.5), and get a joint path $\beta * \gamma$. Then we have

$$\left[\frac{\vartheta_{\gamma,j}(1) - \vartheta_{\gamma,j}(0)}{\pi}\right] = \left[\frac{\vartheta_{\beta*\gamma,j}(1) - \vartheta_{\beta*\gamma,j}(0)}{\pi}\right].$$

But there holds

$$\frac{\vartheta_{\beta*\gamma,j}(1) - \vartheta_{\beta*\gamma,j}(0)}{\pi} = \left[\frac{\vartheta_{\beta*\gamma,j}(1) - \vartheta_{\beta*\gamma,j}(0)}{\pi}\right] + \frac{1}{2}.$$

As in the definition of $\tilde{\gamma}$, we have join a rotation path $\xi(t) = I\cos t + J\sin t,\quad t : \frac{\pi}{2} \to 0$. By direct computation, we have

$$\frac{\vartheta_{\xi,j}(0) - \vartheta_{\xi,j}(\frac{\pi}{2})}{\pi} = -\frac{1}{2}.$$

Thus as in (5.5), we have

$$\frac{\vartheta_{\tilde{\gamma},j}(1)-\vartheta_{\tilde{\gamma},j}(0)}{\pi}=\left[\frac{\vartheta_{\gamma,j}(1)-\vartheta_{\gamma,j}(0)}{\pi}\right]=\left[\frac{\theta_j(1)-\theta_j(0)}{\pi}\right].$$

Taking summation from $j=1$ to n, we get (5.9). □

Remark 1.4 For the non-degenerate case, if we choose $\vartheta_{\gamma,j}(0)=\vartheta_{\beta*\gamma,j}(0)=0$ and $\vartheta_{\gamma,j}(1)=\vartheta_{\beta,j}(0)$, then by the above computations, we have

$$-\frac{1}{2}<\frac{\vartheta_{\beta,j}(1)-\vartheta_{\beta,j}(0)}{\pi}<\frac{1}{2}$$

and thus

$$-\frac{n}{2}<\frac{\Delta_\beta(1)-\Delta_\beta(0)}{\pi}=\sum_{j=1}^{n}\frac{\vartheta_{\beta,j}(1)-\vartheta_{\beta,j}(0)}{\pi}<\frac{n}{2}.$$

Note that

$$i_{L_0}(\gamma)=-\frac{n}{2}+\frac{\Delta_\gamma(1)-\Delta_\gamma(0)}{\pi}+\frac{\Delta_\beta(1)-\Delta_\beta(0)}{\pi},$$

we get

$$-n<i_{L_0}(\gamma)-\frac{\Delta_\gamma(1)-\Delta_\gamma(0)}{\pi}<0.$$

By the arguments in Theorem 1.21 below, for any $\gamma\in\mathcal{P}_1(2n)$,

$$-n\le i_{L_0}(\gamma)-\frac{\Delta_\gamma(1)-\Delta_\gamma(0)}{\pi}<0.$$

The left side equality holds only for $\gamma(1)=I_{2n}$.

For a L_0-degenerate symplectic path $\gamma\in\mathcal{P}_\tau(2n)^0_{L_0}$, its L_0-index is defined by the infimum of the indices of the nearby nondegenerate symplectic paths.

Definition 1.5 ([189]) For a symplectic path $\gamma\in\mathcal{P}_\tau(2n)^0_{L_0}$, we define the L_0-*index of* γ by

$$i_{L_0}(\gamma)=\inf\{i_{L_0}(\tilde{\gamma})\,|\,\tilde{\gamma}\in\mathcal{P}_\tau(2n)^*_{L_0},\text{ and }\tilde{\gamma}\text{ is sufficiently close to }\gamma\}.\qquad(5.10)$$

In the general situation, let L be any linear Lagrangian subspace of $\mathbb{R}^{2n}$. Now we are ready to define the index for any symplectic path $\gamma\in\mathcal{P}_\tau(2n)$ with L-boundary condition. We know from Lemma I.2.14 that $\Lambda(n)=U(n)/O(n)$, this means that for any linear subspace $L\in\Lambda(n)$, there is an orthogonal symplectic matrix $P=\begin{pmatrix}A&-B\\B&A\end{pmatrix}$ with $A\pm\sqrt{-1}B\in U(n)$, such that $PL_0=L$. P is uniquely determined by L up to an orthogonal matrix $C \in O(n)$. It means that for any other choice P' satisfying above conditions, there exists a matrix $C \in O(n)$ such that $P' = P\begin{pmatrix} C & 0 \\ 0 & C \end{pmatrix}$. We define the conjugated symplectic path $\gamma_c \in \mathcal{P}_\tau(2n)$ of γ by $\gamma_c(t) = P^{-1}\gamma(t)P$.

Definition 1.6 (**[189]**) We define the *L-nullity of any symplectic path* $\gamma \in \mathcal{P}_\tau(2n)$ by

$$\nu_L(\gamma) \equiv \dim \ker_L(\gamma(1)) := \dim \ker V_c(1) = n - \operatorname{rank} V_c(1). \tag{5.11}$$

The $n \times n$ matrix function $V_c(t)$ is defined in (5.1) with the symplectic path γ replaced by γ_c, i.e.,

$$\gamma_c(t) = \begin{pmatrix} S_c(t) & V_c(t) \\ T_c(t) & U_c(t) \end{pmatrix}. \tag{5.12}$$

The L-nullity $\nu_L(\gamma)$ is well defined. In fact, for any $C \in O(n)$, we have

$$\begin{pmatrix} C^{-1} & 0 \\ 0 & C^{-1} \end{pmatrix}\begin{pmatrix} S_c(t) & V_c(t) \\ T_c(t) & U_c(t) \end{pmatrix}\begin{pmatrix} C & 0 \\ 0 & C \end{pmatrix} = \begin{pmatrix} C^{-1}S_c(t)C & C^{-1}V_c(t)C \\ C^{-1}T_c(t)C & C^{-1}U_c(t)C \end{pmatrix}, \tag{5.13}$$

and $\dim\ker V_c(1) = \dim\ker C^{-1}V_c(1)C$.

For any $L \in \Lambda(n)$, we define the following two subspaces of $\mathrm{Sp}(2n)$ by

$$\mathrm{Sp}(2n)_L^* = \{M \in \mathrm{Sp}(2n) \mid \det V_c \neq 0\},$$

$$\mathrm{Sp}(2n)_L^0 = \{M \in \mathrm{Sp}(2n) \mid \det V_c = 0\},$$

where V_c is defined by

$$M_c = P^{-1}MP = \begin{pmatrix} S_c & V_c \\ T_c & U_c \end{pmatrix}.$$

The space $\mathrm{Sp}(2n)_L^*$ has two path connected components as well. We denote the two components by

$$\mathrm{Sp}(2n)_L^{\pm} = \{M \in \mathrm{Sp}(2n) \mid \pm \det V_c > 0\},$$

then we have $\mathrm{Sp}(2n)_L^* = \mathrm{Sp}(2n)_L^+ \cup \mathrm{Sp}(2n)_L^-$. We call $\mathrm{Sp}(2n)_L^0$ the L-degenerate subspace of $\mathrm{Sp}(2n)$ and $\mathrm{Sp}(2n)_L^*$ the L-non-degenerate subspace of $\mathrm{Sp}(2n)$. We denote the corresponding symplectic path spaces by

$$\mathcal{P}_\tau(2n)_L^* = \{\gamma \in \mathcal{P}_\tau(2n) \mid \gamma(1) \in \mathrm{Sp}(2n)_L^*\}$$

and

$$\mathcal{P}_\tau(2n)_L^0 = \{\gamma \in \mathcal{P}_\tau(2n) |\, \gamma(1) \in \mathrm{Sp}(2n)_L^0\}.$$

If $\gamma \in \mathcal{P}_\tau(2n)_L^0$, we call it the L-degenerate symplectic path, otherwise, if $\gamma \in \mathcal{P}_\tau(2n)_L^*$, we call it the L-nondegenerate symplectic path.

Definition 1.7 ([189]) For a symplectic path $\gamma \in \mathcal{P}_\tau(2n)$, we define the *$L$-index of* γ by

$$i_L(\gamma) = i_{L_0}(\gamma_c). \tag{5.14}$$

By the arguments (5.12) and (5.13), we see that the L-index $i_L(\gamma)$ is well defined. In fact, for any other choice P', the conjugated symplectic path associated it is

$$\begin{aligned}\gamma_c'(t) = \begin{pmatrix} S_c'(t) & V_c'(t) \\ T_c'(t) & U_c'(t) \end{pmatrix} &= \begin{pmatrix} C^{-1} & 0 \\ 0 & C^{-1} \end{pmatrix} \gamma_c(t) \begin{pmatrix} C & 0 \\ 0 & C \end{pmatrix} \\ &= \begin{pmatrix} C^{-1}S_c(t)C & C^{-1}V_c(t)C \\ C^{-1}T_c(t)C & C^{-1}U_c(t)C \end{pmatrix}.\end{aligned}$$

The associated unitary matrix defined in (2.3) becomes

$$\begin{aligned}\mathcal{Q}_c'(t) &= [U_c'(t) - \sqrt{-1}V_c'(t)][U_c'(t) + \sqrt{-1}V_c'(t)]^{-1} \\ &= C^{-1}[U_c(t) - \sqrt{-1}V_c(t)][U_c(t) + \sqrt{-1}V_c(t)]^{-1}C.\end{aligned}$$

Thus for any symplectic path $\gamma \in \mathcal{P}_\tau(2n)$, the L-index pair

$$(i_L(\gamma), \nu_L(\gamma)) \in \mathbb{Z} \times \{0, 1, \cdots, n\}$$

is well defined.

5.1.1 The Properties of the L-Indices

We only consider the properties of the L-indices with $L = L_0$ in this section. All of the properties are valid for the general cases.

Definition 1.8 ([189]) For two symplectic paths $\gamma_0, \gamma_1 \in \mathcal{P}_\tau(2n)$, we say that they are *$L_0$-homotopic* and denoted by $\gamma_0 \sim_{L_0} \gamma_1$, if there is a map $\delta : [0, 1] \to \mathcal{P}_\tau(2n)$ such that $\delta(j) = \gamma_j$ for $j = 0, 1$, and $\nu_{L_0}(\delta(s))$ is constant for $s \in [0, 1]$.

Theorem 1.9 ([189]) *If $\gamma_0, \gamma_1 \in \mathcal{P}_\tau(2n)_{L_0}^*$, then $i_{L_0}(\gamma_0) = i_{L_0}(\gamma_1)$ if and only if $\gamma_0 \sim_{L_0} \gamma_1$.*

Proof Connecting $\gamma_j(1)$ to M_+ or M_- by a path $\beta_j : [0,1] \to \mathrm{Sp}(2n)^*_{L_0}$ for $j = 0, 1$, we have

$$\gamma_j \sim_{L_0} \beta_j * \gamma_j \text{ along } \beta_j$$

and by definition

$$i_{L_0}(\gamma_j) = i_{L_0}(\beta_j * \gamma_j)$$

where the joint path $\beta * \gamma$ is defined in (5.5). Thus $\gamma_0 \sim_{L_0} \gamma_1$ if and only if $\beta_0 * \gamma_0 \sim_{L_0} \beta_1 * \gamma_1$. By Definition 1.2, it is equivalent to $i_{L_0}(\gamma_0) = i_{L_0}(\gamma_1)$. □

For any $C \in O(n)$, from the definition, we have

$$i_{L_0}(\gamma) = i_{L_0}(\gamma^C), \quad \nu_{L_0}(\gamma) = \nu_{L_0}(\gamma^C), \ \forall \gamma \in \mathcal{P}_\tau(2n), \tag{5.15}$$

where the symplectic path γ^C is defined by

$$\gamma^C(t) = \begin{pmatrix} C & 0 \\ 0 & C \end{pmatrix} \gamma(t) \begin{pmatrix} C^{-1} & 0 \\ 0 & C^{-1} \end{pmatrix}.$$

Lemma 1.10 ([189]) *If $\gamma \in \mathcal{P}_\tau(2n)^0_{L_0}$, and $\nu_{L_0}(\gamma) = k$, there is matrix $C \in O(n)$ and $\theta_0 > 0$ sufficient small satisfying $\nu_{L_0}(\gamma_s^1) = k - 1$ for $s \in [-1, 1]$ and $s \neq 0$, where*

$$\gamma_s^1(t) := R_h(st\theta_0) \begin{pmatrix} C & 0 \\ 0 & C \end{pmatrix} \gamma(t) \begin{pmatrix} C^{-1} & 0 \\ 0 & C^{-1} \end{pmatrix} = R_h(st\theta_0)\gamma^C(t), \tag{5.16}$$

and $R_h(\theta)$ is a rotation matrix defined in (5.18) *below.*

Furthermore, if $k = 1$, then $\gamma_s(1)$ and $\gamma_{-s}(1)$ belong to different connected component of $\mathrm{Sp}(2n)^*_{L_0}$, *i.e., there holds*

$$\det V_s^1(1) \det V_{-s}^1(1) < 0. \tag{5.17}$$

Proof Suppose $\gamma(t) = \begin{pmatrix} S(t) & V(t) \\ T(t) & U(t) \end{pmatrix}$ and $\dim\ker V(1) = k$, $0 < k \leq n$. There exists an orthogonal matrix $C \in O(n)$ such that

$$CV(1)C^{-1} = \begin{pmatrix} A & 0 \\ B & 0 \end{pmatrix}, \quad A \in \mathcal{L}(\mathbb{R}^{n-k}, \mathbb{R}^{n-k}), \quad \det A \neq 0,$$

and with the same block form decomposition, we have

$$CU(1)C^{-1} = \begin{pmatrix} U_{11} & U_{12} \\ U_{21} & U_{22} \end{pmatrix}.$$

Since $\gamma \in \mathcal{P}_\tau(2n)$, we have $U^T(1)V(1) = V^T(1)U(1)$, and then $U_{12}^T A + U_{22}^T B = 0$. By the rank k condition, we have $U_{22} \neq 0$ (otherwise, by $\det A \neq 0$, we have $U_{12} = 0$). Suppose $U_{22} = (u_{i,j})_{k\times k}$ and $u_{i,j} \neq 0$ for some $1 \leq i, j \leq k$.

Now we construct the symplectic matrices (rotation matrix) $R_m(\theta) = (\alpha_{j,l})$ by

$$\begin{cases} \alpha_{m,m} = \alpha_{n+m,n+m} = \cos\theta, \\ \alpha_{n+m,m} = -\alpha_{m,n+m} = \sin\theta, \\ \alpha_{j,j} = 1, \quad \text{for } k \neq m, n+m, \\ \alpha_{j,l} = 0, \quad \text{for any other cases.} \end{cases} \tag{5.18}$$

Thus we have

$$R_{n-k+i}(\theta)\begin{pmatrix} C & 0 \\ 0 & C \end{pmatrix}\begin{pmatrix} S(1) & V(1) \\ T(1) & U(1) \end{pmatrix}\begin{pmatrix} C^{-1} & 0 \\ 0 & C^{-1} \end{pmatrix} = \begin{pmatrix} S_\theta & V_\theta \\ T_\theta & U_\theta \end{pmatrix}$$

with

$$V_\theta = \begin{pmatrix} V_{n-k} & 0 \\ * & -U_{22}(\theta) \end{pmatrix}, \tag{5.19}$$

and $U_{22}(\theta)$ is a matrix function induced from U_{22} by multiplying its i-th row by $\sin\theta$ and by multiplying other rows by 0. It is clear that $\dim\ker V_\theta = \dim\ker V(1) - 1$ for $|\theta| \neq 0$ small. Thus (5.16) follows by choosing $h = n - k + i$.

If $k = 1$, then $U_{22}(\theta) = c\sin\theta$ for some constant $c \neq 0$. Equation (5.17) follows from (5.19).

□

From Lemma 1.10, by induction, we have

$$\nu_{L_0}(\gamma_s^j) - \nu_{L_0}(\gamma_s^{j+1}) = 1, \quad j = 0, 1, \cdots, k-1, \quad \text{and} \quad \nu_{L_0}(\gamma_s^k) = 0, \quad s \neq 0$$

where

$$\gamma_s^{j+1}(t) := R_{h_j}(st\theta_0)\begin{pmatrix} C_j & 0 \\ 0 & C_j \end{pmatrix}\gamma_s^j(t)\begin{pmatrix} C_j^{-1} & 0 \\ 0 & C_j^{-1} \end{pmatrix}, \quad \gamma_s^0 = \gamma. \tag{5.20}$$

We denote by

$$\gamma_s(t) = \gamma_s^k(t), \quad k = \nu_{L_0}(\gamma). \tag{5.21}$$

By Theorem 1.9, we have the following

$$i_{L_0}(\gamma_s) = \text{constant for } s \in (0, 1] \text{ or } s \in [-1, 0).$$

Theorem 1.11 ([189]) *Suppose* $\gamma \in \mathcal{P}_\tau(2n)^0_{L_0}$ *and* $\gamma_s \in \mathcal{P}_\tau(2n)^*_{L_0}$ *defined in Lemma* 1.10. *We have*

$$i_{L_0}(\gamma_s) - i_{L_0}(\gamma_{-s}) = \nu_{L_0}(\gamma), \quad \forall\, s \in (0, 1]. \tag{5.22}$$

Proof We follow the ideas of the proof of Theorem 5.4.1 of [223]. For simplicity, we omit the constant matrix from (5.20), so we write γ_s in following form

$$\gamma_s(t) = R_{h_k}(st\theta_0)R_{h_{k-1}}(st\theta_0)\cdots R_{h_1}(st\theta_0)\gamma^C(t).$$

Fix $0 < s \le 1$. We define symplectic paths by

$$\xi_j(t) = R_{h_k}(st\theta_0)\cdots R_{h_{j+1}}(st\theta_0)R_{h_j}(-st\theta_0)\cdots R_{h_1}(-st\theta_0)\gamma^C(t), \; j = 0, \cdots, k. \tag{5.23}$$

We note that $\xi_j \in \mathcal{P}_\tau(2n)^*_{L_0}$ and $\xi_0 = \gamma_s$, $\xi_k = \gamma_{-s}$. We now only need to prove

$$i_{L_0}(\xi_j) - i_{L_0}(\xi_{j+1}) = 1. \tag{5.24}$$

We take a path

$$\eta_j(t) = R_{h_k}(s\theta_0)\cdots R_{h_{j+1}}(s\theta_0)R_{h_{j+1}}(-2st\theta_0)R_{h_j}(-s\theta_0)\cdots R_{h_1}(-s\theta_0)\gamma^C(1).$$

We have $\eta_j(0) = \xi_j(1)$, $\eta_j(1) = \xi_{j+1}(1)$, $\eta_j(1/2) \in \mathrm{Sp}(2n)^0_{L_0}$ and furthermore $\dim\ker_{L_0}\eta_j(1/2) = 1$, $\eta_j(t) \in \mathrm{Sp}(2n)^*_{L_0}$ for all $t \in [0, 1] \setminus \{1/2\}$, where the dimension of the L_0 kernel is defined in (5.2). So by definition, we have $\xi_{j+1} \sim_{L_0} \eta_j * \xi_j$, the joint path $\alpha * \beta$ is defined in (5.5). So by Theorem 1.9, we need to prove

$$i_{L_0}(\xi_j) - i_{L_0}(\eta_j * \xi_j) = 1. \tag{5.25}$$

We define a path $\bar{\xi} \in \mathcal{P}_\tau(2n)^0_{L_0}$ by

$$\bar{\xi}(t) = \begin{cases} \xi_j(2t), & t \in [0, 1/2], \\ \eta_j(\frac{2t-1}{2}), & t \in [1/2, 1]. \end{cases}$$

The path ξ_j can be taken as γ^1_s in Lemma 1.10 for $\bar{\xi}$, and $\eta_j * \xi_j$ can be taken as γ^1_{-s} for $\bar{\xi}$. So the two paths ξ_j and $\eta_j * \xi_j$ is located in different connected component. Note that $s,\ \theta_0 > 0$, so we have

$$i_{L_0}(\xi_j) - i_{L_0}(\eta_j * \xi_j) > 0. \tag{5.26}$$

But by definition and $\theta_0 > 0$ is small, it is clear that

$$i_{L_0}(\xi_j) - i_{L_0}(\eta_j * \xi_j) \le 1. \tag{5.27}$$

Equation (5.25) follows from (5.26) and (5.27). □

Remark In fact there holds

$$i_{L_0}(\gamma_{-s}) = \inf\{i_{L_0}(\tilde{\gamma}) | \tilde{\gamma} \in \mathcal{P}_\tau(2n)^*_{L_0}, \text{ and it is sufficiently close to } \gamma\}, \tag{5.28}$$

$$i_{L_0}(\gamma_s) = \sup\{i_{L_0}(\tilde{\gamma}) | \tilde{\gamma} \in \mathcal{P}_\tau(2n)^*_{L_0}, \text{ and it is sufficiently close to } \gamma\}. \tag{5.29}$$

We will prove these statements in (5.143).

Theorem 1.12 ([189]) *For any symplectic path $\gamma \in \mathcal{P}_1(2n)$, by using the notations in Proposition 1.3, there holds*

$$i_{L_0}(\gamma) = \sum_{j=1}^{n} E\left(\frac{\theta_j(1) - \theta_j(0)}{\pi}\right), \tag{5.30}$$

where $E(a) = \max\{k \in \mathbb{Z} | \ k < a\}$.

Proof From the arguments of the proof of Proposition 1.3 and the result of Theorem 1.11, we see that there exist exact $k = \nu_{L_0}(\gamma)$ eigenvalues $\theta_j(t)$ with $\theta_j(1) = 2p\pi$ for $p \in \mathbb{Z}$, and

$$i_{L_0}(\gamma_{-s}) = \sum_{j=1}^{n} \left[\frac{\theta_j(1) - \theta_j(0)}{\pi}\right] - k, \quad i_{L_0}(\gamma_s) = \sum_{j=1}^{n} \left[\frac{\theta_j(1) - \theta_j(0)}{\pi}\right], \quad 0 < s \le 1. \tag{5.31}$$

It is easy to see

$$\sum_{j=1}^{n} \left[\frac{\theta_j(1) - \theta_j(0)}{\pi}\right] - k = \sum_{j=1}^{n} E\left(\frac{\theta_j(1) - \theta_j(0)}{\pi}\right). \tag{5.32}$$

Equation (5.30) follows from (5.10), (5.28), (5.31) and (5.32). □

Theorem 1.13 (Symplectic addivity, [189]) *For two path $\gamma_j \in \mathcal{P}(n_j), \ j = 1, 2$, there holds*

$$i_{L_0}(\gamma_1 \diamond \gamma_2) = i_{L_0}(\gamma_1) + i_{L_0}(\gamma_2). \tag{5.33}$$

Proof (5.33) follows from definition and (5.28), (5.29). We omit the details. □

Theorem 1.14 ([189]) *For two path $\gamma_j \in \mathcal{P}_\tau(2n), \ j = 0, 1$ with $\gamma_0(1) = \gamma_1(1)$, it holds that $i_{L_0}(\gamma_0) = i_{L_0}(\gamma_1)$ if and only if $\gamma_0 \sim_{L_0} \gamma_1$ with fixed endpoints.*

Proof The proof follows from (5.28) and (5.29) and the definition of the index theory. We note that $\gamma_0 \sim_{L_0} \gamma_1$ with fixed endpoints if and only if $\gamma_{s,0} \sim_{L_0} \gamma_{s,1}$ with fixed endpoints. □

In the Sect. 5.5 below (see the proof of (5.145)), we shall prove the following result.

Theorem 1.15 (Homotopy invariant, [189]) *For two path* $\gamma_j \in \mathcal{P}_\tau(2n),\ j = 0, 1$, *if* $\gamma_0 \sim_L \gamma_1$, *there hold*

$$i_L(\gamma_0) = i_L(\gamma_1), \quad \nu_L(\gamma_0) = \nu_L(\gamma_1). \tag{5.34}$$

Suppose $\gamma(t) = \begin{pmatrix} C(t) & E(t) \\ E^T(t) & D(t) \end{pmatrix} \in \mathcal{P}_1(2n)$ is a symmetry positive definite symplectic path. From $C(t) > 0$ and $D(t) > 0$ (positive definite matrices), we have

$$\begin{aligned} &\det(D(t) - \sqrt{-1}E(t))(D(t) + \sqrt{-1}E(t))^{-1} \\ &= \det(I - \sqrt{-1}E(t)D^{-1}(t))(I + \sqrt{-1}E(t)D^{-1}(t))^{-1}. \end{aligned} \tag{5.35}$$

Since $D^T(t)E(t) = E^T(t)D(t)$, $E(t)D^{-1}(t)$ is a symmetric $n \times n$ matrix path. Suppose $\lambda_1(t) \leq \cdots \leq \lambda_n(t)$ are the eigenvalues of $E(t)D^{-1}(t)$. We set $1 + \lambda_j\sqrt{-1} = r_j(t)e^{\sqrt{-1}\theta_j(t)/2}$ for $j = 1, \cdots, n$. So we have

$$\det(I - \sqrt{-1}E(t)D^{-1}(t))(I + \sqrt{-1}E(t)D^{-1}(t))^{-1} = e^{2\sqrt{-1}\theta(t)}, \quad \theta(t) = \sum_{j=1}^{n}\theta_j(t).$$

It is easy to see that

$$-\frac{\pi}{2} < \theta_j(t) - \theta_j(0) < \frac{\pi}{2}, \quad t \in [0, 1]. \tag{5.36}$$

So we have

$$-\frac{n\pi}{2} < \sum_{j=1}^{n}(\theta_j(t) - \theta_j(0)) < \frac{n\pi}{2}.$$

By an easy computation(see (5.30)), we have

$$-n \leq i_{L_0}(\gamma) \leq 0. \tag{5.37}$$

Thus we have the following result.

Proposition 1.16 *For a symmetrical positive definite path* $\gamma \in \mathcal{P}_\tau(2n)$, *we have*

$$i_L(\gamma) \in [-n, 0], \quad \forall L \in \Lambda(n). \tag{5.38}$$

Proof γ is a symmetrical positive definite path implies that γ_c defined in Definition 1.7 is also a symmetrical positive definite path. So (5.38) follows from (5.37) and Definition 1.7. □

5.1.2 The Relations of $i_L(\gamma)$ and $i_1(\gamma)$

For a symplectic path $\gamma \in \mathcal{P}_\tau(2n)$, there exist an orthogonal symplectic path $O \in \mathcal{P}_\tau(2n)$ and a symmetric positive definite path $P \in \mathcal{P}_\tau(2n)$ such that $\gamma = OP$. It is called the polar decomposition of γ. Suppose

$$\gamma = \begin{pmatrix} S & V \\ T & U \end{pmatrix}, \quad O = \begin{pmatrix} A & B \\ -B & A \end{pmatrix}, \quad P = \begin{pmatrix} C & E \\ E^T & D \end{pmatrix}.$$

Then we have

$$V = AE + BD, \quad U = -BE + AD.$$

So by direct computations we have

$$\begin{aligned} &\det(U - \sqrt{-1}V)(U + \sqrt{-1}V)^{-1} \\ &= \det(A - \sqrt{-1}B)(A + \sqrt{-1}B)^{-1}\det(D - \sqrt{-1}E)(D + \sqrt{-1}E)^{-1}. \end{aligned} \tag{5.39}$$

Theorem 1.17 ([189]) *For $\gamma \in \mathcal{P}_\tau(2n) \cap O(2n)$, there holds*

$$0 \le i_1(\gamma) - i_{L_0}(\gamma) \le n. \tag{5.40}$$

For any $\gamma \in \mathcal{P}_\tau(2n)$ there holds

$$-n \le i_1(\gamma) - i_{L_0}(\gamma) \le n. \tag{5.41}$$

Proof When $\gamma \in \mathcal{P}_\tau(2n) \cap O(2n)$, the functions θ_j in (5.30) and (2.5) are the same. By the inequality

$$2E(x) \le E(2x) \le 2E(x) + 1$$

we have that

$$i_1(\gamma) - n \le i_{L_0}(\gamma) \le i_1(\gamma).$$

This is exactly (5.40) and (5.41) follows from (5.39) and the estimate (5.36). □

As in the Morse theory of closed geodesics on Riemannian or Finsler Manifolds (cf. [161]), we have

Definition 1.18 ([189]) For any $\gamma \in \mathcal{P}_\tau(2n)$ and $L \in \Lambda(n)$, we call $i_1(\gamma) - i_L(\gamma)$ *the concavity of* γ *w.r.t. the Lagrangian subspace* L and denoted it by

$$\text{concav}_L(\gamma) = i_1(\gamma) - i_L(\gamma). \tag{5.42}$$

If c is a closed geodesic on a Riemannian manifold M, we know from [190] that $\text{ind}(c) = i_1(c)$. In this case, the energy functional is defined in the loop space $\Omega = \{x \in W^{1,2}([0, 1], M) | x(0) = p_1, x(1) = p_2\}$ for $p_1 = p_2$, by linearized the geodesic equation, and translate it to a Hamiltonian system, then the boundary condition is that $z(0), z(1) \in L_0$. So $i_{L_0}(c) = \text{ind}_\Omega(c)$. In this case $\text{concav}_{L_0}(c)$ is just the concavity defined by Morse as in [242] (see also [161]).

Theorem 1.19 ([189]) *The concavity of* $\gamma \in \mathcal{P}_\tau(2n) \cap O(2n)$ *w.r.t.* L *satisfies*

$$0 \le \text{concav}_L(\gamma) \le n. \tag{5.43}$$

The concavity of $\gamma \in \mathcal{P}_\tau(2n)$ *w.r.t.* L *satisfies*

$$-n \le \text{concav}_L(\gamma) \le n. \tag{5.44}$$

Proof From Definition 1.7, we have

$$i_L(\gamma) = i_{L_0}(\gamma_c).$$

By the homotopic invariant of Maslov-type index theory, there holds

$$i_1(\gamma) = i_1(\gamma_c).$$

□

We know that any Lagrangian path $\beta(t)$ starting from L_0 can be written as $\beta(t) = \gamma(t)L_0$ for $\gamma \in \mathcal{P}_\tau(2n) \cap O(2n)$. $i_{L_0}(\gamma)$ in some sense is the algebraic intersection number (with signs determined by orientation) of the Lagrangian path β with the constant path L_0 (cf. [77]). So the special case $\gamma \in \mathcal{P}_\tau(2n) \cap O(2n)$ is important in our index theory.

For a solution $z = z(t)$ of a nonlinear Hamiltonian system with periodic boundary condition:

$$\begin{cases} \dot{z}(t) = JH'(t, z(t)), & z(t) \in \mathbb{R}^{2n} \\ z(1) = z(0), \end{cases} \tag{5.45}$$

by choosing a Lagrangian subspace L such that $z(0) = z(1) \in L$, we can understand it as a special solution of the following problem with Lagrangian boundary condition

$$\begin{cases} \dot{z}(t) = JH'(t, z(t)), \quad z(t) \in \mathbb{R}^{2n} \\ z(1), z(0) \in L. \end{cases} \tag{5.46}$$

Linearizing (5.45) at z, we get a linear Hamiltonian system with periodic boundary condition

$$\begin{cases} \dot{y} = JB(t)y \\ y(0) = y(1), \end{cases} \tag{5.47}$$

where $B(t) = H''(t, z(t))$. Suppose the fundamental solution of the linear system

$$y = JB(t)y \tag{5.48}$$

is $\gamma(t) \in \mathcal{P}_\tau(2n)$. The Maslov-type index $i_1(z) = i_1(\gamma)$ of z as a periodic solution is just an index theory of the problem (5.47). The L-index $i_L(z) = i_L(\gamma)$ of z as a special solution of (5.46) with Lagrangian boundary condition is just an index theory of the following problem

$$\begin{cases} \dot{y} = JB(t)y \\ y(0), \ y(1) \in L, \end{cases} \tag{5.49}$$

which is the linearized form of the problem (5.46).

For the periodic solution z of the Hamiltonian system (5.45), $\gamma \in \mathcal{P}_\tau(2n)$ is the fundamental solution of (5.48). The concavity of z w.r.t. the Lagrangian subspace L with $z(0) \in L$ is defined by

$$\text{concav}_L(z) = \text{concav}_L(\gamma). \tag{5.50}$$

Corollary 1.20 ([189])

(1) *For any* $\gamma \in \mathcal{P}_\tau(2n) \cap O(2n)$ *and* $L \in \Lambda(n)$, *there holds*

$$|i_{L_0}(\gamma) - i_L(\gamma)| \leq n. \tag{5.51}$$

(2) *For any* $\gamma \in \mathcal{P}_\tau(2n)$ *and* $L \in \Lambda(n)$, *there holds*

$$|i_{L_0}(\gamma) - i_L(\gamma)| \leq 2n. \tag{5.52}$$

(3) *For any* $\gamma \in \mathcal{P}_\tau(2n)$ *and* $L \in \Lambda(n)$, $|i_{L_0}(\gamma) - i_L(\gamma)|$ *depends only on the end matrix* $\gamma(\tau)$ *and* L.

Proof From (5.40), (5.41) and (5.43), (5.44) one gets (5.51), (5.52) directly. The third part is a direct consequence of the following result. □

Theorem 1.21 ([189]) *For any* $L \in \Lambda(n)$, *and* $\gamma \in \mathcal{P}_\tau(2n)$, $\mathrm{concav}_L(\gamma)$ *depends only on the end matrix* $\gamma(\tau)$ *and the Lagrangian subspace* L. *Thus for any symplectic matrix* $M \in \mathrm{Sp}(2n)$, *we can define the L-concavity of* M *by*

$$\mathrm{concav}_L(M) := \mathrm{concav}_L(\gamma), \quad \forall \gamma \in \mathcal{P}_\tau(2n), \ \gamma(\tau) = M.$$

Proof We fix $\tau = 1$. Firstly, assume $L = L_0$ and $\gamma \in \mathcal{P}_\tau(2n)^* \cap \mathcal{P}_\tau(2n)^*_{L_0}$. By definition there exist $\beta : [0, 1] \to \mathrm{Sp}^*(2n)$ and $\beta_{L_0} : [0, 1] \to \mathrm{Sp}^*(2n)_{L_0}$ such that

$$i_1(\gamma) = \frac{\theta_\gamma(1) - \theta_\gamma(0)}{\pi} + \frac{\theta_\beta(1) - \theta_\beta(0)}{\pi}, \tag{5.53}$$

where θ_γ and θ_β is defined as $\tilde{\theta}$ just before Definition II.1.1 with respect to γ and β, respectively.

$$i_{L_0}(\gamma) = \frac{\Delta_\gamma(1) - \Delta_\gamma(0)}{\pi} + \frac{\Delta_{\beta_{L_0}}(1) - \Delta_{\beta_{L_0}}(0)}{\pi} - \frac{n}{2}, \tag{5.54}$$

where Δ_γ and $\Delta_{\beta_{L_0}}$ is defined as $\bar{\Delta}$ in (5.7) with respect to γ and β_{L_0}, respectively. We note that (by (5.39))

$$\frac{\Delta_\gamma(1) - \Delta_\gamma(0)}{\pi} = \frac{\theta_\gamma(1) - \theta_\gamma(0)}{\pi} + \frac{\Delta_{\gamma_P}(1) - \Delta_{\gamma_P}(0)}{\pi},$$

where γ_P is the positive definite actor of γ in its polar decomposition. By (5.36), we know that $\frac{\Delta_{\gamma_P}(1) - \Delta_{\gamma_P}(0)}{\pi}$ depends only on its end $\gamma_P(1)$. Thus we have the concavity

$$\begin{aligned} i_1(\gamma) - i_{L_0}(\gamma) = & \tfrac{\theta_\beta(1)-\theta_\beta(0)}{\pi} - \tfrac{\Delta_{\beta_{L_0}}(1)-\Delta_{\beta_{L_0}}(0)}{\pi} \\ & - \tfrac{\Delta_{\gamma_P}(1)-\Delta_{\gamma_P}(0)}{\pi} + \tfrac{n}{2}. \end{aligned} \tag{5.55}$$

Since the symplectic path β and β_{L_0} is determined only by $\gamma(1)$ and L_0, we see that the result of Theorem 1.21 is true in this case. For the degenerate cases, by the homotopic invariant, we can choose a very short rotation path η and η_{L_0} depending only on $\gamma(1)$ such that $\eta * \gamma \in \mathcal{P}_\tau(2n)^*$, $\eta_{L_0} * \gamma \in \mathcal{P}_\tau(2n)^*_{L_0}$ and

$$i_1(\gamma) = i_1(\eta * \gamma), \quad i_{L_0}(\gamma) = i_{L_0}(\eta_{L_0} * \gamma).$$

Then by the same computations as in (5.53) and (5.54), we still get

$$i_1(\gamma) - i_{L_0}(\gamma) = \text{something determined only by } \gamma(1) \text{ and } L_0.$$

For the general case, for any $P \in \mathrm{Sp}(2n) \cap O(2n)$ such that $L = PL_0$, there holds

$$i_1(\gamma) = i_1(P^{-1}\gamma P), \quad i_L(\gamma) = i_{L_0}(P^{-1}\gamma P).$$

□

In the Chap. 7 below, we will revisit the topic of concavity by using Hörmander index theory.

5.1.3 L-Index for General Symplectic Paths

For a general continuous symplectic path $\rho : [a, b] \to \mathrm{Sp}(2n)$ and the Lagrangian subspace L_0, we define its Maslov type L_0-index as follow.

Definition 1.22 We define

$$\hat{i}_{L_0}(\rho) = i_{L_0}(\gamma_b) - i_{L_0}(\gamma_a), \tag{5.56}$$

where $\gamma_a \in \mathcal{P}(2n)$ is a symplectic path ended at $\rho(a)$ and $\gamma_b \in \mathcal{P}(2n)$ is the composite of symplectic path γ_a and ρ, i.e., $\gamma_b = \rho * \gamma_a$.

In general, for any Lagrangian subspace L, we define

$$\hat{i}_L(\rho) = i_L(\gamma_b) - i_L(\gamma_a). \tag{5.57}$$

We remind that for the constant path $\gamma = I$ there holds $i_{L_0}(I) = -n$ and $i_{L_1}(I) = -n$, so $\hat{i}_{L_0}(\gamma) = i_{L_0}(\gamma) + n$ and $\hat{i}_{L_1}(\gamma) = i_{L_1}(\gamma) + n$ for $\gamma \in \mathcal{P}(2n)$.

Lemma 1.23 *The index $\hat{i}_L(\rho)$ is well defined, i.e., it is independent of the choice of γ_a.*

Proof We only prove the case $L = L_0$. For two symplectic paths $\gamma_a,\ \gamma_a' \in \mathcal{P}(2n)$, we denote $\gamma_b,\ \gamma_b' \in \mathcal{P}(2n)$ correspondingly as in Definition 1.22. We should prove

$$i_{L_0}(\gamma_b) - i_{L_0}(\gamma_b') = i_{L_0}(\gamma_a) - i_{L_0}(\gamma_a'). \tag{5.58}$$

We suppose

$$\gamma_b = \begin{pmatrix} S_b & V_b \\ T_b & U_b \end{pmatrix},\quad \gamma_a = \begin{pmatrix} S_a & V_a \\ T_a & U_a \end{pmatrix},$$

and

$$\gamma_b' = \begin{pmatrix} S_b' & V_b' \\ T_b' & U_b' \end{pmatrix},\quad \gamma_a' = \begin{pmatrix} S_a' & V_a' \\ T_a' & U_a' \end{pmatrix}.$$

For these four matrices, as in Definition 1.2, we suppose

$$\det(\bar{U}_a - \sqrt{-1}\bar{V}_a) = \rho_a e^{\sqrt{-1}\bar{\Delta}_a},\ \det(\bar{U}_b - \sqrt{-1}\bar{V}_b) = \rho_b e^{\sqrt{-1}\bar{\Delta}_b}$$

and

$$\det(\bar{U}'_a - \sqrt{-1}\bar{V}'_a) = \rho'_a e^{\sqrt{-1}\bar{\Delta}'_a}, \det(\bar{U}'_b - \sqrt{-1}\bar{V}'_b) = \rho'_b e^{\sqrt{-1}\bar{\Delta}'_b}.$$

By definition, the functions $\bar{\Delta}$'s have the path additivity, we prove the equality (5.58) in the non-degenerate cases. For the degenerate cases, since the two paths have the same end points, we can prove the equality (5.58) by taking non-degenerate paths close to the degenerate ones and taking the infimum in every terms. □

Theorem 1.24 *The index $\hat{i}_{L_0}$ has the following properties*

(1) *(Affine Scale Invariance). For $k > 0, l \geq 0$, we have the affine map $\varphi : [a, b] \to [ka + l, kb + l]$ defined by $\varphi(t) = kt + l$. For a given continuous path $\rho : [ka + l, kb + l] \to \mathrm{Sp}(2n)$, there holds*

$$\hat{i}_{L_0}(\rho) = \hat{i}_{L_0}(\rho \circ \varphi). \tag{5.59}$$

(2) *(Homotopy Invariance rel. End Points). If $\delta : [0, 1] \times [a, b] \to \mathrm{Sp}(2n)$ is a continuous map with $\delta(0, t) = \rho_1(t)$, $\delta(1, t) = \rho_2(t)$, $\delta(s, a) = \rho_1(a) = \rho_2(a)$ and $\delta(s, b) = \rho_1(b) = \rho_2(b)$ for $s \in [0, 1]$, then*

$$\hat{i}_{L_0}(\rho_1) = \hat{i}_{L_0}(\rho_2). \tag{5.60}$$

(3) *(Path Additivity). If $a < b < c$, and $\rho_{[a,c]} : [a, c] \to \mathrm{Sp}(2n)$ is concatenate path of $\rho_{[a,b]}$ and $\rho_{[b,c]}$, then there holds*

$$\hat{i}_{L_0}(\rho_{[a,c]}) = \hat{i}_{L_0}(\rho_{[a,b]}) + \hat{i}_{L_0}(\rho_{[b,c]}). \tag{5.61}$$

(4) *(Symplectic Additivity). Let L_0^k, $\rho_k : [a, b] \to \mathrm{Sp}(2n_k)$, $k = 1, 2$, $L_0 = L_0^1 \diamond L_0^2$, $\rho = \rho_1 \diamond \rho_2$. Then we have*

$$\hat{i}_{L_0}(\rho) = \hat{i}_{L_0^1}(\rho_1) + \hat{i}_{L_0^2}(\rho_2). \tag{5.62}$$

Here the symplectic direct sum of two Lagrangian subspaces L' and L'' is defined by

$$L' \diamond L'' = \{(x', x'', y', y'')^T \mid (x', y')^T \in L',\ (x'', y'')^T \in L''\}.$$

(5) *(Symplectic Invarince) Let matrix $P \in \mathrm{Sp}(2n)$ be a symplectic matrix. We have*

$$\hat{i}_{PL_0}(P\rho P^{-1}) = \hat{i}_{L_0}(\rho). \tag{5.63}$$

(6) *(Normalization). For $L_0 = \mathbb{R}$ and $\rho : [-\varepsilon, \varepsilon] \to \mathrm{Sp}(2)$ with $\varepsilon > 0$ small and $\rho(t) = e^{Jt}$, we have*

$$\begin{aligned} &(i)\ \hat{i}_{L_0}(\rho) = 1; \\ &(ii)\ \hat{i}_{L_0}(\rho_{[-\varepsilon,0]}) = 0; \\ &(iii)\ \hat{i}_{L_0}(\rho_{[0,\varepsilon]}) = 1. \end{aligned} \tag{5.64}$$

Proof We prove the statements (5) and (6) only. The remainders are direct consequence of the definition. The proof of the statement (6) is very easy. By definition, it is easy to see $i_{L_0}(\gamma_{-\varepsilon}) = 0$ and $i_{L_0}(\gamma_\varepsilon) = 1$. This proves statement (i). (ii) and (iii) are similar.

For the proof of the statement (5), we first prove the following result.

Lemma 1.25 *If $P \in \mathrm{Sp}(2n)$ is a symplectic matrix satisfying $PL_0 = L_0$, then we have*

$$i_{L_0}(\gamma) = i_{L_0}(P\gamma P^{-1}) \tag{5.65}$$

for any $\gamma \in \mathcal{P}_1(2n)$.

It is easy to see that we can write the matrix P in the form

$$P = \begin{pmatrix} A & 0 \\ C & (A^T)^{-1} \end{pmatrix}$$

with $A^T C = C^T A$ and so

$$P^{-1} = \begin{pmatrix} A^{-1} & 0 \\ -C^T & A^T \end{pmatrix}.$$

So the set of all $P \in \mathrm{Sp}(2n)$ satisfying $PL_0 = L_0$ is a subgroup of $\mathrm{Sp}(2n)$, we denote it by $\mathcal{F}(L_0)$.

Suppose the end point $\gamma(1)$ of $\gamma : [0,1] \to \mathrm{Sp}(2n)$ is

$$\begin{pmatrix} S & V \\ T & U \end{pmatrix}.$$

Thus we have

$$P\gamma(1)P^{-1} = \begin{pmatrix} ASA^{-1} & AVA^T \\ a_{21} & a_{22} \end{pmatrix}$$

where $a_{21} = CSA^{-1} + (A^{-1})^T TA^{-1} - CVC^T - (A^{-1})^T UC^T$ and $a_{22} = CVA^T + (A^{-1})^T UA^T$. It is clear that $\dim\ker(V) = \dim\ker(AVA^T)$. We can connect the two matrices $I_\pm = \begin{pmatrix} C_\pm & 0 \\ 0 & C_\pm \end{pmatrix}$, $C_\pm = \begin{pmatrix} \pm 1 & 0 \\ 0 & I_{n-1} \end{pmatrix}$ and P in $\mathcal{F}(L_0)$ by a continuous path $P(s)$ with $P(0) = I_\pm$ and $P(1) = P$(in fact, we can get this path in two steps. We first connect C in the presentation of matrix P to 0 by sC, correspondingly we get a path $A(s)$ with $A(s) \equiv A$. Since A is nondegenerate, then we can connect A to $\begin{pmatrix} \pm 1 & 0 \\ 0 & I_{n-1} \end{pmatrix}$). So $\delta(s,t) = P(s)\gamma(t)P(s)^{-1}$ is a homotopy between the two paths γ and $P\gamma P^{-1}$. The statement (5.65) follows from Theorem 1.15. □

Proof of the statement (5) Suppose $L = PL_0$. For L, there is $M \in \mathrm{Osp}(2n)$ such that $ML_0 = L$. By definition, there holds

$$i_L(\gamma) = i_{L_0}(M^{-1}\gamma M) \text{ for } \gamma \in \mathcal{P}(2n).$$

Since $P^{-1}ML_0 = L_0$, by Lemma 1.25 there holds

$$i_L(\gamma) = i_{L_0}(M^{-1}\gamma M) = i_{L_0}(P^{-1}MM^{-1}\gamma MM^{-1}P) = i_{L_0}(P^{-1}\gamma P). \quad (5.66)$$

Now (5.63) follows from the definition. □

For the index $\hat{i}_L$, we have a similar result as Theorem 1.24. The only difference is that the statement (6) in Theorem 1.24 should replaced by

(6)′ For dimension 1 subspace L of $\mathbb{R}^2$, and $\rho : [-\varepsilon, \varepsilon] \to \mathrm{Sp}(2)$ with $\varepsilon > 0$ small and $\rho(t) = Pe^{Jt}P^{-1}$ with $P \in \mathrm{Osp}(2n)$ satisfying $PL_0 = L$, we have

$$\begin{aligned} &(i)\ \hat{i}_L(\rho) = 1; \\ &(ii)\ \hat{i}_L(\rho_{[-\varepsilon,0]}) = 0; \\ &(iii)\ \hat{i}_L(\rho_{[0,\varepsilon]}) = 1. \end{aligned} \quad (5.67)$$

We will see that the six axioms in Theorem 1.24 character the index $\hat{i}_{L_0}$ uniquely. In the next chapter, we will define an index $\mu_{L_0}(\gamma)$ as the Maslov type index of a pair of Lagrangian paths and prove that this index also satisfies the six axioms of Theorem 1.24. So we have

$$\hat{i}_{L_0}(\gamma) = i_{L_0}(\gamma) + n = \mu_{L_0}(\gamma)$$

for every $\gamma \in \mathcal{P}(2n)$.

Remark 1.26 A path $\rho : [0, 1] \to \mathrm{Sp}(2n)$ is called L-nondegenerate if $\nu_L(\rho(0)) = \nu_L(\rho(1)) = 0$. For any path ρ, there exists $\varepsilon > 0$ such that $e^{-\theta J}\rho$ with $0 < |\theta| < \varepsilon$ is L-nondegenerate and there holds

$$\hat{i}_L(\rho) = \hat{i}_L(e^{-\theta J}\rho), \ 0 < \theta < \varepsilon.$$

For any L_0-nondegenerate path ρ, we can connect by a path α from $\rho(0)$ (and β from $\rho(1)$ respectively) to M_+ or M_- in the connected component of $\mathrm{Sp}(2n)^*_{L_0}$. Then we get a composite path $\tilde{\rho} = \beta * \rho * \alpha^{-1}$. We have

$$\hat{i}_{L_0}(\rho) = \hat{i}_{L_0}(\tilde{\rho}) = \frac{\tilde{\Delta}(1) - \tilde{\Delta}(0)}{\pi},$$

where $\tilde{\Delta}$ is defined by $\det \tilde{\mathcal{Q}}(t) = e^{2\sqrt{-1}\tilde{\Delta}(t)}$ with $\tilde{\mathcal{Q}}(t)$ defined from $\tilde{\rho}$ in (5.6) and (5.7). We omit the proofs of these statements and leave them to the readers.

We are ready to define an index for a pair of a continuous Lagrangian path $f: [0,1] \to \Lambda(n) \times \Lambda(n)$ with $f(t) = (L_1(t), L_2(t)), 0 \le t \le 1$. According to (1.20), we know that there are $U_1(t),\ U_2(t) \in \mathrm{Osp}(2n)$ such that $L_j(t) = U_j(t)L_0$, then we have the following definition.

Definition 1.27

$$\hat{\imath}_0(f) = \hat{\imath}_{L_0}(\gamma_{12}), \tag{5.68}$$

where $\gamma_{12}(t) = U_1(t)^{-1}U_2(t),\ 0 \le t \le 1$.

For a general symplectic space $V = (\mathbb{R}^{2n}, \omega)$ and a path of Lagrangian pair $f: [a,b] \to \mathrm{Lag}(V) \times \mathrm{Lag}(V)$, by choosing a linear symplectic map $T: (\mathbb{R}^{2n}, \omega) \to (\mathbb{R}^{2n}, \omega_0)$, we can define

$$\hat{\imath}(f) = \hat{\imath}_0(TfT^{-1}). \tag{5.69}$$

Theorem 1.28 *The definition* (5.68) *is well defined. Furthermore* $\hat{\imath}(f)$ *has the following properties*

(1) *For* $k > 0,\ l \ge 0$, *we have the affine map* $\varphi: [a,b] \to [ka+l, kb+l]$ *defined by* $\varphi(t) = kt + l$. *For a given continuous path* $f: [ka+l, kb+l] \to \mathrm{Lag}(V) \times \mathrm{Lag}(V)$, *there holds*

$$\hat{\imath}(f) = \hat{\imath}(f \circ \varphi). \tag{5.70}$$

(2) *If* $\delta: [0,1] \times [a,b] \to \mathrm{Lag}(V) \times \mathrm{Lag}(V)$ *is a continuous map with* $\delta(0,t) = f_1(t)$, $\delta(1,t) = f_2(t)$, $\delta(s,a) = f_1(a) = f_2(a)$ *and* $\delta(s,b) = f_1(b) = f_2(b)$ *for* $s \in [0,1]$, *then*

$$\hat{\imath}(f_1) = \hat{\imath}(f_2). \tag{5.71}$$

(3) *If* $a < b < c$, *and* $f_{[a,c]}: [a,c] \to \mathrm{Lag}(V) \times \mathrm{Lag}(V)$ *is concatenate path of* $f_{[a,b]}$ *and* $f_{[b,c]}$, *then there holds*

$$\hat{\imath}(f_{[a,c]}) = \hat{\imath}(f_{[a,b]}) + \hat{\imath}(f_{[b,c]}). \tag{5.72}$$

(4) *Let* $f_k: [a,b] \to \mathrm{Lag}(V_k) \times \mathrm{Lag}(V_k)$, $k = 1,2$, $V = V_1 \oplus V_2$ $f = f_1 \oplus f_2$. *Then we have*

$$\hat{\imath}(f) = \hat{\imath}(f_1) + \hat{\imath}(f_2). \tag{5.73}$$

(5) *Let* $P(t) \in \mathrm{Sp}(2n)$ *be a symplectic path and* $f(t) = (L_1(t), L_2(t)) \in \Lambda(n) \times \Lambda(n)$. *We define* $(P_*f)(t) = (P(t)L_1(t), P(t)L_2(t))$. *Then we have*

$$\hat{\imath}_0(P_*f) = \hat{\imath}_0(f). \tag{5.74}$$

(6) *Let* $V_0 = (\mathbb{R}^2, \omega_0)$. *Define* $f : [-\varepsilon, \varepsilon] \to \Lambda(1) \times \Lambda(1)$ *with* $\varepsilon > 0$ *small as a pair Lagrangian path:*

$$f(t) = (\mathbb{R}, \mathbb{R}(\cos t, sint)), \ t \in [-\varepsilon, \varepsilon].$$

Then

$$\begin{aligned} &(i) \ \hat{i}_0(f) = 1, \\ &(ii) \ \hat{i}_0(f_{[-\varepsilon,0]}) = 0, \\ &(iii) \ \hat{i}_0(f_{[0,\varepsilon]}) = 1. \end{aligned} \tag{5.75}$$

Proof We only need to prove the case of $V = (\mathbb{R}^{2n}, \omega_0)$. For that the definition (5.68) is well defined, if $L_1(t) = U_1'(t)L_0 = U_1(t)L_0$ and $L_2(t) = U_2'(t)L_0 = U_2(t)L_0$, there exist two orthogonal matrices $C_j(t) \in O(n), \ j = 1, 2$ such that $U_j'(t) = U_j(t)O_j(t)$ with $O_j(t) = \begin{pmatrix} C_j(t) & 0 \\ 0 & C_j(t) \end{pmatrix}$. Then we have $U_1'(t)^{-1}U_2'(t) = O_1(t)^{-1}U_1(t)^{-1}U_2(t)O_2(t)$. Set $U_1(t)^{-1}U_2(t) = \begin{pmatrix} S(t) & V(t) \\ T(t) & U(t) \end{pmatrix}$. By direct computation, we arrive at

$$\begin{aligned} U_1'(t)^{-1}U_2'(t) &= O_1(t)^{-1}U_1(t)^{-1}U_2(t)O_2(t) \\ &= \begin{pmatrix} C_1(t)^{-1}S(t)C_2(t) & C_1(t)^{-1}V(t)C_2(t) \\ C_1(t)^{-1}T(t)C_2(t) & C_1(t)^{-1}U(t)C_2(t) \end{pmatrix}. \end{aligned}$$

$\dim\ker C_1(t)^{-1}V(t)C_2(t) = \dim\ker V(t)$. We define the homotopy

$$\delta(s,t) = O_1(st)^{-1}U_1(t)^{-1}U_2(t)O_2(st), \ s \in [0,1]$$

and note that

$$\hat{i}_{L_0}(U_1'^{-1}U_2') = \hat{i}_{L_0}(O_1(0)^{-1}U_1^{-1}U_2O_2(0)) = \hat{i}_{L_0}(U_1^{-1}U_2O_2(0)O_1(0)^{-1}).$$

We can connect the matrix $P = O_2(0)O_1(0)^{-1}$ to $I_\pm = \begin{pmatrix} D_\pm & 0 \\ 0 & C_\pm \end{pmatrix}$ in the subgroup $O(n) := \{O; O = \begin{pmatrix} C & 0 \\ 0 & C \end{pmatrix}, \ C$ is orthogonal matrix$\}$ with $C_\pm = \begin{pmatrix} \pm 1 & 0 \\ 0 & I_{n-1} \end{pmatrix}$. So we have

$$\hat{i}_{L_0}(U_1'^{-1}U_2') = \hat{i}_{L_0}(U_1^{-1}U_2C_\pm) = \hat{i}_{L_0}(C_\pm U_1^{-1}U_2). \tag{5.76}$$

We only need to consider the situation of C_-. We recall that $M_+ = -J = \begin{pmatrix} 0 & I \\ -I & 0 \end{pmatrix}$ and $M_- = \begin{pmatrix} 0 & D_n \\ -D_n & 0 \end{pmatrix}$, $D_n = C_-$. It is clear that $M_- = C_- M_+$ and $M_+ = C_- M_-$. In Remark 1.26, we take $\rho(t) = e^{\theta J} U_1(t)^{-1} U_2(t)$ and $C_-\rho(t) = C_- e^{\theta J} U_1(t)^{-1} U_2(t)$, $0 < \theta < \varepsilon$. Then we get the corresponding path $\tilde{\rho}$ of ρ and $\hat{\rho} = C_-\tilde{\rho}$ of $C_-\rho$. So we have

$$\hat{i}_{L_0}(\rho) = \frac{\tilde{\Delta}(1) - \tilde{\Delta}(0)}{\pi}$$

and

$$\hat{i}_{L_0}(C_-\rho) = \frac{\hat{\Delta}(1) - \hat{\Delta}(0)}{\pi} = \frac{\tilde{\Delta}(1) - \tilde{\Delta}(0)}{\pi} = \hat{i}_{L_0}(\rho).$$

Thus we have

$$\hat{i}_{L_0}(U_1^{-1} U_2) = i_{L_0}(U_1^{-1} U_2 C_-) = i_{L_0}(C_- U_1^{-1} U_2),$$

Therefore there holds

$$\hat{i}_{L_0}((U_1')^{-1} U_2') = \hat{i}_{L_0}(U_1^{-1} U_2).$$

The Properties (1)–(6) can be proved as in that of Theorem 1.24. We omit the details. □

5.2 The (L, L')-Index Theory

For any $\gamma \in \mathcal{P}_1(2n)$, we write it in the form $\gamma(t) = \begin{pmatrix} S(t) & V(t) \\ T(t) & U(t) \end{pmatrix}$, where $S(t), T(t), V(t), U(t)$ are $n \times n$ continuous matrix functions. We note that for the linear Lagrangian subspaces L_0, L' there exists an orthogonal symplectic matrix P such that $PL_0 = L'$ and P has the form $P = \begin{pmatrix} A & -B \\ B & A \end{pmatrix}$ with $A \pm \sqrt{-1}B \in U(n)$. So $P^{-1} = \begin{pmatrix} A^T & B^T \\ -B^T & A^T \end{pmatrix}$. Here P is uniquely determined by L' up to an orthogonal matrix $C \in O(n)$. Also we know the fact that the symplectic group is a Lie group and its Lie algebra is $sp(2n) = \{M \in \mathcal{L}(\mathbb{R}^{2n}, \mathbb{R}^{2n}) | JM + M^T J = 0\}$. It is easy to see that for $M \in sp(2n)$ there holds $Je^{tM} = e^{-tM^T} J$. It is well known that for a connected compact Lie group G with its Lie algebra $\mathcal{G}$, there holds $\exp \mathcal{G} = G$.

For any $M \in sp(2n)$, it is well known that the one parameter curve $\exp(tM) = e^{tM}$ in $\mathrm{Sp}(2n)$ is a Lie subgroup of $\mathrm{Sp}(2n)$, and $SO(2n) \cap \mathrm{Sp}(2n) \subseteq \exp(sp(2n))$. So for the above $P \in \mathrm{Sp}(2n)$, there is a matrix $M \in sp(2n)$ such that $M + M^T = 0$, $P = e^M$ and $e^{tM} \in \mathrm{Sp}(2n)$.

Definition 2.1 ([207]) For any symplectic path $\gamma \in \mathcal{P}_1(2n)$ with $\gamma(t) = \begin{pmatrix} S(t) & V(t) \\ T(t) & U(t) \end{pmatrix}$, we define the (L_0, L')-nullity to be

$$\nu_{L_0}^{L'}(\gamma) \equiv \nu_{L_0}(\gamma^M) = \dim\ker(A^T V(1) + B^T U(1)) = \dim\left(\gamma(1)L_0 \cap L'\right), \tag{5.77}$$

where $\gamma^M(t) = e^{-tM}\gamma(t)$ and $\nu_{L_0}(\gamma)$ was defined in Definition 1.1 for any $\gamma \in \mathcal{P}_\tau(2n)$.

Remark 2.2 The (L_0, L')-nullity in (5.77) is well defined. Since for another choice P', we have

$$P' = P\begin{pmatrix} C & 0 \\ 0 & C \end{pmatrix} = \begin{pmatrix} A & -B \\ B & A \end{pmatrix}\begin{pmatrix} C & 0 \\ 0 & C \end{pmatrix} = \begin{pmatrix} AC & -BC \\ BC & AC \end{pmatrix}, \tag{5.78}$$

where $C \in O(n)$, then

$$\dim\ker((AC)^T V(1) + (BC)^T U(1)) = \dim\ker(A^T V(1) + B^T U(1)). \tag{5.79}$$

In general, $i_{L_0}(\gamma^M)$ depends on the choice of P. It is well known that $O(n)$ has two connected components, one contains $C_1 = I_n$ and another contains $C_2 = \begin{pmatrix} -1 & 0 \\ 0 & I_{n-1} \end{pmatrix}$.

Let $P = \begin{pmatrix} A & -B \\ B & A \end{pmatrix}$, we can choose $C = C_1$ (or C_2) such that $\det(AC + \sqrt{-1}BC) = e^{\sqrt{-1}\vartheta}$ with $\vartheta \in (-\frac{\pi}{2}, \frac{\pi}{2}]$. In the following we always choose P satisfying the condition $\det(A + \sqrt{-1}B) = e^{\sqrt{-1}\vartheta}$ with $\vartheta \in (-\frac{\pi}{2}, \frac{\pi}{2}]$.

For P and P' satisfy the relation (5.78) and $\det(A + \sqrt{-1}B) = \det(AC + \sqrt{-1}BC) = e^{\sqrt{-1}\vartheta}$, with $\vartheta \in (-\frac{\pi}{2}, \frac{\pi}{2}]$. Then there holds $\det(C) = 1$, so we have $C \sim C_1$, i.e., there exists $\eta \in C([0,1], O(n))$ such that $\eta(0) = C_1$, $\eta(1) = C$. Then there exists $M \in C([0,1], \mathrm{Sp}(2n))$, satisfies $M(s) + M^T(s) = 0$ and $P(s) = P\eta(s) = e^{M(s)}$. So from Theorem 1.15, we have $i_{L_0}(\gamma^{M(0)}) = i_{L_0}(\gamma^{M(1)})$. We now choose M satisfying $P = e^M$. We write the orthogonal symplectic path in the following form

$$\gamma_0(t) = e^{tM} = \begin{pmatrix} A(t) & B(t) \\ -B(t) & A(t) \end{pmatrix}$$

with $w_0(t) := A(t) - \sqrt{-1}B(t) \in U(n)$. If replacing M by $M + 2k\pi J,\ k \in \mathbb{Z}$, we get another symplectic path $\gamma_k(t) = e^{t(M+2k\pi J)}$ with $w_k(t) = e^{2\sqrt{-1}k\pi t} w_0(t)$. We choose continuous function $\Delta_k : [0, 1] \to \mathbb{R}$, $\Delta_k(0) = 0$ such that

$$\det w_k(t) = e^{\sqrt{-1}\pi \Delta_k(t)}.$$

So by choosing a suitable $k \in \mathbb{Z}$, we can assume that

$$\Delta_k(1) \in (-n, n].$$

In the following we will always choose the matrix M such that $\Delta_0(1) \in (-n, n]$. For example if $L' = L_0$, we have $M = 0$, $P = I_{2n}$ and $\Delta_0(1) = 0$. In the following definition, the matrices P and M are chosen in this way.

Definition 2.3 ([207]) For the symplectic path $\gamma \in \mathcal{P}_\tau(2n)$, we use the L_0-index to define the (L_0, L')-index of γ by

$$i_{L_0}^{L'}(\gamma) \equiv i_{L_0}(\gamma^M),\ \gamma^M(t) = e^{-tM}\gamma(t). \tag{5.80}$$

From the discussion above, we see that the index $i_{L_0}^{L'}(\gamma)$ is well defined and if $L' = L_0$, we have $i_{L_0}^{L'}(\gamma) = i_{L_0}(\gamma)$.

We now give another description of the index via the following definition.

Definition 2.4 For $\gamma \in \mathcal{P}_\tau(2n)$, $L' \in \Lambda(2n)$, the (L_0, L')-index of γ is defined by

$$\mu_{L_0}^{L'}(\gamma) = i_{L_0}(P^{-1}\gamma * \xi) - i_{L_0}(\xi), \tag{5.81}$$

where $\xi \in \mathcal{P}_\tau(2n)$ satisfies $\xi(\tau) = P^{-1}$.

We note that $\mu_{L_0}^{L'}(I) = 0$ for the constant path $\gamma(t) = I$.

Theorem 2.5 *The indices $i_{L_0}^{L'}(\gamma)$ and $\mu_{L_0}^{L'}(\gamma)$ have the following relation*

$$\mu_{L_0}^{L'}(\gamma) = i_{L_0}^{L'}(\gamma) - i_{L_0}^{L'}(I). \tag{5.82}$$

We will restate the result in (5.82) as a special case of Theorem VII.1.3 and prove it later.

For any Lagrangian pair $(L, L') \in \Lambda(n) \times \Lambda(n)$, we now define the index pair $(i_L^{L'}(\gamma), \nu_L^{L'}(\gamma))$ for any $\gamma \in \mathcal{P}_\tau(2n)$. We can choose an orthogonal symplectic matrix O satisfying $OL_0 = L$, then we define a new symplectic path $\gamma_O(t) = O^{-1}\gamma(t)O$. Denoting by $\tilde{L}' = O^{-1}L'$

Definition 2.6 ([207]) The (L, L')-index pair $(i_L^{L'}(\gamma), \nu_L^{L'}(\gamma))$ for any $\gamma \in \mathcal{P}_\tau(2n)$ is defined by

$$\nu_L^{L'}(\gamma) = \nu_{L_0}^{\tilde{L}'}(\gamma_O) = \dim\left(\gamma(1)L \cap L'\right), \tag{5.83}$$

$$i_L^{L'}(\gamma) = i_{L_0}^{\tilde{L}'}(\gamma_O), \; O^{-1}L' = \tilde{L}'. \tag{5.84}$$

Remark 2.7 It is easy to see that the index pair is well defined by the homotopic invariance of the L_0-index theory, and we note that $i_L^{L'}(\gamma)$ can be defined directly as that of $i_{L_0}^{L'}(\gamma)$ in Definition 2.3 with the L-index, i.e., $i_L^{L'}(\gamma) = i_L(\gamma^M)$. We also note that the (L, L')-index theory reserves some properties as the L-index theory in [189], we derive some of them such as the Galerkin approximation and the saddle point reduction in the next two sections. The homotopic invariance of this index theory is still true, but we don't discuss it here. If $L = L'$, we have

$$(i_L^{L'}(\gamma), \nu_L^{L'}(\gamma)) = (i_L(\gamma), \nu_L(\gamma)).$$

We denote by

$$i_L^{L'}(B) = i_L^{L'}(\gamma_B), \nu_L^{L'}(B) = \nu_L^{L'}(\gamma_B), \tag{5.85}$$

where $B \in L^2([0, 1], \mathcal{L}_s(\mathbb{R}^{2n}))$, and γ_B is the fundamental solution of the linear system $\dot{z}(t) = JB(t)z(t)$. If $B_1, B_2 \in L^2([0, 1], \mathcal{L}_s(\mathbb{R}^{2n}))$, with $B_1(t) < B_2(t)$, a.e.$t \in [0, 1]$, we can show as in [189] that

$$i_L^{L'}(B_2) - i_L^{L'}(B_1) = I_L^{L'}(B_1, B_2), \tag{5.86}$$

where $I_L^{L'}(B_1, B_2) = \sum_{s\in[0,1)} \nu_L^{L'}((1 - s)B_1 + sB_2)$ is well defined, that is to say $\nu_L^{L'}((1 - s)B_1 + sB_2) > 0$ only happens at finite points $s \in [0, 1)$.

As in Sect. 1.2, the general symplectic space $(\mathbb{R}^{2n}, \omega)$ is accompanied by the symplectic transform group which is the $\mathcal{J}$-symplectic matrix group $Sp_{\mathcal{J}}(2n)$. We denote by $\mathcal{P}_{\tau,\mathcal{J}}(2n) = \{\gamma \in C^0([0, \tau], Sp_{\mathcal{J}}(2n)) | \gamma(0) = I\}$ the set of $\mathcal{J}$-symplectic paths. From Proposition I.2.2 of Chap. 1, we have

$$T\mathcal{P}_{\tau,\mathcal{J}}(2n)T^{-1} = \mathcal{P}_\tau(2n),$$

where the symplectic transform $T : (\mathbb{R}^{2n}, \omega) \to (\mathbb{R}^{2n}, \omega_0)$ defined in Sect. 1.1 corresponds to a matrix T satisfying $T^T JT = \mathcal{J}^{-1}$. The set of Lagrangian subspaces of the symplectic space $(\mathbb{R}^{2n}, \omega)$ is denoted by $\Lambda(2n, \omega)$. Then we have

$$T : \Lambda(2n, \omega) \simeq \Lambda(2n, \omega_0) = \Lambda(2n).$$

Definition 2.8 For $L, L' \in \Lambda(2n, \omega)$, $\gamma \in \mathcal{P}_{\tau,\mathcal{J}}(2n)$, the (L, L')-index of γ is defined as

$$i_L^{L'}(\gamma) = i_{TL}^{TL'}(\gamma_0), \quad \nu_L^{L'}(\gamma) = \nu_{TL}^{TL'}(\gamma_0), \tag{5.87}$$

where $\gamma_0(t) = T\gamma(t)T^{-1}$.

5.3 Understanding the Index $i^P(\gamma)$ in View of the Lagrangian Index $i_L^{L'}(\gamma)$

From the symplectic space $(\mathbb{R}^{2n}, \omega_0)$, we can construct a dimensional $4n$ symplectic space $(\mathbb{R}^{4n}, \tilde{\Omega}_0) = (\mathbb{R}^{2n} \oplus \mathbb{R}^{2n}, \omega_0 \oplus (-\omega_0))$ in the sense that

$$\omega_0 \oplus (-\omega_0)(x_1 \oplus x_2, y_1 \oplus y_2) = \omega_0(x_1, y_1) - \omega_0(x_2, y_2). \tag{5.88}$$

For a symplectic matrix $P \in \mathrm{Sp}(2n)$, the generated dimensional $2n$ subspace $L_P = \{(x \oplus Px) | x \in \mathbb{R}^{2n}\}$ is a Lagrangian subspace of $(\mathbb{R}^{4n}, \tilde{\Omega}_0)$. Particularly, for $\gamma \in \mathcal{P}_\tau(2n)$, the graph of γ in $\mathbb{R}^{4n}$, $L(t) = \{(x \oplus \gamma(t)x) | x \in \mathbb{R}^{2n}\}$, $t \in [0, \tau]$ is a Lagrangian path of $(\mathbb{R}^{4n}, \tilde{\Omega}_0)$. If we write $x \oplus y \in \mathbb{R}^{2n} \oplus \mathbb{R}^{2n}$ in the form $(x, y)^T$, then any two symplectic matrices $M_1, \ M_2 \in \mathrm{Sp}(2n)$ determine a symplectic transformation of $(\mathbb{R}^{4n}, \tilde{\Omega}_0)$ as

$$\begin{pmatrix} M_1 & 0 \\ 0 & M_2 \end{pmatrix}. \tag{5.89}$$

Particularly, from $\gamma \in \mathcal{P}_\tau(2n)$, we get a symplectic path of $(\mathbb{R}^{4n}, \tilde{\Omega}_0)$ as

$$\Gamma(t) = \begin{pmatrix} I_{2n} & 0 \\ 0 & \gamma(t) \end{pmatrix}.$$

Taking $L_I = \{(x, x)^T \mid x \in \mathbb{R}^{2n}\}$, which is a Lagrangian subspace of $(\mathbb{R}^{4n}, \tilde{\Omega}_0)$, we see that $\Gamma(t)L_I$ is a Lagrangian path of $(\mathbb{R}^{4n}, \tilde{\Omega}_0)$.

Definition 3.1 For $P \in \mathrm{Sp}(2n)$ and $\gamma \in \mathcal{P}_\tau(2n)$, we define

$$\mu_P(\gamma) = i_{L_I}^{L_P}(\Gamma), \quad n_P(\gamma) = \nu_{L_I}^{L_P}(\Gamma). \tag{5.90}$$

Theorem 3.2 *For any $P \in \mathrm{Sp}(2n)$, $\gamma \in \mathcal{P}_\tau(2n)$, there hold*

$$\mu_P(\gamma) = i^P(\gamma), \quad n_P(\gamma) = \nu^P(\gamma). \tag{5.91}$$

We will restate this theorem as a special case of Theorem VII.1.3 and prove it later.

5.4 The Relation with the Morse Index in Calculus Variations

We consider the following problem

$$\begin{cases} [P(t)x'(t) - Q(t)x(t)]' + Q^T(t)x'(t) + R(t)x = 0, \\ x(0) = x(1) = 0, \end{cases} \tag{5.92}$$

where P and R are symmetrial $n \times n$ matrix function, we suppose $-P > 0$ (positive definite). For simplicity, we assume P, Q are smooth and R is continuous. The equations in (5.92) was studied by M.Morse in chapter IV of [242]. We turn it into a first order system with Lagrangian boundary condition by setting $z(t) = (x(t), y(t))^T \in \mathbb{R}^{2n}$ with $y = P(t)x'(t) - Q(t)x(t)$:

$$\begin{cases} \dot{z} = JB(t)z \\ z(0),\ z(1) \in L_0, \end{cases} \tag{5.93}$$

where $B = B(t)$ is defined by

$$B(t) = \begin{pmatrix} -R(t) - Q^T(t)P^{-1}(t)Q(t) & -Q^T(t)P^{-1}(t) \\ -P^{-1}(t)Q(t) & -P^{-1}(t) \end{pmatrix}. \tag{5.94}$$

We take the space $W = W_0^{1,2}([0,1], \mathbb{R}^n)$, the subspace of $W^{1,2}([0,1], \mathbb{R}^n)$ with the elements x satisfying $x(0) = x(1) = 0$. Define the following functional on W

$$\begin{aligned} \varphi(x) = -\tfrac{1}{2} \textstyle\int_0^1 \big(&\langle P^{-1}(t)(P(t)x'(t) - Q(t)x(t)), P(t)x'(t) - Q(t)x(t)\rangle \\ &- \langle (R(t) + Q^T(t)P^{-1}(t)Q(t))x(t), x(t)\rangle \big)\, dt. \end{aligned} \tag{5.95}$$

The critical point of φ is a solution of the problem (5.92), and so we get a solution of the problem (5.93). Denote the Morse index of the functional φ at $x = 0$ by $m^{L_0}(B)$, which is the total multiplicity of the negative eigenvalues of the Hessian of φ at $x = 0$, and the nullity of φ at $x = 0$ by $n^{L_0}(B)$. Assume that the fundamental solution of the linear system in (5.93) is γ_B. We denote by

$$i_{L_0}(B) = i_{L_0}(\gamma_B), \quad \nu_{L_0}(B) = \nu_{L_0}(\gamma_B).$$

Theorem 4.1 ([189]) *There holds*

$$i_{L_0}(B) = m^{L_0}(B), \quad \nu_{L_0}(B) = n^{L_0}(B). \tag{5.96}$$

Proof It is easy to see for any $x, \xi \in W$ there holds

$$\varphi'(x)\xi = \int_0^1 \langle (P(t)x'(t) - Q(t)x(t))' + Q^T(t)x'(t) + R(t)x(t), \xi(t)\rangle\, dt. \tag{5.97}$$

and for $\xi, \zeta \in W$,

$$(\varphi''(0)\xi, \zeta) = \int_0^1 \langle (P(t)\xi'(t) - Q(t)\xi(t))' + Q^T(t)\xi'(t) + R(t)\xi(t), \zeta(t)\rangle \, dt, \tag{5.98}$$

where $\langle\cdot,\cdot\rangle$ is the Euclidean inner product in $\mathbb{R}^{2n}$. From (5.98), we see that the nullity space of this quadratic form is just the space of the solution of problem (5.92), which is one to one corresponding to the space of solutions of problem (5.93). A solution z of (5.93) should satisfy

$$z(t) = \gamma(t)z(0), \quad z(0) \in L_0, \quad z(1) \in L_0. \tag{5.99}$$

From (5.99) we get that

$$n^{L_0}(B) = \dim\ker V(1) = \nu_{L_0}(B), \tag{5.100}$$

$V(t)$ is defined in (5.1). So we get the second result of (5.96). In order to prove the first result of (5.96), following the ideas of [125], we set

$$L(x) = (P(t)x'(t) - Q(t)x(t))' + Q^T(t)x'(t) + R(t)x(t), \quad x \in W. \tag{5.101}$$

Then $L : W \subset L^2 \to L^2 := L^2([0,1], \mathbb{R}^{2n})$ is self-adjoint in L^2. The negative definite subspace W^- is a finite dimensional space. By definition, there holds

$m^{L_0}(B) =$ The number of negative eigenvalues of L counting by multiplicity.

Let $e_j = (a_1, \cdots, a_n)^T \in \mathbb{R}^n$ with $a_j = 1$ and $a_k = 0$ for $k \neq j$. Let x_j be a solution of the first equation of (5.92) satisfying the initial condition $x_j(0) = 0 \in \mathbb{R}^n$ and $\dot{x}_j(0) = P^{-1}e_j$. Then by setting $y(t) = P(t)x'(t) - Q(t)x(t)$ and $z = (x_j, y_j)^T \in \mathbb{R}^{2n}$ we get z a solution of the linear Hamiltonian system in (5.93) with the initial condition

$$(z_1(0), \cdots, z_n(0)) = \begin{pmatrix} 0 \\ I \end{pmatrix}. \tag{5.102}$$

Let $\gamma(t) = \begin{pmatrix} S(t) & V(t) \\ T(t) & U(t) \end{pmatrix} \in \mathcal{P}_\tau(2n)$ be the fundamental solution of the linear Hamiltonian system in (5.93). There holds

$$(z_1(t), \cdots, z_n(t)) = \begin{pmatrix} V(t) \\ U(t) \end{pmatrix}. \tag{5.103}$$

The solution space of the differential equation in (5.92) with the initial data $x(0) = 0$ is that spaned by the special solutions $x_1, \cdots, x_n$. □

We first prove the following result.

Lemma 4.2 *For $V(t)$ defined in* (5.103), *let*

$$\nu(s) = \dim\ker V(s) = n - \mathrm{rank} V(s).$$

There are only finitely many points $s \in (0, 1)$ such that $\nu(s) > 0$ and the following result is true

$$m^{L_0}(B) = \sum_{0<s<1} \nu(s). \tag{5.104}$$

Lemma 4.3 *For $0 < s \le 1$, we consider the following eigenvalue problem:*

$$\begin{cases} [P(t)x'(t) - Q(t)x(t)]' + Q^T(t)x'(t) + R(t)x = \lambda x(t) \\ x(0) = x(s) = 0. \end{cases} \tag{5.105}$$

Let $\lambda_1(s)$ be the smallest eigenvalue of the problem (5.105). *Then*

$$\lim_{s\to 0} \lambda_1(s) = +\infty. \tag{5.106}$$

Proof For $x \in W_0^{1,2}([0, s], \mathbb{R}^n)$, we consider the energy functional

$$\begin{aligned} E(x, x) &= \int_0^s \langle (P(t)x'(t) - Q(t)x(t))' + Q^T(t)x'(t) + R(t)x(t), x(t)\rangle\, dt \\ &= \int_0^s \{-\langle P(t)x'(t), x'(t)\rangle + 2\langle Q(t)x(t), x'(t)\rangle + \langle R(t)x(t), x(t)\rangle\}\, dt. \end{aligned} \tag{5.107}$$

Since for a function $f \in W^{1,2}([0, s], \mathbb{R})$ with $f(0) = 0$,

$$|f(t)| = |f(t) - f(0)| = \left|\int_0^t f'(u)\, du\right| \le \int_0^s |f'(u)|\, du \le \sqrt{s}\|f'\|_2,$$

we have

$$\|x\|_2 \le s\|x'\|_2,\ \|x\|_2 := \left(\int_0^s |x|^2\, du\right)^{1/2}.$$

Thus there exist some constant $a_k > 0,\ k = 1, 2, 3$ such that

$$\begin{aligned} E(x, x) &\ge a_1\|x'\|_2^2 - a_2\|x\|_2\|x'\|_2 - a_3\|x\|_2^2 \\ &\ge a_1\|x'\|_2^2 - \tfrac{a_2\|x\|_2^2}{s} - \tfrac{a_2 s\|x'\|_2^2}{2} - a_3\|x\|_2^2 \\ &\ge \left(a_1 - \tfrac{a_2 s}{2}\right)\|x'\|_2^2 - \left(\tfrac{a_2}{s} + a_3\right)\|x\|_2^2 \\ &\ge \left(a_1 - \tfrac{a_2 s}{2}\right)\tfrac{\|x\|_2^2}{s^2} - \left(\tfrac{a_2}{s} + a_3\right)\|x\|_2^2 \end{aligned}$$

This implies that

$$\lim_{s \to 0} \frac{E(x,x)}{\|x\|_2^2} = +\infty.$$

□

Proof of Lemma 4.2 For $0 < s \leq 1$, let $\lambda_1(s) \leq \lambda_2(s) \leq \cdots$ be the eigenvalues of the problem (5.105). It is well known that $\lambda_j(s)$ is a strictly decreasing function. (see, for example, [291]). Lemma 4.3 implies that $\lambda_j(s)$ decreases from $+\infty$ to $\lambda_j(1)$.

The given number $\lambda = 0$ is an eigenvalue of (5.105) if and only if there are not all zero constants $c_1, \cdots, c_n$ such that $x(t) = c_1 x_1(t) + \cdots + c_n x_n(t)$ satisfies $x(s) = 0 \in \mathbb{R}^n$. This happens if and only if $V(s)c = 0$ for some non-zero vector $c \in \mathbb{R}^n$. Thus the condition for $\lambda = 0$ to be an eigenvalue of (5.105) is that $\det V(s) = 0$. Furthermore, the multiplicity of the eigenvalue $\lambda = 0$ is $\nu(s)$.

On the other hand, $\lambda = 0$ is an eigenvalue of (5.105) if and only if there is λ_j such that $\lambda_j(s) = 0$. The multiplicity of the eigenvalue 0 is the number of j such that $\lambda_j(s) = 0$.

Since $\lambda_j(s)$ is strictly decreasing, for each j there can be at most one s such that $\lambda_j(s) = 0$. Clearly such a value $s \in (0,1)$ exists if and only if $\lambda_j(1) < 0$. Of course, the $\lambda_j(1)$ are the eigenvalues (5.105) with $s = 1$, and only finitely many of these eigenvalues are less than 0. Therefore there are only finitely many points $s \in (0,1)$ such that $\nu(s) > 0$, and $\sum\limits_{0<s<1} \nu(s)$ equals the number of negative eigenvalues of (5.105) with $s = 1$. □

Continue the proof of Theorem 4.1 We recall that

$$\mathcal{Q}(t) = [U(t) - \sqrt{-1}V(t)][U(t) + \sqrt{-1}V(t)]^{-1} \in U(n),$$

and $\mathcal{Q}(0) = I_n$. Let $\chi_j(t) = e^{2\sqrt{-1}\theta_j(t)}$ be the eigenvalues of $\mathcal{Q}(t)$ satisfying

$$\theta_1(t) \leq \theta_2(t) \leq \cdots \leq \theta_n(t), \quad \theta_j(0) = 0.$$

$\theta_j(t)$ have continuous right hand side derivatives. (see, for example, [55])

Since $-P(t) > 0$, if $\theta_j(s_0) = k\pi,\ k \in \mathbb{Z}$, by a result of [55], $\theta'(s_0) > 0$. So by continuity, we have

$$\theta_j(s_1) < \theta_j(s_0) < \theta_j(s_2), \ 0 \leq s_1 < s_0 < s_2 \leq 1.$$

From the proof of Lemma 4.3, $\nu(s) > 0$ only when $\det V(s) = 0$, and this happens when some phase angle $\theta_j(s)$ is a multiple of π. The multiplicity $\nu(s)$ equals the number of phase angles $\theta_j(s)$ that are multiples of π. So the number

$$E\left(\frac{\theta_j(1)}{\pi}\right) = E\left(\frac{\theta_j(1)-\theta_j(0)}{\pi}\right)$$

equals the number of times that is a multiple of π, for $0 < s < 1$. This implies that

$$\sum_{0<s<1} \nu(s) = \sum_{j=1}^{n} E\left(\frac{\theta_j(1)-\theta_j(0)}{\pi}\right) = i_{L_0}(\gamma). \tag{5.108}$$

Comparing (5.30) (note where θ_j should be replaced by $2\theta_j$) and (5.104), we get the first result of (5.96). □

From (5.108), we see that $i_{L_0}(\gamma)$ in some sense is the intersection number (with signs) of the Lagrangian path $\omega(t) = \gamma(t)L_0$ with the constant Lagrangian path L_0. For the Lagrangian intersection index, one can refer to the paper [77]. We note that the definition of the index here is essentially different from those in [77, 259] and [260]. Some similar results of Theorem 4.1 with the periodic boundary condition were proved in [7] and [228].

Theorem 4.4 ([189]) *Suppose $\gamma \in \mathcal{P}_\tau(2n)$ is a fundamental solution of the following linear Hamiltonian system*

$$\dot{z}(t) = JB(t)z(t), \quad z(t) \in \mathbb{R}^{2n}, \tag{5.109}$$

where $B(t) = \begin{pmatrix} S_{11} & S_{12} \\ S_{21} & S_{22} \end{pmatrix}$ is symmetric with $n \times n$ blocks S_{jk}. If $S_{22} > 0$ (positive definite), there holds

$$i_{L_0}(\gamma) \geq 0. \tag{5.110}$$

Proof We choose P, Q, R in (5.94) such that

$$P = -S_{22}^{-1}, \quad Q = S_{22}^{-1}S_{21}, \quad R = S_{12}S_{22}^{-1}S_{21} - S_{11},$$

and define a quadratic form (5.98) by P, Q, R. By Theorem 4.1, we have

$$i_{L_0}(\gamma) = m^{L_0}(B) \geq 0.$$

□

Corollary 4.5 ([189]) *Suppose $\gamma \in \mathcal{P}_\tau(2n)$ is a fundamental solution of* (5.109) *with $B(t) > 0$. There holds*

$$i_L(\gamma) \geq 0. \tag{5.111}$$

Proof There exists an orthogonal symplectic matrix M such that

$$i_L(\gamma) = i_{L_0}(M\gamma M^T).$$

Equation (5.111) follows from that $M\gamma M^T$ is the fundamental solution of $\dot{w}(t) = JMB(t)M^T w(t)$, and $MB(t)M^T > 0$. □

5.5 Saddle Point Reduction Formulas

We consider the Hamiltonian system with lagrangian boundary conditions

$$\begin{cases} \dot{z} = JH'(t, z), & z \in \mathbb{R}^{2n}, \\ z(0) \in L, & z(1) \in L, \end{cases} \tag{5.112}$$

where L is a Lagrangian subspace of $\mathbb{R}^{2n}$, $H \in C^2([0, 1] \times \mathbb{R}^{2n}, \mathbb{R})$ satisfies the following condition

$$|H''(t, z)| \leq C(H), \quad \forall (t, z) \in [0, 1] \times \mathbb{R}^{2n}, \quad C(H) \text{ is a constant}. \tag{5.113}$$

The linear Hamiltonian system

$$\begin{cases} \dot{z} = JB(t)z, & x \in \mathbb{R}^{2n}, \\ z(0) \in L, & z(1) \in L \end{cases} \tag{5.114}$$

is a special case of (5.112) with $H(t, z) = \frac{1}{2}(B(t)z, z)$, for any symmetrical matrix function $B \in C([0, 1], \mathcal{L}_s(\mathbb{R}^{2n}))$. Here we use $\mathcal{L}_s(\mathbb{R}^{2n})$ to denote the set of symmetrical matrices.

We note that there is an orthogonal symplectic matrix $P \in \mathrm{Sp}(2n) \cap O(2n)$ such that $PL_0 = L$. By changing the variables in (5.112) by $z = Pw$, we get the following problem

$$\begin{cases} \dot{z} = JH_L'(t, z), & z \in \mathbb{R}^{2n}, \\ z(0) \in L_0, & z(1) \in L_0, \end{cases} \tag{5.115}$$

where $H_L(t, z) = H(t, Pz)$. H_L still satisfies the condition (5.113). In the special case (5.114), we get

$$\begin{cases} \dot{z} = JB_L(t)z, & z \in \mathbb{R}^{2n}, \\ z(0) \in L_0, & z(1) \in L_0 \end{cases} \tag{5.116}$$

with $B_L(t) = P^T B(t) P$.

We consider the following functional

$$f(z)=\int_0^1\left[\frac{1}{2}(-J\dot{z},z)-H(t,z)\right]dt,\quad z\in W_L,\tag{5.117}$$

where $W_L=\{z=(x,y)^T\in W^{1,2}([0,1],\mathbb{R}^{2n})|\ z(0),\ z(1)\in L\}\subset L^2$. It is a classical result that any critical point of f is a solution of the problem (5.112). We denote the norm and inner product in L^2 by $\|\cdot\|_2$ and $\langle\cdot,\cdot\rangle_2$, respectively. In L^2 we define a self-adjoint operator A by

$$\langle Az,z\rangle_2=\int_0^1(-J\dot{z},z)\,dt,\quad\forall\, z\in\text{dom}A=W_L,\tag{5.118}$$

and define a functional g by

$$g(z)=\int_0^1 H(t,z)\,dt.\tag{5.119}$$

Thus there holds

$$f(z)=\frac{1}{2}\langle Az,z\rangle_2-g(z),\quad\forall\, z\in\text{dom}A=W_L.\tag{5.120}$$

It is clear that

$$g'(z)=H'(t,z(t)),\quad dg'(z)\xi=H''(t,z(t))\xi,$$

and there exists a constant $c(H)>0$ such that

$$\|dg'(z)\|_2:=\|dg'(x)\|_{L^2}\le c(H).\tag{5.121}$$

The kernel of A is the space $E_0=L$ of the constant curves. The range of A is closed and its resolution is compact. We defined an invertible operator A_0 in L^2 by

$$A_0z=Az+P_0z,\quad z\in W_L,\tag{5.122}$$

where $P_0:L^2\to E_0=L$ is the projection map. The spectrum $\sigma(A_0)$ of A_0 is a point spectrum. $\sigma(A)=\{k\pi|\,k\in\mathbb{Z}\}=\pi\mathbb{Z}$. For every eigenvalue $k\pi$, the eigensubspace E_k are a n-dimensional subspace. If $L=L_0$, the eigensubspace

$$\begin{aligned}E_k=\text{span}\{\alpha_1,\cdots,\alpha_n\},\ \ \alpha_j=(a_1,\cdots,a_n,b_1,\cdots,b_n)^T,\\ a_i=b_i=0,i\ne j,\ a_j=\sin k\pi t,\ b_j=-\cos k\pi t.\end{aligned}\tag{5.123}$$

$\sigma(A_0)=1\cup\pi\mathbb{Z}\setminus\{0\}$. The eigensubspace of $1\in\sigma(A_0)$ is dimensional n space E_0, and any other eigenvalue $k\pi$ possesses an n-dimensional eigensubspace E_k. For the general L, the numbers of dimensions are the same, but the eigensubspces should be transformed by a suitable orthogonal symplectic matrix.

We choose $c(H) > \pi$ satisfies $c(H) \notin \sigma(A_0)$. Denote by E_λ the spectral resolution of the self-adjoint operator A_0, we define the projections on L^2 by

$$\mathcal{P} = \int_{-c(H)}^{c(H)} dE_\lambda, \quad \mathcal{P}^+ = \int_{c(H)}^{+\infty} dE_\lambda, \quad \mathcal{P}^- = \int_{-\infty}^{-c(H)} dE_\lambda$$

The Hilbert space L^2 possesses an orthogonal decomposition

$$L^2 = L^+ \oplus L^- \oplus X,$$

where $L^\pm = \mathcal{P}^\pm L^2$, and $X = \mathcal{P}L^2$ is a finite dimensional space with $\dim X = 2d + n = (2m+1)n$ for some $m \in \mathbb{N} \cup \{0\}$.

Theorem 5.1 *There exist a function $a \in C^2(X, \mathbb{R})$ and an injection map $u \in C^1(X, L^2)$ such that $u : X \to W_L$satisfies the following conditions:*

1° *The map u has the form $u(x) = w(x) + x$ with $\mathcal{P}w(x) = 0$.*
2° *The function a satisfies*

$$a(x) = f(u(x)) = \tfrac{1}{2}\langle Au(x), u(x)\rangle_2 - g(u(x)),$$
$$a'(x) = Ax - \mathcal{P}g'(u(x)) = Au(x) - g'(u(x)),$$
$$a''(x) = A\mathcal{P} - \mathcal{P}dg'(u(x))u'(x) = [A - dg'(u(x))]u'(x).$$

And a' is globally Lipschitz continuous.
3° *$x \in X$ is a critical point of a, if and only if $z = u(x)$ is a critical point of f, i.e., $z = u(x)$ is a solution of the problem* (5.112).
4° *If $g'(z) = Bz$ for all $z \in L^2$, where B is the induced linear operator on L^2 from a constant symmetrical matrix $B(t) \equiv B$ defined on $\mathbb{R}^{2n}$, then $a(x) = \frac{1}{2}\langle (A - B)x, x\rangle_2$.*
5° *If Σ is a topological space, for any $\sigma \in \Sigma$ the functional $g : \Sigma \times L^2 \to \mathbb{R}$ satisfies $g(\sigma, \cdot) \in C^1(L^2, \mathbb{R})$, $g' \in C(\Sigma \times L^2, \mathbb{R})$, and the inequality* (5.121) *with the constant $c(H)$ being independent from $\sigma \in \Sigma$. Then the corresponding map $u = u(\sigma, x)$ and its derivative $u'_x(\sigma, x)$ with respect to x are all continuous.*
6° *There holds*

$$\dim\ker a''(x) = \text{the nullity of } f''(u(x)) = \nu_L(\gamma),$$

where γ is the fundamental solution of the linearized system (5.114) *with $B(t) = H''(t, z(t))$ with $z(t) = u(x)(t)$.*

Proof Up to the author's knowledge, the saddle point reduction method was introduced by H.Amann in [4]. For the periodic boundary condition case, a theorem like Theorem 5.1 was proved by H. Amann and E. Zehnder in their celebrated paper [5] by using the monotone operator theory. Then K.C. Chang gave a simple and more direct proof in [33]. A proof of combination of their ideas was given in [223] and [227]. The key ingredient of the proof is to use the contraction mapping theorem and the implicit function theorem. Define operators by

$$S^+ = \int_{c(H)}^{+\infty} \lambda^{-1/2} dE_\lambda, \quad S^- = \int_{-\infty}^{-c(H)} (-\lambda)^{-1/2} dE_\lambda, \quad R = \int_{-c(H)}^{+c(H)} |\lambda|^{-1/2} dE_\lambda.$$

By noting that $S^\pm g_0'(v)$ are contraction mappings, where $g_0(z) = g(z) + \frac{1}{2}\langle P_0 z, z\rangle_2$, one can solve the equations

$$z^\pm = \pm S^\pm g_0'(S^+ z^+ + S^- z^- + Rx), \quad \forall x \in X$$

and get the mappings $z^\pm = z^\pm(x)$. Define

$$u(x) = w(x) + x, \quad w(x) = S^+ z^+(R^{-1}x) + S^- z^-(R^{-1}x).$$

The mapping $u : X \to W_L$ satisfies all conditions $1^o - 6^o$ in Theorem 5.1. We omit the details here. One can find the details in the references mentioned above (for example, see [223]). □

For the special case (5.114), it induces a symmetric operator on L^2 by

$$\langle Bz, w\rangle_2 = \int_0^1 (B(t)z, w)\, dt, \quad \forall z, w \in L^2.$$

Then the functional $f(z)$ in (5.117) is the following one:

$$f(z) = \frac{1}{2}\langle (A - B)z, z\rangle_2, \quad \forall z \in W_L. \tag{5.124}$$

By Theorem 5.1, we obtain an injection map $u : X \to W_L$ and a smooth functional $a \in C^\infty(X, \mathbb{R})$ defined by

$$a(x) = f(u(x)), \quad x \in X. \tag{5.125}$$

Let $\dim X = 2d + n$. Note that the origin of X as a critical point of a corresponds to the origin of W_L as a critical point of f. Denote by $m^*(B)$ for $* = +, 0, -$ the positive, null, and negative Morse indices of the functional a at the origin respectively, i.e., the total multiplicities of positive, zero, and negative eigenvalues of the matrix $a''(0)$ respectively. The following is the main result of this section.

Theorem 5.2 ([189]) *For any $L \in \Lambda(n)$,*

$$\begin{cases} m^-(B) = d + i_L(B) + n, \\ m^0(B) = \nu_L(B), \\ m^+(B) = d - i_L(B) - \nu_L(B). \end{cases} \tag{5.126}$$

Proof We only prove the first result in (5.126) in the special case $L = L_0$. For any C^1 curve $\gamma \in \mathcal{P}_\tau(2n)$, the matrix $J\dot{\gamma}\gamma^{-1}$ is symmetric. We denote by $\mathcal{L}_s(\mathbb{R}^{2n})$ the set of symmetric matrices. We let $B(t) = -J\dot{\gamma}(t)\gamma(t)^{-1} \in \mathcal{L}_s(\mathbb{R}^{2n})$. Then γ is a fundamental solution of the following linear Hamiltonian system

$$\dot{z} = JB(t)z.$$

□

By the same arguments as in the proof of Lemma 5.2.2 of [223], we have the following result.

Lemma 5.3 *Suppose $\gamma_0, \gamma_1 \in \mathcal{P}_1(2n)$ possess common end point $\gamma_0(1) = \gamma_1(1)$. Suppose $\gamma_0 \sim_{L_0} \gamma_1$ in the sense of Definition* 1.8 *via a L_0-homotopy $\delta : [0,1] \times [0,1] \to \mathrm{Sp}(2n)$ with $\delta(s,t) = \delta(s)(t)$ as defined in Definition* 1.8 *such that $\delta(\cdot,1)$ is contractible in* Sp(2*n*). *Then the homotopy can be modified to fix the end points all the time, i.e., $\delta(s,1) = \gamma_0(1)$ for all $0 \le s \le 1$.*

Lemma 5.4 *For the matrix function $Q(t)$ defined in* (5.3) *with $\gamma \in \mathcal{P}_\tau(2n)$, let $\Delta : [0,1] \to \mathbb{R}$ be a continuous function satisfying*

$$\det Q(t) = \exp(2\sqrt{-1}\Delta(t)), \quad \forall t \in [0,1]. \tag{5.127}$$

We define the rotation number of γ by

$$r(\gamma) = \Delta(1) - \Delta(0). \tag{5.128}$$

Then $r(\gamma)$ depends only on γ but not on the choice of the function Δ.

If $\gamma_0, \gamma_1 \in \mathcal{P}_\tau(2n)$ possess common end point $\gamma_0(1) = \gamma_1(1)$, then $r(\gamma_0) = r(\gamma_1)$ if and only if $\gamma_0 \sim_{L_0} \gamma_1$ with fixed end points.

Proof We only prove the second part of Lemma 5.4. We note that $\gamma_0 \sim_{L_0} \gamma_1$ with fixed end points if and only if $\gamma = \gamma_1^{-1} * \gamma_0$ is contractible in $\mathrm{Sp}(2n)$. The latter holds if and only if $\Delta(\gamma) = 0$. Here we define the path γ_1^{-1} by $\gamma_1^{-1}(t) = \gamma_1(1-t)$.

□

The following result was proved in [223].

Lemma 5.5 *Let $B \in C([0,1] \times [0,1], \mathcal{L}_s(\mathbb{R}^{2n}))$ and denote by $B_s(\cdot) = B(s,\cdot)$. Denote the Morse indices of the functional a_s on X corresponding to the system* (5.114) *with $B(t)$ replaced by $B_s(t)$ at $x = 0$ by $m^-(B_s)$, $m^0(B_s)$ and $m^+(B_s)$, where a_s is defined in Theorem* 5.1 *with the same X. Suppose*

$$m^0(B_s) = m^0(B_0), \quad \forall\, s \in [0,1]. \tag{5.129}$$

Then

$$m^-(B_s) = m^-(B_0), \quad m^+(B_s) = m^+(B_0), \quad \forall\, s \in [0, 1]. \tag{5.130}$$

Remark By Theorem 5.1(5°) and the compactness, we can choose the same X for all a_s with $0 \le s \le 1$ in Lemma 5.5.

Continue the proof of Theorem 5.2 Firstly, we assume $\nu_{L_0}(\gamma) = 0$ and $i_{L_0}(\gamma) = k$. By choosing a path β in $\mathrm{Sp}(2n)^*_{L_0}$ connects $\gamma(1)$ with M^+ or M^-, we get a path $\gamma_0 = \beta * \gamma$, and there holds $\gamma \sim_{L_0} \gamma_0$. By definition, we have

$$i_{L_0}(\gamma) = i_{L_0}(\gamma_0) = \frac{r(\gamma_0)}{\pi} - \frac{n}{2}. \tag{5.131}$$

Assume $\delta_s(\cdot) \in \mathcal{P}_\tau(2n)$ is the homotopy map between γ_0 and γ. We can perturb it slightly so that $\delta_s(t)$ is C^1. Then by setting $B_s(t) = -J\dot{\delta}_s(t)\delta_s(t)^{-1}$ as in Lemma 5.5, we have $m^0(B) = m^0(B_0) = 0$ and

$$m^-(B) = m^-(B_0), \quad m^+(B) = m^+(B_0), \tag{5.132}$$

where B and B_0 corresponding to γ and γ_0 respectively. We construct a path $\gamma_k \in \mathcal{P}_\tau(2n)^*$ with $i_{L_0}(\gamma_k) = k$ by

$$\begin{aligned}\gamma^k(t) = &\begin{pmatrix} \cos(k\pi + \frac{\pi}{2})t & -\sin(k\pi + \frac{\pi}{2})t \\ \sin(k\pi + \frac{\pi}{2})t & \cos(k\pi + \frac{\pi}{2})t \end{pmatrix} \diamond \\ &\diamond \begin{pmatrix} \cos\frac{\pi t}{2} & -\sin\frac{\pi t}{2} \\ \sin\frac{\pi t}{2} & \cos\frac{\pi t}{2} \end{pmatrix} \diamond \cdots \diamond \begin{pmatrix} \cos\frac{\pi t}{2} & -\sin\frac{\pi t}{2} \\ \sin\frac{\pi t}{2} & \cos\frac{\pi t}{2} \end{pmatrix}.\end{aligned}$$

Corresponding to this path, the symmetric coefficient matrix of the linear Hamiltonian system is

$$B^k(t) = (k + \frac{1}{2})\pi I_2 \diamond \frac{\pi}{2} I_2 \diamond \cdots \diamond \frac{\pi}{2} I_2.$$

By Lemma 5.4, we can construct a C^1 homotopy map $\delta^k_s(t)$ with fixed end points between γ_0 and γ^k. Then by setting $B^k_s(t) = -J\dot{\delta}^k_s(t)\delta^k_s(t)^{-1}$ as in Lemma 5.5, we get $m^0(B^k) = m^0(B_0) = 0$ and

$$m^-(B^k) = m^-(B_0), \quad m^+(B^k) = m^+(B_0). \tag{5.133}$$

We now only need to consider the case when $n = 1$ and $B(t) = \mu I_2$ with constant $\mu = (k + 1/2)\pi$ for $k \in \mathbb{Z}$. By Theorem 5.1(4°), we have

$$a^\mu(x) = \frac{1}{2}\langle (A - \mu I_2)x, x\rangle_2, \quad x \in X. \tag{5.134}$$

We can choose

$$X = \bigoplus_{j=-d_1}^{d_1} E_j$$

with $E_j = (\sin j\pi t, \cos j\pi t)^T \mathbb{R}$. By direct computations, we have

$$m^-(B^\mu) = d_1 + k + 1. \tag{5.135}$$

Now by Theorem 1.13 and (5.135), we have

$$m^-(B^k) = d + i_{L_0}(\gamma) + n. \tag{5.136}$$

Equation (5.126) follows from (5.132), (5.133) and (5.136).

For the general case, we assume $\nu_{L_0}(\gamma) > 0$. Then by Lemma 1.10 and Theorem 1.11, we have the perturbed paths $\gamma_s \in \mathcal{P}_\tau(2n)$ with $\gamma_s \in \mathcal{P}_\tau(2n)^*_{L_0}$ for $s \neq 0$. Setting $B_s(t) = -J\dot{\gamma}_s(t)\gamma_s(t)^{-1}$. Note that $B_0(t) = B(t)$. By Theorem 5.1, we get a functional a_s on a finite dimensional space X. It is easy to see $a_s \to a$ under the C^2 topology. By Lemma 5.5 and the above case, we have

$$m^+(B_s) = d - i_{L_0}(\gamma_s), \quad m^0(B_s) = 0, \quad m^-(B_s) = d + i_{L_0}(\gamma_s) + n, \quad s \neq 0.$$

By the perturbed theory and Theorem 1.11, if $0 < s \leq 1$,

$$m^+(B) \geq m^+(B_{-s}) - m^0(B) = d - i_{L_0}(\gamma_{-s}) - \nu_{L_0}(\gamma), \tag{5.137}$$

$$m^+(B) \leq m^+(B_s) = d - i_{L_0}(\gamma_s) = d - i_{L_0}(\gamma_{-s}) - \nu_{L_0}(\gamma), \tag{5.138}$$

$$m^-(B) \leq m^-(B_{-s}) = d + i_{L_0}(\gamma_{-s}) + n, \tag{5.139}$$

$$m^-(B) \geq m^-(B_s) - m^0(B) = d + i_{L_0}(\gamma_s) + n - \nu_{L_0}(\gamma) = d + i_{L_0}(\gamma_{-s}) + n. \tag{5.140}$$

From (5.137), (5.138), (5.139) and (5.140), we have

$$m^-(B) = d + i_{L_0}(\gamma_{-s}) + n, \quad m^+(B) = d - i_{L_0}(\gamma_{-s}) - \nu_{L_0}(\gamma). \tag{5.141}$$

If we replace γ_{-s} by any path $\tilde{\gamma} \in \mathcal{P}_\tau(2n)^*_{L_0}$ sufficiently closed to γ and γ_s by any path $\bar{\gamma} \in \mathcal{P}_\tau(2n)^*_{L_0}$, by the estimates (5.139) and (5.140), we get

$$|i_{L_0}(\bar{\gamma}) - i_{L_0}(\tilde{\gamma})| \leq \nu_{L_0}(\gamma). \tag{5.142}$$

But by Theorem 1.11, we have

$$i_{L_0}(\gamma_s) - i_{L_0}(\gamma_{-s}) = \nu_{L_0}(\gamma).$$

Thus there holds

$$\begin{aligned} i_{L_0}(\gamma_{-s}) &= \min\{i_{L_0}(\tilde{\gamma});\ \tilde{\gamma} \in \mathcal{P}_\tau(2n)^*_{L_0} \text{ sufficiently closed to } \gamma\}, \\ i_{L_0}(\gamma_s) &= \max\{i_{L_0}(\tilde{\gamma});\ \tilde{\gamma} \in \mathcal{P}_\tau(2n)^*_{L_0} \text{ sufficiently closed to } \gamma\}. \end{aligned} \tag{5.143}$$

So we have proved (5.28), (5.29) and

$$i_{L_0}(\gamma) = i_{L_0}(\gamma_{-s}). \tag{5.144}$$

Equation (5.126) follows from (5.141) and (5.144). □

Proof of Theorem 1.15 We only need to prove the first result of (5.34). We recall that the condition is that $\gamma_0 \sim_L \gamma_1$. We shall to prove

$$i_L(\gamma_0) = i_L(\gamma_1). \tag{5.145}$$

By perturbed slightly with fixed end points, we assume $\gamma_j \in C^1([0,1], \mathrm{Sp}(2n))$ and the homotopy $\delta_s(\cdot) \in C^1([0,1], \mathrm{Sp}(2n))$ for all $0 \le s \le 1$. As in the proof of Theorem 5.2, we take $B_s(t) = -J\dot{\delta}_s(t)\delta_s(t)^{-1}$. Then by the condition and Lemma 5.5, we have $m^0(B_s) = m^0(B_0)$ for all $s \in [0,1]$. Thus we have

$$m^-(B_s) = m^-(B_0), \quad \forall\, s \in [0,1].$$

So by the result of Theorem 5.2, we have

$$d + n + i_L(B_s) = d + n + i_L(B_0).$$

Take $s = 1$, we get (5.145). The proof is completed. □

5.6 Galerkin Approximation Formulas for L-Index

The eigenspace E_k of the operator $A = -J\frac{d}{dt}$ in the domain $W^{1,2}_{L_0}([0,1], \mathbb{R}^{2n}) := \{z \in W^{1,2}([0,1], \mathbb{R}^{2n}) : z(0) \in L_0, z(1) \in L_0\}$ can be written as

$$\begin{aligned} E_k = -J\exp(k\pi t J)a_k &= -J(\cos(k\pi t)I_{2n} + J\sin(k\pi t))a_k, \\ a_k &= (a_{k1}, \cdots, a_{kn}, 0, \cdots, 0) \in \mathbb{R}^{2n}. \end{aligned}$$

We define a Hilbert space $\mathcal{W}_{L_0} = W^{1/2,2}_{L_0}([0,1], \mathbb{R}^{2n}) \subset \bigoplus_{k\in\mathbb{Z}} E_k$ with L_0 boundary conditions by

$$\mathcal{W}_{L_0} = \{z \in L^2 \mid z(t) = \sum_{k\in\mathbb{Z}} -J\exp(k\pi t J)a_k,\ \|z\|^2 := \sum_{k\in\mathbb{Z}}(1+|k|)|a_k|^2 < \infty\}.$$

We denote its inner product by $\langle\cdot,\cdot\rangle$. By the well-known Sobolev embedding theorem, for any $s \in [1,+\infty)$, there is a constant $C_s > 0$ such that

$$\|z\|_{L^s} \leq C_s\|z\|, \quad \forall z \in \mathcal{W}_{L_0}.$$

For any Lagrangian subspace $L \in \Lambda(n)$, suppose $P \in \mathrm{Sp}(2n) \cap O(2n)$ such that $L = PL_0$. Then we define $\mathcal{W}_L = P\mathcal{W}_{L_0}$.

We denote by $\mathcal{W}_{L_0}^m = \bigoplus\limits_{k=-m}^{m} E_k = \left\{z|z(t) = \sum\limits_{k=-m}^{m} -J\exp(k\pi tJ)a_k\right\}$ the finite dimensional truncation of $\mathcal{W}_{L_0}$, and $\mathcal{W}_L^m = P\mathcal{W}_{L_0}^m$.

Let $P^m = P_L^m : \mathcal{W}_L \to \mathcal{W}_L^m$ the orthogonal projection for $m \in \mathbb{N}$. Then $\Gamma = \{P^m;\ m \in \mathbb{N}\}$ is a Galerkin approximation scheme with respect to A defined in (5.148) below, i.e., there hold

$$P^m \to I \text{ strongly as } m \to \infty$$

and

$$P^m A = AP^m.$$

In this section we still consider the following problem

$$\begin{cases} \dot{z} = JH'(t,z), & z \in \mathbb{R}^{2n}, \\ z(0) \in L, & z(1) \in L, \end{cases} \tag{5.146}$$

with H satisfying

$$|H''(t,z)| \leq a(1+|z|^p), \quad \forall\,(t,z) \in \mathbb{R}\times\mathbb{R}^{2n}, \text{ for some } a > 0, \quad p > 1. \tag{5.147}$$

We consider the functional on $\mathcal{W}_L$

$$f(z) = \int_0^1 \left[\frac{1}{2}(-J\dot{z},z) - H(t,z)\right]dt = \frac{1}{2}\langle Az,z\rangle - g(z), \quad z \in \mathcal{W}_L, \tag{5.148}$$

A critical point of f on $\mathcal{W}_L$ is a solution of (5.146). For a critical point $z = z(t)$, we denote $B(t) = H''(t,z(t))$ and define an operator B on $\mathcal{W}_L$ by

$$\langle Bz,w\rangle = \int_0^1 (B(t)z,w)dt.$$

Using the Floquet theory we have

$$\nu_L(B) = \dim\ker(A-B). \tag{5.149}$$

For $\delta > 0$, we denote by $m_\delta^*(\cdot)$, $* = +, 0, -$ the dimension of the total eigenspace corresponding to the eigenvalue λ belonging to $[\delta, +\infty)$, $(-\delta, \delta)$ and $(-\infty, -\delta]$ resp, and denote by $m^*(\cdot)$, $* = +, 0, -$ the dimension of the total eigenspace corresponding to the eigenvalue λ belonging to $(0, +\infty)$, $\{0\}$ and $(-\infty, 0)$ resp. For any adjoint operator Q, we denote $Q^\sharp = (Q|_{ImQ})^{-1}$, and we also denote $P^m Q P^m = (P^m Q P^m)|_{\mathcal{W}_L^m}$. The following result is adapted from [95] where the periodic boundary condition was considered (see also [227]).

Theorem 6.1 ([192]) *For any $B(t) \in C([0, 1], \mathcal{L}_s(\mathbb{R}^{2n}))$ with the L-index pair $(i_L(B), \nu_L(B))$ and any constant $0 < \delta \le \frac{1}{4}\|(A - B)^\sharp\|$, there exists $m_0 > 0$ such that for $m \ge m_0$, we have*

$$\begin{aligned} m_\delta^+(P^m(A-B)P^m) &= mn - i_L(B) - \nu_L(B) \\ m_d^-(P^m(A-B)P^m) &= mn + i_L(B) + n \\ m_\delta^0(P^m(A-B)P^m) &= \nu_L(B). \end{aligned} \tag{5.150}$$

Proof We follow the ideas of [95].

Step 1. There is $m_1 > 0$ such that for $m \ge m_1$

$$\dim\ker(P^m(A-B)P^m) \le \dim\ker(A-B). \tag{5.151}$$

In fact, by contradiction it is easy to show that there is a constant $m_2 > 0$ such that for $m \ge m_2$

$$\dim P^m \ker(A-B) = \dim\ker(A-B). \tag{5.152}$$

Since B is compact, there is $m_1 \ge m_2$ such that for $m \ge m_1$

$$\|(I - P^m)B\| \le 2\delta.$$

Take $m \ge m_1$, let $\mathcal{W}_L^m = P^m\ker(A-B) \oplus Y^m$, then $Y^m \subset Im(A-B)$. For $y \in Y^m$ we have

$$y = (A-B)^\sharp(A-B)y = (A-B)^\sharp(P^m(A-B)P^m y + (P^m - I)By).$$

This implies

$$\|y\| \le \frac{1}{2\delta}\|P^m(A-B)P^m y\|, \quad \forall\ y \in Y^m. \tag{5.153}$$

By (5.152) and (5.153) we have (5.151).

Step 2. We distinguish two cases.

Case 1. $\nu_L(B) = 0$. By (5.149) and step 1, for $m \ge m_1$ we obtain that

$$m^0(P^m(A-B)P^m) = \dim\ker(A-B) = 0.$$

Since B is compact, there exists $m_3 \geq m_1$ such that for $m \geq m_3$

$$\|(I - P^m)B\| \leq \frac{1}{2}\|(A - B)^\sharp\|^{-1}.$$

Then $P^m(A - B)P^m = (A - B)P^m + (I - P^m)BP^m$ implies that

$$\|P^m(A - B)P^m z\| \geq \frac{1}{2}\|(A - B)^\sharp\|^{-1}\|z\|, \quad \forall\, z \in \mathcal{W}_L^m.$$

Thus the eigen-subspace $M_\delta^*(P^m(A - B)P^m)$ corresponding to the eigenvalue λ belonging to the intervals defined as the symbol $m_\delta^*(P^m(A - B)P^m)$ mentioned above, and the eigen-subspace $M^*(P^m(A - B)P^m)$ satisfy

$$M_\delta^*(P^m(A - B)P^m) = M^*(P^m(A - B)P^m), \quad \text{for } * = +, 0, -.$$

By Theorem 5.2, there is $m_0 \geq m_3$ such that for $m \geq m_0$ the relation (5.150) holds.

Case 2. $\nu_L(B) > 0$. By step 1, it is easy to show that there exists $m_4 > 0$ such that for $m \geq m_4$

$$m_\delta^0(P^m(A - B)P^m) \leq \nu_L(B). \tag{5.154}$$

Let $\gamma \in \mathcal{P}_\tau(2n)$ be the fundamental solution of the linear Hamiltonian system

$$\dot{z} = JB(t)z.$$

Let $\gamma_s,\ 0 \leq s \leq 1$ be the perturbed path defined by Lemma 1.10. Define

$$B_s(t) = -J\dot{\gamma}_s(t)\gamma_s(t)^{-1}, \ t \in [0, 1].$$

Let B_s be the compact operator defined as B corresponding to $B_s(t)$. For $s \neq 0$, there holds $m^0(A - B_s) = 0$ and $\|B_s - B\| \to 0$ as $s \to 0$. If $s \in (0, 1]$, we have

$$i_L(\gamma_s) - i_L(\gamma_{-s}) = \nu_L(\gamma) = \nu_L(B),\ i_L(\gamma_{-s}) = i_L(B) = i_L(\gamma). \tag{5.155}$$

Choose $0 < s < 1$ such that $\|B - B_{\pm s}\| \leq \frac{\delta}{2}$. By case 1, (5.154) and (5.155) and the fact that

$$P^m(A - B_\pm)P^m = P^m(A - B)P^m + P^m(B - B_\pm)P^m$$

there exists $m_0 \geq m_4$ such that for $m \geq m_0$

$$m_\delta^+(P^m(A - B)P^m) \leq m^+(P^m(A - B_s)P^m) = mn - i_L(B) - \nu_L(B)$$

$$\begin{aligned} m_\delta^+(P^m(A-B)P^m) &\geq m^+(P^m(A-B_{-s})P^m) - m_\delta^0(P^m(A-B)P^m) \\ &\geq mn - i_L(B) - \nu_L(B). \end{aligned}$$

Hence, $m_\delta^0(P^m(A-B)P^m) = \nu_L(B)$ and

$$m_\delta^+(P^m(A-B)P^m) = mn - i_L(B) - \nu_L(B).$$

Note that $\dim \mathcal{W}_L^m = (2m+1)n$, so

$$m_\delta^-(P^m(A-B)P^m) = mn + n + i_L(B).$$

□

Remark From Theorem III.4.3, Theorem III.4.5, and Theorem 5.2, we can get a new proof of Theorem 6.1 easily.

Corollary 6.2 ([192]) *Suppose* $B_j(t) \in C([0,1], \mathcal{L}_s(\mathbb{R}^{2n}))$, $j = 1, 2$ *satisfying*

$$B_1(t) < B_2(t), \text{ i.e., } B_2(t) - B_1(t) \text{ is positive definite for all } t \in [0,1].$$

Then there holds

$$i_L(B_1) + \nu_L(B_1) \leq i_L(B_2). \tag{5.156}$$

Proof Just as done in Theorem 6.1, corresponding to $B_j(t)$, we have the operator B_j. Let $\Gamma = \{P^m\}$ be the approximation scheme with respect to the operator A. Then by (5.150), there exists $m_0 > 0$ such that if $m \geq m_0$ there holds

$$m_\delta^-(P^m(A-B_1)P^m) = mn + n + i_L(B_1),$$

$$m_\delta^-(P^m(A-B_2)P^m) = mn + n + i_L(B_2),$$

where we choose $0 < \delta < \frac{1}{2}\|B_2 - B_1\|$. Since $A - B_2 = (A - B_1) - (B_2 - B_1)$ and $B_2 - B_1$ is positive definite in $\mathcal{W}_L^m = P^m\mathcal{W}_L$ and $\langle (B_2 - B_1)x, x\rangle \geq 2\delta\|x\|$. Thus $\langle (P^m(A-B_2)P^m)x, x\rangle \leq -\delta\|x\|$ with $x \in M_\delta^-(P^m(A-B_1)P^m) \oplus M_\delta^0(P^m(A-B_1)P^m)$. It implies that

$$mn + n + i_L(B_1) + \nu_L(B_1) \leq mn + n + i_L(B_2).$$

□

Remark From the proof of Corollary 6.2, it is easy to show that if $B_1(t) \leq B_2(t)$ for all $0 \leq t \leq 1$, there hold

$$i_L(B_1) \leq i_L(B_2), \quad i_L(B_1) + \nu_L(B_1) \leq i_L(B_2) + \nu_L(B_2). \tag{5.157}$$

Definition 6.3 ([192]) For any two matrix functions $B_j \in C([0,1], \mathcal{L}_s(\mathbb{R}^{2n}))$, $j = 0, 1$ with $B_0(t) < B_1(t)$ for all $t \in \mathbb{R}$, we define

$$I_L(B_0, B_1) = \sum_{s \in [0,1)} \nu_L((1-s)B_0 + sB_1). \tag{5.158}$$

Theorem 6.4 ([192]) *For any two matrix functions $B_j \in C([0,1], \mathcal{L}_s(\mathbb{R}^{2n}))$ with $B_0(t) < B_1(t)$ for all $t \in \mathbb{R}$, we have*

$$I_L(B_0, B_1) = i_L(B_1) - i_L(B_0). \tag{5.159}$$

So we call $I_L(B_0, B_1)$ the relative L-index of the pair (B_0, B_1).

Proof Step 1. By Corollary 6.2, if we denote $i_L(\lambda) = i_L((1-\lambda)B_0 + \lambda B_1)$, $\nu_L(\lambda) = \nu_L((1-\lambda)B_0 + \lambda B_1)$, there holds

$$i_L(\lambda_2) \geq i_L(\lambda_1) + \nu_L(\lambda_1), \text{ for } \lambda_2 > \lambda_1. \tag{5.160}$$

So the function $i_L(\lambda)$ is a monotonic increasing function in $[0, 1]$.

Step 2. We prove that for any $\lambda \in [0, 1)$ there holds

$$i_L(\lambda + 0) = i_L(\lambda) + \nu_L(\lambda), \tag{5.161}$$

where $i_L(\lambda + 0)$ is the right-hand limit of $i_L(s)$ at λ. In fact, by (5.160), we have $i_L(\lambda) + \nu_L(\lambda) \leq i_L(\lambda + 0)$. We now use the saddle point reduction methods to prove the opposite inequality $i_L(\lambda) + \nu_L(\lambda) \geq i_L(\lambda + 0)$. Denote by $B_\lambda(t) = (1-\lambda)B_0(t) + \lambda B_1(t)$. We define in $L^2([0,1], \mathbb{R}^{2n})$

$$f_\lambda(x) = \int_0^1 [(-J\dot{x}(t), x(t)) - (B_\lambda(t)x(t), x(t))]\, dt, \ \forall x \in \mathrm{dom}(A) = W_L.$$

Then by the saddle point reduction methods, we can reduce the functional f_λ in $L^2([0,1], \mathbb{R}^{2n})$ to a finite dimensional subspace X of $L^2([0,1], \mathbb{R}^{2n})$ by

$$a_\lambda(x) = f_\lambda(u_\lambda(x)), \ u_\lambda : X \to L^2([0,1], \mathbb{R}^{2n}) \text{ is injection.}$$

a_λ is continuously depending on λ. Denote the Morse indices of a_λ on X at $x = 0$ by m_λ^-, m_λ^0 and m_λ^-. If $\dim X = 2d + n$ large enough, there holds (see Theorem 5.2)

$$m_\lambda^- = d + n + i_L(\lambda), \ m_\lambda^0 = \nu_L(\lambda), \ m_\lambda^+ = d - i_L(\lambda) - \nu_L(\lambda). \tag{5.162}$$

For any fixed $\lambda \in [0, 1)$, choosing $\mu \in (\lambda, 1) \cup [0, \lambda)$ sufficiently close to λ, we obtain

$$m_\lambda^\pm \leq m_\mu^\pm \leq m_\lambda^\pm + \nu_L(\lambda). \tag{5.163}$$

Then by (5.162), we have $i_L(\lambda) \le i_L(\mu)$ and $i_L(\lambda) + \nu_L(\lambda) \ge i_L(\mu)$. It implies $i_L(\lambda)+\nu_L(\lambda) \ge i_L(\lambda+0)$ and $i_L(\lambda) \le i_L(\lambda-0)$. But by (5.160), we have $i_L(\lambda) \ge i_L(\lambda-0)$, so $i_L(\lambda) = i_L(\lambda-0)$. That is to say the function $i_L(\lambda)$ is left continuous at $(0, 1]$. Moreover if $m_\lambda^0 = m^0$ is constant in some interval $[\lambda_1, \lambda_2]$, then $m_\lambda^- = m^-$ and $m_\lambda^+ = m^+$ are constant in this interval. Thus the function $i_L(\lambda)$ is locally constant at its continuous points, whose all discontinuous points are the points with $\nu_L(\lambda) > 0$, and there holds

$$i_L(1) = i_L(0) + \sum_{0\le\lambda<1} \nu_L(\lambda),$$

which is exactly (5.159). □

Corollary 6.5 ([192]) *If $\gamma \in \mathcal{P}_\tau(2n)$ is the fundamental solution of the linear Hamiltonian system with respect to $B(t) > 0$, there holds*

$$i_L(\gamma) = \sum_{0<t<1} \dim(\gamma(t)L \cap L). \tag{5.164}$$

Thus we can understand the index $i_L(\gamma)$ as a kind of intersection number of the two Lagrangian paths $w(t) = \gamma(t)L$ and $w_0(t) = L$.

Proof We take $B_1(t) = B(t)$ and $B_0(t) = 0$ in Theorem 6.4. We note that the fundamental solution corresponding to $B_0(t) = 0$ is the constant path I. We have

$$I_L(0, B) = i_L(\gamma) - i_L(I).$$

But $i_L(I) = i_{L_0}(I) = -n$ and $B_s(t) = (1-s)B_0(t) + sB_1(t) = sB(t)$. The corresponding fundamental solution corresponding to $B_s(t) = sB(t)$ is $\gamma(st)$. Thus

$$I_L(0, B) = \sum_{s\in[0,1)} \nu_L(sB) = \sum_{s\in[0,1)} \dim[(\gamma(s)L) \cap L].$$

But $\dim[(\gamma(0)L) \cap L] = \dim L = n$, so we have (5.164). □

5.7 Dual L-Index Theory for Linear Hamiltonian Systems

Let $B \in C([0, 1], \mathcal{L}_s(\mathbb{R}^{2n}))$. Recall that $\mathcal{L}_s(\mathbb{R}^{2n})$ is the set of symmetrical $2n \times 2n$ metrics. Consider the linear Hamiltonian system

$$\dot{z} = JB(t)z, \ z \in \mathbb{R}^{2n}. \tag{5.165}$$

We consider in this section the dual Morse index theory of system (5.165) with Lagrangian boundary condition. The dual Morse index theory for periodic boundary condition was studied by M. Girardi and M. Matzeu in [121] for the case of superquadratic Hamiltonian systems, and by the author of this book in [185] for the subquadratic Hamiltonian systems. This theory is an application of the Morse-Ekeland index theory [78]. The dual action principal in Hamiltonian framework was first established by F. Clarke in [49–51] (see also [52, 53]), and been adapted by many mathematicians to the study of various variational problems afterward. The index theory for convex Hamiltonian systems was established by I. Ekeland (cf. [78]), whose works are fundamental and important in the studies of convex Hamiltonian systems.

Let W_L be the Hilbert space defined by $W_L = \{z = (x,y)^T \in W^{1,2}([0,1],\mathbb{R}^{2n}) |\ z(0),\ z(1) \in L\} \subset L^2$. The embedding $j : W_L \to \mathcal{L} = L^2([0,1],\mathbb{R}^{2n})$ is compact. Denote by $\langle\cdot,\cdot\rangle$ and $\langle\cdot,\cdot\rangle_2$ the inner product in W_L and $\mathcal{L}$ respectively. We define an operator $A : \mathcal{L} \to \mathcal{L}$ with domain W_L by $A = -J\frac{d}{dt}$. The spectrum of A is isolated. In fact $\sigma(A) = \pi\mathbb{Z}$. Let $k \notin \sigma(A)$ be so large such that $B(t) + kI > 0$. Then the operator $\Lambda_k = A + kI : W_L \to \mathcal{L}$ is invertible, and its inverse is compact. We define a quadratic form in $\mathcal{L}$ by

$$Q^*_{k,B}(v,u) = \int_0^1 [(C_k(t)v(t),u(t)) - (\Lambda_k^{-1}v(t),u(t))]\,dt,\ \forall\, v,u \in \mathcal{L}, \tag{5.166}$$

where $C_k(t) = (B(t)+kI)^{-1}$. Denote $Q^*_{k,B}(v) = Q^*_{k,B}(v,v)$. Then

$$\langle C_k v, v\rangle_2 = \int_0^1 (C_k(t)v(t),v(t))\,dt$$

define a Hilbert structure in $\mathcal{L}$. $C_k^{-1}\Lambda_k^{-1}$ is a self-adjoint and compact operator under this inter product. By the spectral theory, there exists a basis $e_j,\ j \in \mathbb{N}$ of $\mathcal{L}$, and an eigenvalue sequence $\lambda_j \to 0$ in $\mathbb{R}$ such that

$$\langle C_k e_i, e_j\rangle_2 = \delta_{ij},$$
$$\langle \Lambda_k^{-1} e_j, v\rangle_2 = \langle C_k\lambda_j e_j, v\rangle_2,\quad \forall v \in \mathcal{L}.$$

For any $v \in \mathcal{L}$ with $v = \sum_{j=1}^{\infty}\xi_j e_j$, there holds

$$Q^*_{k,B}(v) = -\int_0^1 (\Lambda_k^{-1}v(t),v(t)) - (C_k(t)v(t),v(t))\,dt = \sum_{j=1}^{\infty}(1-\lambda_j)\xi_j^2.$$

Define

$$\mathcal{L}_k^-(B) = \{\textstyle\sum_{j=1}^{\infty}\xi_j e_j \mid \xi_j = 0 \text{ if } 1-\lambda_j \ge 0\}$$
$$\mathcal{L}_k^0(B) = \{\textstyle\sum_{j=1}^{\infty}\xi_j e_j \mid \xi_j = 0 \text{ if } 1-\lambda_j \ne 0\}$$
$$\mathcal{L}_k^+(B) = \{\textstyle\sum_{j=1}^{\infty}\xi_j e_j \mid \xi_j = 0 \text{ if } 1-\lambda_j \le 0\}.$$

Observe that $\mathcal{L}_k^-(B)$, $\mathcal{L}_k^0(B)$ and $\mathcal{L}_k^+(B)$ are $Q_{k,B}^*$-orthogonal, and $\mathcal{L} = \mathcal{L}_k^-(B) \oplus \mathcal{L}_k^0(B) \oplus \mathcal{L}_k^+(B)$. Since $\lambda_j \to 0$ as $j \to \infty$, both $\mathcal{L}_k^-(B)$ and $\mathcal{L}_k^0(B)$ are finite subspaces. We define the k-dual Morse index of B by

$$i_k^*(B) = \dim \mathcal{L}_k^-(B), \ \ \nu_k^*(B) = \dim \mathcal{L}_k^0(B). \tag{5.167}$$

Theorem 7.1 ([192]) *There hold*

$$i_k^*(B) = i_L(B) + n + n\left[\frac{k}{\pi}\right], \ \ \nu_k^*(B) = \nu_L(B), \tag{5.168}$$

where $[a] = \max\{j \in \mathbb{Z} \mid j \leq a\}$.

Proof We only prove (5.168) for the special case $L = L_0$. We first define a functional on

$$W^m = \{x \mid x(t) = \sum_{j=-m}^{m} -J\exp(j\pi tJ)a_j, \ a_j \in \mathbb{R}^n \oplus \{0\} \subset \mathbb{R}^{2n}\}$$

by

$$\begin{aligned} Q_m(x) &= \int_0^1 [(\Lambda_k x(t), x(t)) - (C_k^{-1}(t)x, x)]\, dt \\ &= \int_0^1 [(-J\dot{x}(t), x(t)) - (B(t)x(t), x(t))]\, dt, \ \forall x \in W^m. \end{aligned}$$

We define two linear operators $A_k, \ B_k$ from W^m onto its dual space $W^{m*} \cong W^m$ such that

$$\langle A_k x, y\rangle_2 = \int_0^1 (\Lambda_k x(t), y(t))\, dt, \ \ \forall x, y \in W^m,$$

$$\langle B_k x, y\rangle_2 = \int_0^1 ((B(t) + kI)x(t), y(t))\, dt, \forall x, y \in W^m.$$

$$\langle \cdot, \cdot\rangle_m := \langle B_k \cdot, \cdot\rangle_2$$

is an inner product in W^m. We consider the eigenvalues $\mu_j \in \mathbb{R}$ of A_k with respect to this inner product, i.e.,

$$A_k x_j = \mu_j B_k x_j$$

for some $x_j \in W^m \backslash \{0\}$. Suppose $\mu_1 \leq \mu_2 \leq \cdots \leq \mu_l$ with $l = \dim W^m = 2mn+n$ (each of the eigenvalues is counted with its multiplicity), and corresponding a basis of W^m consists of eigenvectors $v_1, \cdots, v_l$ such that

$$\begin{aligned}\langle v_i, v_j\rangle_m &= \delta_{ij},\\ \langle A_m v_i, v_j\rangle_m &= \mu_i\delta_{ij},\\ Q_m(v_i, v_j) &= (\mu_i - 1)\delta_{ij},\ i, j = 1, 2, \cdots, l.\end{aligned}$$

The Morse index $m^-(Q_m), m^0(Q_m)$ and $m^+(Q_m)$ of Q_m satisfy

$$\begin{aligned}m^-(Q_m) &=^\sharp \{\mu_j \,|\, 1 \le j \le l,\ \mu_j < 1\},\\ m^+(Q_m) &=^\sharp \{\mu_j \,|\, 1 \le j \le l,\ \mu_j > 1\},\\ m^0(Q_m) &=^\sharp \{\mu_j \,|\, 1 \le j \le l,\ \mu_j = 1\}.\end{aligned}$$

By Theorem 6.1, we have for $m > 0$ large enough,

$$m^-(Q_m) = mn + n + i_L(B),\ m^0(Q_m) = \nu_L(B). \tag{5.169}$$

We denote by $Q^*_{k,m}$ the restriction of the quadratic Q^*_k to the subspace W^m, and $i^*_{k,m}(B) = m^-(Q^*_{k,m}),\ \nu^*_{k,m}(B) = m^0(Q^*_{k,m})$. By the same argument in [GM], we have $i^*_{k,m}(B) \to i^*_k(B),\ \nu^*_{k,m}(B) \to \nu^*_k(B)$ as $m \to \infty$. Let $v'_j = A_m v_j$ for $j = 1, 2, \cdots, l$. It is a basis of W^m and

$$Q^*_{k,m}(v'_i, v'_j) = 0 \text{ for } i \ne j,$$

$$Q^*_{k,m}(v'_j) = \mu_j(\mu_j - 1).$$

$Q^*_{k,m}(v'_j)$ is negative if and only if $0 < \mu_j < 1$. We now deduce the total multiplicity of the negative eigenvalues $\mu_j < 0$. If one replaces the inner product $\langle\cdot,\cdot\rangle_m$ by the usual one, i.e., replaces the matrix B_k by the identity I, the eigenvalues μ_j should be replaced by the eigenvalues η_j of A_m with respect to the standard inner product. It is easy to check that μ_j and η_j possesses the same signs. So the total multiplicity of negative μ_j's equals the total multiplicity of negative η_h's. But we have

$$\eta_h = h\pi + k,\ -m \le h \le m, \tag{5.170}$$

each of them has multiplicity n. Therefore, the total multiplicity of the negative η_h is $n(m - [k/\pi])$. So the total multiplicity of $\mu_j \in (0, 1)$ is $m^-(Q_m) - n(m - [k/\pi])$. By definition we have

$$i^*_{k,m}(B) = m^-(Q_m) - n(m - [k/\pi]).$$

So for $m > 0$ large enough, from (5.167) we get (5.168). □

Corollary 7.2 ([192]) *Under the condition of Theorem* 6.4, *there holds*

$$I_L(B_0, B_1) = i_k^*(B_1) - i_k^*(B_0). \tag{5.171}$$

5.8 The (L, ω)-Index Theory

For $\omega = e^{\sqrt{-1}\theta}$ with $\theta \in \mathbb{R}$, we define a Hilbert space $E^{\omega} = E^{\omega}_{L_0}$ consisting of those $x(t)$ in $L^2([0, 1], \mathbb{C}^{2n})$ such that $e^{-\theta t J}x(t)$ has Fourier expending

$$e^{-\theta t J}x(t) = \sum_{j\in\mathbb{Z}} e^{j\pi t J}\begin{pmatrix} 0 \\ a_j \end{pmatrix}, \quad a_j \in \mathbb{C}^n$$

with

$$\|x\|^2 := \sum_{j\in\mathbb{Z}}(1 + |j|)|a_j|^2 < \infty.$$

For $x \in E^{\omega}$, we can write

$$\begin{aligned} x(t) &= e^{\theta t J}\sum_{j\in\mathbb{Z}} e^{j\pi t J}\begin{pmatrix} 0 \\ a_j \end{pmatrix} = \sum_{j\in\mathbb{Z}} e^{(\theta+j\pi)tJ}\begin{pmatrix} 0 \\ a_j \end{pmatrix} \\ &= \sum_{j\in\mathbb{Z}} e^{(\theta+j\pi)t\sqrt{-1}}\begin{pmatrix} \sqrt{-1}a_j/2 \\ a_j/2 \end{pmatrix} + e^{-(\theta+j\pi)t\sqrt{-1}}\begin{pmatrix} -\sqrt{-1}a_j/2 \\ a_j/2 \end{pmatrix}. \end{aligned} \tag{5.172}$$

So we can write

$$x(t) = \xi(t) + N\xi(-t), \;\; \xi(t) = \sum_{j\in\mathbb{Z}} e^{(\theta+j\pi)t\sqrt{-1}}\begin{pmatrix} \sqrt{-1}a_j/2 \\ a_j/2 \end{pmatrix}. \tag{5.173}$$

For $\omega = e^{\sqrt{-1}\theta}$, $\theta \in [0, \pi)$, we define two self-adjoint operators $A^{\omega}, B^{\omega} \in \mathcal{L}(E^{\omega})$ by

$$(A^{\omega}x, y) = \int_0^1 \langle -J\dot{x}(t), y(t)\rangle dt, \quad (B^{\omega}x, y) = \int_0^1 \langle B(t)x(t), y(t)\rangle dt$$

on E^{ω}. Then B^{ω} is also compact.

Definition 8.1 ([209]) We define the index function

$$i_{\omega}^{L_0}(B) = I(A^{\omega},\; A^{\omega}-B^{\omega}), \;\; \nu_{\omega}^{L_0}(B) = m^0(A^{\omega}-B^{\omega}), \; \forall\, \omega = e^{\sqrt{-1}\theta}, \;\; \theta \in (0, \pi). \tag{5.174}$$

We remind that the relative index $I(A^\omega,\quad A^\omega - B^\omega)$ is defined in Definition III.1.5. From Lemma III.1.7, we see that

$$i_\omega^{L_0}(B) = -\mathrm{sf}\{A^\omega - sB^\omega,\ 0 \le s \le 1\}. \tag{5.175}$$

By the Floquet theory, we have $M^0(A^\omega, B^\omega)$ is isomorphic to the solution space of the following linear Hamiltonian system

$$\dot{x}(t) = JB(t)x(t)$$

satisfying the following boundary condition

$$x(0) \in L_0,\quad x(1) \in e^{\theta J} L_0.$$

If $m^0(A^\omega, B^\omega) > 0$, there holds

$$\gamma(1)L_0 \cap e^{\theta J} L_0 \ne \{0\}$$

which is equivalent to

$$\omega^2 = e^{2\theta\sqrt{-1}} \in \sigma\left([U(1) - \sqrt{-1}V(1)][U(1) + \sqrt{-1}V(1)]^{-1}\right).$$

This claim follows from the fact that if $\gamma(1)L_0 \cap e^{\theta J} L_0 \ne \{0\}$, there exist $a, b \in \mathbb{C}^n \setminus \{0\}$ such that

$$[U(1) + \sqrt{-1}V(1)]a = \omega^{-1}b,\quad [U(1) - \sqrt{-1}V(1)]a = \omega b.$$

So we have

$$\nu_\omega^{L_0}(B) = \dim(\gamma(1)L_0 \cap e^{\theta J} L_0),\quad \forall\, \omega = e^{\sqrt{-1}\theta},\ \theta \in (0, \pi). \tag{5.176}$$

Lemma 8.2 ([209]) *The index function $i_\omega^{L_0}(B)$ is locally constant. For $\omega_0 = e^{\sqrt{-1}\theta_0}$, $\theta_0 \in (0, \pi)$ is a point of discontinuity of $i_\omega^{L_0}(B)$, then $\nu_{\omega_0}^{L_0}(B) > 0$ and so $\dim(\gamma(1)L_0 \cap e^{\theta_0 J} L_0) > 0$. Moreover there hold*

$$|i_{\omega_0+}^{L_0}(B) - i_{\omega_0-}^{L_0}(B)| \le \nu_{\omega_0}^{L_0}(B),\ |i_{\omega_0+}^{L_0}(B) - i_{\omega_0}^{L_0}(B)| \le \nu_{\omega_0}^{L_0}(B),$$
$$|i_{\omega_0-}^{L_0}(B) - i_{\omega_0}^{L_0}(B)| \le \nu_{\omega_0}^{L_0}(B),\quad |i_{L_0}(B) + n - i_{1+}^{L_0}(B)| \le \nu_{L_0}(B), \tag{5.177}$$

where $i_{\omega_0+}^{L_0}(B)$, $i_{\omega_0-}^{L_0}(B)$ are the limits on the right and left respectively of the index function $i_\omega^{L_0}(B)$ at $\omega_0 = e^{\sqrt{-1}\theta_0}$ as a function of θ.

Proof For $x(t) = e^{\theta t J} u(t)$, $u(t) = \sum_{j\in\mathbb{Z}} e^{j\pi t J}\begin{pmatrix}0\\a_j\end{pmatrix}$, we have

$$((A^\omega - B^\omega)x, x) = \int_0^1 \langle -J\dot{u}(t), u(t)\rangle dt + \int_0^1 \langle (\theta - e^{-\theta t J} B(t) e^{\theta t J}) u(t), u(t)\rangle dt.$$

So we have

$$((A^\omega - B^\omega)x, x) = (q_\omega u, u)$$

with

$$(q_\omega u, u) = \int_0^1 \langle -J\dot{u}(t), u(t)\rangle dt + \int_0^1 \langle (\theta - e^{-\theta t J} B(t) e^{\theta t J}) u(t), u(t)\rangle dt.$$

Since $\dim(\gamma(1)L_0 \cap e^{\theta J} L_0) > 0$ at only finite (up to n) points $\theta \in (0, \pi)$, for the point $\theta_0 \in (0, \pi)$ such that $\nu_{\omega_0}^{L_0}(B) = 0$, then $\nu_\omega^{L_0}(B) = 0$ for $\omega = e^{\sqrt{-1}\theta}$, $\theta \in (\theta_0 - \delta, \theta_0 + \delta)$, $\delta > 0$ small enough. By using the notation as in Lemma III.1.4, we have

$$(P_m^\omega (A^\omega - B^\omega) P_m^\omega x, x) = (P_m q_\omega P_m u, u).$$

By Lemma III.1.2, we have

$$m_d^0(P_m^\omega (A^\omega - B^\omega) P_m^\omega) = m^0(A^\omega - B^\omega) = \nu_\omega^{L_0}(B) = 0.$$

So by the continuity of the eigenvalue of a continuous family of operators we have that

$$m_d^-(P_m^\omega (A^\omega - B^\omega) P_m^\omega)$$

must be constant for $\omega = e^{\sqrt{-1}\theta}$, $\theta \in (\theta_0 - \delta, \theta_0 + \delta)$. Since $m_d^-(P_m^\omega A^\omega P_m^\omega)$ is constant for $\omega = e^{\sqrt{-1}\theta}$, $\theta \in (\theta_0 - \delta, \theta_0 + \delta)$, we have $i_\omega^{L_0}(B)$ is constant for $\omega = e^{\sqrt{-1}\theta}$, $\theta \in (\theta_0 - \delta, \theta_0 + \delta)$.

The results in (5.177) now follow from some standard arguments. □

It is easy to see that

$$i_1^{L_0}(B) = I(A^1, A^1 - B^1) = i_{L_0}(B) + n. \tag{5.178}$$

By Definition 8.1 and Lemma 8.2, we see that for any $\omega_0 = e^{\sqrt{-1}\theta_0}$, $\theta_0 \in (0, \pi)$, there holds

$$i_{\omega_0}^{L_0}(B) \geq i_{L_0}(B) + n - \sum_{\omega=e^{\sqrt{-1}\theta},\ 0\leq\theta\leq\theta_0} \nu_{\omega}^{L_0}(B). \tag{5.179}$$

We note that

$$\sum_{\omega=e^{\sqrt{-1}\theta},\ 0\leq\theta\leq\theta_0} \nu_{\omega}^{L_0}(B) \leq n. \tag{5.180}$$

So we have

$$i_{L_0}(B) \leq i_{\omega_0}^{L_0}(B) \leq i_{L_0}(B) + 2n. \tag{5.181}$$

5.9 The Bott Formulas of L-Index

In this section, we establish the Bott-type iteration formula for the L_j-index theory with $j = 0, 1$. Without loss of generality, we assume $\tau = 1$. Suppose the continuous symplectic path $\gamma : [0, 1] \to \mathrm{Sp}(2n)$ is the fundamental solution of the following linear Hamiltonian system

$$\dot{z}(t) = JB(t)z(t), \quad t \in \mathbb{R} \tag{5.182}$$

with $B(t)$ satisfying $B(t+2) = B(t)$ and $B(1+t)N = NB(1-t)$ for $t \in \mathbb{R}$. This implies $B(t)N = NB(-t)$ for $t \in \mathbb{R}$. By the unique existence theorem of the linear differential equations, we get

$$\gamma(1+t) = N\gamma(1-t)\gamma(1)^{-1}N\gamma(1), \gamma(2+t) = \gamma(t)\gamma(2). \tag{5.183}$$

For $j \in \mathbb{N}$, we define the j-times iteration path $\gamma^j : [0, j] \to \mathrm{Sp}(2n)$ of γ by

$$\gamma^1(t) = \gamma(t), \ t \in [0, 1],$$

$$\gamma^2(t) = \begin{cases} \gamma(t), \ t \in [0, 1], \\ N\gamma(2-t)\gamma(1)^{-1}N\gamma(1), \ t \in [1, 2], \end{cases} \tag{5.184}$$

and in general, for $k \in \mathbb{N}$, we define $\gamma(2) = N\gamma(1)^{-1}N\gamma(1)$ and

$$\gamma^{2k-1}(t) = \begin{cases} \gamma(t), \ t \in [0, 1], \\ N\gamma(2-t)\gamma(1)^{-1}N\gamma(1), \ t \in [1, 2], \\ \cdots\cdots \\ N\gamma(2k-2-t)N\gamma(2)^{k-1}, \ t \in [2k-3, 2k-2], \\ \gamma(t-2k+2)\gamma(2)^{k-1}, \ t \in [2k-2, 2k-1], \end{cases} \tag{5.185}$$

$$\gamma^{2k}(t) = \begin{cases} \gamma(t), \ t \in [0,1], \\ N\gamma(2-t)\gamma(1)^{-1}N\gamma(1), \ t \in [1,2], \\ \cdots\cdots \\ \gamma(t-2k+2)\gamma(2)^{k-1}, \ t \in [2k-2, 2k-1], \\ N\gamma(2k-t)N\gamma(2)^k, \ t \in [2k-1, 2k]. \end{cases} \tag{5.186}$$

For $\gamma \in \mathcal{P}_\tau(2n)$, we define

$$\gamma^k(\tau t) = \tilde{\gamma}^k(t) \text{ with } \tilde{\gamma}(t) = \gamma(\tau t). \tag{5.187}$$

For the L_0-index of the iteration path γ^k, we have the following Bott-type formulas.

Theorem 9.1 ([209]) *Suppose $\omega_k = e^{\pi\sqrt{-1}/k}$. For odd k we have*

$$i_{L_0}(\gamma^k) = i_{L_0}(\gamma^1) + \sum_{i=1}^{(k-1)/2} i_{\omega_k^{2i}}(\gamma^2),$$

$$\nu_{L_0}(\gamma^k) = \nu_{L_0}(\gamma^1) + \sum_{i=1}^{(k-1)/2} \nu_{\omega_k^{2i}}(\gamma^2),$$

and for even k, we have

$$i_{L_0}(\gamma^k) = i_{L_0}(\gamma^1) + i^{L_0}_{\sqrt{-1}}(\gamma^1) + \sum_{i=1}^{k/2-1} i_{\omega_k^{2i}}(\gamma^2),$$

$$\nu_{L_0}(\gamma^k) = \nu_{L_0}(\gamma^1) + \nu^{L_0}_{\sqrt{-1}}(\gamma^1) + \sum_{i=1}^{k/2-1} \nu_{\omega_k^{2i}}(\gamma^2).$$

We remind that $(i^{L_0}_{\sqrt{-1}}(\gamma), \nu^{L_0}_{\sqrt{-1}}(\gamma))$ is the (L_0, ω) index of γ with $\omega = \sqrt{-1}$.

Before proving Theorem 9.1, we give some notations and definitions.

We define the Hilbert space

$$E^k_{L_0} = \left\{ x \in L^2([0,k], \mathbb{C}^{2n}) \,|\, x(t) = \sum_{j\in\mathbb{Z}} e^{jt\pi/kJ} \begin{pmatrix} 0 \\ a_j \end{pmatrix}, \ a_j \in \mathbb{C}^n, \ \|x\|^2 < \infty \right\},$$

where we still denote $L_0 = \{0\} \times \mathbb{C}^n \subset \mathbb{C}^{2n}$ which is the Lagrangian subspace of the linear complex symplectic space $(\mathbb{C}^{2n}, \tilde{\omega}_0)$ and $|x\|^2 := \sum_{j\in\mathbb{Z}} (1+|j|)|a_j|^2$. For $x \in E^k_{L_0}$, we can write

$$x(t) = \sum_{j\in\mathbb{Z}} e^{jt\pi/kJ}\begin{pmatrix}0\\ a_j\end{pmatrix} = \sum_{j\in\mathbb{Z}}\begin{pmatrix}-\sin(jt\pi/k)a_j\\ \cos(jt\pi/k)a_j\end{pmatrix}$$
$$= \sum_{j\in\mathbb{Z}}\left\{e^{j\pi t\sqrt{-1}/k}\begin{pmatrix}\sqrt{-1}a_j/2\\ a_j/2\end{pmatrix} + e^{-j\pi t\sqrt{-1}/k}\begin{pmatrix}-\sqrt{-1}a_j/2\\ a_j/2\end{pmatrix}\right\} \tag{5.188}$$

On $E^k_{L_0}$ we define two self-adjoint operators and a quadratical form by

$$(A_k x,\, y) = \int_0^k \langle -J\dot{x}(t),\, y(t)\rangle dt, \quad (B_k x,\, y) = \int_0^k \langle B(t)x(t), y(t)\rangle dt, \tag{5.189}$$
$$Q^k_{L_0}(x, y) = ((A_k - B_k)x, y), \tag{5.190}$$

where in this section $\langle\cdot,\cdot\rangle$ is the standard Hermitian inner product in $\mathbb{C}^{2n}$.

Lemma 9.2 ([209]) *$E^k_{L_0}$ has the following natural decomposition*

$$E^k_{L_0} = \bigoplus_{l=0}^{k-1} E^{\omega_k^l}_{L_0}, \tag{5.191}$$

here we have extended the domain of functions in $E^{\omega_k^l}_{L_0}$ from $[0, 1]$ to $[0, k]$ in the obvious way, i.e.,

$$E^{\omega_k^l}_{L_0} = \left\{x \in E^k_{L_0} \mid x(t) = e^{l\pi tJ/k}\sum_{j\in\mathbb{Z}} e^{j\pi tJ}\begin{pmatrix}0\\ a_j\end{pmatrix}\right\}.$$

Proof Any element $x \in E^k_{L_0}$ can be written as

$$x(t) = \sum_{j\in\mathbb{Z}}\left\{e^{j\pi t\sqrt{-1}/k}\begin{pmatrix}\sqrt{-1}a_j/2\\ a_j/2\end{pmatrix} + e^{-j\pi t\sqrt{-1}/k}\begin{pmatrix}-\sqrt{-1}a_j/2\\ a_j/2\end{pmatrix}\right\}$$
$$= \sum_{l=0}^{k-1}\sum_{j\equiv l\,(mod k)}\left\{e^{j\pi t\sqrt{-1}/k}\begin{pmatrix}\sqrt{-1}a_j/2\\ a_j/2\end{pmatrix} + e^{-j\pi t\sqrt{-1}/k}\begin{pmatrix}-\sqrt{-1}a_j/2\\ a_j/2\end{pmatrix}\right\}$$
$$= \sum_{l=0}^{k-1}\sum_{j\in\mathbb{Z}}\left\{e^{l\pi t\sqrt{-1}/k}e^{j\pi t\sqrt{-1}}\begin{pmatrix}\sqrt{-1}b_j/2\\ b_j/2\end{pmatrix}\right.$$
$$\left. + e^{-l\pi t\sqrt{-1}/k}e^{-j\pi t\sqrt{-1}}\begin{pmatrix}-\sqrt{-1}b_j/2\\ b_j/2\end{pmatrix}\right\}$$
$$:= \xi_x(t) + \eta_x(t), \tag{5.192}$$

where $\xi_x(t) = \sum_{l=0}^{k-1}\sum_{j\in\mathbb{Z}} e^{l\pi t\sqrt{-1}/k} e^{j\pi t\sqrt{-1}} \begin{pmatrix} \sqrt{-1}b_j/2 \\ b_j/2 \end{pmatrix}$, $\eta_x(t) = N\xi_x(-t)$ and $b_j = a_{jk+l}$. By setting $\omega_k = e^{\pi\sqrt{-1}/k}$, and comparing (5.172) and (5.192), we obtain (5.191). □

Note that the natural decomposition (5.191) is not orthogonal under the quadratical form $Q^k_{L_0}$ defined in (5.190). So the type of the iteration formulas in Theorem 9.1 is somewhat different from the original Bott formulas in [28] of the Morse index theory for closed geodesics and of Maslov-type index theory for periodic solutions of Hamiltonian systems and the Bott-type formulas in [78]. This is also our main difficulty in the proof of Theorem 9.1. However, after recombining the terms in the decomposition of Lemma 9.2, we can obtain an orthogonal decomposition under the quadratical form $Q^k_{L_0}$.

For $1 \le l < \frac{k}{2}$ and $l \in \mathbb{N}$, we set

$$E^{\omega_k,l}_{L_0} = E^{\omega_k^l}_{L_0} + E^{\omega_k^{k-l}}_{L_0}.$$

Then we have the following

Lemma 9.3 *$E^k_{L_0}$ has the following $Q^k_{L_0}$ orthogonal decompositions:*

$$E^k_{L_0} = E^1_{L_0} \oplus \bigoplus_{l=1}^{(k-1)/2} E^{\omega_k,l}_{L_0} \quad \forall k \in 2\mathbb{N}+1, \tag{C_{odd}}$$

$$E^k_{L_0} = E^1_{L_0} \oplus E^{\sqrt{-1}}_{L_0} \oplus \bigoplus_{l=1}^{\frac{k}{2}-1} E^{\omega_k,l}_{L_0} \quad \forall k \in 2\mathbb{N}. \tag{C_{even}}$$

Proof For any $x \in E^k_{L_0}$, in the following argument we always extend it naturally to an element in $L^2([0,2k],\mathbb{C}^{2n})$ by $x(t) = \sum_{j\in\mathbb{Z}} e^{jt\pi/kJ} \begin{pmatrix} 0 \\ a_j \end{pmatrix}$, $t \in [0,2k]$. By direct computation, for any $x,\ y \in E^k_{L_0}$ we have

$$((A_k - B_k)x,\ y) = \frac{1}{2}\int_0^{2k} \langle -J\dot{x}(t),\ y(t)\rangle dt - \frac{1}{2}\int_0^{2k} \langle B(t)x(t),\ y(t)\rangle dt. \tag{5.193}$$

We denote simply by $\frac{1}{2}Q_k(x,y) = ((A_k - B_k)x,\ y)$. For any $0 \le l \le k$, define the Hilbert space

$$E^l_k = \left\{ x \in L^2([0,2k],\mathbb{C}^{2n}) \mid x(t) = e^{lt\pi/kJ} \sum_{j\in\mathbb{Z}} e^{jt\pi J} c_j,\ c_j \in \mathbb{C}^{2n},\ \|x\|^2 < \infty \right\}.$$

Thus for any $x \in E_k^l$ we have $x(t+2) = w_k^{2l}x(t)$. Then by the argument of Lemma I.5.2 of [78], we have E_k^i and E_k^j are Q_k orthogonal for any $i \neq j$.

For any $x \in E_{L_0}^{\omega_k^l}$ with $1 \le l \le k-1$, by the definition of ξ_x and η_x in the proof of Lemma 9.2 we have $\xi_x \in E_k^l$ and $\eta_x \in E_k^{k-l}$. Hence $x \in E_k^l + E_k^{k-l}$. So for any $x \in E_{L_0}^{\omega_k,i}$ with $1 \le i < \frac{k}{2}$ we have $x \in E_k^i + E_k^{k-i}$. Note that for any $x \in E_{L_0}^{\omega_k,0}$ we have $x \in E_k^0$, if k is even, then for any $x \in E_{L_0}^{\omega_k,k/2}$ we have $x \in E_k^{k/2}$. Hence the $Q_{L_0}^k$ orthogonal decompositions (C_{odd}) and (C_{even}) hold by (5.193) and the Q_k orthogonality of E_k^i and E_k^j with $i \neq j$. The proof of Lemma 9.3 is complete. □

Let $1 \le l < \frac{k}{2}, l \in \mathbb{N}$. For any $z = x + y \in E_{L_0}^{\omega_k,l}$ with $x \in E_{L_0}^{\omega_k^l}$ and $y \in E_{L_0}^{\omega_k^{k-l}}$, by the arguments in the proofs of Lemmas 9.2 and 9.3 we have

$$z = \xi_x + \eta_x + \xi_y + \eta_y$$

with $\xi_x + \eta_y := z_1 \in E_k^l$ and $\eta_x + \xi_y := z_2 \in E_k^{k-l}$. Then by $B(-t) = NB(t)N$ and B being 2-periodic we have

$$\begin{aligned}
&\int_0^{2k} \langle B(t)z_2(t), z_2(t)\rangle \\
&= \int_0^{2k} \langle B(t)(y(t) + Nx(-t)), (y(t) + Nx(-t))\rangle dt \\
&= \int_0^{2k} \langle B(2k-t)(y(2k-t) + Nx(t-2k)), (y(2k-t) + Nx(t-2k))\rangle dt \\
&= \int_0^{2k} \langle NB(t)N(y(-t) + Nx(t)), (y(-t) + Nx(t))\rangle dt \\
&= \int_0^{2k} \langle B(t)(Ny(-t) + x(t)), N(y(-t) + x(t))\rangle dt \\
&= \int_0^{2k} \langle B(t)z_1(t), z_1(t)\rangle \qquad (5.194)
\end{aligned}$$

So by the Q_k orthogonality of E_k^i and E_k^j with $i \neq j$, (5.193), (5.194), and formula (34) on page 37 of [78] we have

$$\begin{aligned}
(B_k z, z) &= \frac{1}{2}\int_0^{2k} \langle B(t)(z_1(t) + z_2(t)), (z_1(t) + z_2(t))\rangle \\
&= \frac{1}{2}\int_0^{2k} \langle B(t)z_1(t), z_1(t)\rangle + \frac{1}{2}\int_0^{2k} \langle B(t)z_2(t), z_2(t)\rangle
\end{aligned}$$

$$= \int_0^{2k} \langle B(t)z_1(t), z_1(t)\rangle$$

$$= k\int_0^2 \langle B(t)z_1(t),\ z_1(t)\rangle dt. \tag{5.195}$$

Similarly we have

$$(A_k z, z) = k\int_0^2 \langle -J\dot{z}_1(t),\ z_1(t)\rangle dt. \tag{5.196}$$

Note that for $z \in E^1_{L_0}$ and $z \in E^{\sqrt{-1}}_{L_0}$ we have

$$(B_k z, z) = \frac{k}{2}\int_0^2 \langle B(t)z(t), z(t)\rangle dt = k\int_0^1 \langle B(t)z(t), z(t)\rangle dt, \tag{5.197}$$

$$(A_k z, z) = \frac{k}{2}\int_0^2 \langle -J\dot{z}(t), z(t)\rangle dt = k\int_0^1 \langle -J\dot{z}(t), z(t)\rangle dt. \tag{5.198}$$

We also note that

$$u(t) = \xi_x(t) + \eta_y(t) = \sum_{j\in\mathbb{Z}} e^{l\pi\sqrt{-1}t/k} e^{j\pi\sqrt{-1}t}\begin{pmatrix}\sqrt{-1}(\alpha_j - \beta_j)\\ (\alpha_j + \beta_j)\end{pmatrix}$$

$$= \sum_{j\in\mathbb{Z}} e^{l\pi\sqrt{-1}t/k} e^{j\pi\sqrt{-1}t} u_j,\ \ u_j \in \mathbb{C}^{2n}. \tag{5.199}$$

We set

$$E_{\omega_k^{2l}} = \left\{ u \in L^2([0,2], \mathbb{C}^{2n}) \mid u(t) \right.$$

$$\left. = e^{l\pi\sqrt{-1}t/k}\sum_{j\in\mathbb{Z}} e^{j\pi\sqrt{-1}t}u_j,\ \|u\|^2 := \sum_{j\in\mathbb{Z}}(1+|j|)|u_j|^2 < +\infty \right\}.$$

Proof We define self-adjoint operators on $E_{\omega_k^{2l}}$ by

$$(A_{\omega_k^{2l}}u, v) = \int_0^2 \langle -J\dot{u}(t),\ v(t)\rangle dt,\ (B_{\omega_k^{2l}}u, v) = \int_0^2 \langle B(t)u(t),\ v(t)\rangle dt$$

and a quadratic form

$$Q_{\omega_k^{2l}}(u) = ((A_{\omega_k^{2l}} - B_{\omega_k^{2l}})u, u),\ u \in E_{\omega_k^{2l}}.$$

Here Q_ω is just the quadratic form f_ω defined on p.133 of [223]. In order to complete the proof of Theorem 9.1, we need the following result. □

Lemma 9.4 *For a symmetric 2-periodic matrix function B and $\omega \in \mathbf{U} \setminus \{1\}$, there hold*

$$I(A_\omega, A_\omega - B_\omega) = i_\omega(\gamma^2), \tag{5.200}$$

$$m^0(A_\omega - B_\omega) = \nu_\omega(\gamma^2). \tag{5.201}$$

Proof In fact, (5.200) follows directly from Definition 2.3 and Corollary 2.1 of [234] and Lemma III.1.7, (5.201) follows from the Floquet theory. We note also that (5.200) is the eventual form of the Galerkin approximation formula. We can also prove it step by step as the proof of Theorem 3.1 of [192] by using the saddle point reduction formula in Theorem 6.1.1 of [223]. □

Continue the proof of Theorem 9.1 By Lemma 9.4, for $1 \le l < \frac{k}{2}$, $l \in \mathbb{N}$ we have

$$I(A_{\omega_k^{2l}}, A_{\omega_k^{2l}} - B_{\omega_k^{2l}}) = i_{\omega_k^{2l}}(\gamma^2), \quad m^0(A_{\omega_k^{2l}} - B_{\omega_k^{2l}}) = \nu_{\omega_k^{2l}}(\gamma^2). \tag{5.202}$$

By Definition 8.1, we have

$$I(A^{\sqrt{-1}}, A^{\sqrt{-1}} - B^{\sqrt{-1}}) = i^{L_0}_{\sqrt{-1}}(\gamma), \quad m^0(A^{\sqrt{-1}} - B^{\sqrt{-1}}) = \nu^{L_0}_{\sqrt{-1}}(\gamma). \tag{5.203}$$

We also have

$$I(A^1, A^1 - B^1) = i_{L_0}(\gamma) + n, \quad m^0(A^1 - B^1) = \nu_{L_0}(\gamma), \tag{5.204}$$

and

$$I(A_k, A_k - B_k) = i_{L_0}(\gamma^k) + n, \quad m^0(A_k - B_k) = \nu_{L_0}(\gamma^k). \tag{5.205}$$

By (5.195)–(5.199), Lemma III.1.4, Definition III.1.5, Remark III.1.6 and Lemma 9.3, for odd k, sum the first equality in (5.202) for $l = 1, 2, \cdots, \frac{k-1}{2}$ and the first equality of (5.204) correspondingly. By comparing with the first equality of (5.205) we have

$$i_{L_0}(\gamma^k) = i_{L_0}(\gamma) + \sum_{l=1}^{\frac{k-1}{2}} i_{\omega_k^{2l}}(\gamma^2), \tag{5.206}$$

and for even k, sum the first equality in (5.202) for $l = 1, 2, \cdots, \frac{k}{2} - 1$ and the first equalities of (5.203) and (5.204) correspondingly. By comparing with the first equality of (5.205) we have

$$i_{L_0}(\gamma^k) = i_{L_0}(\gamma) + i^{L_0}_{\sqrt{-1}}(\gamma) + \sum_{l=1}^{\frac{k}{2}-1} i_{\omega_k^{2l}}(\gamma^2). \tag{5.207}$$

Similarly we have

$$\nu_{L_0}(\gamma^k) = \nu_{L_0}(\gamma) + \sum_{l=1}^{\frac{k-1}{2}} \nu_{\omega_k^{2l}}(\gamma^2), \quad \text{if k is odd}, \tag{5.208}$$

$$\nu_{L_0}(\gamma^k) = \nu_{L_0}(\gamma) + \nu^{L_0}_{\sqrt{-1}}(\gamma) + \sum_{l=1}^{\frac{k}{2}-1} \nu_{\omega_k^{2l}}(\gamma^2), \quad \text{if k is even}. \tag{5.209}$$

Then Theorem 9.1 holds from (5.206)–(5.209). □

From the formulas in Theorem 9.1, we note that

$$i_{L_0}(\gamma^2) = i_{L_0}(\gamma^1) + i^{L_0}_{\sqrt{-1}}(\gamma^1), \quad \nu_{L_0}(\gamma^2) = \nu_{L_0}(\gamma^1) + \nu^{L_0}_{\sqrt{-1}}(\gamma^1).$$

Definition 9.5 The mean L_0-index of γ is defined by

$$\bar{i}_{L_0}(\gamma) = \lim_{k\to+\infty} \frac{i_{L_0}(\gamma^k)}{k}.$$

By definitions of $\bar{i}_{L_0}(\gamma)$ and $\bar{i}(\gamma^2)$(cf. [223] for example), the following result is obvious.

Proposition 9.6 *The mean L_0-index of γ is well defined, and*

$$\bar{i}_{L_0}(\gamma) = \frac{1}{2\pi}\int_0^{\pi} i_B(e^{\sqrt{-1}\theta})d\theta = \frac{\bar{i}(\gamma^2)}{2}, \tag{5.210}$$

here we have written $i_B(\omega) = i_\omega(B) = i_\omega(\gamma_B)$.

For $L_1 = \mathbb{R}^n \times \{0\}$, we have the L_1-index theory established in [189]. Similarly as in Definition 8.1, for $\omega = e^{\theta\sqrt{-1}}$, $\theta \in (0, \pi)$, we define

$$E^\omega_{L_1} = \left\{ x \in L^2([0,1], \mathbb{C}^{2n}) \mid x(t) = e^{\theta t J}\sum_{j\in\mathbb{Z}} e^{j\pi t J}\begin{pmatrix} a_j \\ 0 \end{pmatrix},\ a_j \in \mathbb{C}^n,\ \|x\| < +\infty \right\}.$$

In $E^\omega_{L_1}$ we define two operators $A^\omega_{L_1}$ and $B^\omega_{L_1}$ by the same way as the definitions of operators A^ω and B^ω in Sect. 5.8, but the domain is $E^\omega_{L_1}$. We define

$$i^{L_1}_\omega(B) = I(A^\omega_{L_1}, A^\omega_{L_1} - B^\omega_{L_1}),\ \nu^{L_1}_\omega(B) = m^0(A^\omega_{L_1} - B^\omega_{L_1}).$$

Theorem 9.7 ([209]) *Suppose $\omega_k = e^{\pi\sqrt{-1}/k}$. For odd k we have*

$$i_{L_1}(\gamma^k) = i_{L_1}(\gamma^1) + \sum_{i=1}^{\frac{k-1}{2}} i_{\omega_k^{2i}}(\gamma^2),$$

$$\nu_{L_1}(\gamma^k) = \nu_{L_1}(\gamma^1) + \sum_{i=1}^{\frac{k-1}{2}} \nu_{\omega_k^{2i}}(\gamma^2). \tag{5.211}$$

For even k, we have

$$i_{L_1}(\gamma^k) = i_{L_1}(\gamma^1) + i_{\sqrt{-1}}^{L_1}(\gamma^1) + \sum_{i=1}^{k/2-1} i_{\omega_k^{2i}}(\gamma^2),$$

$$\nu_{L_1}(\gamma^k) = \nu_{L_1}(\gamma^1) + \nu_{\sqrt{-1}}^{L_1}(\gamma^1) + \sum_{i=1}^{k/2-1} \nu_{\omega_k^{2i}}(\gamma^2).$$

Proof The proof is almost the same as that of Theorem 9.1. The only thing different from that is that the matrix N should be replaced by $N_1 = -N$. □

Proposition 9.8 ([209]) *There hold*

$$i(\gamma^2) = i_{L_0}(\gamma^1) + i_{L_1}(\gamma^1) + n, \tag{5.212}$$

$$\nu_1(\gamma^2) = \nu_{L_0}(\gamma^1) + \nu_{L_1}(\gamma^1), \tag{5.213}$$

$$i_{-1}(\gamma^2) = i_{\sqrt{-1}}^{L_1}(\gamma^1) + i_{\sqrt{-1}}^{L_1}(\gamma^1), \tag{5.214}$$

$$\nu_{-1}(\gamma^2) = \nu_{\sqrt{-1}}^{L_1}(\gamma^1) + \nu_{\sqrt{-1}}^{L_1}(\gamma^1). \tag{5.215}$$

Proof As in the proof of Theorem 9.1, we set

$$E_1 = \{z \in L^2([0,2], \mathbb{C}^{2n}) | \ z(t) = \sum_{j\in\mathbb{Z}} e^{jt\pi J}\alpha_j, \ \ \alpha_j \in \mathbb{C}^{2n}, \ \ \|z\| < \infty\}.$$

Then we have $E_1 = \mathcal{W}_{L_0} \oplus \mathcal{W}_{L_1}$. So from Theorem 6.1 we have

$$I(A_1, A_1 - B_1) = i(\gamma^2) + n = i_{L_0}(\gamma^1) + n + i_{L_1}(\gamma^1) + n,$$

which implies (5.212) and (5.213) is also true. By the Bott-type index theory of the ω-index theory for symplectic paths we have

$$i(\gamma^4) = i(\gamma^2) + i_{-1}(\gamma^2),$$

$$\nu(\gamma^4) = \nu(\gamma^2) + \nu_{-1}(\gamma^2).$$

Hence (5.214) and (5.215) hold from Theorems 9.1–9.7.

□

5.10 Iteration Inequalities of L-Index

5.10.1 Precise Iteration Index Formula

From the Bott-type formulas in Theorem 9.1, we prove the abstract precise iteration index formula of i_{L_0}.

Theorem 10.1 *Let $\gamma \in \mathcal{P}_\tau(2n)$, γ^k be defined by (5.185)–(5.187), and $M = \gamma^2(2\tau)$. Then for every $k \in 2\mathbb{N}-1$, there holds*

$$
\begin{aligned}
i_{L_0}(\gamma^k) &= i_{L_0}(\gamma^1) + \frac{k-1}{2}(i(\gamma^2) + S_M^+(1) - C(M)) \\
&\quad + \sum_{\theta \in (0,2\pi)} E\left(\frac{k\theta}{2\pi}\right) S_M^-(e^{\sqrt{-1}\theta}) - C(M),
\end{aligned} \tag{5.216}
$$

where $C(M)$ is defined by

$$C(M) = \sum_{\theta \in (0,2\pi)} S_M^-(e^{\sqrt{-1}\theta})$$

and

$$S_M^{\pm}(\omega) = \lim_{\varepsilon \to 0+} i_{\omega exp(\pm\sqrt{-1}\varepsilon)}(\gamma^2) - i_\omega(\gamma^2)$$

is the splitting number of the symplectic matrix M at ω for $\omega \in \mathbf{U}$.

For every $k \in 2\mathbb{N}$, there holds

$$
\begin{aligned}
i_{L_0}(\gamma^k) = i_{L_0}(\gamma^2) + \left(\frac{k}{2} - 1\right)\left(i(\gamma^2) + S_M^+(1) - C(M)\right) - C(M) \\
- \sum_{\theta \in (\pi,2\pi)} S_M^-(e^{\sqrt{-1}\theta}) + \sum_{\theta \in (0,2\pi)} E\left(\frac{k\theta}{2\pi}\right) S_M^-(e^{\sqrt{-1}\theta}).
\end{aligned} \tag{5.217}
$$

Proof By the definition of the splitting number, we have

$$i_{\omega_0}(\gamma^2) = i(\gamma^2) + \sum_{0 \le \theta < \theta_0} S_M^+(e^{\sqrt{-1}\theta}) - \sum_{0 < \theta \le \theta_0} S_M^-(e^{\sqrt{-1}\theta}),$$

where $\omega_0 = e^{\sqrt{-1}\theta_0}$. So for $k \in 2\mathbb{N} - 1$, let $m = \frac{k-1}{2}$, we have

$$\sum_{i=1}^{m} i_{\omega_k^{2i}}(\gamma^2) = mi(\gamma^2) + \sum_{i=1}^{m}\left(\sum_{0\le\theta<\frac{2i\pi}{k}} S_M^+(e^{\sqrt{-1}\theta}) - \sum_{0<\theta\le\frac{2i\pi}{k}} S_M^-(e^{\sqrt{-1}\theta})\right)$$

$$= m(i(\gamma^2) + S_M^+(1)) + \sum_{\theta\in(0,\pi)}\left(\sum_{\frac{k\theta}{2\pi}<i\le m} S_M^+(e^{\sqrt{-1}\theta}) - \sum_{\frac{k\theta}{2\pi}\le i\le m} S_M^-(e^{\sqrt{-1}\theta})\right)$$

$$\begin{aligned}
&= m(i(\gamma^2) + S_M^+(1)) \\
&\quad + \sum_{\theta\in(0,\pi)}\left(\left(m - \left[\frac{k\theta}{2\pi}\right]\right) S_M^+(e^{\sqrt{-1}\theta}) - \left[m + 1 - \frac{k\theta}{2\pi}\right] S_M^-(e^{\sqrt{-1}\theta})\right) \\
&= m(i(\gamma^2) + S_M^+(1)) \\
&\quad + \sum_{\theta\in(0,\pi)}\left(\left(m - \left[\frac{k\theta}{2\pi}\right]\right) S_M^-(e^{\sqrt{-1}(2\pi-\theta)}) - \left(m + 1 - E\left(\frac{k\theta}{2\pi}\right)\right) S_M^-(e^{\sqrt{-1}\theta})\right) \\
&= m(i(\gamma^2) + S_M^+(1)) + \sum_{\theta\in(\pi,2\pi)}\left(m - \left[\frac{k(2\pi-\theta)}{2\pi}\right]\right) S_M^-(e^{\sqrt{-1}\theta}) \\
&\quad - \sum_{\theta\in(0,\pi)}\left(m + 1 - E\left(\frac{k\theta}{2\pi}\right)\right) S_M^-(e^{\sqrt{-1}\theta}) \\
&= m(i(\gamma^2) + S_M^+(1)) + \sum_{\theta\in(0,\pi)\cup(\pi,2\pi)}\left(-(m+1) + E\left(\frac{k\theta}{2\pi}\right)\right) S_M^-(e^{\sqrt{-1}\theta}) \\
&= m(i(\gamma^2) + S_M^+(1)) - (m+1)C(M) + \sum_{\theta\in(0,2\pi)} E\left(\frac{k\theta}{2\pi}\right) S_M^-(e^{\sqrt{-1}\theta}) \\
&= m(i(\gamma^2) + S_M^+(1) - C(M)) + \sum_{\theta\in(0,2\pi)} E\left(\frac{k\theta}{2\pi}\right) S_M^-(e^{\sqrt{-1}\theta}) - C(M),
\end{aligned}$$

where in the fourth equality and sixth equality we have used the facts that

$$S_M^+(e^{\sqrt{-1}\theta}) = S_M^-(e^{\sqrt{-1}(2\pi-\theta)}),$$

$k = 2m + 1$ and $E(a) + [b] = a + b$ if $a,\ b \in \mathbb{R}$ and $a + b \in \mathbb{Z}$, especially $E(-a)+[a] = 0$ for any $a \in \mathbb{R}$. By using Theorem 9.1 and $m = \frac{k-1}{2}$ we get (5.216). Similarly we obtain (5.217).

□

Corollary 10.2 *For mean L_0-index, there holds*

$$\bar{i}_{L_0}(\gamma) = \frac{1}{2}\bar{i}(\gamma^2) = \frac{1}{2}(i(\gamma^2) + S_M^+(1) - C(M)) + \sum_{\theta\in(0,2\pi)} \frac{\theta}{2\pi} S_M^-(e^{\sqrt{-1}\theta}).$$

Proof The above equality follows from Theorem 9.1 and the definition of the mean L_0-index

$$\bar{i}_{L_0}(\gamma) = \lim_{k\to\infty} \frac{i_{L_0}(\gamma^k)}{k}.$$

□

5.10.2 Iteration Inequalities

Theorem 10.3

1° *For any $\gamma \in \mathcal{P}(2n)$ and $k \in \mathbb{N}$, there holds*

$$\begin{aligned} &i_{L_0}(\gamma^1) + \left[\tfrac{k}{2}\right](i_1(\gamma^2) + \nu_1(\gamma^2) - n) \le i_{L_0}(\gamma^k)\\ &\le i_{L_0}(\gamma^1) + \left[\tfrac{k}{2}\right](i_1(\gamma) + n) - \tfrac{1}{2}\nu_1(\gamma^{2k}) + \tfrac{1}{2}\nu_1(\gamma^2), \ \text{if } k \in 2\mathbb{N}-1, \end{aligned} \tag{5.218}$$

$$\begin{aligned} &i_{L_0}(\gamma^1) + i_{\sqrt{-1}}^{L_0}(\gamma^1) + \left(\tfrac{k}{2}-1\right)(i_1(\gamma^2) + \nu_1(\gamma^2) - n) \le i_{L_0}(\gamma^k) \le i_{L_0}(\gamma^1)\\ &+ i_{\sqrt{-1}}^{L_0}(\gamma^1) + \left(\tfrac{k}{2}-1\right)(i_1(\gamma) + n) - \tfrac{1}{2}\nu_1(\gamma^{2k}) + \tfrac{1}{2}\nu_1(\gamma^2) + \tfrac{1}{2}\nu_{-1}(\gamma^2),\\ &\text{if } k \in 2\mathbb{N}. \end{aligned} \tag{5.219}$$

The index $(i_\omega^{L_0}(\gamma), \nu_\omega^{L_0}(\gamma))$ is defined in Definition 8.1 *for $\omega \in \mathbb{U} = \{z \in \mathbb{C} \mid |z| = 1\}$.*

2° *The left equality of* (5.218) *holds for some $k \ge 3$ and of* (5.219) *holds for some $k \ge 4$ if and only if there holds $I_{2p}\diamond N_1(1,-1)^{\diamond q}\diamond K \in \Omega_0(\gamma^2(2))$ for some non-negative integers p and q satisfying $p+q \le n$ and some $K \in \mathrm{Sp}(2(n-p-q))$ satisfying $\sigma(K) \subset \mathbf{U}\backslash\mathbb{R}$. If $r = n-p-q > 0$, then $R(\theta_1)\diamond\cdots\diamond R(\theta_r) \in \Omega_0(K)$ for some $\theta_j \in (0,\pi)$. In this case, all eigenvalues of K on $\mathbf{U}^+$ (on $\mathbf{U}^-$) are located on the arc between* 1 *and* $\exp(2\pi\sqrt{-1}/k)$ *(and* $\exp(-2\pi\sqrt{-1}/k)$*) in $\mathbf{U}^+$ (in $\mathbf{U}^-$) and are all Krein negative (positive) definite.*

3° *The right equality of* (5.218) *holds for some $k \ge 3$ and of* (5.219) *holds for some $k \ge 4$ if and only if there holds $I_{2p} \diamond N_1(1,1)^{\diamond r} \in \Omega_0(\gamma^2(2))$ for some non-negative integers p and r satisfying $p + r = n$.*

4° *Both equalities of* (5.218)*, and also of* (5.219)*, hold for some $k > 2$ if and only if $\gamma^2(2) = I_{2n}$.*

Proof By Theorem 9.1, summing the inequalities of (2.17) with $\omega = \omega_k^{2i}$, $1 \le i < k/2$, $i \in \mathbb{N}$, we obtain the inequalities (5.218) for odd k and (5.219) for even k. We remind that here we have used the Bott-type formula

$$\nu_1(\gamma^k) = \sum_{\omega^k=1} \nu_\omega(\gamma).$$

The equality conditions follow from 2° and 4° of Proposition II.2.8 together with Lemma II.3.14. We note that from Lemma II.3.14, no eigenvalue on $\mathbf{U}^+$ is Krein positive definite. □

Chapter 6
Maslov Type Index for Lagrangian Paths

6.1 Lagrangian Paths

Let $(\mathbb{R}^{2n}, \tilde{\omega}_0)$ be the standard symplectic space with $\tilde{\omega}_0 = \sum_{i=1}^{n} dx_i \wedge dy_i$. For $z_i = (x_i, y_i) \in \mathbb{R}^n \times \mathbb{R}^n,\ i = 1, 2$, there holds

$$\tilde{\omega}_0(z_1, z_2) = \langle x_1, y_2\rangle - \langle x_2, y_1\rangle.$$

A Lagrangian subspace $L \subset (\mathbb{R}^{2n}, \tilde{\omega}_0)$ is a dimensional n subspace with $\tilde{\omega}_0(z_1, z_2) = 0$ for all $z_1,\ z_2 \in L$. A Lagrangian frame for a Lagrangian subspace L is an injection linear map $Z : \mathbb{R}^n \to \mathbb{R}^{2n}$ whose image is L. Such a frame has the following form

$$Z = \begin{pmatrix} X \\ Y \end{pmatrix} \tag{6.1}$$

where X, Y are $n \times n$ matrices with $Y^T X$ being a symmetric matrix. The n vectors $v_1, \cdots, v_n$ of the columns in Z form a basis of L, i.e., $L = \text{span}\{v_1, \cdots, v_n\}$. For any symplectic matrix $M = \begin{pmatrix} A & B \\ C & D \end{pmatrix}$, the first and last n columns form a Lagrangian frame.

A quadratic form on a vector space V can be view as a map from V to the dual space V^*. If $\mathbb{R}^{2n} = L_1 \oplus L_2$ is a Lagrangian splitting then L_2 can be identifies with L_1^* via $\tilde{\omega}_0$, i.e., for any $v_2 \in L_2$, it defines a linear function v_2^* on L_1 by $v_2^*(v_1) = \tilde{\omega}_0(v_1, v_2)$. So every Lagrangian subspace of V transverse L_2 is the graph of a quadratic form $A : L_1 \to L_1^* = L_2$. We denote the set of all Lagrangian subspaces of $(\mathbb{R}^{2n}, \tilde{\omega}_0)$ by $\Lambda(n)$. It is a submanifold of the Grassmann $G_{2n,n}(\mathbb{R})$. A smooth Lagrangian path is a smooth map $L : [a, b] \to \Lambda(n)$.

Lemma 1.1 *Suppose* $L : [0, 1] \to \Lambda(n)$ *is a Lagrangian path and the Lagrangian frame at* $L(t)$ *is* $\begin{pmatrix} X(t) \\ Y(t) \end{pmatrix}$. *By setting* $X(0) = X,\ Y(0) = Y,\ \dot{X}(0) = \hat{X},\ \dot{Y}(0) = \hat{Y}$, *then the matrix* $S(X, Y, \hat{X}, \hat{Y}) := \hat{X}^T Y - \hat{Y}^T X$ *is symmetric.*

Proof In fact, for any $u_1,\ u_2 \in \mathbb{R}^n$, $z_i(t) = \begin{pmatrix} X(t)u_i \\ Y(t)u_i \end{pmatrix} \in L(t)$, there holds

$$\tilde{\omega}_0(z_1(t), z_2(t)) = \langle X(t)u_1, Y(t)u_2\rangle - \langle X(t)u_2, Y(t)u_1\rangle \equiv 0. \tag{6.2}$$

By taking derivative of (6.2) at $t = 0$, we see that the matrix $\hat{X}^T Y - \hat{Y}^T X$ is symmetric. □

With this observation, for any $L \in \Lambda(n)$ with its Lagrangian frame $Z = \begin{pmatrix} X \\ Y \end{pmatrix}$ and a tangent vector $\hat{L} \in T_L\Lambda(n)$ with its frame $\hat{Z} = \begin{pmatrix} \hat{X} \\ \hat{Y} \end{pmatrix}$ we can define a quadratic form on L.

Definition 1.2 For $v \in L$, we define

$$Q_{L,\hat{L}}(v) \equiv Q(v) = u^T(\hat{X}^T Y - \hat{Y}^T X)u, \tag{6.3}$$

where $v = Zu$.

For this quadratic form, we have the following result.

Lemma 1.3 ([259]) *Let* $L(t) \in \Lambda(n),\ \ t \in [0, 1]$ *be a Lagrangian path with* $L(0) = L$ *and* $\dot{L}(0) = \hat{L}$.

(1) *Let* W *be a fixed Lagrangian complement of* L *and for* $v \in L$ *and a small* t *define* $w(t) \in W$ *by* $v + w(t) \in L(t)$. *Then there holds*

$$Q(v) = \frac{d}{dt}\bigg|_{t=0} \tilde{\omega}_0(v, w(t)),$$

which means that the derivative is independent of the choice of W.

(2) *The form* Q *is natural in the sense that*

$$Q_{\varphi L, \varphi\hat{L}}(\varphi(v)) = Q_{L,\hat{L}}(v)$$

for a symplectic matrix φ.

Proof Choose coordinates so that $L = L(0) = \mathbb{R}^n \times 0$. Then any Lagrangian complement W of L is the graph of a symmetric matrix $B \in \mathcal{L}(\mathbb{R}^n)$:

$$W = \{(By, y) |\ y \in \mathbb{R}^n\}$$

and for small t, the Lagrangian subspace $L(t)$ is the graph of a symmetric matrix $A(t) \in \mathcal{L}(\mathbb{R}^n)$:

$$L(t) = \{(x, A(t)x) | \ x \in \mathbb{R}^n\}.$$

Hence $v = (x, 0)$, $w(t) = (B(y(t)), y(t))$, and $y(t) = A(t)(x + By(t))$. From definition, there holds $y(0) = 0$ and $A(0) = 0$. So we have $\tilde{\omega}_0(v, w(t)) = \langle x, y(t)\rangle$ and

$$\left.\frac{d}{dt}\right|_{t=0} \tilde{\omega}_0(v, w(t)) = \langle x, \dot{y}(0)\rangle = \langle x, \dot{A}(0)x\rangle.$$

This shows that $\left.\frac{d}{dt}\right|_{t=0} \tilde{\omega}_0(v, w(t))$ is independent of B, and as a result it is independent of the choice of W.

With this observation, we now prove Lemma 1.3 by choosing $W = 0 \times \mathbb{R}^n$. Suppose the Lagrangian frame of Ł(t) is $Z(t) = \begin{pmatrix} X(t) \\ Y(t) \end{pmatrix}$, $Z(0) = \begin{pmatrix} X \\ Y \end{pmatrix}$ and $\dot{Z}(0) = \begin{pmatrix} \hat{X} \\ \hat{Y} \end{pmatrix}$. Then $v = \begin{pmatrix} Xu \\ Yu \end{pmatrix}$ and $w(t) = \begin{pmatrix} 0 \\ y(t) \end{pmatrix}$, where $Yu + y(t) = Y(t)X^{-1}(t)Xu$. Hence $\tilde{\omega}_0(v, w(t)) = \langle Xu, y(t)\rangle$. Thus by using $X^tY = Y^TX$ we have

$$\begin{aligned} \left.\frac{d}{dt}\right|_{t=0} \tilde{\omega}_0(v, w(t)) &= \langle Xu, \dot{y}(0)\rangle \\ &= \langle Xu, \hat{Y}u\rangle - \langle Xu, YX^{-1}\hat{X}u\rangle \\ &= \langle Xu, \hat{Y}u\rangle - \langle \hat{X}u, Yu\rangle \\ &= Q(v). \end{aligned}$$

This proves the statement (1). The statement (2) is obvious consequence of the definition. □

Remark 1.4

(1) For any $L \in \Lambda(n)$ with frame $\begin{pmatrix} X \\ Y \end{pmatrix}$, the path $L(t) = e^{Jt}L$ satisfies $L(0) = L$ and the frame of $\dot{L}(0) = \hat{L}$ is $\begin{pmatrix} -Y \\ X \end{pmatrix}$. So

$$Q(v) = u^T(X^XX + Y^TY)u, \quad v = (Xu, Yu)^T.$$

Thus Q is positive definite in L.

(2) If $L = \mathbb{R}^n \times \{0\}$ and $M = \{0\} \times \mathbb{R}^n$, then for any $L' \in \Lambda^0(V, M)$ which is close to L, we can write $L' = \{(x, Ax) | \ x \in L\}$, $A : L \to M$ is a linear map. As in (1.28), we can define a quadratic form

$$Q(L')(x, y) = \tilde{\omega}_0(Ax, y), \quad x, y \in L. \tag{6.4}$$

It is easy to see that the frame of L is $\begin{pmatrix} I \\ 0 \end{pmatrix}$ and the frame of L' is $\begin{pmatrix} I \\ A \end{pmatrix}$. So we have $Q(L') = Q_{L,L'}$ as defined in Definition 1.2, where we identity $A : L \to M$ with $\begin{pmatrix} 0 & 0 \\ A & 0 \end{pmatrix}$.

Lemma 1.5 *For a continuous symmetric matrix function $B(t)$, $t \in [0, 1]$, suppose $\gamma_B(t)$ is the fundamental solution of the linear Hamiltonian system*

$$\dot{\gamma}_B(t) = JB(t)\gamma_B(t).$$

For any Lagrangian subspace L, we define a Lagrangian path $L(t) = \gamma_B(t)L$. We assume $B(t) > 0$ (positively definite). Then for any $t_0 \in [0, 1)$, there holds

$$Q_{L(t_0),\dot{L}(t_0)}(v) > 0, \ \forall\, v \in L(t_0) \setminus \{0\}. \tag{6.5}$$

Proof We set $\gamma_B(t) = \begin{pmatrix} \gamma_{11}(t) & \gamma_{12}(t) \\ \gamma_{21}(t) & \gamma_{22}(t) \end{pmatrix}$. By Lemma 1.3(2), we only need to prove the result for $L = L_0 = \{0\} \times \mathbb{R}^n$ and $t_0 = 0$. In this case, we have the frame of $L(t) = \begin{pmatrix} \gamma_{12}(t) \\ \gamma_{22}(t) \end{pmatrix}$ and $L(0) = \begin{pmatrix} 0 \\ I \end{pmatrix}$. So

$$Q_{L(0),\dot{L}(0)}(v) = -\langle u, \dot{\gamma}_{12}(0)u\rangle = \langle u, B_{22}(0)u\rangle, \ v = \begin{pmatrix} 0 \\ u \end{pmatrix}.$$

□

Remark 1.6 Together with Remark 1.4 and Lemma 1.5, we know that if we orient the manifold $\Lambda(n) \setminus \Lambda^0(\mathbb{R}^{2n}, L) = \{L' \in \Lambda(n) | \dim L \cap L' > 0\}$ in $e^{\theta J}L'$, $\theta > 0$, $L' \in \Lambda(n) \setminus \Lambda^0(\mathbb{R}^{2n}, L)$, then the path $L(t) = \gamma_B(t)L$ crosses the manifold $\Lambda(n) \setminus \Lambda^0(\mathbb{R}^{2n}, L)$ transversally and in positive direction provided $B(t) > 0$ for $t \in [0, 1]$.

6.2 Maslov Type Index for a Pair of Lagrangian Paths

Let $V = (\mathbb{R}^{2n}, \tilde{\omega})$ be a symplectic vector space. The set of its Lagrangian space is $\mathrm{Lag}(V)$. A path of Lagrangian pair f is a continuous map:

$$f : [a, b] \to \mathrm{Lag}(V) \times \mathrm{Lag}(V)$$

for some interval $[a, b]$, $a < b$. So $f(t) = (L_1(t), L_2(t)) \in \mathrm{Lag}(V) \times \mathrm{Lag}(V)$. The set of all paths of Lagrangian pair is denote by $P(V)$.

Theorem 2.1 ([32]) *There is a unique function* $\mu_V : P(V) \to \mathbb{Z}$ *satisfying the following axioms.*

(1) *(Affine Scale Invariance). For* $k > 0, l \geq 0$, *we have the affine map* $\varphi : [a, b] \to [ka + l, kb + l]$ *defined by* $\varphi(t) = kt + l$. *For a given continuous path* $f : [ka + l, kb + l] \to \mathrm{Lag}(V) \times \mathrm{Lag}(V)$, *there holds*

$$\mu_V(f) = \mu_V(f \circ \varphi). \tag{6.6}$$

(2) *(Homotopy Invariance rel. End Points). If* $\delta : [0, 1] \times [a, b] \to \mathrm{Lag}(V) \times \mathrm{Lag}(V)$ *is a continuous map with* $\delta(0, t) = f_1(t)$, $\delta(1, t) = f_2(t)$, $\delta(s, a) = f_1(a) = f_2(a)$ *and* $\delta(s, b) = f_1(b) = f_2(b)$ *for* $s \in [0, 1]$, *then*

$$\mu_V(f_1) = \mu_V(f_2). \tag{6.7}$$

(3) *(Path Additivity). If* $a < b < c$, *and* $f_{[a,c]} : [a, c] \to \mathrm{Lag}(V) \times \mathrm{Lag}(V)$ *is concatenate path of* $f_{[a,b]}$ *and* $f_{[b,c]}$, *then there holds*

$$\mu_V(f_{[a,c]}) = \mu_V(f_{[a,b]}) + \mu_V(f_{[b,c]}). \tag{6.8}$$

(4) *(Symplectic Additivity). Let* $f_k : [a, b] \to \mathrm{Lag}(V_k) \times \mathrm{Lag}(V_k)$, $k = 1, 2$, $V = V_1 \oplus V_2$ $f = f_1 \oplus f_2$. *Then we have*

$$\mu_V(f) = \mu_{V_1}(f_1) + \mu_{V_2}(f_2). \tag{6.9}$$

(5) *(Symplectic Invarince) Let matrix* $P(t) \in \mathrm{Sp}(2n)$ *be a symplectic path. We define* $(P_* f)(t) = (P(t)L_1(t), P(t)L_2(t))$. *Then we have*

$$\mu_V(P_* f) = \mu_V(f). \tag{6.10}$$

(6) *(Normalization). Let* $V = (\mathbb{R}^2, \tilde{\omega}_0)$. *Define* $f : [-\varepsilon, \varepsilon] \to \Lambda(1) \times \Lambda(1)$ *with* $\varepsilon > 0$ *small by the formula*

$$f(t) = (\mathbb{R}, \mathbb{R}(\cos t, sint)), \ \ t \in [-\varepsilon, \varepsilon].$$

Then

$$\begin{aligned} &(i) \ \ \mu_V(f) = 1; \\ &(ii) \ \ \mu_V(f_{[-\varepsilon,0]}) = 0; \\ &(iii) \ \mu_V(f_{[0,\varepsilon]}) = 1. \end{aligned} \tag{6.11}$$

Proof The existence is established in Definition V.1.27 and Theorem V.1.28.

Now we follow the ideas of [32] to prove the uniqueness.

We say that a path of Lagrangian pair $f(t) = (L_1(t), L_2(t)),\ t \in [a, b]$ is proper if $L_1(a) \cap L_2(a) = \{0\}$ and $L_1(b) \cap L_2(b) = \{0\}$.

Suppose $\mu_V(f)$ is the index for $f(t) = (L_1(t), L_2(t)), a \le t \le b$ satisfying the six axioms in Theorem 2.1. We prove that $\mu_V(f) = \hat{i}(f)$. By choosing a symplectic basis of V, we can present the data as $V = (\mathbb{R}^{2n}, \tilde{\omega}_0)$ with the standard symplectic matrix J as the complex structure. Taking $L = \mathbb{R}^n \times \{0\}$, we suppose $L_1(t) = \phi_1(t)L$ and $L_2(t) = \phi_2(t)L$ with $\phi_1(t), \phi_2(t) \in \mathrm{Osp}(2n)$ by (1.20) and the bijection $\phi(L)$ defined after that. Then by the axiom (5)

$$\mu_V(f) = \mu_V(f'),\ \ f'(t) = (L, \phi_1^{-1}(t)\phi_2(t)L).$$

So we can in further suppose $f(t) = (L, L_2(t)), a \le t \le b$. Up to a symplectic matrix of the form $\begin{pmatrix} A & B \\ 0 & D \end{pmatrix}$, we can assume $L \cap L_2(a) = \mathbb{R}^\alpha$ and $L_2(a) = \mathbb{R}^\alpha \oplus J\mathbb{R}^{n-\alpha}$. We introduce the path

$$\gamma_1(t) = (\mathbb{R}^n, e^{Jt}\mathbb{R}^\alpha \oplus J\mathbb{R}^{n-\alpha}),\ \ t \in [0, \frac{\pi}{4}].$$

Then we define the "tail" consists of first traveling along $\gamma_1(t-(a-\frac{\pi}{2}))$ for $a-\frac{\pi}{2} \le t \le a - \frac{\pi}{4}$ and then along the reverse path for $a - \frac{\pi}{4} \le t \le a$. Denote by Δ the composite of the "tail" with the original path $f(t) = (L, L_2(t)), a \le t \le b$, and $\hat{\Delta}$ the part of Δ from $a - \frac{\pi}{4}$ to b. Then we have

$$\mu_V(f) = \mu_V(\Delta) = \mu_V(\gamma_1) + \mu_V(\hat{\Delta}).$$

Note that at the beginning point $t = a - \frac{\pi}{4}$ the path $\hat{\Delta}$ is proper.

In a similar manner, at $t = b$, we may arrange $L_2(b) = \mathbb{R}^\beta \oplus J\mathbb{R}^{n-\beta}$, and consider the path

$$\gamma_2(t) = (\mathbb{R}^n, e^{Jt}\mathbb{R}^\beta \oplus J\mathbb{R}^{n-\beta}),\ 0 \le t \le \frac{\pi}{4}$$

and the composite path $\tilde{\Delta}$ of $\hat{\Delta}$ and $\gamma_2(\frac{\pi}{4} - t + b), b \le t \le b + \frac{\pi}{4}$. Then we have

$$\mu_V(f) = \mu_V(\gamma_1) + \mu_V(\gamma_2) + \mu_V(\tilde{\Delta}),$$

where $\tilde{\Delta}$ is proper.

Next we modify $\tilde{\Delta}(t) = (L, \tilde{L}_2(t)), a - \frac{\pi}{4} \le t \le b + \frac{\pi}{4}$ by continuously deforming $\tilde{L}_2(t)$ to a smooth path which intersects L transversally. At these finite intersection points t_i, we may assume that $\tilde{\Delta}$ is locally isomorphic to one of the following two types:

(1) $\gamma^+(t) = \{\mathbb{R}^n, e^{J(t-t_i)}\mathbb{R}^1 \oplus \mathbb{R}^{n-1}, \quad |t - t_i| < \delta\}$,
(2) $\gamma^-(t) = \{\mathbb{R}^n, e^{-J(t-t_i)}\mathbb{R}^1 \oplus \mathbb{R}^{n-1}, \quad |t - t_i| < \delta\}$.

Outside these intervals $|t - t_i| < \delta$, the two Lagrangian paths have trivial intersection. We assume $\mu_V(\gamma^+|_{[t_i-\delta,t_i]}) = y$ and $\mu_V(\gamma^+|_{[t_i,t_i+\delta]}) = x$. So $\mu_V(\gamma^+) = x + y$, $\mu_V(\gamma^-) = -(x + y)$ and furthermore $\mu_V(\gamma_1) = \alpha x$, $\mu_V(\gamma_2) = \beta y$. So if after making $\tilde{\Delta}$ transverse to L there are p intersection points of type (1) and q intersection points of type (2), then it yields the computation:

$$\mu_V(f) = (p - q)(x + y) + \alpha x + \beta y. \tag{6.12}$$

By using conditions in (6.11), we see that $x = 1$ and $y = 0$. So $\mu_V(f) = p+q-\alpha$. From the definition and the properties of the index $\hat{i}$, we also have $\hat{i}(f) = p+q-\alpha$. Thus $\mu_V(f) = \hat{i}(f)$ for any path f. □

In [32] the authors defined four indices for paths of Lagrangian pair satisfying the six axioms in Theorem 2.1, two of them are geometrical ones $\mu_{geo,1}$ and $\mu_{geo,2}$, the others are analytical ones $\mu_{anal,1}$ and $\mu_{anal,2}$. In the following we introduce the first geometrical Maslov index $\mu_{geo,1}$ and simply denote it by $\mu^{CLM}(f)$.

We first define the Maslov index for a proper path of Lagrangian pair (cf. [127]). For a proper path, say $h(t) = (\hat{L}_1(t), \hat{L}_2(t))$, we will define the Maslov index

$$\mu_{proper}(h) \in \mathbb{Z}.$$

This index is to count with signs and multiplicities the number of times that $\hat{L}_1(t) \cap \hat{L}_2(t) \neq \{0\}$ at t ranges from $t = a$ to $t = b$. More precisely, we first assume $(\hat{L}_1(t), \hat{L}_2(t))$ is a smooth path. Let $\mathcal{Z}$ be the subspace in $[a, b] \times \mathrm{Lag}(V)$ consisting of all pairs (t, L) which has the property

$$\hat{L}_1(t) \cap L \neq 0. \tag{6.13}$$

It is well known that $\mathcal{Z} \cap (\{t\} \times \mathrm{Lag}(V))$ is a codimension one subvariety of $\{t\} \times \mathrm{Lag}(V)$ and has singularities of codimension 3 in $\{t\} \times \mathrm{Lag}(V)$ (see Theorem I.3.3). The top stratum of $\mathcal{Z} \cap (\{t\} \times \mathrm{Lag}(V))$ has a canonical transverse orientation (see [10], and Lemma I.3.1). Indeed, if $\{t\} \times L$ is a point on this top stratum, then the path of Lagrangian $\{t\} \times e^{J\theta}L$ crosses the stratum transversally as θ increases. It defines the desired transverse orientation.

Hence for the path $h(t) = (\hat{L}_1(t), \hat{L}_2(t))$, $a \leq t \leq b$, we may, by a slight perturbation keeping the endpoints fixed, modify the oriented path

$$\gamma = (t, \hat{L}_2(t)), \quad a \leq t \leq b$$

to a new path γ' intersecting $\mathcal{Z}$ only at points of top smooth stratum and crossing them transversely.

Definition I Define $\mu_{proper}(h)$ as

$$\mu_{proper}(h) = \sharp(\mathcal{Z} \cap \gamma') \text{ in } [a, b] \times \text{Lag}(V). \tag{6.14}$$

This intersection number is counted with signs. Since $h(t)$ and so $L_1(t)$ is assumed to be smooth, the union of strata

$$\coprod_{a<t<b} \{\text{top strata of} \mathcal{Z} \cap (\{t\} \times \text{Lag}(V))\}$$

forms a smooth open manifold in $\mathcal{Z}$ and the singularities in $\mathcal{Z}$ are of codimension at least 3. Because of this codimension property, two different choices of γ' may be deformed from one to another avoiding the singularities. Thus the intersection number of (6.14) is well defined and independent of the choice of γ'.

Similarly, if $h(t) = (L_1(t), L_2(t))$ is continuous and piecewise smooth, then we can approximate $h(t)$ by a smooth path and define $\mu_{proper}(h)$ by (6.14) using this smooth approximation. Again the codimension 3 property of the singularities guaranties the independence from the choice of approximation.

From the properties of geometric intersection number, the function

$$\mu_{proper} : \{Proper\ continuous\ and\ piecewise\ smooth\ maps\} \to \mathbb{Z} \tag{6.15}$$

satisfies the properties (1)–(6) as stated in the theorem.

For a continuous and piecewise smooth path of Lagrangian pair

$$f(t) = (L_1(t), L_2(t)),\ t \in [a, b],$$

we have by Lemma I.3.1 an $\varepsilon \in (0, \pi)$ such that for all θ with $0 < |\theta| < \varepsilon$

$$L_1(a) \cap e^{J\theta} L_2(a) = \{0\},\ \ L_1(b) \cap e^{J\theta} L_2(b) = \{0\}. \tag{6.16}$$

In particular, the perturbed path with $0 < \theta' < \varepsilon$ the path

$$g_{\theta'}(t) = (L_1(t), e^{-J\theta'} L_2(t)) \tag{6.17}$$

is proper.

Definition II We define the index $\mu^{CLM}(f)$ by

$$\mu^{CLM}(f) = \mu_{proper}(g_{\theta'}). \tag{6.18}$$

This index was first defined in [32] by Cappell, Lee and Miller. To prove that this index is well defined, we need the following properties of the proper index μ_{proper}.

Property I ([32]) If $h_s(t) = (L_1(s,t), L_2(s,t))$, $a \le t \le b$, $0 \le s \le 1$, defines a continuous mapping

$$[a,b] \times [0,1] \to \mathrm{Lag}(V) \times \mathrm{Lag}(V)$$

such that $\forall s \in [0,1]$, $h_s : [a,b] \to \mathrm{Lag}(V) \times \mathrm{Lag}(V)$ is a proper continuous and piecewise smooth path, then

$$\mu_{proper}(h_0) = \mu_{proper}(h_1). \tag{6.19}$$

Proof The endpoints of $h_s(t)$ may move so long as the properness condition is satisfied. This ensures that all intersections are confined to a compact subset of the open interval (a,b) away from the endpoints. By a small perturbation, the above mapping images in disjoint from the codimension 3 singular strata and is transversal to the top smooth stratrum. From this, it follows (6.19). □

Property II ([32]) If $h(t) = (L_1(t), L_2(t))$, $a \le t \le b$, satisfies the condition $L_1(t) \cap L_2(t) = \{0\}$ for all $t \in [a,b]$, then $\mu_{proper}(h) = 0$.

Proof This is obvious because h gives rise to a path in $\mathrm{Lag}(V) \times \mathrm{Lag}(V)$ which does not intersect $\mathcal{Z}$. □

Proof of the well definedness of $\mu^{CLM}(f)$ First, as θ' varies with $0 < \theta' < \varepsilon$, the corresponding path $g_{\theta'} = \{(L_1(t), e^{-J\theta'}L_2(t)),\ a \le t \le b\}$ is a continuously varying, piecewise smooth path which is proper. Hence, by Property I of the proper index, $\mu_{proper}(g_{\theta'})$ is independent of θ'. Similarly by choosing $0 < \theta' < \varepsilon$ small, we ensure that the properness holds as we smoothly interpolate between any two choice of J and $(\cdot,\cdot)$. We know that the set of complex structures and the set of Hermitian inner products are contractible. Again by Property I, this proves that $\mu_{proper}(g_{\theta'})$ is independent of the choice of J and $(\cdot,\cdot)$. Thus μ^{CLM} is well defined and depends only on the symplectic structure $\tilde{\omega}$ of V. □

Proof of properties (1)–(6) in Theorem 2.1 for $\mu^{CLM}(f)$ By the above discussion of properties (1), (2), (4) of $\mu_{proper}(h)$, the analogous properties of $\mu^{CLM}(f)$ follow immediately from the definition. For example, in (2), the endpoint conditions ensure that once θ is chosen $0 < \theta < \varepsilon$ satisfying the proper condition for the path $f(0)$, it also satisfies the same condition for all $f(s)$, $s \in [0,1]$. Hence $\mu_{proper}(g(0)_{\theta'}) = \mu_{proper}(g(1)_{\theta'})$, so we have $\mu^{CLM}(f(0)) = \mu^{CLM}(f(1))$.

To prove the property (3) for $f(t) = (L_1(t), L_2(t))$, $a \le t \le c$, and $a < b < c$, we choose $\varepsilon \in (0,\pi)$ such that $L_1(t) \cap e^{J\theta}L_2(t) = 0$ for all θ with $0 < |\theta| < \varepsilon$ and for $t = a$, $t = b$ and $t = c$. Hence for $0 < \theta' < \varepsilon$,

$$\begin{aligned} g_{\theta'}^{[a,b]} &= \{(L_1(t), e^{-J\theta'}L_2(t),\ \ a \le t \le b)\}, \\ g_{\theta'}^{[b,c]} &= \{(L_1(t), e^{-J\theta'}L_2(t),\ \ b \le t \le c)\}, \\ g_{\theta'}^{[a,c]} &= \{(L_1(t), e^{-J\theta'}L_2(t),\ \ a \le t \le c)\} \end{aligned}$$

are proper paths, and by proper path additivity,

$$\mu_{proper}(g_{\theta'}^{[a,c]}) = \mu_{proper}(g_{\theta'}^{[a,b]}) + \mu_{proper}(g_{\theta'}^{[b,c]}). \tag{6.20}$$

So we have $\mu^{CLM}(f^{[a,c]}) = \mu^{CLM}(f^{[a,b]}) + \mu^{CLM}(f^{[b,c]})$.

In similar fashion, property (5) is proved. Given $\phi : [a,b] \to \mathrm{Sp}(V)$, we choose smooth map $\psi : [a,b]\times[0,1] \to \mathrm{Sp}(V)$ such that $\psi(t,1) = \phi(t)$ and $\psi(t,0) = Id$ for $t \in [a,b]$. For $t = a$ or $t = b$, we consider the largest $\varepsilon_s \in (0, \frac{\pi}{2}]$ such that

$$\psi(t,s)L_1(t) \cap e^{-J\theta}\psi(t,s)L_2(t) = \{0\} \tag{6.21}$$

for all θ with $0 < |\theta| < \varepsilon_s$. Since ε_s varies in an upper-semicontinuous manner on s, we may choose ε such that (6.21) holds for all s, all θ with $0 < |\theta| < \varepsilon$ and for $t = a,\ b$. With this choice ε and any $\theta \in (0, \varepsilon)$, the path

$$(\psi(t,s)L_1(t), e^{-J\theta}\psi(t,s)L_2(t)),\quad t \in [a,b]$$

is a proper path for all s. By Property I of μ_{proper}, we have $\mu^{CLM}(\phi_* f) = \mu^{CLM}(f)$ because $\psi(t,1) = \phi(t),\ \psi(t,0) = Id$.

The property (6) is easily verified. After a perturbation, the path $f(t) = (\mathbb{R}, \mathbb{R}\{(\cos t, \sin t)\})$, $t \in [-\varepsilon, \varepsilon]$, become $g_{\theta'}(t) = (\mathbb{R}, \mathbb{R}\{(\cos(t-\theta'), \sin(t-\theta'))\})$, $t \in [-\varepsilon, \varepsilon]$. The path $(t, \mathbb{R})$ and $(t, \mathbb{R}\{(\cos(t-\theta'), \sin(t-\theta'))\})$ have no intersection between $-\varepsilon \le t \le 0$, and have intersection $+1$ in $0 \le t \le \varepsilon$. This proves (i)–(iii) in (6.11). □

Remark 2.2 From (6.12), it is easy to see for a collection of functions $\hat{\mu}_V(f)$ satisfies properties (1)–(5) in Theorem 2.1, then

$$\hat{\mu}_V(f) = (x+y)\mu^{CLM}(f) + y(\cap_b - \cap_a),$$

where x, y are some fixed integers and $\cap_t = \dim L_1(t) \cap L_2(t)$.

From Definition V.1.27, Theorem V.1.28 and Theorem 2.1, there holds

$$\hat{i}(f) = \mu^{CLM}(f) \tag{6.22}$$

for any Lagrangian pair path f. We note that our definition of $\hat{i}(f)$ is algebraic and a more simple one.

Here are some additional properties of μ_V (the readers can prove them as exercises):

(7) For a path $f(t) = (L_1(t), L_2(t)),\quad t \in [a,b]$ with $\cap_t = \dim L_1(t) \cap L_2(t)$ independent of t, then

$$\mu_V(f) = 0. \tag{6.23}$$

(8) Let $\psi : [c,d] \to [a,b]$ be a continuous and piecewise smooth function with $\psi(c) = a$, $\psi(d) = b$. Then for $f(t) = (L_1(t), L_2(t))$, $t \in [a,b]$, we have

$$\mu_V(f \circ \psi) = \mu_V(f). \tag{6.24}$$

(9) For $f(t) = (L_1(t), L_2(t))$, $t \in [a,b]$, we define its reverse path $f^-(t) = (L_1(-t), L_2(-t))$, $-b \le t \le -a$. Then

$$\mu_V(f) = -\mu_V(f^-). \tag{6.25}$$

(10) For $f(t) = (L_1(t), L_2(t))$, $t \in [a,b]$, we define $f_c(t) = (L_2(t), L_1(t))$, $t \in [a,b]$. Then we have

$$\mu_V(f) + \mu_V(f_c) = \cap_a - \cap_b. \tag{6.26}$$

6.3 Hörmander Index Theory

To measure the difference of two indices, we introduce the Hörmander index $s(L_1, L_2, M_1, M_2)$ for four Lagrangian subspaces L_1, L_2, M_1, M_2. We recall that for $L \in \mathrm{Lag}(V)$, $\Lambda^k(V,L) = \{L' \in \mathrm{Lag}(V) \mid \dim L \cap L' = k\}$.

Theorem 3.1 ([9, 32, 143]) $\mathrm{Lag}(V) \setminus \Lambda^0(V,L)$ *defines an oriented cycle* μ *of codimension* 1 *in* $\mathrm{Lag}(V)$. $\pi_1(\mathrm{Lag}(V)) \simeq \mathbb{Z}$. *The mapping*

$$\alpha : \gamma \to \gamma \cdot \mu = intersection\, number\, of\, \gamma\, and\, \mu, \quad \forall \gamma \in \pi_1(\mathrm{Lag}(V))$$

defines a generator of $H^1(\mathrm{Lag}(V), \mathbb{Z})$. α *is called the Maslov-Arnold index (or class) of* $(V, \tilde{\omega})$.

Proof Since only a variety of codimension 3 in $\mathrm{Lag}(V)$ is attached to its regular part $\Lambda^k(V,L)$, $\overline{\Lambda^1(V,L)} = \mathrm{Lag}(V) \setminus \Lambda^0(V,L)$ defines a chain of codimension 1 in $\mathrm{Lag}(V)$ without boundary. Moreover, it is oriented by defining as the positive side at $M \in \Lambda^1(V,L)$ the set of all $L' \in T_M\mathrm{Lag}(V)$ such that the corresponding quadratic form $Q(L')$ defined in (1.28) is positive definite on the line $L \cap M$. The orientation is also determined by the direction of $e^{Jt}M$ as t increasing.

To prove that $\pi_1(\mathrm{Lag}(V)) \simeq \mathbb{Z}$, and α is a generator of $H^1(\mathrm{Lag}(V), \mathbb{Z})$, we prove that α is bijective and exhibit an element γ_0 of $\pi_1(\mathrm{Lag}(V))$ such that $\gamma_0 \cdot \mu = 1$.

Suppose $\gamma \in \pi_1(\mathrm{Lag}(V))$ such that $\gamma \cdot \mu = 0$. Represent γ by a closed differentiable curve $\gamma(t)$, intersecting $\mathrm{Lag}(V) \setminus \Lambda^0(V,L)$ only at finitely often and only at $\Lambda^1(V,L)$ and transversally. Then

$$\gamma \cdot \mu = \sum_{\gamma(t) \in \Lambda^1(V,L)} \pm 1$$

with 1 if $\gamma(t)$ crosses $\Lambda^1(V, L)$ from the negative side to the positive side, and -1 otherwise. Now let $\gamma(t_1)$ be a positive and $\gamma(t_2)$ a subsequent negative crossing, that is, $\gamma(t) \in \Lambda^0(V, L)$ if $t_1 < t < t_2$. Connect $\gamma(t_1)$ and $\gamma(t_2)$ by a differentiable curve δ in $\Lambda^1(V, L)$, which can be accompanied by nearby curves δ_+, respectively, δ_- at the positive and negative side of $\Lambda^1(V, L)$, connecting $\gamma(t_1+\varepsilon)$ with $\gamma(t_2-\varepsilon)$ and $\gamma(t_1-\varepsilon)$ with $\gamma(t_2+\varepsilon)$, respectively. Here $\varepsilon > 0$ is small. Because $\Lambda^0(V, L)$ is simply connected, the curve $\gamma(t)$, $t_1+\varepsilon \le t \le t_2-\varepsilon$, followed by $(\delta_+)^{-1}$ is a contractible closed curve. A simple geometric argument shows that the curve $\gamma(t)$, $t_1-\varepsilon \le t \le t_1+\varepsilon$, followed by δ_+, then $\gamma(t)$, $t_2-\varepsilon \le t \le t_2+\varepsilon$ and then by $(\delta_-)^{-1}$ also is a contractible closed curve. So γ is homotopic to γ, $t \le t_1-\varepsilon$, followed by $(\delta_-)^{-1}$, and then by $\gamma(t)$, $t \ge t_2+\varepsilon$. However this curve has two intersection less with $\Lambda^1(V, L)$. Eliminating in this way all intersections, we arrive at the conclusion that γ is homotopic to a point. This implies that α is injective.

Now we construct γ_0. Let P be a nonisotropic two-dimensional subspace of $(V, \tilde{\omega})$. Then $(P, \tilde{\omega})$ is a symplectic vector space and $V = P \oplus P^{\perp_{\tilde{\omega}}}$. On canonical coordinates in P, define

$$\gamma_P(\theta) = \left\{ \begin{pmatrix} x\cos\theta \\ x\sin\theta \end{pmatrix}; x \in \mathbb{R} \right\}, \quad \theta \text{ from } 0 \text{ to } \pi.$$

γ_P is a closed curve in $\operatorname{Lag}(P)$ and intersects the ξ-axis only for $\theta = \frac{\pi}{2}$. At that point, its derivative $\dot{\gamma}_P(\frac{\pi}{2})$ is represented by the quadratic form

$$\xi \to \frac{d}{d\theta}\tilde{\omega}(\begin{pmatrix} \xi\cos\theta/\sin\theta \\ 0 \end{pmatrix}, \begin{pmatrix} 0 \\ \xi \end{pmatrix})_{\theta=\frac{\pi}{2}} = \xi^2,$$

so the intersection is positive. To lift this example to V, let L_0, M_0 be transeval Lagrangian subspaces of the symplectic vector space $P^{\perp_{\tilde{\omega}}}$. Define

$$\gamma_0(\theta) = \gamma_P(\theta) + M_0, \quad L = \gamma_P(\frac{\pi}{2}) + L_0.$$

Then $\gamma_0(\theta)$ intersects $\operatorname{Lag}(V) \setminus \Lambda^0(V, L)$ only for $\theta = \frac{\pi}{2}$, in $\Lambda^1(V, L)$ and in the positive direction, so $\gamma_0 \cdot \mu = 1$. □

It should be remarked that the definition of $\alpha = \alpha_L$ does not depend on the choice of $L \in \operatorname{Lag}(V)$, because the connected group $\operatorname{Sp}(V, \mathbb{R})$ acts transitively on $\operatorname{Lag}(V)$ and the intersection number is homotopy invariant.

Now we are ready to define the Hörmander index.

Definition 3.2 (**[143]**) Let L_1, L_2, M_1, M_2 be four Lagrangian subspaces of $V = (\mathbb{R}^{2n}, \tilde{\omega})$ such that L_j is transversal to M_k for all $j,\ k = 1,\ 2$. Then the Hörmander index is defined by

$$s(L_1, L_2; M_1, M_2) = \alpha(\gamma) = \gamma \cdot \mu, \tag{6.27}$$

where γ is a closed curve in $\mathrm{Lag}(V)$ consisting of an arc of Lagrangian subspaces from M_1 to M_2 transversal to L_1, followed by an arc of Lagrangian subspaces from M_2 to M_1 transversal to L_2.

Remark 3.3

(1) Since $\Lambda^0(V, L_1)$ and $\Lambda^0(V, L_2)$ are connected and simply connected, the Hörmander index $s(L_1, L_2; M_1, M_2)$ is well defined. In fact, If γ_1 and γ_1' are two paths in $\Lambda^0(V, L_1)$ from M_1 to M_2, and γ_2 and γ_2' are two paths in $\Lambda^0(V, L_2)$ from M_2 to M_1, $\gamma = \gamma_1 * \gamma_2$, $\gamma' = \gamma_1' * \gamma_2'$, then $\gamma_1 \sim \gamma_1'$ in $\Lambda^0(V, L_1)$ and $\gamma_2 \sim \gamma_2'$ in $\Lambda^0(V, L_2)$. By definition, there hold

$$\langle \gamma, \alpha \rangle = \gamma \cdot \mu_{L_2} = \gamma_1 \cdot \mu_{L_2},$$

$$\langle \gamma', \alpha \rangle = \gamma' \cdot \mu_{L_2} = \gamma_1' \cdot \mu_{L_2}.$$

Thus $\langle \gamma, \alpha \rangle = \langle \gamma', \alpha \rangle$.

(2) If M_1 and M_2 are in the same connected component of $\Lambda^0(V, L_1) \cap \Lambda^0(V, L_2)$, we can choose $\gamma = \gamma_1 * \gamma_2$ to be the composite path of γ_1 and γ_2 with γ_1 from M_1 to M_2 and γ_2 from M_2 to M_1 such that γ_1 and γ_2 are in the same connected component of $\Lambda^0(V, L_1) \cap \Lambda^0(V, L_2)$. Thus we have $s(L_1, L_2; M_1, M_2) = 0$. The inverse is also true (see [76] for a proof). Clearly s is integer-valued and continuous (locally constant) in all variables. Moreover there hold

$$s(L_1, L_2; M_1, M_2) = -s(L_1, L_2; M_2, M_1) = s(L_2, L_1; M_2, M_1) \tag{6.28}$$

and

$$s(L_1, L_2; M_1, M_2) + s(L_1, L_2; M_2, M_3) = s(L_2, L_2; M_1, M_3). \tag{6.29}$$

(3) With the same notation as in (2), we see that

$$s(L_1, L_2; M_1, M_2) = \mu^{CLM}(L_2, \gamma_1), \ \mu^{CLM}(L_1, \gamma_1) = 0. \tag{6.30}$$

Thus there holds

$$s(L_1, L_2; M_1, M_2) = \mu^{CLM}(L_2, \gamma_1) - \mu^{CLM}(L_1, \gamma_1) \tag{6.31}$$

as the difference of the index of path γ_1 with two different "boundary" Lagrangian subspaces L_2 and L_1.

(4) From Theorem V.1.21, Theorem V.1.24, Theorem 2.1, Theorem VII.1.3 and Theorem VII.3.2, we can translate the Hörmander index $s(L, L'; M_1, M_2)$ as the (L, L')-concavity of any symplectic path $\gamma : [0, 1] \to \mathrm{Sp}(2n)$ with $\gamma(0)L'' = M_1$, $\gamma(1)L'' = M_2$, where $L'' \in \Lambda(n)$ such that

$$\hat{i}_{L'}(\gamma) - \hat{i}_L(\gamma) = s(L, L'; M_1, M_2). \tag{6.32}$$

If $\gamma \in \mathcal{P}_1(2n)$, then there holds

$$i_{L'}(\gamma) - i_{L'}(I) - i_L(\gamma) + i_L(I) = s(L, L'; M, \gamma(1)M), \tag{6.33}$$

for some $M \in \Lambda(n)$ such that $s(L, L'; M, \gamma(1)M)$ is well defined. Especially, for $L = L_0 = \{0\} \oplus \mathbb{R}^n$ and $L' = L_1 = \mathbb{R}^n \oplus \{0\}$, we have that $i_{L_0}(I) = i_{L_1}(I) = -n$, so there holds

$$i_{L_1}(\gamma) - i_{L_0}(\gamma) = s(L_0, L_1; M, \gamma(1)M). \tag{6.34}$$

Definition 3.4 ([143]) If L_1, L_2 are transversal Lagrangian subspaces, $L' \in \mathrm{Lag}(V)$ transversal to L_2, then $\mathrm{sgn}(L_1, L_2; L')$ is the signature of the symmetric bilinear form $Q(L') : (x, y) \to \tilde{\omega}(Ax, y)$ on L_1, if $L' = \{x + Ax|\ x \in L_1\}$, $A : L_1 \to L_2$.

Lemma 3.5 ([143]) *If L_1, L_2 are transversal Lagrangian subspaces, and M_j is transversal to L_k for $M_1, M_2 \in \mathrm{Lag}(V)$, then*

$$s(L_1, L_2; M_1, M_2) = \frac{1}{2}[\mathrm{sgn}(L_1, L_2; M_1) - \mathrm{sgn}(L_1, L_2; M_2)]. \tag{6.35}$$

Proof $s(L_1, L_2; M_1, M_2)$ is clearly equal to the intersection number of a differentiable curve γ from M_2 to M_1 in $\Lambda^0(V, L_2)$ with $\mathrm{Lag}(V) \setminus \Lambda^0(V, L_1)$, the curve intersects the latter only in its regular part(top stratum) $\Lambda^1(V, L_1)$ and transversally. Now $\mathrm{sgn}(L_1, L_2; \gamma(t))$ only changes when $\gamma(t)$ crosses $\Lambda^1(V, L_1)$. We claim that it changes by $+2$, respectively, -2, if the crossing is positive, respectively, negative.

Suppose the crossing at t_0 is positive. Choose coordinates in L_1 such that $\gamma(t_0) \cap L_1$ is the x_1-axis and

$$Q(\gamma(t_0)) = \begin{pmatrix} 0 & 0 \\ 0 & Q_0 \end{pmatrix}$$

for some symmetric nonsingular $(n-1) \times (n-1)$ matrix Q_0. Now if D is nonsingular,

$$\mathrm{sgn} \begin{pmatrix} B & C \\ C^T & D \end{pmatrix} = \mathrm{sgn}D + \mathrm{sgn}[B - CD^{-1}C^T]. \tag{6.36}$$

Here $CD^{-1}C^T$ vanishes of second order as $C = 0$, so

$$\mathrm{sgn}[B(t) - C(t)D(t)^{-1}C(t)^T]$$

jumps by $+2$ if t passes t_0, $B(t_0) = 0$, $c(t_0) = 0$, $D(t_0) = Q_0$, $B'(t_0) > 0$. Note that $\mathrm{sgn}D(t)$ remains constant.

Finally we remark that

$$\begin{aligned}\gamma(t) &= \{x + A(t)x | x \in L_1\},\ A(t) : L_1 \to L_2\\ &= \{y + \tilde{A}(t)y |\ y \in \gamma(t_0)\},\ \tilde{A}(t) : \gamma(t_0) \to L_2.\end{aligned}$$

So $A(t) = A(t_0) + \tilde{A}(t)(I + A(t_0))$ on L_1, hence

$$\frac{d}{dt}\tilde{\omega}(A(t)x, y)_{t=t_0} = \frac{d}{dt}\tilde{\omega}(\tilde{A}(t)x, y)_{t=t_0} \tag{6.37}$$

if $x, y \in \gamma(t_0) \cap L_1$. So $B'(t_0) > 0$ if and only if $\gamma(t)$ crosses $\Lambda^1(V, L_1)$ in the positive direction. □

Lemma 3.6 ([143]) $\operatorname{sgn}(L_1, L_2; M) = -\operatorname{sgn}(L_1, M; L_2)$ *if* $L_1,\ L_2,\ M$ *are mutually transversal.*

Proof We can write

$$M = \{x + Ax |\ x \in L_1\},\quad A : L_1 \to L_2,$$

$$L_2 = \{y + By |\ y \in L_1\},\quad B : L_1 \to M.$$

If $x \in L_1$, then $(x + Ax) + Bx = (x + Bx) + Ax$ belongs to both M and L_2, hence $x + Ax + Bx = 0$. So

$$\tilde{\omega}(Ax, y) = -\tilde{\omega}(Bx, y),\quad \forall\, x, y \in L_1.$$

□

Lemma 3.7 ([143]) *If* $L_1, L_2, M_1, M_2 \in \operatorname{Lag}(V)$, L_j *transverses to* M_k *for* $j,\ k = 1,\ 2$, *then*

$$s(L_1, L_2; M_1, M_2) = \frac{1}{2}[\operatorname{sgn}(L_1, M_2; L_2) - \operatorname{sgn}(L_1, M_1; L_2)]. \tag{6.38}$$

Proof Note that the right hand side is well defined and (6.38) holds if L_1, L_2 are transversal in view of Lemmas 3.5 and 3.6. Now assume $\dim L_1 \cap L_2 = k$. Let $\gamma(t)$ be a curve in $\operatorname{Lag}(V)$ such that $\gamma(0) = L_2$ and $\gamma'(0)$, regarded as a quadratic form on L_2, is positive definite on $L_1 \cap L_2$. Then, using formulas analogous to (6.36) and (6.37) we obtain that $\gamma(t)$ is transversal to L_1,

$$\operatorname{sgn}(L_1, M_2; \gamma(t)) = \operatorname{sgn}(L_1, M_2; L_2) + k,$$

$$\operatorname{sgn}(L_1, M_1; \gamma(t)) = \operatorname{sgn}(L_1, M_2; L_2) + k$$

if $t > 0$ is small. This proves (6.38) because it holds when L_2 is replaced by $\gamma(t)$, $t > 0$, small, and the left-hand side is locally constant in L_2. □

For any four Lagrangian subspaces L_1, L_2, M_1, M_2, there is an $\varepsilon > 0$ such that L_j is transversal to $e^{J\theta}M_k$, $0 < |\theta| < \varepsilon$, $j, k = 1, 2$.

Definition 3.8 For $L_1, L_2, M_1, M_2 \in \mathrm{Lag}(V)$, we define the Hörmander index as

$$s(L_1, L_2; M_1, M_2) = s(L_1, L_2; e^{-J\theta}M_1, e^{-J\theta}M_2), \quad 0 < \theta < \varepsilon, \tag{6.39}$$

the right hand side of (6.39) is defined in Definition 3.2.

It is easy to see that the Hörmander index $s(L_1, L_2; M_1, M_2)$ is well-defined, i.e., it is independent of $\theta \in (0, \varepsilon)$ and the Eqs. (6.31), (6.32), (6.33) and (6.34) hold in this case.

Chapter 7
Revisit of Maslov Type Index for Symplectic Paths

7.1 Maslov Type Index for Symplectic Paths

We recall that $(\mathbb{R}^{2n}, \tilde{\omega})$ is a symplectic space, and $\mathrm{Sp}(2n, \tilde{\omega})$ is the symplectic group of $(\mathbb{R}^{2n}, \tilde{\omega})$. That is

$$\mathrm{Sp}(2n, \tilde{\omega}) = \{M \in \mathcal{L}(\mathbb{R}^{2n}) |\ M^*\tilde{\omega} = \tilde{\omega}\}.$$

We denote by $\mathcal{P}(2n, \tilde{\omega}) = \{\gamma \in C([0, 1], \mathrm{Sp}(2n, \tilde{\omega}))|\ \gamma(0) = I\}$ the set of continuous and piecewise smooth symplectic paths starting from I and $\Lambda(n, \tilde{\omega})$ the set of Lagrangian subspaces of $(\mathbb{R}^{2n}, \tilde{\omega})$. We recall that $\mathcal{P}(2n) = \mathcal{P}(2n, \tilde{\omega}_0)$. We also denote by $\tilde{\mathcal{P}}(2n, \tilde{\omega}) = \{\gamma |\ \gamma \in C([a, b], \mathrm{Sp}(2n, \tilde{\omega}))\}$ the set of continuous and piecewise smooth symplectic paths.

Suppose $L \in \Lambda(n, \tilde{\omega})$. For $\gamma \in \tilde{\mathcal{P}}(2n, \tilde{\omega})$, we now define the L-index $\mu_L(\gamma)$ of γ.

Definition 1.1 We define

$$\mu_L(\gamma) = \mu^{CLM}(f), \quad f(t) = (L, \gamma(t)L), \ 0 \le t \le 1. \tag{7.1}$$

From Lemma VI.1.5 and Remark VI.1.6, we see that (compare with Corollary V.6.5)

$$\mu_L(\gamma_B) = \sum_{t\in[0,1)} \dim(\gamma_B(t)L \cap L) \tag{7.2}$$

for $B(t) > 0$, $t \in [0, 1]$, and specially, we have $\mu_L(\gamma_B) \ge n$.

We note that the symplectic structure $\tilde{\omega}$ may not be the standard one. We will unify various indices as spacial cases of $\hat{\mu}_{\hat{L}}(\gamma)$ as defined in (7.3). For this purpose, we consider the linear symplectic spaces $(\mathbb{R}^{2n} \oplus \mathbb{R}^{2n}, -\tilde{\omega}_0 \oplus \tilde{\omega}_0)$ with $-\tilde{\omega}_0 \oplus$

$\tilde{\omega}_0((x_1, y_1), (x_2, y_2)) = -\tilde{\omega}_0(x_1, x_2) + \tilde{\omega}_0(y_1, y_2)$ for all $(x_j, y_j) \in \mathbb{R}^{2n} \oplus \mathbb{R}^{2n}$, $j = 1, 2$. For a symplectic matrix $M \in \mathrm{Sp}(2n)$, the graph $\mathrm{Gr}(M)$ of $M : \mathbb{R}^{2n} \to \mathbb{R}^{2n}$ is defined by $\mathrm{Gr}(M) = \{(x, Mx) |\ x \in \mathbb{R}^{2n}\}$. It is obvious that $\mathrm{Gr}(M)$ is a Lagrangian subspace of $V = (\mathbb{R}^{2n} \oplus \mathbb{R}^{2n}, -\tilde{\omega}_0 \oplus \tilde{\omega}_0)$. Let $\hat{L} \in \mathrm{Lag}(V)$, and $\gamma : [a, b] \to \mathrm{Sp}(2n)$ be a continuous symplectic paths.

Definition 1.2 We define

$$\hat{\mu}_{\hat{L}}(\gamma) = \mu^{CLM}(f), \quad f(t) = (\hat{L}, \mathrm{Gr}(\gamma(t))), \ a \le t \le b. \tag{7.3}$$

We have the following four spacial cases:

(1) $\hat{L} = L \oplus L$ for $L \in \Lambda(n)$, we have the L-index: $\hat{\mu}_L(\gamma)$.
(2) $\hat{L} = L \oplus L'$ for $L,\ L' \in \Lambda(n)$, we have the (L, L')-index: $\hat{\mu}_L^{L'}(\gamma)$.
(3) $\hat{L} = \mathrm{Gr}(P)$ for $P \in \mathrm{Sp}(2n)$, we have the P-index: $\hat{\mu}_P(\gamma)$.
(4) $\hat{L} = \mathrm{Gr}(I)$, I is the $2n \times 2n$ identity matrix, we have the I-index: $\hat{\mu}_I(\gamma)$.

We recall that the L-index $\hat{i}_L(\gamma), \gamma \in \tilde{\mathcal{P}}(2n)$, is defined in (5.57), the index $i^P(\gamma)$ is defined in Definition IV.1.7, $\mu_P(\gamma)$ is defined in Definition V.3.1, and $i_{L_0}^L(\gamma)$ defined in Definition V.2.6, $\mu_{L_0}^L(\gamma)$ defined in Definition V.2.4. The following result shows that indices defined in different manner in fact are the same.

Theorem 1.3 *We have the following equalities*

(1) $\hat{i}_L(\gamma) = \hat{\mu}_L(\gamma)$ *for* $\gamma \in \tilde{\mathcal{P}}(2n)$;
(2) $i_P(\gamma) = \mu_P(\gamma) = \hat{\mu}_P(\gamma)$, $\gamma \in \mathcal{P}(2n)$, $P \in \mathrm{Sp}(2n)$;
(3) $i_{L_0}^L(\gamma) - i_{L_0}^L(I) = \mu_{L_0}^L(\gamma) - \mu_{L_0}^L(I) = \hat{\mu}_{L_0}^L(\gamma)$, $\gamma \in \mathcal{P}(2n)$;
(4) $i_1(\gamma) + n = i_I(\gamma) = \hat{\mu}_I(\gamma)$, $\gamma \in \mathcal{P}(2n)$.

Proof By Theorem VI.2.1, we only need to prove the various indices satisfy the axioms (1)–(6). The equality in (1) above is a consequence of Theorem V.1.24. The equality (2) is a consequence of Theorem IV.1.21. And equality (3) can be proved in the same manner. Equality (4) is a special case of equality (2). □

Let $B \in C([0, 1], \mathcal{L}_s(\mathbb{R}^{2n}))$, and $\gamma_B : [0, 1] \to \mathrm{Sp}(2n)$ be the fundamental solution of the following linear Hamiltonian system

$$\dot{y}(t) = JB(t)y(t).$$

We take a Lagrangian subspace $V_\omega = L_0 \oplus (e^{\theta J} L_0)$ for $\omega = e^{\sqrt{-1}\theta}, \theta \in [0, \pi),$.

Definition 1.4 We define the ω-index $\mu_\omega^{L_0}(B)$ by

$$\mu_\omega^{L_0}(B) = \mu^{CLM}(V_\omega, \mathrm{Gr}(\gamma_B)). \tag{7.4}$$

Lemma 1.5 *There holds*

$$i_\omega^{L_0}(B) = \mu_\omega^{L_0}(B), \quad \omega = e^{\sqrt{-1}\theta}, \quad \theta \in [0, \pi). \tag{7.5}$$

Proof It is a direct consequence of Theorem 1.3(3). □

7.2 The ω-Index Function for P-Index

Definition 2.1 A complex symplectic space $V = \mathbb{C}^{2n}$ is a complex vector space, together with a prescribed symplectic form Ω, namely a sesqui-linear (or conjugate bilinear) complex-valued function $\Omega : V \times V \to \mathbb{C}$

$$u, v \to \Omega(u, v) \tag{7.6}$$

satisfying the following axioms, for all $u, v, w \in V$ and all complex numbers $c_1, c_2 \in \mathbb{C}$,

(1) $\Omega(c_1 u + c_2 v, w) = c_1\Omega(u, w) + c_2\Omega(v, w)$ (linearity property in first argument),
(2) $\Omega(u, v) = -\overline{\Omega(v, u)}$ (skew-Hermitian property),
(3) $\Omega(u, V)$ implies $u = 0$ (non-degeneracy property).

Properties (1) and (2) of Definition 2.1 imply $\Omega(u, c_1 v + c_2 w) = \bar{c}_1\Omega(u, v) + \bar{c}_2\Omega(u, w)$. The subspace L with maximum dimension is called Lagrangian subspace if $\Omega|_L = 0$.

As an example, we complexification the symplectic space $(\mathbb{R}^{2n}, \tilde{\omega}_0, J)$ as $(\mathbb{C}^{2n}, \tilde{\omega}_0, J)$. In $(\mathbb{R}^{2n}, \tilde{\omega}_0, J)$, we have $\tilde{\omega}_0(u, v) = \langle Ju, v\rangle = v^T Ju$ where $\langle\cdot,\cdot\rangle$ is the standard inner product. In $(\mathbb{C}^{2n}, \tilde{\omega}_0, J)$, we have $\tilde{\omega}_0(u, v) = (Ju, v) = \bar{v}^T Ju$ where $(\cdot,\cdot)$ is the standard Hermitian inner product. We know that $(\mathbb{C}^{2n}, \tilde{\omega}_0, J)$ is a complex symplectic vector space. The linear bijection $M : \mathbb{C}^{2n} \to \mathbb{C}^{2n}$ satisfying $\tilde{\omega}_0(Mu, Mv) = \tilde{\omega}_0(u, v)$ for all $u, v \in \mathbb{C}^{2n}$ is called complex symplectic map, the corresponding matrix M satisfies $M^* JM = J$, where M^* is the conjugate transpose of M, which is called complex symplectic matrix. We denote by $\mathrm{Sp}(2n, \mathbb{C})$ the set of all complex symplectic matrices. It is a Lie group and $\mathrm{Sp}(2n) = \mathrm{Sp}(2n, \mathbb{R})$ is its a Lie subgroup. For any $M \in \mathrm{Sp}(2n, \mathbb{C})$, we have $e^{\sqrt{-1}\theta} M \in \mathrm{Sp}(2n, \mathbb{C})$. The dimension n subspace L of $\mathbb{C}^{2n}$ is called Lagrangian subspace if $\tilde{\omega}_0(x, y) = 0$ for all $x, y \in L$. We note that the complexification of any Lagrangian subspace L of $(\mathbb{R}^{2n}, \tilde{\omega}_0, J)$ is a Lagrangian subspace.

We now define $(\mathbb{C}^{2n} \times \mathbb{C}^{2n}, \Omega, \mathcal{J})$ with $\Omega = -\tilde{\omega}_0 \oplus \tilde{\omega}_0$ and $\mathcal{J} = \begin{pmatrix} -J & 0 \\ 0 & J \end{pmatrix}$. For any $M \in \mathrm{Sp}(2n, \mathbb{C})$, the graph $\mathrm{Gr}(M) = \{(x, Mx) |\ x \in \mathbb{C}^{2n}\}$ of $M : \mathbb{C}^{2n} \to \mathbb{C}^{2n}$ is a Lagrangian subspace of $(\mathbb{C}^{2n} \times \mathbb{C}^{2n}, \Omega, \mathcal{J})$.

For $P \in \mathrm{Sp}(2n, \mathbb{R})$, $\omega = e^{\theta J}$ and $\gamma \in \mathcal{P}(2n, \mathbb{R})$, we have two Lagrangian subspaces $\mathrm{Gr}(\omega P) = \{\begin{pmatrix} x \\ \omega Px \end{pmatrix} : x \in \mathbb{C}^{2n}\}$ and $\mathrm{Gr}(\gamma) = \{\begin{pmatrix} x \\ \gamma(t)x \end{pmatrix} : x \in \mathbb{C}^{2n}, t \in [0, 1]\}$.

Definition 2.2 For $P \in \mathrm{Sp}(2n, \mathbb{R})$, $\omega = e^{\theta J}$ and $\gamma \in \mathcal{P}(2n, \mathbb{R})$, we define

$$\mu_\omega^P(\gamma) = \mu_{\mathbb{C}}^{CLM}(\mathrm{Gr}(\omega P), \mathrm{Gr}(\gamma)). \tag{7.7}$$

Theorem 2.3 *For $P \in \mathrm{Sp}(2n, \mathbb{R})$, $\omega = e^{\theta J}$ and $\gamma \in \mathcal{P}(2n, \mathbb{R})$, we have*

$$i_\omega^P(\gamma) = \mu_\omega^P(\gamma). \tag{7.8}$$

Proof The proof is the same as uniqueness arguments as before, we omit it here. □

7.3 The Concavity of Symplectic Paths and (ε, L_0, L_1)-Signature

From Theorem V.1.21, we know that the concavity defined by

$$\mathrm{concav}_L(M) := \mathrm{concav}_L(\gamma) = i_1(\gamma) - i_L(\gamma)$$

depends only on the end matrix $M = \gamma(1)$ and the Lagrangian subspace L.

Definition 3.1 ([189, 209]) For two Lagrangian subspaces $L,\ L' \in \Lambda(n)$, we define the (L, L')-concavity of γ by

$$\mathrm{concav}_{(L,L')}(\gamma) = i_{L'}(\gamma) - i_L(\gamma),\ \ \gamma \in \mathcal{P}(2n), \tag{7.9}$$

and the $*$-(L, L')-concavity of γ by

$$\mathrm{concav}^*_{(L,L')}(\gamma) = i_{L'}(\gamma) + \nu_{L'}(\gamma) - i_L(\gamma) - \nu_L(\gamma),\ \ \gamma \in \mathcal{P}(2n). \tag{7.10}$$

Theorem 3.2 *The concavity* $\mathrm{concav}_{(L,L')}(\gamma)$ *depends only on the end matrix $M = \gamma(1)$. So we denote it by* $\mathrm{concav}_{(L,L')}(M)$ *and*

$$\mathrm{concav}^0_{(L,L')}(\gamma) = \hat{i}_{L'}(\gamma) - \hat{i}_L(\gamma),\ \ \gamma \in \tilde{\mathcal{P}}(2n) \tag{7.11}$$

depends only on the end matrices $M_1 = \gamma(0)$ and $M_2 = \gamma(1)$. So we denote it by $\mathrm{concav}^0_{(L,L')}(M_1, M_2)$.

Proof From Theorem V.1.21, we see that

$$i_{L'}(\gamma) - i_L(\gamma) = \mathrm{concav}_L(\gamma) - \mathrm{concav}_{L'}(\gamma)$$

depends only on the end matrix $M = \gamma(1)$. And

$$\hat{i}_{L'}(\gamma) - \hat{i}_L(\gamma) = i_{L'}(\gamma) - i_L(\gamma) - i_{L'}(I) + i_L(I)$$

depends only on the end matrix $M = \gamma(1)$ for $\gamma \in \mathcal{P}(2n)$. Now by definition, if $\gamma \in \tilde{\mathcal{P}}(2n)$, we have γ_0 and $\gamma_1 \in \mathcal{P}(2n)$ such that $\gamma_1 = \gamma * \gamma_0$. Thus

$$\hat{i}_{L'}(\gamma) - \hat{i}_L(\gamma) = i_{L'}(\gamma_1) - i_L(\gamma_1) - i_{L'}(\gamma_0) + i_L(\gamma_0)$$

depends only on $M_1 = \gamma_0(1) = \gamma(0)$ and $M_2 = \gamma_1(1) = \gamma(1)$. □

From Remark VI.3.3(3), Definition VI.3.8, Theorem 1.3 and Theorem 3.2, we see that

$$\mathrm{concav}^0_{(L,L')}(M_1, M_2) = s(L \oplus L, L' \oplus L'; \mathrm{Gr}(M_1), \mathrm{Gr}(M_2)). \tag{7.12}$$

As a special case, for $\gamma \in \mathcal{P}(2n)$, there holds

$$i_{L'}(\gamma) - i_L(\gamma) = s(L \oplus L, L' \oplus L'; \mathrm{Gr}(I), \mathrm{Gr}(M)) + i_{L'}(I) - i_L(I). \tag{7.13}$$

If $L_0 = \{0\} \times \mathbb{R}^n$ and $L_1 = \mathbb{R}^n \times \{0\}$, then we have

$$i_{L_1}(\gamma) - i_{L_0}(\gamma) = s(L_0 \oplus L_0, L_1 \oplus L_1; \mathrm{Gr}(I), \mathrm{Gr}(M)) \tag{7.14}$$

since there hold $i_{L_0}(I) = i_{L_1}(I) = n$ in this case. We further have the following results.

Theorem 3.3 *For $\gamma \in \mathcal{P}(2n)$ with $M = \gamma(1) \in \mathrm{Osp}(2n)$, there holds*

$$\mathrm{concav}_{(L_0,L_1)}(M) = s(L_0 \oplus L_0, L_1 \oplus L_1; \mathrm{Gr}(I), \mathrm{Gr}(M)) = 0. \tag{7.15}$$

Proof Since $M = \gamma(1) \in \mathrm{Osp}(2n)$ can be written as $M = e^P$ with $JP = PJ$. So $JM = MJ$ and by choosing $\gamma(t) = e^{Pt}$, we obtain the result (7.15) from Definition III.1.7 of $i_{L_1}(\gamma)$ with $L_1 = JL_0$. □

For any $P \in \mathrm{Sp}(2n)$ and $\varepsilon \in \mathbb{R}$, we set $N = \begin{pmatrix} -I & 0 \\ 0 & I \end{pmatrix}$ and

$$\begin{aligned} M_\varepsilon(P) &= e^{-\varepsilon J}\begin{pmatrix} 0 & I \\ I & 0 \end{pmatrix} e^{\varepsilon J} - P^T e^{\varepsilon J}\begin{pmatrix} 0 & I \\ I & 0 \end{pmatrix} e^{-\varepsilon J} P \\ &= P^T J e^{2\varepsilon J} N P - J e^{-2\varepsilon J} N \\ &= P^T \begin{pmatrix} \sin 2\varepsilon I & -\cos 2\varepsilon I \\ -\cos 2\varepsilon I & -\sin 2\varepsilon I \end{pmatrix} P + \begin{pmatrix} \sin 2\varepsilon I & \cos 2\varepsilon I \\ \cos 2\varepsilon I & -\sin 2\varepsilon I \end{pmatrix}. \end{aligned} \tag{7.16}$$

Theorem 3.4 ([307]) *For $\gamma \in \mathcal{P}(2n)$, there holds*

$$\mathrm{concav}_{(L_0,L_1)}(\gamma(1)) = \frac{1}{2}\mathrm{sgn} M_\varepsilon(\gamma(1)) \tag{7.17}$$

for $\varepsilon > 0$ being sufficiently small. Moreover, there holds

$$\mathrm{concav}^*_{(L_0,L_1)}(\gamma(1)) = \frac{1}{2}\mathrm{sgn} M_\varepsilon(\gamma(1)) \tag{7.18}$$

for $\varepsilon < 0$ and $|\varepsilon|$ being sufficiently small.

We call $\mathrm{sgn} M_\varepsilon(A)$ the of the symplectic matrix A.

Proof We set $V_0 = L_0 \oplus L_0$ and $V_1 = L_1 \oplus L_1$. By definition and above arguments, we have

$$i_{L_1}(\gamma) - i_{L_0}(\gamma) = s(V_0, V_1; \mathrm{Gr}(I), \mathrm{Gr}(M)) = s(V_0, V_1; e^{-\tilde{J}\varepsilon}\mathrm{Gr}(I), e^{-\tilde{J}\varepsilon}\mathrm{Gr}(M)),$$

here $\tilde{J} = -J \oplus J = \begin{pmatrix} -J & 0 \\ 0 & J \end{pmatrix}$ and $M = \gamma(1)$. By (6.38), we have

$$\begin{aligned} &s(V_0, V_1; e^{-\tilde{J}\varepsilon}\mathrm{Gr}(I), e^{-\tilde{J}\varepsilon}\mathrm{Gr}(M)) \\ &\quad = \frac{1}{2}[\mathrm{sgn}(V_0, e^{-\tilde{J}\varepsilon}\mathrm{Gr}(M); V_1) - \mathrm{sgn}(V_0, e^{-\tilde{J}\varepsilon}\mathrm{Gr}(I); V_1)]. \end{aligned} \tag{7.19}$$

Claim 1. For $\varepsilon > 0$ small enough,

$$\mathrm{sgn}(V_0, e^{-\tilde{J}\varepsilon}\mathrm{Gr}(I); V_1) = 0. \tag{7.20}$$

In fact,

$$e^{-\tilde{J}\varepsilon}\mathrm{Gr}(I) = \left\{ \begin{pmatrix} e^{\varepsilon J}x \\ e^{-\varepsilon J}x \end{pmatrix};\ x \in \mathbb{R}^{2n} \right\}.$$

Suppose $V_1 = (\mathbb{R}^n \times \{0\}) \times (\mathbb{R}^n \times \{0\}) = \{y + Ay |\ y \in V_0\}$, $A : V_0 \to e^{-\tilde{J}\varepsilon}\mathrm{Gr}(I)$. If $y = (0, a, 0, b)^T \in V_0$ satisfies $Ay = \begin{pmatrix} e^{\varepsilon J}x \\ e^{-\varepsilon J}x \end{pmatrix}$, $x = \begin{pmatrix} p \\ q \end{pmatrix}$, then we have $p = \frac{b-a}{2\sin\varepsilon}$ and $q = -\frac{a+b}{2\cos\varepsilon}$. So there holds

$$Ay = (\frac{\alpha}{2}a + \frac{\beta}{2}b, -a, -\frac{\beta}{2}a - \frac{\alpha}{2}b, -b),$$

where $\alpha = \frac{\sin\varepsilon}{\cos\varepsilon} - \frac{\cos\varepsilon}{\sin\varepsilon}$, $\beta = \frac{\sin\varepsilon}{\cos\varepsilon} + \frac{\cos\varepsilon}{\sin\varepsilon}$. Setting $\Omega = -\tilde{\omega}_0 \oplus \tilde{\omega}_0$, then we have

$$\Omega(Ay, y) = -\langle \frac{\alpha}{2}a + \frac{\beta}{2}b, a\rangle - \langle -\frac{\beta}{2}a - \frac{\alpha}{2}b, b\rangle = \sum_{j=1}^{n} \left[-\frac{\alpha}{2}(x_j^2 + y_j^2) - \beta x_j y_j \right],$$

where $a = (x_1, \cdots, x_n)$, $b = (y_1, \cdots, y_n)$.

Now it is easy to see that $\alpha < 0$, $\beta > 0$ and the signature of the above quadratic form is zero. Thus we complete the proof of Claim 1.

Claim 2. For $\varepsilon > 0$ small enough,

$$\mathrm{sgn}(V_0, e^{-\tilde{J}\varepsilon}\mathrm{Gr}(\gamma(1)); V_1) = \mathrm{sgn}(M_\varepsilon(\gamma(1))). \tag{7.21}$$

Proof of Claim 2. $\mathrm{Gr}(\gamma(1)) = \left\{\begin{pmatrix} x \\ \gamma(1)x \end{pmatrix}; x = \begin{pmatrix} p \\ q \end{pmatrix} \in \mathbb{R}^{2n}\right\}$ and $e^{-\tilde{J}\varepsilon}\mathrm{Gr}(\gamma(1)) = \left\{\begin{pmatrix} e^{\varepsilon J}x \\ e^{-\varepsilon J}\gamma(1)x \end{pmatrix}; x \in \mathbb{R}^{2n}\right\}$.

Suppose $V_1 = \{y + Ay \mid y \in V_0\}$, $A : V_0 \to e^{-\tilde{J}\varepsilon}\mathrm{Gr}(\gamma(1))$. If $y = (0, a, 0, b)^T \in V_0$ satisfies $Ay = \begin{pmatrix} e^{\varepsilon J}x \\ e^{-\varepsilon J}x \end{pmatrix}$, $x \in \mathbb{R}^{2n}$, then $y = \begin{pmatrix} -P_2 e^{\varepsilon J}x \\ -P_2 e^{-\varepsilon J}x \end{pmatrix}$, where P_2 is the projection $P_2 : \mathbb{R}^{2n} \to L_0$, i.e., $P_2 = \begin{pmatrix} 0 & 0 \\ 0 & I \end{pmatrix}$. So A is a bijection and the quadratic form $\Omega(Ay, y)$ is non-degenerate.

$$\begin{aligned} \Omega(Ay, y) &= \Omega\left(\begin{pmatrix} e^{\varepsilon J}x \\ e^{-\varepsilon J}x \end{pmatrix}, \begin{pmatrix} -P_2 e^{\varepsilon J}x \\ -P_2 e^{-\varepsilon J}x \end{pmatrix}\right) \\ &= \tilde{\omega}_0(e^{\varepsilon J}x, P_2 e^{\varepsilon J}x) - \tilde{\omega}_0(e^{-\varepsilon J}\gamma(1)x, P_2 e^{-\varepsilon J}\gamma(1)x). \end{aligned}$$

Since $\tilde{\omega}_0(u, v) = v^T J u$, we have

$$\begin{aligned} \Omega(Ay, y) &= x^T e^{-\varepsilon J} P_2^T J e^{\varepsilon J} x - x^T \gamma(1)^T e^{\varepsilon J} P_2^T J e^{-\varepsilon J}\gamma(1)x \\ &= x^T \tilde{M}_\varepsilon(\gamma(1))x \\ &= \tfrac{1}{2} x^T [\tilde{M}_\varepsilon(\gamma(1)) + \tilde{M}_\varepsilon(\gamma(1))^T]x, \end{aligned}$$

where $\tilde{M}_\varepsilon(\gamma(1)) = \begin{pmatrix} \sin\varepsilon\cos\varepsilon I & -\sin^2\varepsilon I \\ \cos^2\varepsilon I & -\sin\varepsilon\cos\varepsilon I \end{pmatrix}$ $+\gamma(1)^T \begin{pmatrix} \sin\varepsilon\cos\varepsilon I & \sin^2\varepsilon I \\ -\cos^2\varepsilon I & -\sin\varepsilon\cos\varepsilon I \end{pmatrix}\gamma(1)$. We see that $\tilde{M}_\varepsilon(\gamma(1)) + \tilde{M}_\varepsilon(\gamma(1))^T = M_\varepsilon(\gamma(1))$.

Equation (7.17) follows from (7.19), (7.20) and (7.21).

For $\varepsilon < 0$, $|\varepsilon|$ small enough, it is clear that

$$i_{L_0}(\gamma) = \mu^{CLM}(V_0, e^{-\varepsilon\tilde{J}}\mathrm{Gr}(\gamma)) - \nu_{L_0}(\gamma),$$

$$i_{L_1}(\gamma) = \mu^{CLM}(V_1, e^{-\varepsilon\tilde{J}}\mathrm{Gr}(\gamma)) - \nu_{L_1}(\gamma).$$

By the same arguments as above, we see that (7.18) holds. □

If $P \in \mathrm{Osp}(2n)$, then $JP = PJ$, $P^T J = J P^T$, and so we have $JM_\varepsilon(P)J^T = -M_\varepsilon(P)$. Thus we have

$$\mathrm{sgn}(M_\varepsilon(P)) = \mathrm{sgn}(JM_\varepsilon(P)J^T) = -\mathrm{sgn}(M_\varepsilon(P)),$$

it implies $\mathrm{sgn}(M_\varepsilon(P)) = 0$ and give a new proof of Theorem 3.3.

Remark 3.5 We list some basic facts about the signature of $M_\varepsilon(P)$ and leave the proofs to readers as exercises.

(1) $\mathrm{sgn}(M_\varepsilon(P(s)))$ is constant for continuous path $P : [0, 1] \to \mathrm{Sp}(2n)$ with $\nu_{L_j}(P(s)) = \text{constant}$ for $j = 0,\ 1$ (cf. Lemma 2.2 of [307] for a proof);
(2) We have

$$M_\varepsilon(P_1 \diamond P_2) = M_\varepsilon(P_1) \diamond M_\varepsilon(P_2),$$
$$\mathrm{sgn}M_\varepsilon(P_1 \diamond P_2) = \mathrm{sgn}M_\varepsilon(P_1) + \mathrm{sgn}M_\varepsilon(P_2) :$$

(3) $\mathrm{sgn}(M_\varepsilon(P) = 0$ for $P = \begin{pmatrix} \cos\theta & -\sin\theta \\ \sin\theta & \cos\theta \end{pmatrix}, \begin{pmatrix} a & 0 \\ 0 & a^{-1} \end{pmatrix}, \pm\begin{pmatrix} 1 & b \\ 0 & 1 \end{pmatrix}$ or $\pm\begin{pmatrix} 1 & 0 \\ -b & 1 \end{pmatrix}$, $\theta \in [0, 2\pi]$, $a \neq 0$, $b > 0$;
(4) $\mathrm{sgn}(M_\varepsilon(P) = 2$ for $P = \pm\begin{pmatrix} 1 & -b \\ 0 & 1 \end{pmatrix}$ or $\pm\begin{pmatrix} 2 & -1 \\ -1 & 1 \end{pmatrix}$, $b > 0$;
(5) $\mathrm{sgn}(M_\varepsilon(P) = -2$ for $P = \pm\begin{pmatrix} 1 & 0 \\ b & 1 \end{pmatrix}$, $b > 0$.

Definition 3.6 ([308]) We call two symplectic matrices M_1 and M_2 in $\mathrm{Sp}(2n)$ are *special homotopic*(or (L_0, L_1)-homotopic) and denote by $M_1 \overset{s}{\sim} M_2$, if there are $P_j \in \mathrm{Sp}(2n)$ with $P_j = \mathrm{diag}(Q_j, (Q_j^T)^{-1})$, where Q_j is a $n \times n$ invertible real matrix, and $\det(Q_j) > 0$ for $j = 1, 2$, such that

$$M_1 = P_1 M_2 P_2.$$

It is clear that $\overset{s}{\sim}$ is an equivalent relation.

Remark 3.7 Let $M_i = \begin{pmatrix} A_i & B_i \\ C_i & D_i \end{pmatrix} \in \mathrm{Sp}(2k_i)$, $i = 0, 1, 2$, $k_1 = k_2$ and $M_1 \overset{s}{\sim} M_2$(in this time), then $A_1^T C_1$, $B_1^T D_1$ are congruent to $A_2^T C_2$, $B_2^T D_2$ respectively. So $m^*(A_1^T C_1) = m^*(A_2^T C_2)$ and $m^*(B_1^T D_1) = m^*(B_2^T D_2)$ for $* = \pm,\ 0$. Furthermore, if $M_0 = M_1 \diamond M_2$(here $k_1 = k_2$ is not necessary), then

$$m^*(A_0^T C_0) = m^*(A_1^T C_1) + m^*(A_2^T C_2),$$
$$m^*(B_0^T D_0) = m^*(B_1^T D_1) + m^*(B_2^T D_2). \tag{7.22}$$

So $m^*(A^T C)$ and $m^*(B^T D)$ are (L_0, L_1)-homotopic invariant. The following formula will be used frequently

$$N_k M_1^{-1} N_k M_1 = I_{2k} + 2\begin{pmatrix} B_1^T C_1 & B_1^T D_1 \\ A_1^T C_1 & C_1^T B_1 \end{pmatrix}. \tag{7.23}$$

We recall that $\Omega_0(M)$ is the path connected component of $\Omega(M)$ containing M (cf. Definition I.5.2).

Definition 3.8 For any M_1,$M_2 \in \mathrm{Sp}(2n)$, we call $M_1 \approx M_2$ if $M_1 \in \Omega_0(M_2)$.

Remark 3.9 It is easy to check that $\approx$ is an equivalent relation. If $M_1 \approx M_2$, we have $M_1^k \approx M_2^k$ for any $k \in \mathbb{N}$ and $M_1 \diamond M_3 \approx M_2 \diamond M_4$ for $M_3 \approx M_4$. Also we have $PMP^{-1} \approx M$ for any $P, M \in \mathrm{Sp}(2n)$.

Lemma 3.10 *Assume $M_1 \in \mathrm{Sp}(2(k_1+k_2))$, $M_2 \in \mathrm{Sp}(2k_3)$ and $M_3 \in \mathrm{Sp}(2(k_1 + k_2 + k_3))$ have the following block forms*

$$M_1 = \begin{pmatrix} A_1 & A_2 & B_1 & B_2 \\ A_3 & A_4 & B_3 & B_4 \\ C_1 & C_2 & D_1 & D_2 \\ C_3 & C_4 & D_3 & D_4 \end{pmatrix}, \quad M_2 = \begin{pmatrix} A_5 & B_5 \\ C_5 & D_5 \end{pmatrix}, \quad M_3 = \begin{pmatrix} A_1 & 0 & A_2 & B_1 & 0 & B_2 \\ 0 & A_5 & 0 & 0 & B_5 & 0 \\ A_3 & 0 & A_4 & B_3 & 0 & B_4 \\ C_1 & 0 & C_2 & D_1 & 0 & D_2 \\ 0 & C_5 & 0 & 0 & D_5 & 0 \\ C_3 & 0 & C_4 & D_3 & 0 & D_4 \end{pmatrix}$$

with $A_1, B_1, C_1, D_1 \in \mathcal{L}(\mathbb{R}^{k_1})$, $A_4, B_4, C_4, D_4 \in \mathcal{L}(\mathbb{R}^{k_2})$, $A_5, D_5 \in \mathcal{L}(\mathbb{R}^{k_3})$. Then

$$M_3 \approx M_1 \diamond M_2. \tag{7.24}$$

Proof Let $P = \mathrm{diag}\left(\begin{pmatrix} I_{k_1} & 0 & 0 \\ 0 & 0 & I_{k_2} \\ 0 & I_{k_3} & 0 \end{pmatrix}, \begin{pmatrix} I_{k_1} & 0 & 0 \\ 0 & 0 & I_{k_2} \\ 0 & I_{k_3} & 0 \end{pmatrix} \right)$. It is east to verify that $P \in \mathrm{Sp}(2(k_1 + k_2 + k_3))$ and $M_3 = P(M_1 \diamond M_2)P^{-1}$. Then (7.24) holds from Remark 3.9 and the proof of Lemma 3.10 is completed. □

Lemma 3.11 ([308]) *Let $k \in \mathbb{N}$ and any symplectic matrix $P = \begin{pmatrix} I_k & 0 \\ C & I_k \end{pmatrix}$. Then $P \approx I_2^{\diamond p} \diamond N_1(1,1)^{\diamond q} \diamond N_1(1,-1)^{\diamond r}$ with $p = m^0(C)$, $q = m^-(C)$, $r = m^+(C)$.*

Proof It is clear that

$$P \approx \begin{pmatrix} I_k & 0 \\ B & I_k \end{pmatrix},$$

where $B = \mathrm{diag}(0, -I_{m^-(C)}, I_{m^+(C)})$. Since $J_1 N_1(1, \pm 1)(J_1)^{-1} = \begin{pmatrix} 1 & 0 \\ \mp 1 & 1 \end{pmatrix}$, by Remark 3.9 we have $N_1(1, \pm 1) \approx \begin{pmatrix} 1 & 0 \\ \mp 1 & 1 \end{pmatrix}$. Then

$$P \approx I_2^{\diamond m^0(C)} \diamond N_1(1,1)^{\diamond m^-(C)} \diamond N_1(1,-1)^{\diamond m^+(C)}.$$

By Lemma II.3.14 we have

$$S_P^+(1) = m^0(C) + m^-(C) = p + q. \tag{7.25}$$

By the definition of the relation $\approx$, we have

$$2p + q + r = \nu_1(P) = 2m^0(C) + m^+(C) + m^-(C). \tag{7.26}$$

Also we have

$$p + q + r = m^0(C) + m^+(C) + m^-(C) = k. \tag{7.27}$$

By (7.25)–(7.27) we have

$$m^0(C) = p, \quad m^-(C) = q, \quad m^+(C) = r.$$

The proof of Lemma 3.11 is complete. □

Lemma 3.12 ([308]) *For $\tilde{M}_1, \tilde{M}_2 \in \mathrm{Sp}(2n)$, if $\tilde{M}_1 \overset{s}{\sim} \tilde{M}_2$, then*

$$\mathrm{sgn} M_\varepsilon(\tilde{M}_1) = \mathrm{sgn} M_\varepsilon(\tilde{M}_2), \quad 0 \le |\varepsilon| \ll 1, \tag{7.28}$$

$$N\tilde{M}_1^{-1} N\tilde{M}_1 \approx N\tilde{M}_2^{-1} N\tilde{M}_2. \tag{7.29}$$

Proof By Definition 3.6, there are $P_j \in \mathrm{Sp}(2n)$ with $P_j = \mathrm{diag}(Q_j, (Q_j^T)^{-1})$, Q_j being $n \times n$ invertible real matrix, and $\det(Q_j) > 0$ such that

$$\tilde{M}_1 = P_1 \tilde{M}_2 P_2.$$

Since $\det(Q_j) > 0$ for $j = 1, 2$, we can joint Q_j to I_n by invertible matrix path. Hence we can joint $P_1\tilde{M}_2P_2$ to $\tilde{M}_2$ by symplectic path preserving the nullity ν_{L_0} and ν_{L_1}. By the basic fact (1) in Remark 3.5, (7.28) holds. Since $P_jN = NP_j$ for $j = 1, 2$. Direct computation shows that

$$N\tilde{M}_1^{-1}N\tilde{M}_1 = N(P_1\tilde{M}_2P_2)^{-1}N(P_1\tilde{M}_2P_2) = P_2^{-1}N\tilde{M}_2^{-1}N\tilde{M}_2P_2. \tag{7.30}$$

Thus (7.29) holds by Remark 3.9. The proof of Lemma 3.12 is complete. □

Lemma 3.13 ([308]) *Let $P = \begin{pmatrix} A & B \\ C & D \end{pmatrix} \in \mathrm{Sp}(2n)$, where A, B, C, D are all $n \times n$ matrices. Then*

(i) $\frac{1}{2}\mathrm{sgn} M_\varepsilon(P) \le n - \nu_{L_0}(P)$, *for* $0 < \varepsilon \ll 1$. *If* $B = 0$, *we have* $\frac{1}{2}\mathrm{sgn} M_\varepsilon(P) \le 0$ *for* $0 < \varepsilon \ll 1$.
(ii) *Let* $m^+(A^TC) = q$, *we have*

$$\frac{1}{2}\mathrm{sgn} M_\varepsilon(P) \le n - q, \quad 0 \le |\varepsilon| \ll 1. \tag{7.31}$$

(iii) $\frac{1}{2}\mathrm{sgn}M_\varepsilon(P) \geq \dim\ker C - n$ *for* $0 < \varepsilon \ll 1$, *If* $C = 0$, *then* $\frac{1}{2}\mathrm{sgn}M_\varepsilon(P) \geq 0$ *for* $0 < \varepsilon \ll 1$

(iv) *If both* B *and* C *are invertible, we have*

$$\mathrm{sgn}M_\varepsilon(P) = \mathrm{sgn}M_0(P), \quad 0 \leq |\varepsilon| \ll 1.$$

Proof Since P is symplectic, so is for P^T. From $P^T J P = J$ and $P J P^T = J$ we get $A^T C, B^T D, AB^T, CD^T$ are all symmetric matrices and

$$AD^T - BC^T = I_n, \quad A^T D - C^T B = I_n. \tag{7.32}$$

We denote by $s = \sin 2\varepsilon$ and $c = \cos 2\varepsilon$. By definition of $M_\varepsilon(P)$, we have

$$\begin{aligned} M_\varepsilon(P) &= \begin{pmatrix} A^T & C^T \\ B^T & D^T \end{pmatrix}\begin{pmatrix} sI_k & -cI_k \\ -cI_k & -sI_k \end{pmatrix}\begin{pmatrix} A & B \\ C & D \end{pmatrix} + \begin{pmatrix} sI_k & cI_k \\ cI_k & -sI_k \end{pmatrix} \\ &= \begin{pmatrix} sA^T A - 2cA^T C - sC^T C + sI_k & sA^T B - 2cC^T B - sC^T D \\ sB^T A - 2cB^T C - sD^T C & sB^T B - 2cB^T D - sD^T D - sI_k \end{pmatrix}. \end{aligned}$$

So

$$M_0(P) = -2\begin{pmatrix} A^T C & C^T B \\ B^T C & B^T D \end{pmatrix} = -2\begin{pmatrix} C^T & 0 \\ 0 & B^T \end{pmatrix}\begin{pmatrix} A & B \\ C & D \end{pmatrix},$$

where we have used $A^T C$ is symmetric. So if both B and C are invertible, $M_0(P)$ is invertible and symmetric, its signature is invariant under small perturbation, so (iv) holds.

If $\nu_{L_0}(P) = \dim\ker B > 0$, since $B^T D = D^T B$, for any $x \in \ker B \subseteq \mathbb{R}^n$, $x \neq 0$, and $0 < \varepsilon \ll 1$, we have

$$\begin{aligned} M_\varepsilon(P)\begin{pmatrix} 0 \\ x \end{pmatrix} \cdot \begin{pmatrix} 0 \\ x \end{pmatrix} &= (sB^T B - 2cD^T B - sD^T D - sI_n)x \cdot x \\ &= -s(D^T D + I_n)x \cdot x < 0. \end{aligned} \tag{7.33}$$

So $M_\varepsilon(P)$ is negative definite on $(0 \oplus \ker B) \subseteq \mathbb{R}^{2n}$. Hence $m^-(M_\varepsilon(p) \geq \dim\ker B$ which yields that $\frac{1}{2}\mathrm{sgn}M_\varepsilon(P) \leq n - \dim\ker B = n - \nu_{L_0}(P)$, for $0 < \varepsilon \ll 1$. Thus (i) holds. Similarly we can prove (iii).

If $m^+(A^T C) = q > 0$, let $A^T C$ is positive definite on $E \subseteq \mathbb{R}^n$, then for $0 \leq |s| \ll 1$, similar to (7.33) we have $M_\varepsilon(P)$ is negative on $E \oplus 0 \subseteq \mathbb{R}^{2n}$. Hence $m^-(M_\varepsilon(P) \geq q$, which yields (7.31). □

Lemma 3.14 ([309]) *Let* $2k \times 2k$ *symmetric real matrix* E *have the following block form* $E = \begin{pmatrix} 0 & E_1 \\ E_1^T & E_2 \end{pmatrix}$. *Then*

$$m^{\pm}(E) \geq \text{rank}E_1. \tag{7.34}$$

Proof We only need to prove Lemma 3.14 in the case $\text{rank}E_1 > 0$. Suppose $\text{rank}E_1 > 0$, there is a linear subspace F of $\mathbb{R}^k$ with $\dim F = \text{rank}E_1$ and a number $\delta > 0$ such that for any $y \in F$ with $|y| = 1$ there holds

$$|E_1 y| \geq \delta. \tag{7.35}$$

So for $0 < \varepsilon \ll 1$ and any $y \in F$ with $|y| = 1$ we have

$$E\begin{pmatrix} E_1 y \\ \varepsilon y \end{pmatrix} \cdot \begin{pmatrix} E_1 y \\ \varepsilon y \end{pmatrix} = 2\varepsilon E_1 y \cdot E_1 y + \varepsilon^2 E_2 y \cdot y = \varepsilon(2|E_1 y|^2 + \varepsilon E_2 y \cdot y)$$

$$\geq \varepsilon(2\delta^2 - \varepsilon||E_2||) > 0. \tag{7.36}$$

where $||E_2||$ is the operator norm of E_2 on $\mathbb{R}^k$. So E is positive definite on the $\text{rank}E_1$-dimensional linear subspace $\left\{\begin{pmatrix} E_1 y \\ \varepsilon y \end{pmatrix} | y \in F\right\}$ of $\mathbb{R}^{2k}$. Hence we have $m^+(E) \geq \text{rank}E_1$. By the same argument we have $m^-(E) = m^+(-E) \geq \text{rank}(-E_1) = \text{rank}E_1$. The proof of Lemma 3.14 is complete. □

Lemma 3.15 ([210]) *Let A_1 and A_3 be $k \times k$ real matrices. Assume both A_1 and $A_1 A_3$ are symmetric and $\sigma(A_3) \subset (-\infty, 0)$. Then*

$$\text{sgn}A_1 + \text{sgn}(A_1 A_3) = 0. \tag{7.37}$$

Proof It is clear that A_3 is invertible. We prove Lemma 3.15 in the following two steps.

Step 1. We assume that A_1 is invertible and proceed by induction on $k \in \mathbb{N}$.

If $k = 1$, then $A_1, A_3 \in \mathbb{R}$ and (7.37) holds obviously. Now assume (7.37) holds for $1 \leq k \leq l$. If we can prove (7.37) for $k = l + 1$, then by the mathematical induction (7.37) holds for any $k \in \mathbb{N}$ and Lemma 3.15 is proved in the case A_1 is invertible.

In view of the real Jordan canonical form decomposition of A_3, we only need to prove (7.37) for $k = l + 1$ in the following Case 1 and Case 2.

Case 1. There is an invertible $(l+1) \times (l+1)$ real matrix such that $Q^{-1}A_3 Q$ is the $(l+1)$-order Jordan form $\begin{pmatrix} \lambda\ 1\ 0 \cdots 0\ 0 \\ 0\ \lambda\ 1 \cdots 0\ 0 \\ \vdots\ \vdots\ \vdots\ \cdots\ \vdots\ 0 \\ 0\ 0\ 0 \cdots \lambda\ 1 \\ 0\ 0\ 0 \cdots 0\ \lambda \end{pmatrix} := \tilde{A}_3$ with $\lambda < 0$.

Denoting by $\tilde{A}_1 = Q^T A_1 Q$. We have

$$\tilde{A}_1 \tilde{A}_3 = Q^T A_1 Q\, Q^{-1} A_3 Q = Q^T A_1 A_3 Q.$$

Hence both matrices $\tilde{A}_1$ and $\tilde{A}_1\tilde{A}_3$ are symmetric and

$$\mathrm{sgn}A_1 + \mathrm{sgn}(A_1A_3) = \mathrm{sgn}\tilde{A}_1 + \mathrm{sgn}(\tilde{A}_1\tilde{A}_3). \tag{7.38}$$

Since $\tilde{A}_1 = (a_{i,j})_{1\le i,j\le l+1}$ and $\tilde{A}_1\tilde{A}_3 = (c_{i,j})_{1\le i,j\le l+1}$ are symmetric, $a_{i,j} = a_{j.i}$ and $c_{i,j} = c_{j,i}$ for $1 \le i, j \le l+1$.

Claim C.1. $a_{i,j} = 0$ for $i + j \le l + 1$ and $a_{i,j} = a_{l+1,1}$ for $i + j = l + 2$ with $1 \le i, j \le l + 1$.

For $2 \le j \le l + 1$, since $c_{1,j} = c_{j,1}$,

$$\lambda a_{1,j} + a_{1,j-1} = \lambda a_{j,1} = \lambda a_{1,j}.$$

Thus

$$a_{1,j-1} = 0, \quad 2 \le j \le l + 1. \tag{7.39}$$

For $2 \le i, j \le l + 1$, since $c_{i,j} = c_{j,i}$ we have

$$\lambda a_{i,j} + a_{i,j-1} = \lambda a_{j,i} + a_{j,i-1} = \lambda a_{i,j} + a_{i-1,j}.$$

So

$$a_{i,j-1} = a_{i-1,j}, \quad 2 \le i, j \le l + 1. \tag{7.40}$$

By (7.39) and (7.40) we have

$$a_{i,j} = a_{i-1,j+1} = \cdots = a_{2,i+j-2} = a_{1,i+j-1} = 0,$$
$$1 \le i, j \text{ and } i + j \le l + 1, \tag{7.41}$$
$$a_{l+1,1} = a_{l,2} = a_{l-1,3} = \cdots = a_{2,l} = a_{1,l+1}. \tag{7.42}$$

Hence, by (7.41) and (7.42), Claim C.1 is proved.

By Claim C.1, let $a = a_{1,l+1}$, then

$$\tilde{A}_1 = \begin{pmatrix} 0&0&0&0&0&0&a \\ 0&0&0&0&0&a&* \\ 0&0&0&0&\cdot&*&* \\ 0&0&0&\cdot&*&*&* \\ 0&0&\cdot&*&*&*&* \\ 0&a&*&*&*&*&* \\ a&*&*&*&*&*&* \end{pmatrix}, \quad \tilde{A}_1\tilde{A}_3 = \begin{pmatrix} 0&0&0&0&0&0&\lambda a \\ 0&0&0&0&0&\lambda a&* \\ 0&0&0&0&\cdot&*&* \\ 0&0&0&\cdot&*&*&* \\ 0&0&\cdot&*&*&*&* \\ 0&\lambda a&*&*&*&*&* \\ \lambda a&*&*&*&*&*&* \end{pmatrix}. \tag{7.43}$$

It is easy to see that $\tilde{A}_1\tilde{A}_3$ is congruent to $\lambda\tilde{A}_1$. Since $\lambda < 0$,

$$\text{sgn}(\tilde{A}_1\tilde{A}_3) = \text{sgn}(\lambda\tilde{A}_1) = -\text{sgn}(\tilde{A}_1),$$

$$\text{sgn}(\tilde{A}_1\tilde{A}_3) + \text{sgn}\tilde{A}_1 = 0. \tag{7.44}$$

Equations (7.38) and (7.44) imply (7.37). Hence Step 1 is proved in Case 1.
Case 2. There exists an invertible $(l+1)\times(l+1)$ real matrix Q such that $Q^{-1}A_3Q = \text{diag}(A_4, A_5)$, where A_4 is a $k_1 \times k_1$ real matrix with $\sigma(A_4) \subset (-\infty, 0)$ and A_5 is a k_2-order Jordan form

$$A_5 = \begin{pmatrix} \lambda & 1 & 0 & \cdots & 0 & 0 \\ 0 & \lambda & 1 & \cdots & 0 & 0 \\ \vdots & \vdots & \vdots & \cdots & \vdots & 0 \\ 0 & 0 & 0 & \cdots & \lambda & 1 \\ 0 & 0 & 0 & \cdots & 0 & \lambda \end{pmatrix}$$

with $\lambda < 0$, $1 \le k_1, k_2 \le l$ and $k_1 + k_2 = l + 1$.
We still denote by $\tilde{A}_1 = Q^T A_1 Q$, then

$$\tilde{A}_1\tilde{A}_3 = Q^T A_1 Q\, Q^{-1} A_3 Q = Q^T A_1 A_3 Q.$$

So both $\tilde{A}_1$ and $\tilde{A}_1\tilde{A}_3$ are symmetric and

$$\text{sgn}A_1 + \text{sgn}(A_1A_3) = \text{sgn}\tilde{A}_1 + \text{sgn}(\tilde{A}_1\tilde{A}_3). \tag{7.45}$$

Correspondingly we can write $\tilde{A}_1$ in the block form decomposition $\tilde{A}_1 = \begin{pmatrix} E_1 & E_2 \\ E_2^T & E_4 \end{pmatrix}$, where E_1 is a $k_1 \times k_1$ real symmetric matric and E_4 is a $k_2 \times k_2$ real symmetric matrix. Then

$$\tilde{A}_1\tilde{A}_3 = \begin{pmatrix} E_1A_4 & E_2A_5 \\ E_2^T A_4 & E_4A_5 \end{pmatrix}$$

is symmetric.
Subcase 1. E_4 is invertible.
In this case we have

$$\begin{pmatrix} I_{k_1} & -E_2E_4^{-1} \\ 0 & I_{k_2} \end{pmatrix} \begin{pmatrix} E_1 & E_2 \\ E_2^T & E_4 \end{pmatrix} \begin{pmatrix} I_{k_1} & 0 \\ -E_4^{-1}E_2^T & I_{k_2} \end{pmatrix}$$
$$= \begin{pmatrix} E_1 - E_2E_4^{-1}E_2^T & 0 \\ 0 & E_4 \end{pmatrix} \tag{7.46}$$

and

$$\begin{pmatrix} I_{k_1} & -E_2E_4^{-1} \\ 0 & I_{k_2} \end{pmatrix} \begin{pmatrix} E_1A_4 & E_2A_5 \\ E_2^TA_4 & E_4A_5 \end{pmatrix} \begin{pmatrix} I_{k_1} & 0 \\ -E_4^{-1}E_2^T & I_{k_2} \end{pmatrix}$$
$$= \begin{pmatrix} E_1A_4 - E_2E_4^{-1}E_2^TA_4 & 0 \\ 0 & E_4A_5 \end{pmatrix}$$
$$= \begin{pmatrix} (E_1 - E_2E_4^{-1}E_2^T)A_4 & 0 \\ 0 & E_4A_5 \end{pmatrix}. \tag{7.47}$$

Since the matrices $\tilde{A}_1$ and $\tilde{A}_1\tilde{A}_3$ are symmetric and invertible, by (7.46) and (7.47), both $E_1 - E_2E_4^{-1}E_2^T$ and $(E_1 - E_2E_4^{-1}E_2^T)A_4$ are symmetric and invertible. Hence from $1 \le k_1 \le l$, $\sigma(A_4) \subset (-\infty, 0)$ and our induction hypothesis we obtain

$$\mathrm{sgn}((E_1 - E_2E_4^{-1}E_2^T)A_4) + \mathrm{sgn}(E_1 - E_2E_4^{-1}E_2^T) = 0. \tag{7.48}$$

By (7.47), E_4A_5 is symmetric. Since E_4 is symmetric and invertible, $\sigma(A_5) \subset (-\infty, 0)$ and $1 \le k_2 \le l$, by our induction hypothesis we have

$$\mathrm{sgn}(E_4A_5) + \mathrm{sgn}E_4 = 0. \tag{7.49}$$

From (7.46) we obtain

$$\mathrm{sgn}\tilde{A}_1 = \mathrm{sgn}(E_1 - E_2E_4^{-1}E_2^T) + \mathrm{sgn}E_4. \tag{7.50}$$

By (7.47) there holds

$$\mathrm{sgn}(\tilde{A}_1\tilde{A}_3) = \mathrm{sgn}((E_1 - E_2E_4^{-1}E_2^T)A_4) + \mathrm{sgn}(E_4A_5). \tag{7.51}$$

Then by (7.48)–(7.51) we have

$$\mathrm{sgn}(\tilde{A}_1\tilde{A}_3) + \mathrm{sgn}\tilde{A}_1 = 0. \tag{7.52}$$

Therefore, (7.45) and (7.52) imply (7.37).

Subcase 2. E_4 is not invertible.

In this case we define k_2-order real invertible matrix

$$E_0 = \begin{pmatrix} 0&0&0&0&0&0&1 \\ 0&0&0&0&0&1&0 \\ 0&0&0&0&\cdot&0&0 \\ 0&0&0&\cdot&0&0&0 \\ 0&0&\cdot&0&0&0&0 \\ 0&1&0&0&0&0&0 \\ 1&0&0&0&0&0&0 \end{pmatrix}.$$

Then it is easy to verify that E_0A_5 is symmetric and $E_4 + \varepsilon E_0$ is invertible for $0 < \varepsilon \ll 1$. Define $A_\varepsilon = \begin{pmatrix} E_1 & E_2 \\ E_2^T & E_4 + \varepsilon E_0 \end{pmatrix}$. Since $\tilde{A}_1$ and $\tilde{A}_1\tilde{A}_3$ are invertible, we have both A_ε and $A_\varepsilon\tilde{A}_3$ are symmetric and invertible. Thus

$$\mathrm{sgn}\tilde{A}_1 = \mathrm{sgn}A_\varepsilon, \quad \mathrm{sgn}(\tilde{A}_1\tilde{A}_3) = \mathrm{sgn}(A_\varepsilon\tilde{A}_3), \quad \text{for } 0 < \varepsilon \ll 1. \tag{7.53}$$

By the proof of Subcase 1, we have

$$\mathrm{sgn}(A_\varepsilon\tilde{A}_3) + \mathrm{sgn}A_\varepsilon = 0. \tag{7.54}$$

So from (7.53) we obtain

$$\mathrm{sgn}(\tilde{A}_1\tilde{A}_3) + \mathrm{sgn}\tilde{A}_1 = 0. \tag{7.55}$$

Then (7.37) holds from (7.55).

So in Case 2 (7.37) holds for $k = l + 1$. Hence in the case A_1 is invertible Lemma 3.15 holds and Step 1 is finished.

Step 2. We assume that A_1 is not invertible.

If $A_1 = 0$, (7.37) holds obviously.

If $1 \leq \mathrm{rank}A_1 = m \leq k - 1$, there is a real orthogonal matrix G such that

$$G^T A_1 G = \begin{pmatrix} 0 & 0 \\ 0 & \hat{A}_1 \end{pmatrix}, \tag{7.56}$$

where $\hat{A}_1$ is an m-order invertible real symmetric matrix. Correspondingly we write

$$G^{-1}A_3G = \begin{pmatrix} F_1 & F_2 \\ F_3 & F_4 \end{pmatrix},$$

where F_1 is a $(k-m) \times (k-m)$ real matrix and F_4 is an $m \times m$ real matrix.

Since A_1A_3 is symmetric, from

$$G^T A_1A_3G = G^T A_1 G G^{-1} A_3 G = \begin{pmatrix} 0 & 0 \\ \hat{A}_1F_3 & \hat{A}_1F_4 \end{pmatrix}$$

we get $\hat{A}_1F_3 = 0$. Hence $F_3 = 0$ by the invertibility of $\hat{A}_1$. Therefore we can write

$$G^{-1}A_3G = \begin{pmatrix} F_1 & F_2 \\ 0 & F_4 \end{pmatrix}. \tag{7.57}$$

Hence

$$G^T A_1 A_3 G^T = \begin{pmatrix} 0 & 0 \\ 0 & \hat{A}_1 F_4 \end{pmatrix}, \tag{7.58}$$

where $\hat{A}_1 F_4$ is symmetric. Also by (7.57) the matrix F_4 is invertible and $\sigma(F_4) \subset (-\infty, 0)$. Thus by the proof of Step 1, there holds

$$\mathrm{sgn}(\hat{A}_1 F_4) + \mathrm{sgn}\hat{A}_1 = 0. \tag{7.59}$$

Identities (7.56) and (7.58) give

$$\mathrm{sgn}(A_1 A_3) + \mathrm{sgn}A_1 = \mathrm{sgn}(\hat{A}_1 F_4) + \mathrm{sgn}\hat{A}_1. \tag{7.60}$$

Then (7.59) and (7.60) give (7.37). Hence the Step 2 is proved.

By Step 1 and Step 2 Lemma 3.15 holds. □

We recall that the elliptic hight $e(P)$ of P is the total algebraic multiplicity of all eigenvalues of P on $\mathbf{U}$ for any $P \in \mathrm{Sp}(2n)$.

Lemma 3.16 ([210]) *Let* $R = \begin{pmatrix} A_1 & I_k \\ A_3 & A_2 \end{pmatrix} \in \mathrm{Sp}(2k)$ *with* A_3 *being invertible. If* $e(N_k R^{-1} N_k R) = 2m$, *where* $0 \leq m \leq k$, *then*

$$m - k \leq \frac{1}{2}\mathrm{sgn}M_\varepsilon(R) \leq k - m, \qquad 0 \leq |\varepsilon| \ll 1. \tag{7.61}$$

Proof Since $e(N_k R^{-1} N_k R) = 2m$, there exists a symplectic matrix $P \in \mathrm{Sp}(2k)$ such that

$$P^{-1}(N_k R^{-1} N_k R)P = Q_1 \diamond Q_2 \tag{7.62}$$

with $\sigma(Q_1) \in \mathbf{U}$, $\sigma(Q_2) \cap \mathbf{U} = \emptyset$, $Q_1 \in \mathrm{Sp}(2m)$, and $Q_2 \in \mathrm{Sp}(2k - 2m)$. By (ii) of Lemma 3.13, since A_3 is invertible we only need to prove (7.61) for $\varepsilon = 0$.

Step 1. Assume A_1 is invertible.

Since R is symplectic, we conclude from $R^T J_k R = J_k$ that $A_1^T A_3$ and A_2 are symmetric and

$$A_1^T A_2 - A_3^T = I_k.$$

Because R^T is also symplectic, A_1 is symmetric. Hence $A_1 A_3$ is symmetric and

$$A_1 A_2 - A_3^T = I_k. \tag{7.63}$$

By definition we have

$$M_0(R) = R^T \begin{pmatrix} 0 & -I_k \\ -I_k & 0 \end{pmatrix} R + \begin{pmatrix} 0 & I_k \\ I_k & 0 \end{pmatrix}$$
$$= -2 \begin{pmatrix} A_1A_3 & A_3^T \\ A_3 & A_2 \end{pmatrix}. \tag{7.64}$$

Since A_1 is invertible, there holds

$$\begin{pmatrix} I_k & 0 \\ -A_1^{-1} & I_k \end{pmatrix} \begin{pmatrix} A_1A_3 & A_3^T \\ A_3 & A_2 \end{pmatrix} \begin{pmatrix} I_k & -A_1^{-1} \\ 0 & I_k \end{pmatrix}$$
$$= \begin{pmatrix} A_1A_3 & 0 \\ 0 & -A_1^{-1}A_3^T + A_2 \end{pmatrix}$$
$$= \begin{pmatrix} A_1A_3 & 0 \\ 0 & A_1^{-1} \end{pmatrix}, \tag{7.65}$$

where in the last equality we have used the equality (7.63). From (7.65) we obtain

$$\frac{1}{2}\mathrm{sgn} M_0(R) = -\frac{1}{2}\mathrm{sgn} \begin{pmatrix} A_1A_3 & 0 \\ 0 & A_1^{-1} \end{pmatrix}. \tag{7.66}$$

By the Jordan canonical form decomposition of complex matrix, there exists a complex invertible k-order matrix G_1 such that

$$G_1^{-1}A_3G_1 = \begin{pmatrix} u_1 & * & * & * & * \\ 0 & u_2 & * & * & * \\ 0 & 0 & \ddots & * & * \\ 0 & 0 & 0 & u_{k-1} & * \\ 0 & 0 & 0 & 0 & u_k \end{pmatrix}$$

with $u_1, u_2, \ldots, u_k \in \mathbb{C}$.

Equation (7.23) gives

$$N_kR^{-1}N_kR = I_{2k} + 2 \begin{pmatrix} A_3 & A_2 \\ A_1A_3 & A_3^T \end{pmatrix}. \tag{7.67}$$

Since

$$\begin{pmatrix} I_k & 0 \\ -A_1 & I_k \end{pmatrix} \begin{pmatrix} A_3 & A_2 \\ A_1A_3 & A_3^T \end{pmatrix} \begin{pmatrix} I_k & 0 \\ A_1 & I_k \end{pmatrix} = \begin{pmatrix} I_k + 2A_3 & A_2 \\ -A_1 & -I_k \end{pmatrix},$$

by (7.67) we have

$$\begin{pmatrix} I_k & 0 \\ A_1 & I_k \end{pmatrix}^{-1} (N_k R^{-1} N_k R) \begin{pmatrix} I_k & 0 \\ A_1 & I_k \end{pmatrix} = \begin{pmatrix} 3I_k + 4A_3 & 2A_2 \\ -2A_1 & -I_k \end{pmatrix} := R_1. \tag{7.68}$$

By (7.68), for any $\lambda \in \mathbb{C}$ we get

$$\lambda I_{2k} - R_1 = \begin{pmatrix} (\lambda - 3)I_k - 4A_3 & -2A_2 \\ 2A_1 & (\lambda + 1)I_k \end{pmatrix}. \tag{7.69}$$

Since A_1 is invertible, by (7.63) there holds

$$\begin{aligned} &\begin{pmatrix} I_k & -\frac{1}{2}((\lambda - 3)I_k - 4A_3)A_1^{-1} \\ 0 & I_k \end{pmatrix} \begin{pmatrix} (\lambda - 3)I_k - 4A_3 & -2A_2 \\ 2A_1 & (\lambda + 1)I_k \end{pmatrix} \\ &= \begin{pmatrix} 0 & -\frac{1}{2}((\lambda^2 - 2\lambda + 1)I_k - 4\lambda A_3)A_1^{-1} \\ 2A_1 & (\lambda + 1)I_k \end{pmatrix}. \end{aligned} \tag{7.70}$$

Then by (7.69)–(7.70) we have

$$\det(\lambda I_{2k} - R_1) = \det((\lambda^2 - 2\lambda + 1)I_k - 4\lambda A_3). \tag{7.71}$$

Denote by $u_1, u_2, \ldots, u_k$ the k complex eigenvalues of A_3, (7.71) gives

$$\begin{aligned} \det(\lambda I_{2k} - R_1) &= \Pi_{i=1}^k (\lambda^2 - 2\lambda + 1 - 4\lambda u_i) \\ &= \Pi_{i=1}^k (\lambda^2 - (2 + 4u_i)\lambda + 1). \end{aligned} \tag{7.72}$$

Thus from (7.68) and (7.72) we get

$$\begin{aligned} \det(\lambda I_{2k} - N_k R^{-1} N_k R) &= \Pi_{i=1}^k (\lambda^2 - 2\lambda + 1 - 4\lambda u_i) \\ &= \Pi_{i=1}^k (\lambda^2 - (2 + 4u_i)\lambda + 1). \end{aligned} \tag{7.73}$$

It is easy to check that the equation $\lambda^2 - (2 + u_i)\lambda + 1 = 0$ has two solutions on $\mathbf{U}$ if and only if $-4 \le u_i \le 0$ for $i = 1, 2, 3 \ldots, k$. So by (7.62) without loss of generality we assume $u_j \in [-4, 0)$ for $1 \le j \le m$ and $u_j \notin [-4, 0)$ for $m + 1 \le j \le k$. Then there exists a real invertible matrix k-order Q such that

$$Q^{-1} A_3 Q = \begin{pmatrix} A_4 & 0 \\ 0 & A_5 \end{pmatrix} := \tilde{A}_3$$

and $\sigma(A_4) \subset [-4, 0)$, $\sigma(A_5) \cap [-4, 0) = \emptyset$, where A_4 is an m-order real invertible matrix and A_5 is a $(k - m)$-order real matrix.

Denote by $\tilde{A}_1 = Q^T A_1 Q$. We have

$$\tilde{A}_1 \tilde{A}_3 = Q^T A_1 Q\, Q^{-1} A_3 Q = Q^T A_1 A_3 Q.$$

Hence both $\tilde{A}_1$ and $\tilde{A}_1\tilde{A}_3$ are symmetric and we conclude that

$$\mathrm{sgn}A_1 + \mathrm{sgn}(A_1A_3) = \mathrm{sgn}\tilde{A}_1 + \mathrm{sgn}(\tilde{A}_1\tilde{A}_3). \tag{7.74}$$

Correspondingly we can write $\tilde{A}_1$ in the block form decomposition $\tilde{A}_1 = \begin{pmatrix} E_1 & E_2 \\ E_2^T & E_4 \end{pmatrix}$, where E_1 is an m-order real symmetric matric and E_4 is a $(k-m)$-order real symmetric matrix. Then

$$\tilde{A}_1\tilde{A}_3 = \begin{pmatrix} E_1A_4 & E_2A_5 \\ E_2^TA_4 & E_4A_5 \end{pmatrix}$$

is symmetric.

By the same argument as the proof of Subcase 2 of Lemma 3.15 without loss of generality we can assume E_1 is invertible (otherwise we can perturb it slightly such that it is invertible). So as in Subcase 1 of the proof of Lemma 3.15 we obtain

$$\begin{aligned} &\begin{pmatrix} I_m & 0 \\ -E_2^TE_1^{-1} & I_{k-m} \end{pmatrix} \begin{pmatrix} E_1 & E_2 \\ E_2^T & E_4 \end{pmatrix} \begin{pmatrix} I_m & -E_1^{-1}E_2 \\ 0 & I_{k-m} \end{pmatrix} \\ &= \begin{pmatrix} E_1 & 0 \\ 0 & E_4 - E_2^TE_1^{-1}E_2 \end{pmatrix} \end{aligned} \tag{7.75}$$

and

$$\begin{aligned} &\begin{pmatrix} I_m & 0 \\ -E_2^TE_1^{-1} & I_{k-m} \end{pmatrix} \begin{pmatrix} E_1A_4 & E_2A_5 \\ E_2^TA_4 & E_4A_5 \end{pmatrix} \begin{pmatrix} I_m & -E_1^{-1}E_2 \\ 0 & I_{k-m} \end{pmatrix} \\ &= \begin{pmatrix} E_1A_4 & 0 \\ 0 & (E_4 - E_2^TE_1^{-1}E_2)A_5 \end{pmatrix}. \end{aligned} \tag{7.76}$$

By (7.76) we also have that E_1A_4 is symmetric. Since E_1 is symmetric and invertible, $\sigma(A_4) \subset [-4, 0)$, by Lemma 3.15 we have

$$\mathrm{sgn}(E_1A_4) + \mathrm{sgn}E_1 = 0. \tag{7.77}$$

By (7.75) and (7.75), there hold

$$\mathrm{sgn}\tilde{A}_1 = \mathrm{sgn}(E_4 - E_2^TE_1^{-1}E_2) + \mathrm{sgn}E_1, \tag{7.78}$$

$$\mathrm{sgn}(\tilde{A}_1\tilde{A}_3) = \mathrm{sgn}((E_4 - E_2^TE_1^{-1}E_2)A_5) + \mathrm{sgn}(E_1A_4). \tag{7.79}$$

Equations (7.77)–(7.79) give

$$\begin{aligned}
&\operatorname{sgn}(\tilde{A}_1\tilde{A}_3) + \operatorname{sgn}\tilde{A}_1 \\
&= \operatorname{sgn}((E_4 - E_2^T E_1^{-1} E_2)A_5) + \operatorname{sgn}(E_4 - E_2^T E_1^{-1} E_2) \\
&\in [-2(k-m), 2(k-m)].
\end{aligned} \tag{7.80}$$

Then (7.61) holds from (7.66), (7.74) and (7.80).
Step 2. Assume A_1 is not invertible.

If $A_1 = 0$, then $A_3 = -I_k$ and $m = k$. It is easy to check that $M_0(R) = 2\begin{pmatrix} 0 & I_k \\ I_k & -A_2 \end{pmatrix}$ is congruent to $2\begin{pmatrix} 0 & I_k \\ I_k & 0 \end{pmatrix}$, so $\operatorname{sgn}M_0(R) = 0$ and (7.61) holds.

If $1 \le \operatorname{rank}A_1 = r \le k-1$, there is a $k \times k$ invertible matrix G with $\det G > 0$ such that

$$(G^{-1})^T A_1 G^{-1} = \operatorname{diag}(0, \Lambda), \tag{7.81}$$

where Λ is an $r \times r$ real invertible matrix. Hence

$$\begin{aligned}
\operatorname{diag}((G^T)^{-1}, G) \cdot R \cdot \operatorname{diag}(G^{-1}, G^T) &= \begin{pmatrix} (G^T)^{-1}A_1G^{-1} & I_k \\ GA_3G^{-1} & GA_2G^T \end{pmatrix} \\
&:= R_2 = \begin{pmatrix} 0 & 0 & I_{k-r} & 0 \\ 0 & \Lambda & 0 & I_r \\ B_1 & B_2 & D_1 & D_2 \\ B_3 & B_4 & D_3 & D_4 \end{pmatrix},
\end{aligned} \tag{7.82}$$

where B_1 and D_1 are $(k-r) \times (k-r)$ matrices, B_4 and D_4 are $r \times r$ matrices.

Since R_2 is symplectic and Λ is invertible, there holds $R_2^T J_k R_2 = J_k$. It implies that $B_3 = 0$, $D_3 = D_2^T$, $B_1 = -I_{k-r}$, and D_1, D_4 are symmetric. Thus

$$R_2 = \begin{pmatrix} 0 & 0 & I_{k-r} & 0 \\ 0 & \Lambda & 0 & I_r \\ B_1 & B_2 & D_1 & D_2 \\ 0 & B_4 & D_2^T & D_4 \end{pmatrix}.$$

For $t \in [0, 1]$, we define

$$\beta(t) = \begin{pmatrix} 0 & 0 & I_{k-r} & 0 \\ 0 & \Lambda & 0 & I_r \\ B_1 & tB_2 & tD_1 & tD_2 \\ 0 & B_4 & tD_2^T & D_4 \end{pmatrix}.$$

It is easy to check that β is a symplectic path and $\nu_{L_j}(\beta(t) = 0$ for all $t \in [0, 1]$ and $j = 0, 1$. We also have $\beta(1) = R_2$ and

$$\beta(0)=\begin{pmatrix}0&0&I_{k-r}&0\\0&\Lambda&0&I_r\\B_1&0&0&0\\0&B_4&0&D_4\end{pmatrix}=-J_{k-r}\diamond\begin{pmatrix}\Lambda&I_r\\B_4&D_4\end{pmatrix}:=R_3.$$

Then by (1) of Remark 3.5 we have

$$\begin{aligned}\frac{1}{2}\mathrm{sgn}M_0(R_2)&=\frac{1}{2}\mathrm{sgn}M_0(-J_{k-r})+\frac{1}{2}\mathrm{sgn}M_0\left(\begin{pmatrix}\Lambda&I_r\\B_4&D_4\end{pmatrix}\right)\\&=\frac{1}{2}\mathrm{sgn}M_0\left(\begin{pmatrix}\Lambda&I_r\\B_4&D_4\end{pmatrix}\right).\end{aligned}\tag{7.83}$$

Since $R_2\overset{s}{\sim}R$, by (7.83) we have

$$\frac{1}{2}\mathrm{sgn}M_0(R)=\frac{1}{2}\mathrm{sgn}M_0\left(\begin{pmatrix}\Lambda&I_r\\B_4&D_4\end{pmatrix}\right).\tag{7.84}$$

By (7.23), there holds

$$N_kR_2^{-1}N_kR_2=I_{2k}+2\begin{pmatrix}B_1&B_2&D_1&D_2\\0&B_4&D_2^T&D_4\\0&0&B_1^T&0\\0&\Lambda B_4&B_2^T&B_4^T\end{pmatrix}.\tag{7.85}$$

By (7.85) for any $\lambda\in\mathbb{C}$, we obtain

$$\begin{aligned}&\det(\lambda I_{2k}-N_kR_2^{-1}N_kR_2)\\&=\det((\lambda-1)I_{k-r}-2B_1)\det((\lambda-1)I_{k-r}-2B_1^T)\cdot\\&\quad\cdot\det\begin{pmatrix}(\lambda-1)I_r-2B_4&-2D_4\\-2\Lambda B_4&(\lambda-1)I_r-2B_4^T\end{pmatrix}\\&=\det(\lambda I_{2k}-N_kR_3^{-1}N_kR_3),\end{aligned}\tag{7.86}$$

where

$$N_kR_3^{-1}N_kR_3=I_{2k}+2\begin{pmatrix}B_1&0&0&0\\0&B_4&0&D_4\\0&0&B_1^T&0\\0&\Lambda B_4&0&B_4^T\end{pmatrix}.$$

So (7.86) gives

$$\sigma(N_k R^{-1} N_k R) = \sigma(N_k R_2^{-1} N_k R_2) = \sigma(N_k R_3^{-1} N_k R_3). \tag{7.87}$$

Since $B_1 = -I_{k-r}$ and $R_3 = (-J_{k-r}) \diamond \begin{pmatrix} \Lambda & I_r \\ B_4 & D_4 \end{pmatrix}$, (7.87) gives

$$e\left(N_r \begin{pmatrix} \Lambda & I_r \\ B_4 & D_4 \end{pmatrix}^{-1} N_r \begin{pmatrix} \Lambda & I_r \\ B_4 & D_4 \end{pmatrix}\right) = 2(m-(k-r)). \tag{7.88}$$

Step 1 implies that

$$\frac{1}{2}\left|\mathrm{sgn} M_0\left(\begin{pmatrix} \Lambda & I_r \\ B_4 & D_4 \end{pmatrix}\right)\right| \leq r-(m-(k-r)) = k-m. \tag{7.89}$$

Then (7.61) follows from (7.84) and (7.89). This finishes the proof of Step 2.

With Step 1 and Step 2, the proof of Lemma 3.16 is completed. □

Lemma 3.17 ([309]) *Let $R \in \mathrm{Sp}(2k)$ has the block form $R = \begin{pmatrix} A & B \\ C & D \end{pmatrix}$ with $1 \leq \mathrm{rank} B = r < k$. We have*

(i) $R \overset{s}{\sim} \begin{pmatrix} A_1 & B_1 & I_r & 0 \\ 0 & D_1 & 0 & 0 \\ A_3 & B_3 & A_2 & 0 \\ C_3 & D_3 & C_2 & D_2 \end{pmatrix}$, *where A_1, A_2, A_3 are $r \times r$ matrices, D_1, D_2, D_3 are $(k-r)\times(k-r)$ matrices, B_1, B_3 are $r \times (k-r)$ matrices, and C_2, C_3 are $(k-r) \times r$ matrices.*

(ii) *If A_3 is invertible, we have*

$$R \overset{s}{\sim} \begin{pmatrix} A_1 & I_r \\ A_3 & A_2 \end{pmatrix} \diamond \begin{pmatrix} D_1 & 0 \\ \tilde{D}_3 & D_2 \end{pmatrix}, \tag{7.90}$$

where $\tilde{D}_3$ is a $(k-r)\times(k-r)$ matrix.

(iii) *If $1 \leq \mathrm{rank} A_3 = \lambda \leq r-1$, then*

$$R \overset{s}{\sim} \begin{pmatrix} U & I_\lambda \\ \Lambda & V \end{pmatrix} \diamond \begin{pmatrix} \tilde{A}_1 & \tilde{B}_1 & I_{r-\lambda} & 0 \\ 0 & D_1 & 0 & 0 \\ 0 & \tilde{B}_3 & \tilde{A}_2 & 0 \\ \tilde{C}_3 & \tilde{D}_3 & \tilde{C}_2 & \tilde{D}_2 \end{pmatrix}, \tag{7.91}$$

where $\tilde{A}_1, \tilde{A}_2$ are $(r-\lambda)\times(r-\lambda)$ matrices, $\tilde{B}_1, \tilde{B}_3$ are $(r-\lambda)\times(k-r)$ matrices, $\tilde{C}_2, \tilde{C}_3$ are $(k-r)\times(r-\lambda)$ matrices, $D_1, \tilde{D}_2, \tilde{D}_3$ are $(k-r)\times(k-r)$ matrices, U, V, Λ are $\lambda \times \lambda$ matrices, and Λ is invertible.

(iv) *If $A_3 = 0$, then $\mathrm{rank} B_3 = \mathrm{rank} C_3$ and $\sigma(N_k R^{-1} N_k R) = \{1\}$. Moreover, if $B_3 = 0$ and $C_3 = 0$ we have*

$$R \overset{s}{\sim} \begin{pmatrix} A_1 & I_r \\ 0 & A_2 \end{pmatrix} \diamond \begin{pmatrix} D_1 & 0 \\ D_3 & D_2 \end{pmatrix}; \tag{7.92}$$

if $k \geq 2r$ *and* $\mathrm{rank} B_3 = \mathrm{rank} C_3 = r$, *then*

$$N_k R^{-1} N_k R \approx \begin{pmatrix} I_r & 2I_r & 2A_2 & 0 \\ 0 & I_r & 0 & 0 \\ 0 & 2A_1 & I_r & 0 \\ 2A_1 & 0 & 2I_r & I_r \end{pmatrix} \diamond \begin{pmatrix} I_{k-2r} & 0 \\ U_4 & I_{k-2r} \end{pmatrix}, \tag{7.93}$$

$$m^*(A^T C) = m^* \begin{pmatrix} 0 & A_1 \\ A_1 & 0 \end{pmatrix} + m^*(U_4), \quad * = +, -, 0, \tag{7.94}$$

where U_4 *is a* $(k-2r) \times (k-2r)$ *symmetric matrix, when* $k = 2r$, *the term* $\begin{pmatrix} I_{k-2r} & 0 \\ U_4 & I_{k-2r} \end{pmatrix}$ *will not appear in the right hand side of* (7.93) *and* $m^*(U_4) = 0$ *for* $* = +, -, 0$ *in* (7.94).

Remark 3.18 Note that R_2 in (iii) satisfies the condition of (iv) above, so there hold

$$R_2 \overset{s}{\sim} \begin{pmatrix} \tilde{A}_1 & I_r \\ 0 & \tilde{A}_2 \end{pmatrix} \diamond \begin{pmatrix} D_1 & 0 \\ \tilde{D}_3 & \tilde{D}_2 \end{pmatrix}$$

and $\mathrm{rank}\tilde{B}_3 = \mathrm{rank}\tilde{C}_3$.

Proof of Lemma 3.17 Since $\mathrm{rank} B = r$, there are two invertible $k \times k$ matrices U and V with $\det U > 0$ and $\det V > 0$ such that

$$UBV = \begin{pmatrix} I_r & 0 \\ 0 & 0 \end{pmatrix}.$$

So there holds

$$R \overset{s}{\sim} \mathrm{diag}(U, (U^T)^{-1})\, R\, \mathrm{diag}((V^T)^{-1}, V) = \begin{pmatrix} A_1 & B_1 & I_r & 0 \\ C_1 & D_1 & 0 & 0 \\ A_3 & B_3 & A_2 & B_2 \\ C_3 & D_3 & C_2 & D_2 \end{pmatrix} := \tilde{R}_1, \tag{7.95}$$

where for $j = 1, 2, 3$, A_j is an $r \times r$ matrix, D_j is a $(k-r) \times (k-r)$ matrix, B_j is an $r \times (k-r)$ matrix, and C_j is $(k-r) \times r$ matrix. Since $\tilde{R}_1$ is still a symplectic matrix, we have $C_1 = 0$, $B_2 = 0$. So

$$\tilde{R}_1 = \begin{pmatrix} A_1 & B_1 & I_r & 0 \\ 0 & D_1 & 0 & 0 \\ A_3 & B_3 & A_2 & 0 \\ C_3 & D_3 & C_2 & D_2 \end{pmatrix}.$$

This proves the statement (i).

Suppose A_3 is invertible. By $\tilde{R}_1$ is symplectic, we have

$$\begin{pmatrix} A_1^T & 0 \\ B_1^T & D_1^T \end{pmatrix}\begin{pmatrix} A_2 & 0 \\ C_2 & D_2 \end{pmatrix} - \begin{pmatrix} A_3^T & C_3^T \\ B_3^T & D_3^T \end{pmatrix}\begin{pmatrix} I_r & 0 \\ 0 & 0 \end{pmatrix} = I_k. \tag{7.96}$$

Hence

$$D_1^T D_2 = I_{k-r}. \tag{7.97}$$

By direct computation we have

$$\begin{pmatrix} A_1 & B_1 & I_r & 0 \\ 0 & D_1 & 0 & 0 \\ A_3 & B_3 & A_2 & 0 \\ C_3 & D_3 & C_2 & D_2 \end{pmatrix}\begin{pmatrix} I_r & -A_3^{-1}B_3 & 0 & 0 \\ 0 & I_{k-r} & 0 & 0 \\ 0 & 0 & I_r & 0 \\ 0 & 0 & B_3^T(A_3^T)^{-1} & I_{k-r} \end{pmatrix} = \begin{pmatrix} A_1 & \tilde{B}_1 & I_r & 0 \\ 0 & D_1 & 0 & 0 \\ A_3 & 0 & A_2 & 0 \\ C_3 & \tilde{D}_3 & \tilde{C}_2 & D_2 \end{pmatrix}. \tag{7.98}$$

So by (7.97) we have

$$\begin{pmatrix} I_r & -\tilde{B}_1 D_2^T & 0 & 0 \\ 0 & I_{k-r} & 0 & 0 \\ 0 & 0 & I_r & 0 \\ 0 & 0 & D_2\tilde{B}_1^T & I_{k-r} \end{pmatrix}\begin{pmatrix} A_1 & \tilde{B}_1 & I_r & 0 \\ 0 & D_1 & 0 & 0 \\ A_3 & 0 & A_2 & 0 \\ C_3 & \tilde{D}_3 & \tilde{C}_2 & D_2 \end{pmatrix} = \begin{pmatrix} A_1 & 0 & I_r & 0 \\ 0 & D_1 & 0 & 0 \\ A_3 & 0 & A_2 & 0 \\ \tilde{C}_3 & \tilde{D}_3 & \tilde{C}_2 & D_2 \end{pmatrix} := \tilde{R}_2. \tag{7.99}$$

Then we have

$$\tilde{R}_2 \overset{s}{\sim} \tilde{R}_1 \overset{s}{\sim} R. \tag{7.100}$$

Since $\tilde{R}_2$ is a symplectic matrix, we have $\tilde{R}_2^T J_k \tilde{R}_2 = J_k$, then it is easy to check that $\tilde{C}_3 = 0$, $\tilde{C}_2 = 0$. Hence we have

$$\tilde{R}_2 = \begin{pmatrix} A_1 & I_r \\ A_3 & A_2 \end{pmatrix} \diamond \begin{pmatrix} D_1 & 0 \\ \tilde{D}_3 & D_2 \end{pmatrix}. \tag{7.101}$$

This proves the statement (ii).

Suppose $A_3 \neq 0$ and A_3 is not invertible. In this case, suppose $\mathrm{rank} A_3 = \lambda$, then $0 < \lambda < r$. There is an invertible $r \times r$ matrix G with $\det G > 0$ such that

$$GA_3G^{-1} = \begin{pmatrix} \Lambda & 0 \\ 0 & 0 \end{pmatrix} \tag{7.102}$$

where Λ is a $\lambda \times \lambda$ invertible matrix. Then we have

$$\begin{pmatrix} (G^T)^{-1} & 0 & 0 & 0 \\ 0 & I_{k-r} & 0 & 0 \\ 0 & 0 & G & 0 \\ 0 & 0 & 0 & I_{k-r} \end{pmatrix} \begin{pmatrix} A_1 & B_1 & I_r & 0 \\ 0 & D_1 & 0 & 0 \\ A_3 & B_3 & A_2 & 0 \\ C_3 & D_3 & C_2 & D_2 \end{pmatrix} \begin{pmatrix} G^{-1} & 0 & 0 & 0 \\ 0 & I_{k-r} & 0 & 0 \\ 0 & 0 & G^T & 0 \\ 0 & 0 & 0 & I-k-r \end{pmatrix}$$

$$= \begin{pmatrix} \tilde{A}_1 & \tilde{B}_1 & I_r & 0 \\ 0 & D_1 & 0 & 0 \\ GA_3G^{-1} & \tilde{B}_3 & \tilde{A}_2 & 0 \\ \tilde{C}_3 & D_3 & \tilde{C}_2 & D_2 \end{pmatrix} := \tilde{R}_3. \tag{7.103}$$

By (7.102) we can write $\tilde{R}_3$ as the following block form

$$\tilde{R}_3 = \begin{pmatrix} U_1 & U_2 & F_1 & I_\lambda & 0 & 0 \\ U_3 & U_4 & F_2 & 0 & I_{r-\lambda} & 0 \\ 0 & 0 & D_1 & 0 & 0 & 0 \\ \Lambda & 0 & E_1 & W_1 & W_2 & 0 \\ 0 & 0 & E_2 & W_3 & W_4 & 0 \\ G_1 & G_2 & D_3 & K_1 & K_2 & D_2 \end{pmatrix}. \tag{7.104}$$

Let $Q_1 = \begin{pmatrix} I_\lambda & 0 & 0 \\ 0 & I_{r-\lambda} & 0 \\ -G_1\Lambda^{-1} & 0 & I_{k-r} \end{pmatrix}$ and $Q_2 = \begin{pmatrix} I_\lambda & 0 & -\Lambda^{-1}E_1 \\ 0 & I_{r-\lambda} & 0 \\ 0 & 0 & I_{k-r} \end{pmatrix}$. By (7.104) we have

$$\mathrm{diag}((Q_1^T)^{-1}, Q_1)\,\tilde{R}_3\,\mathrm{diag}(Q_2, (Q_2^T)^{-1}) = \begin{pmatrix} U_1 & U_2 & \tilde{F}_1 & I_\lambda & 0 & 0 \\ U_3 & U_4 & \tilde{F}_2 & 0 & I_{r-\lambda} & 0 \\ 0 & 0 & D_1 & 0 & 0 & 0 \\ \Lambda & 0 & 0 & W_1 & W_2 & 0 \\ 0 & 0 & E_2 & W_3 & W_4 & 0 \\ 0 & G_2 & \tilde{D}_3 & \tilde{K}_1 & \tilde{K}_2 & D_2 \end{pmatrix} := \tilde{R}_4. \tag{7.105}$$

Since $\tilde{R}_4$ is a symplectic matrix we have $\tilde{R}_4^T J \tilde{R}_4 = J$. Then by direct computation we have $U_2 = 0$, $U_3 = 0$, $W_2 = 0$, $W_3 = 0$, $\tilde{F}_1 = 0$, $\tilde{K}_1 = 0$, and U_1, U_4, W_1, W_4 are all symmetric matrices, moreover there hold

$$U_4W_4 = I_{r-\lambda}, \tag{7.106}$$

$$D_1D_2^T = I_{k-r}, \tag{7.107}$$

$$U_4\tilde{E}_2 = G_2^T D_1. \tag{7.108}$$

So

$$\tilde{R}_4 = \begin{pmatrix} U_1 & 0 & 0 & I_\lambda & 0 & 0 \\ 0 & U_4 & \tilde{F}_2 & 0 & I_{r-\lambda} & 0 \\ 0 & 0 & D_1 & 0 & 0 & 0 \\ \Lambda & 0 & 0 & W_1 & 0 & 0 \\ 0 & 0 & \tilde{E}_2 & 0 & W_4 & 0 \\ 0 & G_2 & \tilde{D}_3 & 0 & K_2 & D_2 \end{pmatrix} = \begin{pmatrix} U & I_\lambda \\ \Lambda & V \end{pmatrix} \diamond \begin{pmatrix} \tilde{A}_1 & \tilde{B}_1 & I_{r-\lambda} & 0 \\ 0 & D_1 & 0 & 0 \\ 0 & \tilde{B}_3 & \tilde{A}_2 & 0 \\ \tilde{C}_3 & \tilde{D}_3 & \tilde{C}_2 & \tilde{D}_2 \end{pmatrix}, \tag{7.109}$$

where $U = U_1$, $V = W_1$, $\tilde{A}_1 = U_4$, $\tilde{B}_1 = \tilde{F}_2$, $\tilde{B}_3 = \tilde{E}_2$, $\tilde{C}_3 = G_2$, $\tilde{A}_2 = W_4$, $\tilde{C}_2 = K_2$, $\tilde{D}_2 = D_2$. This proves statement (iii).

Lastly we prove (iv). If $A_3 = 0$, then $R \overset{s}{\sim} \begin{pmatrix} A_1 & B_1 & I_r & 0 \\ 0 & D_1 & 0 & 0 \\ 0 & B_3 & A_2 & 0 \\ C_3 & D_3 & C_2 & D_2 \end{pmatrix} := R_3$. Since $R_3^T J_k R_3 = J_k$, it is easy to check that A_1, A_2 and $B_3^T B_1 + D_3^T D_1$ are symmetric, $A_1A_2 = I_r$, $D_1D_2^T = I_{k-r}$, $A_1^T B_3 = C_3^T D_1$ and $B_1^T A_2 + D_1^T C_2 = B_3^T$. So $\mathrm{rank} B_3 = \mathrm{rank} C_3$. By direct computation we have

$$N_k R_3^{-1} N_k R_3 - I_{2k} = 2 \begin{pmatrix} 0 & B_3 & A_2 & 0 \\ 0 & 0 & 0 & 0 \\ 0 & A_1^T B_3 & 0 & 0 \\ B_3^T A_1 & B_1^T B_3 + D_1^T D_3 & B_3^T & 0 \end{pmatrix}. \tag{7.110}$$

Then it is easy to check that

$$(N_k R_3^{-1} N_k R_3 - I_{2k})^4 = 0. \tag{7.111}$$

Hence by Lemma 3.12 $\sigma(N_k R^{-1} N_k R) = \sigma(N_k R_3^{-1} N_k R_3) = \{1\}$.

Furthermore if $B_3 = 0$ and $C_3 = 0$, we have

$$\mathrm{diag}\left(\begin{pmatrix} I_r & -B_1 D_1^{-1} \\ 0 & I_{k-r} \end{pmatrix}, \begin{pmatrix} I_r & 0 \\ (D_1^T)^{-1} B_1^T & I_{k-r} \end{pmatrix} \right) \begin{pmatrix} A_1 & B_1 & I_r & 0 \\ 0 & D_1 & 0 & 0 \\ 0 & 0 & A_2 & 0 \\ 0 & D_3 & C_2 & D_2 \end{pmatrix}$$

$$= \begin{pmatrix} A_1 & 0 & I_r & 0 \\ 0 & D_1 & 0 & 0 \\ 0 & 0 & A_2 & 0 \\ 0 & D_3 & \tilde{C}_2 & D_2 \end{pmatrix} := \tilde{R}. \tag{7.112}$$

Since $\tilde{R}$ is symplectic, it is easy to check that $\tilde{C}_2 = 0$, hence by the conclusion of (i) and (7.112), (7.92) holds.

If $k > 2r$ and $\operatorname{rank} B_3 = \operatorname{rank} C_3 = r$, then there are $(k-r) \times (k-r)$ matrices G_1 and G_2 with $\det(G_i) > 0$ for $i = 1, 2$ such that

$$G_1 C_3 = (I_r, 0)^T, \quad B_3 G_2 = (I_r, 0). \tag{7.113}$$

Then we have

$$\operatorname{diag}\left(\begin{pmatrix} I_r & 0 \\ 0 & (G_1^T)^{-1} \end{pmatrix}, \begin{pmatrix} I_r & 0 \\ 0 & G_1 \end{pmatrix}\right) R_3 \left(\begin{pmatrix} I_r & 0 \\ 0 & G_2 \end{pmatrix}, \begin{pmatrix} I_r & 0 \\ 0 & (G_2^T)^{-1} \end{pmatrix}\right)$$

$$= \begin{pmatrix} A_1 & \hat{B}_1 & I_r & 0 \\ 0 & \hat{D}_1 & 0 & 0 \\ 0 & B_3 G_2 & A_2 & 0 \\ G_1 C_3 & \hat{D}_3 & \hat{C}_2 & \hat{D}_2 \end{pmatrix} := R_4. \tag{7.114}$$

By (7.113) there exist $(k-r) \times r$ matrix G_3 and an $r \times (k-r)$ matrix G_4 such that

$$\begin{pmatrix} I_r & 0 \\ G_3 & I_{k-r} \end{pmatrix} \begin{pmatrix} 0 & B_3 G_2 \\ G_1 C_3 & \hat{D}_3 \end{pmatrix} \begin{pmatrix} I_r & G_4 \\ 0 & I_{k-r} \end{pmatrix} = \begin{pmatrix} 0 & I_r & 0 \\ I_r & 0 & 0 \\ 0 & 0 & \tilde{D}_3 \end{pmatrix}, \tag{7.115}$$

where $\tilde{D}_3$ is a $(k-2r) \times (k-2r)$ matrix. Then we have

$$\operatorname{diag}\left(\begin{pmatrix} I_r & -G_3^T \\ 0 & I_{k-r} \end{pmatrix}, \begin{pmatrix} I_r & 0 \\ G_3 & I_{k-r} \end{pmatrix}\right) R_4 \operatorname{diag}\left(\begin{pmatrix} I_r & G_4 \\ 0 & I_{k-r} \end{pmatrix}, \begin{pmatrix} I_r & 0 \\ -G_4^T & I_{k-r} \end{pmatrix}\right)$$

$$= \begin{pmatrix} A_1 & U_1 & U_2 & I_r & 0 & 0 \\ 0 & E_1 & E_2 & 0 & 0 & 0 \\ 0 & E_3 & E_4 & 0 & 0 & 0 \\ 0 & I_r & 0 & A_2 & 0 & 0 \\ I_r & 0 & 0 & V_1 & F_1 & F_2 \\ 0 & 0 & \tilde{D}_3 & V_2 & F_3 & F_4 \end{pmatrix} := R_5, \tag{7.116}$$

where E_1 and F_1 are $r \times r$ matrices, E_4, F_4 are $(k-2r)\times(k-2r)$ matrices, others are corresponding matrix blocks. Since R_5 is symplectic, $R_5^T J_k R_5 = J_k$. Then we can check that $E_1 = A_1$, $E_2 = 0$, $F_3 = 0$, $F_1 = A_2$, $U_2 = E_3^T \tilde{D}_3$, U_1, V_1 and $E_4^T \tilde{D}_3$ are symmetric. So we have

$$R_5 = \begin{pmatrix} A_1 & U_1 & U_2 & I_r & 0 & 0 \\ 0 & A_1 & 0 & 0 & 0 & 0 \\ 20 & E_3 & E_4 & 0 & 0 & 0 \\ 0 & I_r & 0 & A_2 & 0 & 0 \\ I_r & 0 & 0 & V_1 & A_2 & F_2 \\ 0 & 0 & \tilde{D}_3 & V_2 & 0 & F_4 \end{pmatrix}. \tag{7.117}$$

Note that by the above construction

$$\begin{pmatrix} 0 & I_r & 0 \\ I_r & 0 & 0 \\ 0 & 0 & \tilde{D}_3 \end{pmatrix}^T \begin{pmatrix} A_1 & U_1 & U_2 \\ 0 & A_1 & 0 \\ 0 & E_3 & E_4 \end{pmatrix} = \begin{pmatrix} 0 & A_1 & 0 \\ A_1 & U_1 & U_2 \\ 0 & U_2^T & \tilde{D}_3^T E_4 \end{pmatrix}.$$

Hence by the Remark 3.7 we have

$$m^*(A^T C) = m^* \begin{pmatrix} 0 & A_1 & 0 \\ A_1 & U_1 & U_2 \\ 0 & U_2^T & \tilde{D}_3^T E_4 \end{pmatrix}, \quad * = +, -, 0. \tag{7.118}$$

Since

$$\begin{pmatrix} I_r & 0 & 0 \\ -\frac{1}{2}U_1 A_1^{-1} & I_r & 0 \\ -U_2^T A_1^{-1} & 0 & I_{k-2r} \end{pmatrix} \begin{pmatrix} 0 & A_1 & 0 \\ A_1 & U_1 & U_2 \\ 0 & U_2^T & \tilde{D}_3^T E_4 \end{pmatrix} \begin{pmatrix} I_r & -\frac{1}{2}A_1^{-1}U_1^T & -A_1^{-1}U_2 \\ 0 & I_r & 0 \\ 0 & 0 & I_{k-2r} \end{pmatrix}$$
$$= \begin{pmatrix} 0 & A_1 & 0 \\ A_1 & 0 & 0 \\ 0 & 0 & \tilde{D}_3^T E_4 \end{pmatrix},$$

by (7.118) we have

$$m^*(A^T C) = m^* \begin{pmatrix} 0 & A_1 \\ A_1 & 0 \end{pmatrix} + m^*(\tilde{D}_3^T E_4), \quad * = +, -, 0. \tag{7.119}$$

By direct computation we have

$$N_kR_5^{-1}N_kR_5 = I_{2k} + 2\begin{pmatrix} 0 & I_r & 0 & A_2 & 0 & 0 \\ 0 & 0 & 0 & 0 & 0 & 0 \\ 0 & 0 & 0 & 0 & 0 & 0 \\ 0 & A_1 & 0 & 0 & 0 & 0 \\ A_1 & U_1 & U_2 & I_r & 0 & 0 \\ 0 & U_2^T & \tilde{D}_3^T E_4 & 0 & 0 & 0 \end{pmatrix}. \tag{7.120}$$

Since A_1 and A_2 is invertible, it is easy to see that $\nu_1(N_kR_5^{-1}N_kR_5) = k - r + \dim\ker(\tilde{D}_3^T E_4)$. For any $t \in \mathbb{R}$, define

$$\beta(t) = \begin{pmatrix} I_r & 2I_r & 0 & 2A_2 & 0 & 0 \\ 0 & I_r & 0 & 0 & 0 & 0 \\ 0 & 0 & I_{k-2r} & 0 & 0 & 0 \\ 0 & 2A_1 & 0 & I_r & 0 & 0 \\ 2A_1 & 2tU_1 & 2tU_2 & 2I_r & I_r & 0 \\ 0 & 2tU_2^T & (1+t)\tilde{D}_3^T E_4 & 0 & 0 & I_r \end{pmatrix}.$$

Since $N_kR_5^{-1}N_kR_5$ is a symplectic matrix, we can check that $\beta(t)$ is a symplectic matrix for each $t \in \mathbb{R}$. Since A_1 is invertible and $\mathrm{rank}(\beta(t) - I_{2k})$ is independent of $t \in [0, 1]$. So we have

$$\sigma(\beta(t)) = \{1\} \text{ and } \nu_1(\beta(t)) = k - r + \dim\ker(\tilde{D}_3^T E_4) = constant. \tag{7.121}$$

So by the definition of R_3, R_4, R_5, Lemma 3.12, and (7.121) we have

$$N_kR^{-1}N_kR \approx N_kR_5^{-1}N_kR_5 = \beta(1) \approx \beta(0) = \begin{pmatrix} I_r & 2I_r & 2A_2 & 0 \\ 0 & I_r & 0 & 0 \\ 0 & 2A_1 & I_r & 0 \\ 2A_1 & 0 & 2I_r & I_r \end{pmatrix}$$

$$\diamond \begin{pmatrix} I_{k-2r} & 0 \\ \tilde{D}_3^T E_4 & I_{k-2r} \end{pmatrix}. \tag{7.122}$$

Denote by $U_4 = \tilde{D}_3^T E_4$. Then (7.93) and (7.94) hold from (7.119) and (7.122). So (iv) is proved. The proof of Lemma 3.17 is complete. □

The following result is about the (L_0, L_1)-normal forms of L_0-degenerate symplectic matrices.

Lemma 3.19 ([210]) *Using the same notations in Lemma* 3.17. *If* $A_3 = 0$, *then* A_1, A_2 *are symmetric and* $A_1A_2 = I_r$. *Suppose* $m^+(A_1) = p$, $m^-(A_1) = r - p$ *and* $0 \le \mathrm{rank} B_3 = \lambda \le \min\{r, k - r\}$, *then*

$$N_k R^{-1} N_k R \approx \begin{pmatrix} 1 & 1 \\ 0 & 1 \end{pmatrix}^{\diamond p+q^-} \diamond \begin{pmatrix} 1 & -1 \\ 0 & 1 \end{pmatrix}^{\diamond(r-p+q^+)} \diamond I_2^{\diamond q^0} \diamond D(2)^{\diamond\lambda}, \quad (7.123)$$

$$m^+(A^T C) = \lambda + q^+, \quad (7.124)$$

$$m^0(A^T C) = r - \lambda + q^0, \quad (7.125)$$

$$m^-(A^T C) = \lambda + q^-, \quad (7.126)$$

where $q^* \geq 0$ for $* = \pm, 0$, $q^+ + q^0 + q^- = k - r - \lambda$, $M^{\diamond 0}$ means correspondent component does not appears at all for M being one of the four matrices on the right hand side of (7.123).

Proof By (i) of Lemma 3.17 and $A_3 = 0$ we have

$$R \overset{s}{\sim} \begin{pmatrix} A_1 & B_1 & I_r & 0 \\ 0 & D_1 & 0 & 0 \\ 0 & B_3 & A_2 & 0 \\ C_3 & D_3 & C_2 & D_2 \end{pmatrix} := R_1. \quad (7.127)$$

Since R_1 is symplectic we have $R_1^T J_k R_1 = J_k$. Then we have A_1, A_2 are symmetric and $A_1 A_2 = I_r$. $D_1 D_2^T = I_{k-r}$ and $A_1^T B_3 = C_3^T D_1$. Equation (7.23) yields

$$N_k R_1^{-1} N_k R_1 = \begin{pmatrix} I_r & 2B_3 & 2A_2 & 0 \\ 0 & I_{k-r} & 0 & 0 \\ 0 & 2A_1^T B_3 & I_r & 0 \\ 2B_3^T A_1 & 2B_1^T B_3 + 2D_1^T D_3 & 2B_3^T & I_{k-r} \end{pmatrix}. \quad (7.128)$$

By Remark 3.7 we obtain

$$m^*(A^T C) = m^*\left(\begin{pmatrix} 0 & A_1^T B_3 \\ B_3^T A_1 & B_1^T B_3 + D_1^T D_3 \end{pmatrix}\right), \qquad * = +, -, 0. \quad (7.129)$$

Since $0 \leq \text{rank} B_3 = \lambda \leq \min\{r, k-r\}$, there exist $r \times r$ and $(k-r) \times (k-r)$ real invertible matrices G_1 and G_2 such that

$$G_1 B_3 G_2 = \begin{pmatrix} I_\lambda & 0 \\ 0 & 0 \end{pmatrix} := F. \quad (7.130)$$

Note that if $\lambda = 0$ then $B_3 = 0$, if $\lambda = \min\{r, k-r\}$ then $G_1 B_3 G_2 = \begin{pmatrix} I_\lambda & 0 \end{pmatrix}$ or $\begin{pmatrix} I_\lambda \\ 0 \end{pmatrix}$, if $\lambda = r = k - r$ then $G_1 B_3 G_2 = I_\lambda$. The proof below can still go through by corresponding adjustment.

By (7.130) we have

$$\begin{pmatrix} G_1A_1^{-1} & 0 \\ 0 & G_2^T \end{pmatrix} \begin{pmatrix} 0 & A_1^TB_3 \\ B_3^TA_1 & B_1^TB_3+D_1^TD_3 \end{pmatrix} \begin{pmatrix} A_1^{-1}G_1^T & 0 \\ 0 & G_2 \end{pmatrix}$$

$$= \begin{pmatrix} 0 & G_1B_3G_2 \\ G_2^TB_3^TG_1^T & U \end{pmatrix} = \begin{pmatrix} 0 & 0 & I_\lambda & 0 \\ 0 & 0 & 0 & 0 \\ I_\lambda & 0 & U_1 & U_2 \\ 0 & 0 & U_2^T & U_4 \end{pmatrix}. \tag{7.131}$$

Then

$$\begin{pmatrix} I_\lambda & 0 & 0 & 0 \\ 0 & I_{r-\lambda} & 0 & 0 \\ -\frac{1}{2}U_1 & 0 & I_\lambda & 0 \\ -U_2^T & 0 & 0 & I_{k-r-\lambda} \end{pmatrix} \begin{pmatrix} 0 & 0 & I_\lambda & 0 \\ 0 & 0 & 0 & 0 \\ I_\lambda & 0 & U_1 & U_2 \\ 0 & 0 & U_2^T & U_4 \end{pmatrix} \begin{pmatrix} I_\lambda & 0 & -\frac{1}{2}U_1 & -U_2 \\ 0 & I_{r-\lambda} & 0 & 0 \\ 0 & 0 & I_\lambda & 0 \\ 0 & 0 & 0 & I_{k-r-\lambda} \end{pmatrix}$$

$$= \begin{pmatrix} 0 & 0 & I_\lambda & 0 \\ 0 & 0 & 0 & 0 \\ I_\lambda & 0 & 0 & 0 \\ 0 & 0 & 0 & U_4 \end{pmatrix}. \tag{7.132}$$

Set

$$q^* = m^*(U_4), \quad * = \pm,\ 0. \tag{7.133}$$

Then $q^+ + q^0 + q^- = k - r - \lambda$ and (7.124)–(7.126) hold from (7.129), (7.131) and (7.132).

Also by (7.132) and Lemma 3.11 we have

$$\begin{pmatrix} I_{k-r-\lambda} & 0 \\ 2U_4 & I_{k-r-\lambda} \end{pmatrix} \approx \begin{pmatrix} 1 & 1 \\ 0 & 1 \end{pmatrix}^{\diamond q^-} \diamond I_2^{\diamond q^0} \diamond \begin{pmatrix} 1 & -1 \\ 0 & 1 \end{pmatrix}^{\diamond q^+}. \tag{7.134}$$

By (7.131), there holds

$$\begin{aligned} &\mathrm{diag}((G_1^T)^{-1}A_1, G_2^{-1}, G_1A_1^{-1}, G_2^T)(N_kR_1^{-1}N_kR_1) \\ &\mathrm{diag}(A_1^{-1}G_1^T, G_2, A_1G_1^{-1}, (G_2^T)^{-1}) \\ &= \begin{pmatrix} I_r & 2E & 2\tilde{A}_1 & 0 \\ 0 & I_{k-r} & 0 & 0 \\ 0 & 2F & I_r & 0 \\ 2F^T & 2U & 2E^T & I_{k-r} \end{pmatrix} := M, \end{aligned} \tag{7.135}$$

where $\tilde{A}_1 = (G_1^T)^{-1}A_1G_1^{-1}$, $E = (G_1^T)^{-1}A_1B_3G_2 = \tilde{A}_1F$.

Since M is symplectic, we have $M^T J_k M = J_k$. Then we have $E = \tilde{A}_1 F$. Since $\tilde{A}_1 = (G_1^T)^{-1} A_1 G_1^{-1}$, it is congruent to $\mathrm{diag}(a_1, a_2, \ldots, a_r)$ with

$$\begin{aligned} &a_i = 1, \ 1 \le i \le p \quad \text{and} \\ &a_j = -1, \ p+1 \le j \le r \text{ for some } 0 \le p \le r. \end{aligned} \tag{7.136}$$

Then there is an invertible $r \times r$ real matrix Q such that $\det Q > 0$ and

$$\begin{aligned} Q\tilde{A}_1 Q^T &= \mathrm{diag}(a_1, a_2, \ldots, a_r) \\ &= \mathrm{diag}(\mathrm{diag}(a_1, a_2, \ldots, a_\lambda), \mathrm{diag}(a_{\lambda+1}, \ldots, a_r)) := \mathrm{diag}(\Lambda_1, \Lambda_2). \end{aligned} \tag{7.137}$$

Since $\det Q > 0$ we can joint it to I_r by invertible continuous matrix path. So there is a continuous invertible symmetric matrix path α_1 such that $\alpha_1(1) = \tilde{A}_1$ and $\alpha_1(0) = \mathrm{diag}(a_1, a_2, \ldots, a_r)$ with

$$m^*(\alpha_1(t)) = m^*(\tilde{A}_1) = m^*(A_1), \quad t \in [0, 1], \ * = +, -.$$

Define symmetric matrix path

$$\alpha_2(t) = \begin{pmatrix} 2tU_1 & 2tU_2 \\ 2tU_2^T & 2U_4 \end{pmatrix}, \quad t \in [0, 1].$$

For $t \in [0, 1]$, define

$$\beta(t) = \begin{pmatrix} I_r & 2\alpha_1(t)F & 2\alpha_1(t) & 0 \\ 0 & I_{k-r} & 0 & 0 \\ 0 & 2F & I_r & 0 \\ 2F^T & \alpha_2(t) & 2F^T\alpha_1(t)^T & I_{k-r} \end{pmatrix}.$$

Then since M is symplectic, it is easy to check that β is a continuous path of symplectic matrices. Since $F = \begin{pmatrix} I_\lambda & 0 \\ 0 & 0 \end{pmatrix}$, and $\alpha_1(t)$ is invertible, by direct computation, we have

$$\mathrm{rank}(\beta(t) - I_{2k}) = 2\lambda + \mathrm{rank}(\alpha_1(t)) + \mathrm{rank}(U_4) = 2\lambda + r + m^+(U_4) + m^-(U_4).$$

Hence

$$\nu_1(\beta(t)) = \nu_1(\beta(1)) = \nu_1(M), \qquad t \in [0, 1].$$

Because $\sigma(\beta(t)) = \{1\}$, by Definition 3.8 and Lemma 3.10

$$M = \beta(1) \approx \beta(0)$$

$$= \begin{pmatrix} I_\lambda & 0 & 2\Lambda_1 & 2\Lambda_1 & 0 & 0 \\ 0 & I_{r-\lambda} & 0 & 0 & 2\Lambda_2 & 0 \\ 0 & 0 & I_\lambda & 0 & 0 & 0 \\ 0 & 0 & 2I_\lambda & I_\lambda & 0 & 0 \\ 0 & 0 & 0 & 0 & I_{r-\lambda} & 0 \\ 2I_\lambda & 0 & 0 & 2\Lambda_1 & 0 & I_\lambda \end{pmatrix} \diamond \begin{pmatrix} I_{k-r-\lambda} & 0 \\ 2U_4 & I_{k-r-\lambda} \end{pmatrix}$$

$$\approx \begin{pmatrix} I_\lambda & 2\Lambda_1 & 2\Lambda_1 & 0 \\ 0 & I_\lambda & 0 & 0 \\ 0 & 2I_\lambda & I_\lambda & 0 \\ 2I_\lambda & 0 & 2\Lambda_1 & I_\lambda \end{pmatrix} \diamond \begin{pmatrix} I_{r-\lambda} & 2\Lambda_2 \\ 0 & I_{r-\lambda} \end{pmatrix} \diamond \begin{pmatrix} I_{k-r-\lambda} & 0 \\ 2U_4 & I_{k-r-\lambda} \end{pmatrix}$$

$$= \begin{pmatrix} I_\lambda & 2\Lambda_1 & 2\Lambda_1 & 0 \\ 0 & I_\lambda & 0 & 0 \\ 0 & 2I_\lambda & I_\lambda & 0 \\ 2I_\lambda & 0 & 2\Lambda_1 & I_\lambda \end{pmatrix} \diamond \Diamond_{j=\lambda+1}^{r} \begin{pmatrix} 1 & 2a_j \\ 0 & 1 \end{pmatrix} \diamond \begin{pmatrix} I_{k-r-\lambda} & 0 \\ 2U_4 & I_{k-r-\lambda} \end{pmatrix}. \tag{7.138}$$

We define continuous symplectic matrix path

$$\psi(t) = \begin{pmatrix} I_\lambda & 2(1-t^2)\Lambda_1 & 2\Lambda_1 & 0 \\ 0 & (1+t)I_\lambda & 0 & 0 \\ 0 & 2(1-t^2)I_\lambda & I_\lambda & 0 \\ 2(1-t)I_\lambda & 0 & 2(1-t)\Lambda_1 & \frac{1}{1+t}I_\lambda \end{pmatrix}, \quad t \in [0, 1].$$

Since Λ_1 is invertible, $\nu(\psi(t)) \equiv \lambda$ for $t \in [0, 1]$. So by $\sigma(\psi(t)) \cap \mathbf{U} = \{1\}$ for $t \in [0, t]$ and Definition 3.8 we obtain

$$\begin{pmatrix} I_\lambda & \Lambda_1 & 2\Lambda_1 & 0 \\ 0 & I_\lambda & 0 & 0 \\ 0 & 2I_\lambda & I_\lambda & 0 \\ 2I_\lambda & 0 & 2\Lambda_1 & I_\lambda \end{pmatrix} = \psi(0) \approx \psi(1) = \begin{pmatrix} I_\lambda & 2\Lambda_1 \\ 0 & I_\lambda \end{pmatrix} \diamond \begin{pmatrix} 2I_\lambda & 0 \\ 0 & \frac{1}{2}I_\lambda \end{pmatrix}$$

$$= \Diamond_{j=1}^{\lambda} \begin{pmatrix} 1 & 2a_j \\ 0 & 1 \end{pmatrix} \diamond D(2)^{\diamond\lambda}. \tag{7.139}$$

Thus by (7.138), (7.139) and Remark 3.9 we get

$$M \approx \left(\Diamond_{j=1}^{r} \begin{pmatrix} 1 & a_j \\ 0 & 1 \end{pmatrix}\right) \diamond D(2)^{\diamond\lambda} \diamond \begin{pmatrix} I_{k-r-\lambda} & 0 \\ U_4 & I_{k-r-\lambda} \end{pmatrix}. \tag{7.140}$$

So by (7.134), (7.136) and Remark 3.9, there holds

$$M \approx \begin{pmatrix} 1 & 1 \\ 0 & 1 \end{pmatrix}^{\diamond(p+q^-)} \diamond \begin{pmatrix} 1 & -1 \\ 0 & 1 \end{pmatrix}^{\diamond(r-p+q^+)} \diamond I_2^{\diamond q^0} \diamond D(2)^{\diamond\lambda}. \tag{7.141}$$

By Lemma 3.12, (7.127) and (7.135), we have

$$N_k R^{-1} N_k R \approx M. \tag{7.142}$$

Then (7.123) holds from (7.141) and (7.142). The proof of Lemma 3.19 is completed. □

7.4 The Mixed (L_0, L_1)-Concavity

Definition 4.1 ([210]) The mixed (L_0, L_1)-concavity and mixed (L_1, L_0)-concavity of a symplectic path $\gamma \in \mathcal{P}_\tau(2n)$ are defined respectively by

$$\mu_{(L_0,L_1)}(\gamma) = i_{L_0}(\gamma) - \nu_{L_1}(\gamma), \quad \mu_{(L_1,L_0)}(\gamma) = i_{L_1}(\gamma) - \nu_{L_0}(\gamma).$$

Proposition V.9.8 and Theorem 3.4, imply the following result.

Proposition 4.2 *There hold*

$$\mu_{(L_0,L_1)}(\gamma) + \mu_{(L_1,L_0)}(\gamma) = i(\gamma^2) - \nu(\gamma^2) - n, \tag{7.143}$$

$$\begin{aligned}\mu_{(L_0,L_1)}(\gamma) - \mu_{(L_1,L_0)}(\gamma) &= concav^*_{(L_0,L_1)}(\gamma) \\ &= \frac{1}{2}\mathrm{sgn} M_\varepsilon(\gamma(\tau)), \quad 0 < -\varepsilon \ll 1.\end{aligned} \tag{7.144}$$

Theorem 4.3 ([210]) *For $\gamma \in \mathcal{P}_\tau(2n)$, let $P = \gamma(\tau)$. If $i_{L_0}(\gamma) \geq 0$, $i_{L_1}(\gamma) \geq 0$, $i(\gamma) \geq n$, $\gamma^2(t) = \gamma(t-\tau)\gamma(\tau)$ for all $t \in [\tau, 2\tau]$, then*

$$\mu_{(L_0,L_1)}(\gamma) + S^+_{P^2}(1) \geq 0, \tag{7.145}$$

$$\mu_{(L_1,L_0)}(\gamma) + S^+_{P^2}(1) \geq 0. \tag{7.146}$$

We recall that γ^2 is the 2-times iteration path $\gamma^2 : [0, 2\tau] \to \mathrm{Sp}(2n)$ of γ defined in (5.184).

Proof The proofs of (7.145) and (7.146) are almost the same. We only prove (7.146).
Claim 1. Under the conditions of Theorem 4.3, if

$$P^2 \approx \begin{pmatrix} 1 & 1 \\ 0 & 1 \end{pmatrix}^{\diamond p_1} \diamond D(2)^{\diamond p_2} \diamond \tilde{P}, \tag{7.147}$$

then

$$i(\gamma^2)+2S^+_{P^2}(1)-\nu(\gamma^2)\geq n+p_1+p_2. \tag{7.148}$$

Proof of Claim 1. By Theorem 7.8 of [226] we have

$$\begin{aligned}P\approx\ & I_2^{\diamond q_1}\diamond\begin{pmatrix}1&1\\0&1\end{pmatrix}^{\diamond q_2}\diamond\begin{pmatrix}1&-1\\0&1\end{pmatrix}^{\diamond q_3}\diamond(-I_2)^{\diamond q_4}\diamond\begin{pmatrix}-1&1\\0&-1\end{pmatrix}^{\diamond q_5}\diamond\begin{pmatrix}-1&-1\\0&-1\end{pmatrix}^{\diamond q_6}\\ &\diamond R(\theta_1)\diamond\cdots\diamond R(\theta_{q_7})\diamond\cdots\diamond R(\theta_{q_7+q_8})\diamond N_2(\omega_1,b_1)\diamond\cdots\diamond N_2(\omega_{q_9},b_{q_9})\\ &\diamond D(2)^{\diamond q_{10}}\diamond D(-2)^{\diamond q_{11}},\end{aligned}\tag{7.149}$$

where $q_i\geq 0$ for $1\leq i\leq 11$ with $q_1+q_2+\cdots+q_8+2q_9+q_{10}+q_{11}=n$, $\theta_j\in(0,\pi)$ for $1\leq j\leq q_7$, $\theta_j\in(\pi,2\pi)$ for $q_7+1\leq j\leq q_7+q_8$, $\omega_j\in(\mathbf{U}\setminus\mathbb{R})$ for $1\leq j\leq q_9$ and $b_j=\begin{pmatrix}b_{j1}&b_{j2}\\b_{j3}&b_{j4}\end{pmatrix}$ satisfying $b_{j2}\neq b_{j3}$ for $1\leq j\leq q_9$.

By (7.149) and Remark 3.9 we obtain

$$\begin{aligned}P^2\approx\ & I_2^{\diamond(q_1+q_4)}\diamond\begin{pmatrix}1&1\\0&1\end{pmatrix}^{\diamond(q_2+q_6)}\diamond\begin{pmatrix}1&-1\\0&1\end{pmatrix}^{\diamond(q_3+q_5)}\\ &\diamond R(2\theta_1)\diamond\cdots\diamond R(2\theta_{q_7})\diamond\cdots\diamond R(2\theta_{q_7+q_8})\\ &\diamond N_2(\omega_1,b_1)^2\diamond\cdots\diamond N_2(\omega_{q_9},b_{q_9})^2\diamond D(2)^{\diamond(q_{10}+q_{11})}.\end{aligned}\tag{7.150}$$

By Theorem 7.8 of [226], (7.147) and (7.150), there hold

$$q_2+q_6\geq p_1,\qquad q_{10}+q_{11}\geq p_2. \tag{7.151}$$

Since $\gamma^2(t)=\gamma(t-\tau)\gamma(\tau)$ for all $t\in[\tau,2\tau]$ we have γ^2 is also the twice iteration of γ in the periodic boundary value case, so by the Bott-type formula (2.13), the proof of Lemma 4.1 of [232], and Lemma II.3.14 we have

$$\begin{aligned}& i(\gamma^2)+2S^+_{P^2}(1)-\nu(\gamma^2)\\ &=2i(\gamma)+2S^+_P(1)+\sum_{\theta\in(0,\pi)}(S^+_P(e^{\sqrt{-1}\theta})\\ &\quad-(\sum_{\theta\in(0,\pi)}(S^-_P(e^{\sqrt{-1}\theta})+(\nu(P)-S^+_P(1))+(\nu_{-1}(P)-S^-_P(-1)))\\ &=2i(\gamma)+2(q_1+q_2)+(q_8-q_7)-(q_1+q_3+q_4+q_5)\\ &\geq 2n+q_1+2q_2+(q_8-q_7)-(q_3+q_4+q_5)\\ &=n+(2q_1+3q_2+q_6+2q_8+2q_9+q_{10}+q_{11})\end{aligned}$$

$$\geq n + 2q_2 + q_6 + q_{10} + q_{11}$$
$$\geq n + p_1 + p_2, \tag{7.152}$$

where in the first equality we have used $S^+_{P^2}(1) = S^+_P(1) + S^+_P(-1)$ and $\nu(\gamma^2) = \nu(\gamma) + \nu_{-1}(\gamma)$, in the first inequality we have used the condition $i(\gamma) \geq n$, in the third equality we have used that $q_1 + q_2 + \cdots + q_8 + 2q_9 + q_{10} + q_{11} = n$, in the last inequality we have used (7.151). By (7.152) Claim 1 holds.

We continue with the proof of Theorem 4.3. We set $\mathcal{A} = \mu_{(L_1,L_0)}(\gamma) + S^+_{P^2}(1)$ and $\mathcal{B} = \mu_{(L_0,L_1)}(\gamma) + S^+_{P^2}(1)$.

By (5.212) and (5.213) we have

$$i_{L_0}(\gamma) + i_{L_1}(\gamma) = i(\gamma^2) - n, \quad \nu_{L_0}(\gamma) + \nu_{L_1}(\gamma) = \nu(\gamma^2). \tag{7.153}$$

From (7.153) or (7.143) we obtain

$$\mathcal{A} + \mathcal{B} = i(\gamma^2) + 2S^+_{P^2}(1) - \nu(\gamma^2) - n. \tag{7.154}$$

Case 1. $\nu_{L_0}(\gamma) = 0$.

In this case

$$i_{L_1}(\gamma) + S^+_{P^2}(1) - \nu_{L_0}(\gamma) \geq 0 + 0 - 0 = 0$$

and (7.146) holds.

Case 2. $\nu_{L_0}(\gamma) = n$.

In this case $P = \begin{pmatrix} A & 0 \\ C & D \end{pmatrix}$, so A is invertible and

$$m^0(A^T C) = \nu_{L_1}(P) = \nu_{L_1}(\gamma). \tag{7.155}$$

By Lemma 3.11 we have

$$NP^{-1}NP = \begin{pmatrix} I_n & 0 \\ 2A^T C & I_n \end{pmatrix}$$
$$\approx I_2^{\diamond m^0(A^T C)} \diamond N_1(1,1)^{\diamond m^-(A^T C)} \diamond N_1(1,-1)^{\diamond m^+(A^T C)}. \tag{7.156}$$

By Claim 1, (7.156) and (7.154), there holds

$$\mathcal{A} + \mathcal{B} \geq m^-(A^T C). \tag{7.157}$$

By Theorem 3.4, Lemma 3.13 and (7.155) we obtain

$$\mathcal{A}-\mathcal{B} \geq m^{+}(A^{T}C)+m^{0}(A^{T}C)-n. \tag{7.158}$$

Then (7.157) and (7.158) give

$$2\mathcal{A} \geq m^{-}(A^{T}C)+(m^{+}(A^{T}C)+m^{0}(A^{T}C))-n=0$$

which yields $\mathcal{A} \geq 0$ and (7.146) holds.
Case 3. $1 \leq \nu_{L_0}(\gamma)=\nu_{L_0}(P) \leq n-1$.

In this case by (i) of Lemma 3.17 we have

$$P:=\begin{pmatrix} A & B \\ C & D \end{pmatrix} \overset{s}{\sim} \begin{pmatrix} A_1 & B_1 & I_r & 0 \\ 0 & D_1 & 0 & 0 \\ A_3 & B_3 & A_2 & 0 \\ C_3 & D_3 & C_2 & D_2 \end{pmatrix},$$

where A_1, A_2, A_3 are $r \times r$ matrices, D_1, D_2, D_3 are $(n-r)\times(n-r)$ matrices, B_1, B_3 are $r\times(n-r)$ matrices, and C_2, C_3 are $(n-r)\times r$ matrices. We divide Case 3 into the following 3 subcases.
Subcase 1. $A_3=0$.

In this subcase let $\lambda=\operatorname{rank} B_3$. Then $0 \leq \lambda \leq \min\{r, n-r\}$, A_1 is invertible, $A_1A_2=I_r$ and $D_1D_2^T=I_{k-r}$, so we have that A is invertible, furthermore there holds $m^0(A^TC)=\dim\ker C=\nu_{L_1}(P)$. Suppose $m^+(A_1)=p, m^-(A_1)=r-p$, then by Lemma 3.19 we have

$$N_kR^{-1}N_kR \approx \begin{pmatrix} 1 & 1 \\ 0 & 1 \end{pmatrix}^{\diamond p+q^-} \diamond \begin{pmatrix} 1 & -1 \\ 0 & 1 \end{pmatrix}^{\diamond(r-p+q^+)} \diamond I_2^{\diamond q^0} \diamond D(2)^{\diamond\lambda}, \tag{7.159}$$

$$m^{+}(A^{T}C)=\lambda+q^{+}, \tag{7.160}$$

$$m^{0}(A^{T}C)=r-\lambda+q^{0}, \tag{7.161}$$

$$m^{-}(A^{T}C)=\lambda+q^{-}, \tag{7.162}$$

where $q^* \geq 0$ for $*=+,-,0$ and $q^++q^0+q^-=n-r-\lambda$.

By (7.159) and Claim 1, there holds

$$i(\gamma^2)+2S_{P^2}^{+}(1)-\nu(\gamma^2) \geq n+p+q^-+\lambda \geq n+q^-+\lambda. \tag{7.163}$$

Equation (7.163) and (7.154) give

$$\mathcal{A}+\mathcal{B} \geq q^-+\lambda. \tag{7.164}$$

By Theorem 3.4, Lemma 3.13, and (7.160)–(7.162), we have

$$
\begin{aligned}
&\mathcal{A} - \mathcal{B} \\
&\geq m^+(A^T C) + m^0(A^T C) - n \\
&= q^+ + \lambda + r - \lambda + q^0 - n \\
&= r + q^+ + q^0 - n.
\end{aligned} \tag{7.165}
$$

Since $q^+ + q^0 + q^- = n - r - \lambda$, (7.164) and (7.165) imply

$$
\begin{aligned}
2\mathcal{A} &\geq q^- + \lambda + (r + q^+ + q^0) - n \\
&= (q^- + q^+ + q^0) - (n - r - \lambda) \\
&= 0
\end{aligned}
$$

which yields (7.146).

Subcase 2. A_3 is invertible.

In this case by (ii) of Lemma 3.17 there holds

$$
P \overset{s}{\sim} \begin{pmatrix} A_1 & I_r \\ A_3 & A_2 \end{pmatrix} \diamond \begin{pmatrix} D_1 & 0 \\ \tilde{D}_3 & D_2 \end{pmatrix} := P_1 \diamond P_2, \tag{7.166}
$$

where $\tilde{D}_3$ is a $(k - r) \times (k - r)$ matrix. Then by (7.166) and Lemma 3.12 we obtain

$$
P^2 \approx (N_r P_1^{-1} N_r P_1) \diamond (N_{n-r} P_2^{-1} N_{n-r} P_2). \tag{7.167}
$$

Let $e(N_r P_1^{-1} N_r P_1) = 2m$, by Lemma 3.16 we have $0 \leq m \leq r$ and

$$
\frac{1}{2}\mathrm{sgn} M_\varepsilon(P_1) \leq r - m, \quad 0 < -\varepsilon \ll 1. \tag{7.168}
$$

Also by (7.167) and (7.150), there exists $\tilde{P}_1 \in \mathrm{Sp}(2m)$ such that

$$
N_r P_1^{-1} N_r P_1 \approx D(2)^{\diamond(r-m)} \diamond \tilde{P}_1. \tag{7.169}
$$

By Lemma 3.11, there holds

$$
\begin{aligned}
N_{n-r}\ P_2^{-1}\ N_{n-r} P_2 &= \begin{pmatrix} I_{n-r} & 0 \\ 2D_1^T \tilde{D}_3 & I_{n-r} \end{pmatrix} \\
&\approx \begin{pmatrix} 1 & 1 \\ 0 & 1 \end{pmatrix}^{\diamond m^-(D_1^T \tilde{D}_3)} \diamond I_2^{\diamond m^0(D_1^T \tilde{D}_3)} \diamond \begin{pmatrix} 1 & -1 \\ 0 & 1 \end{pmatrix}^{\diamond m^+(D_1^T \tilde{D}_3)}.
\end{aligned} \tag{7.170}
$$

So by Claim 1 and (7.169), (7.170), (7.167) and (7.154) we have

$$\mathcal{A}+\mathcal{B}\geq m^{-}(D_1^T\tilde{D}_3)+r-m. \tag{7.171}$$

By Theorem 3.4 and Lemma 3.13 together with Lemma 3.16, for $0<-\varepsilon\ll 1$ we get

$$\begin{aligned}\mathcal{A}-\mathcal{B}&=-\frac{1}{2}\mathrm{sgn}M_\varepsilon(P_1)-\frac{1}{2}\mathrm{sgn}M_\varepsilon(P_2)\\&\geq -r+m-(n-r)+m^{+}(D_1^T\tilde{D}_3)+m^{0}(D_1^T\tilde{D}_3)\\&=m+m^{+}(D_1^T\tilde{D}_3)+m^{0}(D_1^T\tilde{D}_3)-n,\end{aligned} \tag{7.172}$$

where we have used the fact $m^0(D_1^T\tilde{D}_3)=\ker(\tilde{D}_3)=\nu_{L_1}(P_2)$.

Note that

$$m^{+}(D_1^T\tilde{D}_3)+m^{0}(D_1^T\tilde{D}_3)+m^{-}(D_1^T\tilde{D}_3)=n-r. \tag{7.173}$$

Then by (7.171), (7.172) and (7.173) we have

$$\begin{aligned}2\mathcal{A}&\geq m^{-}(D_1^T\tilde{D}_3)+r-m+(m+m^{+}(D_1^T\tilde{D}_3)+m^{0}(D_1^T\tilde{D}_3))-n\\&=m^{+}(D_1^T\tilde{D}_3)+m^{0}(D_1^T\tilde{D}_3)+m^{-}(D_1^T\tilde{D}_3)-(n-r)\\&=0\end{aligned}$$

which yields (7.146).

Subcase 3. $1\leq \mathrm{rank}A_3=l\leq r-1$.

In this case by (iii) of Lemma 3.17 there holds

$$P\overset{s}{\sim}\begin{pmatrix}U & I_l\\ \Lambda & V\end{pmatrix}\diamond\begin{pmatrix}\tilde{A}_1 & \tilde{B}_1 & I_{r-l} & 0\\ 0 & D_1 & 0 & 0\\ 0 & \tilde{B}_3 & \tilde{A}_2 & 0\\ \tilde{C}_3 & \tilde{D}_3 & \tilde{C}_2 & \tilde{D}_2\end{pmatrix}:=P_3\diamond P_4, \tag{7.174}$$

where $\tilde{A}_1,\tilde{A}_2$ are $(r-l)\times(r-l)$ matrices, $\tilde{B}_1,\tilde{B}_3$ are $(r-l)\times(n-r)$ matrices, $\tilde{C}_2,\tilde{C}_3$ are $(n-r)\times(r-l)$ matrices, $D_1,\tilde{D}_2,\tilde{D}_3$ are $(n-r)\times(n-r)$ matrices, U,V,Λ are $l\times l$ matrices, and Λ is invertible.

Let $\lambda=\mathrm{rank}\tilde{B}_3$ and denote $P_4=\begin{pmatrix}\tilde{A} & \tilde{B}\\ \tilde{C} & \tilde{D}\end{pmatrix}$, where $\tilde{A},\tilde{B},\tilde{C},\tilde{D}$ are $(n-l)$-order real matrices. Assume $m^{+}(\tilde{A}_1)=p$, $m^{-}(\tilde{A}_1)=r-l-p$, then by Lemma 3.19 we have

$$N_k P_4^{-1} N_k P_4 \approx \begin{pmatrix} 1 & 1 \\ 0 & 1 \end{pmatrix}^{\diamond(p+q^-)} \diamond \begin{pmatrix} 1 & -1 \\ 0 & 1 \end{pmatrix}^{\diamond(r-l-p+q^+)} \diamond I_2^{\diamond q^0} \diamond D(2)^{\diamond\lambda}, \tag{7.175}$$

$$m^+(\tilde{A}^T\tilde{C}) = \lambda + q^+, \tag{7.176}$$
$$m^0(\tilde{A}^T\tilde{C}) = r - l - \lambda + q^0, \tag{7.177}$$
$$m^-(\tilde{A}^T\tilde{C}) = \lambda + q^-, \tag{7.178}$$

where $q^* \geq 0$ for $* = +, -, 0$ and $q^+ + q^0 + q^- = n - r - \lambda$.
Let $e(N_l P_3^{-1} N_l P_3) = 2m$, by Lemma 3.16 we obtain $0 \leq m \leq l$ and

$$\frac{1}{2}\mathrm{sgn} M_\varepsilon(P_3) \leq l - m, \quad 0 < -\varepsilon \ll 1. \tag{7.179}$$

By similar argumet as in the proof of Subcase 2, there exists $\tilde{P}_3 \in \mathrm{Sp}(2m)$ such that

$$N_r P_3^{-1} N_r P_3 \approx D(2)^{\diamond(l-m)} \diamond \tilde{P}_3. \tag{7.180}$$

So by Claim 1, (7.174), (7.175), (7.180), and (7.154) we have

$$\mathcal{A} + \mathcal{B} \geq q^- + l - m + \lambda. \tag{7.181}$$

By Theorem 3.4, Lemma 3.13, (7.176), (7.177) and (7.179), for $0 \leq -\varepsilon \ll 1$ we obtain

$$\begin{aligned}
\mathcal{A} - \mathcal{B} &= -\frac{1}{2}\mathrm{sgn} M_\varepsilon(P_3) - \frac{1}{2}\mathrm{sgn} M_\varepsilon(P_4) \\
&\geq -\frac{1}{2}\mathrm{sgn} M_\varepsilon(P_3) - (n-l) - m^+(\tilde{A}^T\tilde{C}) + m^0(\tilde{A}^T\tilde{C}) \\
&\geq -l + m - (n-l) + (\lambda + q^+) + (r - l - \lambda + q^0) \\
&= (q^+ + q^0 + r) - n - (l - m).
\end{aligned} \tag{7.182}$$

Since $q^+ + q^0 + q^- = n - r - \lambda$, by (7.181) and (7.182) we have

$$\begin{aligned}
2\mathcal{A} &\geq q^- + l - m + \lambda + (q^+ + q^0 + r) - n - (l - m) \\
&= (q^+ + q^0 + q^-) - (n - r - \lambda) \\
&= 0
\end{aligned}$$

which yields (7.146). Hence (7.146) holds in Cases 1–3 and the proof of Theorem 4.3 is completed. □

Remark 4.4 Both the estimates (7.145) and (7.146) in Theorem 5.1 are optimal. In fact, we can construct a symplectic path satisfying the conditions of Theorem 5.1 such that the equalities in (7.145) and (7.146) hold. Let $\tau = \pi$ and $\gamma(t) = R(t)^{\diamond n}$, $t \in [0, \pi]$. It is easy to see that $i_{L_0}(\gamma) = \sum_{0<t<\pi} \nu_{L_0}(\gamma(t)) = 0$ and also $i_{L_1}(\gamma) = \sum_{0<t<\pi} \nu_{L_1}(\gamma(t)) = 0$, $\nu_{L_0}(\gamma) = \nu_{L_1}(\gamma) = n$, $\gamma^2(t) = \gamma(t-\pi)\gamma(\pi)$ for $t \in [\pi, 2\pi]$, $i(\gamma) = n$ and $P = \gamma(\pi) = -I_{2n}$ hence by Lemma 2.2 $S^+_{P^2}(1) = S^+_{I_{2n}}(1) = n$. Thus

$$\mu_{(L_0,L_1)}(\gamma) + S^+_{P^2}(1) = \mu_{(L_1,L_0)}(\gamma) + S^+_{P^2}(1) = 0 - n + n = 0.$$

Chapter 8
Applications of P-Index

8.1 The Existence of P-Solution of Nonlinear Hamiltonian Systems

One of the most beautiful application of the Maslov-type index theory for the case of periodic Hamiltonian systems is to study the existence and multiplicity of solutions of asymptotically linear Hamiltonian systems. The study on the periodic solutions of asymptotically linear Hamiltonian systems in global sense started from 1980 in [5]. Since then many mathematicians made their contributions on this problem (cf. [33, 38, 94, 177, 223, 224] and so on). In this section, we consider the non-periodic (P-boundary) solutions of asymptotically linear Hamiltonian systems.

We consider the following Hamiltonian systems with P-boundary condition:

$$\begin{cases} \dot{x}(t) = JH'(t, x(t)) \\ x(1) = Px(0) \end{cases}, \tag{8.1}$$

where $P \in \mathrm{Sp}(2n)$, $H \in C^2(\mathbb{R} \times \mathbb{R}^{2n}, \mathbb{R})$ satisfying $H(t+1, Px) = H(t, x)$, $\forall (t, x)$ and $H'(t, x)$ is the gradient of H with respect to the variables x. It is clear that

$$P^T H''(t+1, Px)P = H''(t, x). \tag{8.2}$$

We turn to study the existence and multiplicity of solutions of asymptotically linear Hamiltonian systems. A Hamiltonian system is called asymptotically linear at infinity if

(H_∞) there exists a continuous symmetric matrix function $B_\infty(t)$ such that $P^T B_\infty(t+1)P = B_\infty(t)$ for all $t \in \mathbb{R}$ and

$$\|H'(t, x) - B_\infty(t)x\|_{\mathbb{R}^{2n}} = o(\|x\|_{\mathbb{R}^{2n}}) \tag{8.3}$$

uniformly in t, as $\|x\|_{\mathbb{R}^{2n}} \to \infty$.

We assume the following conditions on H:

(H1) $H \in C^2(\mathbb{R} \times \mathbb{R}^{2n}, \mathbb{R})$ and

$$H(t+1, Px) = H(t, x), \quad \forall (t, x) \in \mathbb{R} \times \mathbb{R}^{2n}. \tag{8.4}$$

(H2) There exist constants $a > 0$ and $p > 1$ such that

$$|H''(t, x)| \le a(1 + |x|^p), \quad \forall (t, x) \in \mathbb{R} \times \mathbb{R}^{2n}. \tag{8.5}$$

(H3) There exists a constant $C(H) > 0$ such that

$$|H''(t, x)| \le C(H), \quad \forall (t, x) \in [0, 1] \times \mathbb{R}^{2n}. \tag{8.6}$$

We remind that for a matrix function $B = B(t)$ satisfying $P^T B(t+1)P = B(t)$, the index pair $(i^P(B), \nu^P(B))$ is defined Chap. 4. In the following result, we suppose $\nu^P(B_\infty) = 0$ and denote $i_\infty = i^P(B_\infty)$.

Theorem 1.1 ([186]) *Suppose H satisfies conditions (H1), (H3), (H_∞). Then the problem* (8.1) *possesses at least one solution x_0. Let $B(t) = H''(t, x_0(t))$ and $(i_0, \nu_0) = (i^P(B), \nu^P(B))$. If*

$$i_\infty \notin [i_0, i_0 + \nu_0], \tag{8.7}$$

the problem (8.1) *possesses at least two solutions. Furthermore, if x_0 is not pseudo-degenerated, and*

$$i_\infty \notin [i_0 - 2n, i_0 + \nu_0 + 2n], \tag{8.8}$$

the problem (8.1) *possesses at least three solutions.*

Remark 1.2 x_0 is a critical point of f defined by

$$f(x) = \frac{1}{2}\langle Ax, x\rangle_L - g(x), \quad \forall x \in \mathrm{dom} A \subset L, \tag{8.9}$$

where L is the space $L = \gamma_P L^2(S^1, \mathbb{R}^{2n}) = \{x \,|\, x(t) = \gamma_P(t)\xi(t), \ \xi \in L^2(S^1, \mathbb{R}^{2n})\}$, $\gamma_P \in \mathcal{P}(2n)$ is defined in Sect. 4.2 of Chap. 4. We set

$$\langle Ax, y\rangle_L = \int_0^1 (-J\dot{x}, y)\, dt, \quad \forall x, \ y \in L \tag{8.10}$$

and

$$g(x) = \int_0^1 H(t, x(t))\, dt, \quad x \in L. \tag{8.11}$$

Using the notations of Theorem III.4.1 (see papers [4, 5, 33, 34] and [223] for nonlinear cases.), let $z_0 = \mathcal{P}x$ be a critical point of the function a. Suppose the critical set of a is isolated. Then z_0 is an isolated invariant set of the gradient flow of a. By Conley homotopic index theory, we get the Conley homotopic index $h(z_0)$, and its poincaré polynomial

$$p(t, h(z_0)) = t^{m^+(z_0)} \sum_{j=0}^{m^0(z_0)} a_j t^j,\ a_j \in \{0, 1, 2, \cdots\}, a_0 = 0 \text{ or } a_{m^0(z_0)} = 0,$$

where $m^*(z_0),\ * = 0, \pm$ are respectively the nullity, positive and negative Morse indices of a at z_0 (cf. Definition III.4.2). We say that x_0 is pseudo-degenerated if $p(t, h(z_0)) = 0$ or contains the factor $(1+t)$. Theorem 1.1 should be compared with the main result of [224] (see also Theorem 7.2.2 of [223] and Theorem 4.1.3 of [34]) where they considered the periodic solutions of asymptotically Hamiltonian systems. The main ingredients of the proof are the Maslov-type index theory, the Poincaré polynomial of the Conley homotopic index of isolated invariant set, and the saddle point reduction method. The Conley homotopic index theory can be used here for the proof of Theorem 1.1. Now by using the index theory and the saddle point method, we can prove Theorem 1.1 as done in [223] and [224]. We omit the details.

By the index theory and the Galerkin approximation methods developed in Chap. 4, similar to [38] and [94], we have the following result.

Theorem 1.3 ([186]) *Suppose H satisfies conditions (H1), (H2), (H_∞) and*

(H4) *There exists continuous symmetric matrix function $B_0(t)$ such that for all $t \in \mathbb{R}$, $P^T B_0(t+1)P = B_0(t)$ and*

$$H'(t, x) = B_0(t)x + o(|x|) \text{ as } |x| \to 0 \text{ uniformly in } t. \tag{8.12}$$

(H5) *For $h(t, x) = H(t, x) - \frac{1}{2}(B_\infty(t)x, x)$ with $(t, x) \in \mathbb{R} \times \mathbb{R}^{2n}$, either*

$$h(t, x) \to 0,\quad |h'(t, x)| \to 0 \text{ as } |x| \to +\infty \text{ uniformly in } t. \tag{8.13}$$

or

$$h(t, x) \to \pm\infty,\quad |h'(t, x)| = 0(1) \text{ as } |x| \to \infty \text{ uniformly in } t. \tag{8.14}$$

Then the problem (8.1) *possesses a nontrivial solution, provided that*

$$[i^P(B_0), i^P(B_0) + \nu^P(B_0)] \cap [i^P(B_\infty), i^P(B_\infty) + \nu^P(B_\infty)] = \emptyset. \tag{8.15}$$

Remark 1.4 Theorem 1.3 should be compared with results in [38] and [94] (see also [223]), where the periodic solutions were studied. The main ingredients of the proof given in [94] are the Maslov-type index theory, the Galerkin approximation methods, and the critical point theory, specially, the saddle point theorem.

Similarly, we have developed the index theory which is parallel to that of Maslov-type index in the periodic case and is suitable to use in our case here. We have also developed the Galerkin approximation methods in Theorem IV.3.3. The critical point theory can also be used here to find as many critical points as that in [94] and [38]. So we can modify the proof given in [94] or [38] to give a proof of Theorem 1.3. We omit the details for simplicity.

8.2 The Existence of Periodic Solutions for Delay Differential Equations

The goal of this section is to consider the existence and multiplicity of periodic solutions for some asymptotically linear delay differential systems and some asymptotically linear delay Hamiltonian systems via the methods of Hamiltonian systems and index theory. For this purpose, we first consider in general the so called $\mathcal{M}$-boundary problem of a Hamiltonian system.

8.2.1 $\mathcal{M}$-Boundary Problem of a Hamiltonian System

We recall that for a skew-symmetric non-degenerate $2N \times 2N$ matrix $\mathcal{J} = (a_{ij})$, it can define a symplectic structure on $\mathbb{R}^{2N}$ by

$$\omega(v, w) = v^T \cdot \mathcal{J}^{-1} \cdot w$$

or $\omega = \frac{1}{2}\sum\limits_{i,j} a^{ij} dx_i \wedge dx_j$ with $\mathcal{J}^{-1} = (a^{ij})$. A $2N \times 2N$ matrix $\mathcal{M}$ is called $\mathcal{J}$-symplectic if there holds $\mathcal{M}^T \cdot \mathcal{J}^{-1} \cdot \mathcal{M} = \mathcal{J}^{-1}$. We denote the set of all $\mathcal{J}$-symplectic matrices by $\mathrm{Sp}_{\mathcal{J}}(2N)$. The usual symplectic group $\mathrm{Sp}(2N)$ is special case with $\mathcal{J} = J_N = \begin{pmatrix} 0 & -I_N \\ I_N & 0 \end{pmatrix}$, i.e., $\mathrm{Sp}(2N) = \mathrm{Sp}_{J_N}(2N)$. Here I_N is the $N \times N$ identity matrix. We will write J for J_N if the dimension $2N$ is clear from the text.

For a $\mathcal{J}$-symplectic matrix $\mathcal{M}$ with $\mathcal{M}^k = I_{2N}$, and a function $H \in C^2(\mathbb{R} \times \mathbb{R}^{2N})$ with $H(t + \tau, \mathcal{M}z) = H(t, z)$, we consider $k\tau$-periodic solution of the following $\mathcal{M}$-boundary value problem

$$\begin{cases} \dot{z}(t) = \mathcal{J}\nabla H(t, z(t)) \\ z(\tau) = \mathcal{M}z(0). \end{cases} \tag{8.16}$$

For the general $k\tau$-periodic problem, the corresponding functional is defined in $E = W^{1/2,2}(S^1, \mathbb{R}^{2N})$ with $S^1 = \mathbb{R}/(k\tau\mathbb{Z})$ by

$$\varphi(z) = \frac{1}{2}\int_0^{k\tau} (\mathcal{J}^{-1}\dot{z}(t), z(t))dt - \int_0^{k\tau} H(t, z(t))dt. \tag{8.17}$$

The critical point of φ in E is a $k\tau$-periodic solution of the nonlinear Hamiltonian system in (8.16). In order to solve the problem (8.16), we define a group action σ on E by

$$\sigma z(t) = \mathcal{M}z(t - \tau).$$

It is clear that $\sigma^k = id$ and φ is σ-invariant, i.e., there holds

$$\varphi(\sigma z) = \varphi(z). \tag{8.18}$$

Setting $E^\sigma = \{z \in E | \sigma z = z\} = fix(\sigma)$, by the well known Palais symmetric principal (see [248]), a critical point of φ in E^σ is a solution of the boundary problem (8.16). For $z \in E^\sigma$, there holds

$$\varphi(z) = \frac{k}{2}\int_0^{\tau} (\mathcal{J}^{-1}\dot{z}(t), z(t))dt - k\int_0^{\tau} H(t, z(t))dt.$$

The linearized system along a solution $z(t)$ of the nonlinear Hamiltonian system in (8.16) is the following linear Hamiltonian system

$$\dot{y}(t) = \mathcal{J}H''(t, z(t))y(t).$$

Its fundamental solution $\gamma_z(t)$ with $\gamma_z(0) = I_{2N}$ should satisfy

$$\dot{\gamma}_z(t) = \mathcal{J}H''(t, z(t))\gamma_z(t).$$

The following result is well known and the proof is omitted.

Lemma 2.1 *γ_z is a $\mathcal{J}$-symplectic path, i.e., $\gamma_z(t)^T\mathcal{J}^{-1}\gamma_z(t) = \mathcal{J}^{-1}$ for all $t \in \mathbb{R}$.*

For simplicity we take $\tau = 1$. For a symmetric continuous matrix function $B(t)$ satisfying $\mathcal{M}^T B(t+1)\mathcal{M} = B(t)$, suppose $\gamma_B(t)$ is the fundamental solution of the linear Hamiltonian system

$$\dot{y}(t) = \mathcal{J}B(t)y(t).$$

Definition 2.2 The $(\mathcal{J}, \mathcal{M})$-nullity of the symmetric matrix function B is defined by

$$\nu_{\mathcal{M}}^{\mathcal{J}}(B) = \dim_{\mathbb{C}} \ker_{\mathbb{C}}(\gamma_B(1) - \mathcal{M}).$$

For the standard case of $\mathcal{J} = J_N$ and a matrix $P \in \mathrm{Sp}_{J_N}(2N)$, the (J, P)-nullity and Maslov-type index of a symmetric matrix function B was defined first in [186] by algebra method (see Chap. 4). By using the notation of Proposition I.2.2, we have the following definition.

Definition 2.3 The $(\mathcal{J}, \mathcal{M})$-index of the symmetric matrix function B is defined by

$$i_{\mathcal{M}}^{\mathcal{J}}(B) = i^P(T\gamma_B T^{-1}), \quad P = T\mathcal{M}T^{-1}.$$

So the index pair $(i_{\mathcal{M}}^{\mathcal{J}}(B), \nu_{\mathcal{M}}^{\mathcal{J}}(B)) \in \mathbb{Z} \times \{0, 1, \cdots, 2N\}$ makes sense for all symmetric continuous matrix function $B(t)$ satisfying $\mathcal{M}^T B(t+1)\mathcal{M} = B(t)$.

8.2.2 Delay Differential Systems

For simplicity, as in [158] we first consider the 4τ-periodic solutions of the following delay differential systems

$$x'(t) = \nabla V(t, x(t-\tau)), \tag{8.19}$$

where the function $V \in C^2(\mathbb{R} \times \mathbb{R}^n, \mathbb{R})$ is τ-periodic in variable t and is even in variables x. To find 4τ-periodic solution $x(t)$, we only need to find solution with $x(t+2\tau) = -x(t)$. If $x(t)$ is so a solution, let $x_1(t) = x(t),\ x_2(t) = x(t-\tau)$ and $z(t) = (x_1(t), x_2(t))^T$, then there holds

$$\begin{cases} x_1'(t) = \nabla V(t, x_2(t)), \\ x_2'(t) = -\nabla V(t, x_1(t)). \end{cases} \tag{8.20}$$

Set $H(t, x_1, x_2) = V(t, x_1) + V(t, x_2)$, then we can rewrite (8.20) as

$$\dot{z}(t) = -J\nabla H(t, z(t)), \quad J = \begin{pmatrix} 0 & -I_n \\ I_n & 0 \end{pmatrix}. \tag{8.21}$$

Moreover, if $z(t) = (x_1(t), x_2(t))^T$ is a 4τ-periodic solution of (8.21) with $z(t) = \sigma z(t)$ for the 4-periodic action

$$\sigma z(t) = -Jz(t+\tau), \tag{8.22}$$

then $x(t) = -x_1(t)$ is a solution of (8.19) with $x(t+2\tau) = -x(t)$. The condition (8.22) is equivalent to

$$z(\tau) = Jz(0). \tag{8.23}$$

So the problem can be transformed to the problem (8.16) with $\mathcal{J} = -J$ and $M = J$.

In general, for a function $V \in C^2(\mathbb{R} \times \mathbb{R}^n, \mathbb{R})$ with period τ in variable t and even in variables x, we consider the $2m\tau$-periodic solutions of the following delay differential system

$$x'(t) = \nabla V(t, x(t-\tau)) + \nabla V(t, x(t-2\tau)) + \cdots + \nabla V(t, x(t-(m-1)\tau)). \tag{8.24}$$

If we get a solution $x(t)$ with $x(t-m\tau) = -x(t)$, then by setting $x_1(t) = x(t), x_2(t) = x(t-\tau), \cdots, x_m(t) = x(t-(m-1)\tau)$ and $H(t, x_1, \cdots, x_m) = V(t, x_1) + \cdots + V(t, x_m)$, $z = (x_1, \cdots, x_m)^T$, we rewrite the system (8.24) as

$$\dot{z}(t) = A_m \nabla H(t, z(t)), \tag{8.25}$$

where the $mn \times mn$ skew symmetric matrix A_m is defined by

$$A_m = \begin{pmatrix} 0 & I_n & \cdots & I_n \\ -I_n & 0 & \cdots & I_n \\ \vdots & \vdots & \cdots & \vdots \\ -I_n & -I_n & \cdots & 0 \end{pmatrix}.$$

We see that $\det A_m \neq 0$ if $m \in 2\mathbb{N}$ and $\det A_m = 0$ if $m \in 2\mathbb{N}+1$. Furthermore, if we take $X = \big(I_n \ -I_n \ I_n \ \cdots \ (-1)^{m-1} I_n)\big)$, there holds $XA_m = 0$ when $m \in 2\mathbb{N}+1$. Setting

$$T_m = \begin{pmatrix} 0 & I_n & 0 & \cdots & 0 \\ 0 & 0 & I_n & \cdots & 0 \\ \vdots & \vdots & \vdots & \cdots & \vdots \\ -I_n & 0 & 0 & \cdots & 0 \end{pmatrix} = \begin{pmatrix} 0 & I_{n(m-1)} \\ -I_n & 0 \end{pmatrix}.$$

It is easy to see $T_m^m = -I_{mn}$, $T_m^{2m} = I_{mn}$ and $(T_m^{-1})^T A_m^{-1} T_m^{-1} = A_m^{-1}$ when $m \in 2\mathbb{N}$, so $G_m = \{g | \ g = T_m^k, \ k = 1, 2, \cdots, 2m\}$ is a discrete group. We see also that

$$T_m^T = T_m^{-1} = \begin{pmatrix} 0 & \cdots & 0 & -I_n \\ I_n & \cdots & 0 & 0 \\ \vdots & \cdots & \vdots & \vdots \\ 0 & \cdots & I_n & 0 \end{pmatrix} = \begin{pmatrix} 0 & -I_n \\ I_{n(m-1)} & 0 \end{pmatrix}.$$

If $z(t)$ is a solution of (8.25) with $z(\tau) = T_m^{-1}z(0)$ then $x(t) = x_1(t)$ is a $2m\tau$-periodic solution of equation (8.24) for $m \in 2\mathbb{N}$. So we also turn our problem into the problem (8.16) with $\mathcal{J} = A_m$ and $\mathcal{M} = T_m^{-1}$. In order to understand the case of $m \in 2\mathbb{N} + 1$, we first recall the notion about Poisson structure.

8.2.3 Poisson Structure

Any $k \times k$ skew symmetric matrix A determines a Poisson structure on $\mathbb{R}^k$. For any functions $F, H \in C^\infty(\mathbb{R}^k, \mathbb{R})$, the Poisson structure defined by

$$\{F, H\} = (\nabla F)^T \cdot A \cdot \nabla H \in C^\infty(\mathbb{R}^k, \mathbb{R}). \tag{8.26}$$

We recall the definition of a Poisson structure $\{\cdot, \cdot\}$ on a manifold M. For any two functions $F, H \in C^\infty(M)$, the Poisson bracket $\{F, H\} \in C^\infty(M)$ is defined by the following properties

(1) $\{c_1F_1 + c_2F_2, H\} = c_1\{F_1, H\} + c_2\{F_2, H\}, \forall c_1, c_2 \in \mathbb{R}$,
(2) $\{F, H\} = -\{H, F\}$,
(3) $\{\{F, H\}, P\} + \{\{P, F\}, H\} + \{\{H, P\}, F\} = 0$,
(4) $\{F, H \cdot P\} = \{F, H\}P + H\{F, P\}$.

A differential manifold M with a Poisson structure is called Poisson manifold. Furthermore if there holds

(5) $\{F, H\} = 0$ for any function F implies $H \equiv c$,

then the Poisson structure $\{\cdot, \cdot\}$ determines a symplectic structure, and M is a symplectic manifold. For example, when $k = 2m$ and $A = J_m = \begin{pmatrix} 0 & -I_m \\ I_m & 0 \end{pmatrix}$, then it defines a Poisson structure on $\mathbb{R}^{2m}$ by

$$\{F, H\} = \sum_{i=1}^{m} \left(\frac{\partial F}{\partial q_i} \frac{\partial H}{\partial p_i} - \frac{\partial F}{\partial p_i} \frac{\partial H}{\partial q_i} \right).$$

This Poisson structure on $\mathbb{R}^{2m}$ is the standard Poisson structure and it determines the standard symplectic structure on $\mathbb{R}^{2m}$.

Let M be a Poisson manifold and $H : M \to \mathbb{R}$ be a smooth function. The Hamiltonian vector field associated with H is the unique smooth vector field V_H on M satisfying

$$V_H(F) = \{F, H\}$$

for every smooth function $F : M \to \mathbb{R}$. If the Poisson structure on $\mathbb{R}^k$ is defined by (8.26), the Hamiltonian vector field is

$$V_H(x) = A\nabla H(x).$$

The Hamiltonian equation becomes

$$\dot{x}(t) = A\nabla H(x(t)).$$

For a time depending function $H_t(x) = H(t, x)$, the Hamiltonian vector field makes sense as $V_{H_t}(x) = A\nabla_x H(t, x)$ and the Hamiltonian equation becomes

$$\dot{x}(t) = A\nabla_x H(t, x(t)).$$

The following Darboux Theorem will be very useful.

Theorem 2.4 (Darboux Theorem) *Suppose the rank of the $k\times k$ matrix A in* (8.26) *is $2m$ with $k = 2m+l$, then there is a coordinates transformation $y = Bx$ such that*

$$\{F, H\}(x) = \sum_{i=1}^{m}\left(\frac{\partial f}{\partial q_i}\frac{\partial h}{\partial p_i} - \frac{\partial f}{\partial p_i}\frac{\partial h}{\partial q_i}\right)(y) = \nabla f(y)^T \cdot J' \cdot \nabla h(y),$$

where $f(y) = F(B^{-1}y)$, $h(y) = H(B^{-1}y)$ and $y = (p, q, z)^T$, $p = (p_1, \cdots, p_m)$, $q = (q_1, \cdots, q_m,)$, $z = (z_1, \cdots, z_l)$. That is to say $J' = BAB^T = \begin{pmatrix} J_m & 0 \\ 0 & 0 \end{pmatrix}$, $J_m = \begin{pmatrix} 0 & -I_m \\ I_m & 0 \end{pmatrix}$.

Particularly, for the matrix A_m defined in (8.25) with $m \in 2\mathbb{N}+1$, and the matrix B_m defined by

$$B_m = \begin{pmatrix} I_n & 0 & \cdots & 0 & 0 \\ 0 & I_n & \cdots & 0 & 0 \\ \vdots & \vdots & \cdots & \vdots & \vdots \\ 0 & 0 & \cdots & I_n & 0 \\ I_n & -I_n & \cdots & -I_n & I_n \end{pmatrix},$$

there holds

$$B_m A_m B_m^T = \begin{pmatrix} A_{m-1} & 0 \\ 0 & 0 \end{pmatrix}.$$

So by choosing matrix C_{m-1} satisfying $C_{m-1}A_{m-1}C_{m-1}^T = J_{nl}$ with $l = \frac{m-1}{2}$ and taking $B'_m = \begin{pmatrix} C_{m-1} & 0 \\ 0 & I_m \end{pmatrix}$ there holds $B'_m B_m A_m (B'_m B_m)^T = \begin{pmatrix} J_{nl} & 0 \\ 0 & 0 \end{pmatrix}$. By taking the coordinates transformation $y = \begin{pmatrix} \tilde{y} \\ y_m \end{pmatrix} = B_m z$ and $h(t, y) = H(t, z)$, the system (8.25) becomes

$$\dot{\tilde{y}}(t) = A_{m-1}\nabla_{\tilde{y}} h(t, y), \quad y_m \equiv c. \tag{8.27}$$

From the coordinates transformation, we see that $y_i = x_i$ for $1 \le i \le m-1$ and $y_m = x_1 - x_2 + \cdots - x_{m-1} + x_m$. So system (8.27) becomes

$$\dot{\tilde{z}}(t) = A_{m-1}\nabla \tilde{H}(t, \tilde{z}(t)), \tag{8.28}$$

where $\tilde{H}(t, \tilde{z}) = H(t, \tilde{z}, c - x_1 + x_2 - \cdots + x_{m-1})$ and $\tilde{z} = (x_1, \cdots, x_{m-1})$. When choosing $c = 0$, any solution $\tilde{z}(t)$ of (8.28) with $\tilde{z}(0) = \tilde{B}_{m-1}\tilde{z}(1)$ determines a solution of system (8.25) with $x_m(t) = -x_1(t) + x_2(t) - \cdots + x_{m-1}(t)$, so $x(t) = x_1(t)$ is a solution of the delay system (8.24), where

$$\tilde{B}_{m-1} = \begin{pmatrix} 0 & I_n & 0 & \cdots & 0 \\ 0 & 0 & I_n & \cdots & 0 \\ \vdots & \vdots & \vdots & \cdots & \vdots \\ 0 & 0 & 0 & \cdots & I_n \\ -I_n & I_n & -I_n & \cdots & I_n \end{pmatrix}.$$

It is easy to see that $\tilde{B}_{m-1}A_{m-1}(\tilde{B}_{m-1})^T = A_{m-1}$ and $(\tilde{B}_{m-1})^{2m} = I_{n(m-1)}$. So in this case, we also transform the problem into the problem (8.16) with $\mathcal{J} = A_{m-1}$ and $\mathcal{M} = \tilde{B}_{m-1}^{-1}$.

8.2.4 First Order Delay Hamiltonian Systems

For a function $G \in C^2(\mathbb{R} \times \mathbb{R}^{2n}, \mathbb{R})$ with $G(t + \tau, x) = G(t, x)$, we consider the 2τ periodic solutions of following first order delay Hamiltonian system

$$\dot{x}(t) = J_n \nabla G(t, x(t - \tau)). \tag{8.29}$$

For a 2τ period solution $x(t)$ of (8.29), by setting $x_1(t) = x(t),\ x_2(t) = x(t-\tau)$ and $z = (x_1, x_2)^T$, the delay Hamiltonian system (8.29) is read as

$$\dot{z}(t) = \tilde{J}_{2n} \nabla H(t, z(t)) \tag{8.30}$$

with $H(t, z) = H(t, x_1, x_2) = G(t, x_1) + G(t, x_2)$ and $\tilde{J}_{2n} = \begin{pmatrix} 0 & J_n \\ J_n & 0 \end{pmatrix}$.

If $z(t)$ is a solution of (8.30) with $z(0) = P_{2n} z(\tau),\ P_{2n} = \begin{pmatrix} 0 & I_{2n} \\ I_{2n} & 0 \end{pmatrix}$, then $x(t) = x_1(t)$ is a 2τ-periodic solution of (8.29). It is easy to see $P_{2n}^{-1} = P_{2n} = P_{2n}^T$ and $P_{2n}^T \tilde{J}_{2n}^{-1} P_{2n} = \tilde{J}_{2n}^{-1}$.

In general, we consider the following first order delay Hamiltonian systems

$$\dot{x}(t) = J_n(\nabla G(t, x(t-\tau)) + \nabla G(t, x(t-2\tau)) + \cdots + \nabla G(t, x(t-(m-1)\tau))). \tag{8.31}$$

For an $m\tau$-periodic solution $x(t)$ of (8.31), by setting $x_1(t) = x(t),\ x_2(t) = x(t-\tau), \cdots, x_m(t) = x(t-(m-1)\tau)$ and $z = (x_1, x_2, \cdots, x_m)^T$, the delay Hamiltonian system (8.31) is read as

$$\dot{z}(t) = J_{n,m} \nabla H(t, z(t)) \tag{8.32}$$

with $H(t, z) = G(t, x_1) + \cdots + G(t, x_m)$ and

$$J_{n,m} = \begin{pmatrix} 0 & J_n & \cdots & J_n & J_n \\ J_n & 0 & \cdots & J_n & J_n \\ \vdots & \vdots & \cdots & \vdots & \vdots \\ J_n & J_n & \cdots & J_n & 0 \end{pmatrix}.$$

Conversely, if $z(t)$ is a solution of (8.32) with $z(0) = P_{n,m} z(\tau)$, where

$$P_{n,m} = \begin{pmatrix} 0 & I_{2n} & 0 & \cdots & 0 \\ 0 & 0 & I_{2n} & \cdots & 0 \\ \vdots & \vdots & \vdots & \cdots & \vdots \\ I_{2n} & 0 & 0 & \cdots & 0 \end{pmatrix} = \begin{pmatrix} 0 & I_{(2n)(m-1)} \\ I_{2n} & 0 \end{pmatrix},$$

then $x(t) = x_1(t)$ is a $m\tau$-periodic solution of (8.31). It is easy to see that $P_{n,m}^{-1} = P_{n,m}^T = \begin{pmatrix} 0 & I_{2n} \\ I_{(2n)(m-1)} & 0 \end{pmatrix}$, $P_{n,m}^m = I_{2nm}$ and $P_{n,m} J_{n,m} = J_{n,m} P_{n,m}$. In this case, we still turn our problem into problem (8.16) with $\mathcal{J} = J_{n,m}$ and $\mathcal{M} = P_{n,m}^{-1}$.

8.2.5 Second Order Delay Hamiltonian Systems

For a function $V \in C^2(\mathbb{R} \times \mathbb{R}^n, \mathbb{R})$ with $V(t+\tau, x) = V(t, x)$, we consider the periodic solutions of following second order delay Hamiltonian system

$$\ddot{x}(t) + \nabla V(t, x(t-2\tau)) = 0.$$

We can turn it into a first order delay Hamiltonian system as (8.29) with $H(t, x, y) = -\frac{|y|^2}{2} - V(t, x)$ and $y(t) = \dot{x}(t+\tau)$. In general, we consider the $m\tau$-periodic solutions of the following second order delay system

$$\ddot{x}(t) = -[\nabla V(t, x(t-\tau)) + \nabla V(t, x(t-2\tau)) + \cdots + \nabla V(t, x(t-(m-1)\tau))]. \tag{8.33}$$

Let $x_1(t) = x(t)$, $x_2(t) = x(t-\tau)$, $x_m(t) = x(t-(m-1)\tau)$ and $z(t) = (x_1(t), \cdots, x_m(t))^T$, then by $x(t+m\tau) = x(t)$, there holds

$$\ddot{z}(t) = -\mathcal{A}_m \nabla H(t, z(t)),$$

where $H(t, z) = V(t, x_1) + \cdots + V(t, x_m)$ and

$$\mathcal{A}_m = \begin{pmatrix} 0 & I_n & \cdots & I_n \\ I_n & 0 & \cdots & I_n \\ \vdots & \vdots & \cdots & \vdots \\ I_n & I_n & \cdots & 0 \end{pmatrix}.$$

It is easy to see that $\det \mathcal{A}_m \neq 0$. By taking $y(t) = \mathcal{A}_m^{-1}\dot{z}(t)$ and $w = (z, y)^T$, there holds

$$\dot{w}(t) = J_{mn} \nabla \mathcal{H}(t, w(t)), \tag{8.34}$$

where $\mathcal{H}(t, w) = -\frac{1}{2}(\mathcal{A}_m y, y) - H(t, z)$. A solution of this Hamiltonian system with the boundary value condition $w(0) = \begin{pmatrix} P_{n,m} & 0 \\ 0 & \mathcal{A}_m^{-1} P_{n,m} \mathcal{A}_m \end{pmatrix} w(\tau)$ determine an $m\tau$-periodic solution $x(t) = x_1(t)$ of the second order delay system. We note that $P_{n,m}\mathcal{A}_m = \mathcal{A}_m P_{n,m}$, so there holds

$$\mathcal{P}_{n,m} := \begin{pmatrix} P_{n,m} & 0 \\ 0 & \mathcal{A}_m^{-1} P_{n,m} \mathcal{A}_m \end{pmatrix} = \begin{pmatrix} P_{n,m} & 0 \\ 0 & P_{n,m} \end{pmatrix} \in \mathrm{Sp}(2mn),$$

$\mathcal{P}_{n,m} J_{mn} \mathcal{P}_{n,m}^T = J_{mn}$, and $\mathcal{P}_{n,m}^m = I_{2nm}$. We also turn our problem into the problem (8.16) with $\mathcal{J} = J_{mn}$ and $\mathcal{M} = \mathcal{P}_{n,m}^{-1}$.

We note that $\mathcal{J} = A_m$ in (8.25) when $m \in 2\mathbb{Z}$, $\mathcal{J} = J_{n,m}$ and $\mathcal{J} = J_{mn}$ all satisfy the property $\mathcal{J} J_N = J_N \mathcal{J}$ for suitable N.

8.2.6 Background and Related Works

In 1974, Kaplan and Yorke in [158] studied the autonomous delay differential equation as (8.19) and introduced a new technique for establishing the existence of periodic solutions. More precisely, the authors of [158] considered the periodic solutions of the following kinds of delay differential equations

$$\dot{x}(t) = f(x(t-1))$$

and

$$\dot{x}(t) = f(x(t-1)) + f(x(t-2))$$

with odd function f. They turned their problems into the problems of periodic solution of autonomous Hamiltonian system and under some twisted condition on the origin and infinity for the function f, it was proved that there exists an energy surface of the Hamiltonian function containing at least one periodic solution. Since then many papers (see [92, 93, 130, 131, 173, 174] and [175] and the references therein) used Kaplan and Yorke's original idea to search for periodic solutions of more general differential delay equations of the following form

$$\dot{x}(t) = f(x(t-1)) + f(x(t-2)) + \cdots + f(x(t-m+1)).$$

The existence of periodic solutions of above delay differential equation has been investigated by Nussbaum in [246] using different techniques. For other related works, the readers may refer to the references [109, 124] and the references therein. The readers can also refer [132, 133, 247] for systematic introduction on delay differential equations.

8.2.7 Main Results

8.2.7.1 Asymptotically Linear Delay Differential Systems

Theorem 2.5 ([188]) *Suppose V satisfies the following conditions*

(V1) *There exists a constant $C > 0$ such that*

$$|V''(t,x)| \le C, \quad (t,x) \in [0,1] \times \mathbb{R}^n.$$

(V_∞) *There exists a continuous symmetric matrix function $C_\infty(t)$ such that*

$$V'(t,x) = C_\infty(t)x + o(|x|)$$

uniformly in $t \in \mathbb{R}$ as $|x| \to \infty$.

Let $C_0(t) = V''(t,0)$. For $m \in 2\mathbb{N}$

$$B_0(t) = \begin{pmatrix} C_0(t) & 0 & \cdots & 0 \\ 0 & C_0(t) & \cdots & 0 \\ \vdots & \vdots & \cdots & \vdots \\ 0 & 0 & \cdots & C_0(t) \end{pmatrix}, \tag{8.35}$$

and

$$B_\infty(t) = \begin{pmatrix} C_\infty(t) & 0 & \cdots & 0 \\ 0 & C_\infty(t) & \cdots & 0 \\ \vdots & \vdots & \cdots & \vdots \\ 0 & 0 & \cdots & C_\infty(t) \end{pmatrix}. \tag{8.36}$$

Set $(i_0, \nu_0) = (i_{\mathcal{M}}^{\mathcal{J}}(B_0), \nu_{\mathcal{M}}^{\mathcal{J}}(B_0))$, $(i_\infty, \nu_\infty) = (i_{\mathcal{M}}^{\mathcal{J}}(B_\infty), \nu_{\mathcal{M}}^{\mathcal{J}}(B_\infty))$ with $\mathcal{J} = A_m$, $\mathcal{M} = T_m^{-1}$ defined before.

For $m \in 2\mathbb{N}+1$,

$$B_0(t) = \begin{pmatrix} 2C_0(t) & -C_0(t) & C_0(t) & \cdots & C_0(t) & -C_0(t) \\ -C_0(t) & 2C_0(t) & -C_0(t) & \cdots & -C_0(t) & C_0(t) \\ \vdots & \vdots & \vdots & \cdots & \vdots & \vdots \\ -C_0(t) & C_0(t) & -C_0(t) & \cdots & -C_0(t) & 2C_0(t) \end{pmatrix} \tag{8.37}$$

and

$$B_\infty(t) = \begin{pmatrix} 2C_\infty(t) & -C_\infty(t) & C_\infty(t) & \cdots & C_\infty(t) & -C_\infty(t) \\ -C_\infty(t) & 2C_\infty(t) & -C_\infty(t) & \cdots & -C_\infty(t) & C_\infty(t) \\ \vdots & \vdots & \vdots & \cdots & \vdots & \vdots \\ -C_\infty(t) & C_\infty(t) & -C_\infty(t) & \cdots & -C_\infty(t) & 2C_\infty(t) \end{pmatrix}. \tag{8.38}$$

Set $(i_0, \nu_0) = (i_{\mathcal{M}}^{\mathcal{J}}(B_0), \nu_{\mathcal{M}}^{\mathcal{J}}(B_0))$, $(i_\infty, \nu_\infty) = (i_{\mathcal{M}}^{\mathcal{J}}(B_\infty), \nu_{\mathcal{M}}^{\mathcal{J}}(B_\infty))$ with $\mathcal{J} = A_{m-1}$, $\mathcal{M} = \tilde{B}_{m-1}^{-1}$ defined before.

If

$$i_\infty \notin [i_0, i_0 + \nu_0], \tag{8.39}$$

the delay differential system (8.24) possesses at least one nontrivial $2m\tau$-periodic solution x with $x(t - m\tau) = -x(t)$. Furthermore, if the trivial solution $z_0 = 0$ of problem (8.16), with $H(t,z) = V(t,x_1) + \cdots + V(t,x_m)$ for even m and $H(t,z) =$

$V(t,x_1)+\cdots+V(t,x_{m-1})+V(t,-x_1+x_2-\cdots+x_{m-1})$ *for odd* m, *is not pseudo-degenerated, and*

$$i_\infty \notin [i_0-2N, i_0+\nu_0+2N], \tag{8.40}$$

the delay differential system (8.24) *possesses at least two nontrivial* $2m\tau$*-periodic solutions as above.*

Here and in the sequel, the number N in (8.40) is the dimension of the space where the corresponding first order Hamiltonian systems are considered.

Proof Since $V(t,-x)=V(t,x)$, $z_0=0$ is the trivial solution of problem (8.16). Here $H(t,z)=V(t,x_1)+\cdots+V(t,x_m)$, $z=(x_1,\cdots,x_m)$ for even m and $H(t,z)=V(t,x_1)+\cdots+V(t,x_{m-1})+V(t,-x_1+x_2-\cdots+x_{m-1})$, $z=(x_1,\cdots,x_{m-1})$ for odd m.

It is clear that the condition (V1) implies (H2) in Theorem 1.1. For $m\in 2\mathbb{N}$,

$$\frac{|H'(t,z)-B_\infty(t)z|}{|z|}=\sum_{j=1}^{m}\frac{|V'(t,x_j)-C_\infty(t)x_j|}{|x_j|}\frac{|x_j|}{|z|}.$$

We only need to deal with all $x_j\neq 0$. Since $V\in C^2$, (V1) implies that $\frac{|V'(t,x_j)-C_\infty(t)x_j|}{|x_j|}$ is bounded. So when $\frac{|V'(t,x_j)-C_\infty(t)x_j|}{|x_j|}\nrightarrow 0$, then $\frac{|x_j|}{|z|}\to 0$ as $|z|\to\infty$. Since $\frac{|x_j|}{|z|}\leq 1$, there holds

$$\frac{|H'(t,z)-B_\infty(t)z|}{|z|}\to 0,\ \text{as } |z|\to 0.$$

The case $m\in 2\mathbb{N}+1$ is similar, we note that $|z|\to\infty$ if and only if $(x_1,\cdots,x_{m-1},-x_1+x_2-\cdots+x_{m-1})\to\infty$. So the condition ($V_\infty$) implies ($H_\infty$) in Theorem 1.1. Now the result follows from Theorem 1.1. □

Remark 2.6

(i) Since $V(t,-x)=V(t,x)$, so the solutions of equation (8.24) appear in pairs. That is to say if x is a solution of (8.24), so is $-x$.
(ii) When $m=n=1$, the problem (8.24) was considered by Kaplan and Yorke in [15] with $V'(t,x)=-f(x)$. Hence $\mathcal{M}=\mathcal{J}=-J$. In this case, for a constant matrix function $\bar{B}(t)=aId_2$, the index $i_{\mathcal{M}}^{\mathcal{J}}(\bar{B})=2j$ if

$$\frac{(4j-1)\pi}{2}<a\leq\frac{(4j+3)\pi}{2},$$

$\nu_{\mathcal{M}}^{\mathcal{J}}(\bar{B})=2$ if $a=\frac{(4i-1)\pi}{2}$ and $\nu_{\mathcal{M}}^{\mathcal{J}}(\bar{B})=0$ otherwise. Since $V'(t,x)=-f(x)$, we should choose $B_\infty=-\alpha Id$ and $B_0=-\beta Id$ to compute the index. If $\alpha<\frac{\pi}{2}<\beta$ or in general $\min\{\alpha,\beta\}<\frac{(1-4j)\pi}{2}<\max\{\alpha,\beta\}$, then

$|i_\infty - i_0| \geq 2$. Furthermore, if $\nu_\infty \neq 0$ or $\nu_0 \neq 0$, then $|i_\infty - i_0| \geq 4$. This means that $i_\infty \notin [i_0, i_0 + \nu_0]$. So when replace the condition $xf(x) > 0$ for $x \neq 0$ with $f'(x)$ is bounded and with all other conditions in Theorem 1.1 of [15], the result is still true, i.e., there exists a pair nontrivial periodic solution x, $-x$ of equation (8.19) with $x(t + 2\tau) = -x(t)$. If $\alpha < \frac{(1-4j)\pi}{2}$ and $\beta > \frac{(5-4j)\pi}{2}$, then it is easy to see (8.40) holds, so there exist at least two pair nontrivial periodic solutions. We note that the function $H(t, z)$ is even in z since $V(t, x)$ is even in x, so the solutions appear in pairs. Moreover, if $\nu_0 = 0$, Eq. (8.24) may possesses at least $|i_\infty - i_0| - 1$ pairs nontrivial solutions under some conditions on the Hessian $H''(t, z)$.

Similarly, as consequences of Theorem 1.3, we have the following two results.

Theorem 2.7 ([188]) *Suppose V satisfies the condition (V_∞) and the following conditions*

(V2) *There exist constants $a > 0$ and $p > 1$ such that*

$$|V''(t, x)| \leq a(1 + |x|^p), \quad \forall (t, x) \in [0, 1] \times \mathbb{R}^n.$$

(V3) *The matrix function $C_0(t) := V''(t, 0)$ satisfies*

$$V'(t, x) = C_0(t)x + o(|x|) \text{ as } |x| \to 0 \text{ uniformly in } t \in [0, 1].$$

(V4) *For $v(t, x) = V(t, x) - \frac{1}{2}(C_\infty(t)x, x)$ with $(t, x) \in \mathbb{R} \times \mathbb{R}^n$, either*

$$v(t, x) \to 0, \quad |v'(t, x)| \to 0 \text{ as } |x| \to +\infty \text{ uniformly in } t, \tag{8.41}$$

or

$$v(t, x) \to \pm\infty, \quad |v'(t, x)| = 0 \text{ as } |x| \to \infty \text{ uniformly in } t. \tag{8.42}$$

Then the delay differential system (8.24) possesses a nontrivial $2m\tau$-periodic solution x with $x(t - m\tau) = -x(t)$, provided that

$$[i_{\mathcal{M}}^{\mathcal{J}}(B_0), i_{\mathcal{M}}^{\mathcal{J}}(B_0) + \nu_{\mathcal{M}}^{\mathcal{J}}(B_0)] \cap [i_{\mathcal{M}}^{\mathcal{J}}(B_\infty), i_{\mathcal{M}}^{\mathcal{J}}(B_\infty) + \nu_{\mathcal{M}}^{\mathcal{J}}(B_\infty)] = \emptyset, \tag{8.43}$$

where in (8.43) when $m \in 2\mathbb{N}$, $\mathcal{J} = A_m$, $\mathcal{M} = T_m^{-1}$, $B_0(t)$ and $B_\infty(t)$ are defined as in (8.35) and (8.36), when $m \in 2\mathbb{N}+1$, $\mathcal{J} = A_{m-1}$, $\mathcal{M} = \tilde{B}_{m-1}^{-1}$, $B_0(t)$ and $B_\infty(t)$ are defined as in (8.37) and (8.38) with $C_0(t)$ and $C_\infty(t)$ defined in (V3) and (V_∞) respectively.

Proof Take $H(t, z) = V(t, x_1) + \cdots + V(t, x_m)$, $z = (x_1, \cdots, x_m)$ for even m and $H(t, z) = V(t, x_1) + \cdots + V(t, x_{m-1}) + V(t, -x_1 + x_2 - \cdots + x_{m-1})$, $z = (x_1, \cdots, x_{m-1})$ for odd m in Theorem 1.3. It is easy to see that V satisfying the condition (V_∞) implies H satisfying the condition (H_∞). Also (V2) implies (H1), (V3) implies (H3) and (H4) follows from (V4). □

8.2.7.2 First Order Delay Hamiltonian Systems

For a function $G \in C^2(\mathbb{R} \times \mathbb{R}^{2n}, \mathbb{R})$ with $G(t+\tau, x) = G(t, x)$, we consider the periodic solutions of following first order delay Hamiltonian system

$$\dot{x}(t) = J_n(\nabla G(t, x(t-\tau)) + \nabla G(t, x(t-2\tau)) + \cdots + \nabla G(t, x(t-(m-1)\tau))). \tag{8.44}$$

Let $z = (x_1, \cdots, x_m)$ and $H(t, z) = G(t, x_1) + \cdots + G(t, x_m)$.

The following results are also consequences of Theorems 1.1 and 1.3.

Theorem 2.8 *Suppose G satisfies the conditions (V1) and (V_∞) in Theorem 2.5. Then the system* (8.44) *possesses an $m\tau$-periodic solution x_0. Suppose $z_0(t)$ is the solution of* (8.32) *corresponding to x_0. Let $B_0(t) = H''(t, z_0(t))$ and*

$$B_\infty(t) = \begin{pmatrix} C_\infty(t) & 0 & \cdots & 0 \\ 0 & C_\infty(t) & \cdots & 0 \\ \vdots & \vdots & \cdots & \vdots \\ 0 & 0 & \cdots & C_\infty(t) \end{pmatrix}.$$

Set $(i_0, \nu_0) = (i_{\mathcal{M}}^{\mathcal{J}}(B_0), \nu_{\mathcal{M}}^{\mathcal{J}}(B_0))$, $(i_\infty, \nu_\infty) = (i_{\mathcal{M}}^{\mathcal{J}}(B_\infty), \nu_{\mathcal{M}}^{\mathcal{J}}(B_\infty))$ with $\mathcal{J} = J_{n,m}$, $\mathcal{M} = P_{n,m}^{-1}$ defined before. If (8.7) *holds, the system* (8.44) *possesses at least two $m\tau$-periodic solutions. Furthermore, if z_0 is not pseudo-degenerated, and* (8.40) *holds, then the system* (8.44) *possesses at least three an $m\tau$-periodic solutions.*

Theorem 2.9 ([188]) *Suppose G satisfies the conditions (V2),(V3), (V4) and (V_∞) in Theorem 2.7. Then the system* (8.44) *possesses a nontrivial solution, provided the twisted condition* (8.43) *holds with $\mathcal{J} = J_{n,m}$, $\mathcal{M} = P_{n,m}^{-1}$ and the matrix functions $B_0(t)$, $B_\infty(t)$ are defined as in* (8.35), (8.36) *with $C_0(t)$ and $C_\infty(t)$ defined in (V3) and (V_∞) respectively.*

The proofs of above two results are the same as in that of Theorems 2.5 and 2.7, respectively. We omit the details.

8.2.7.3 Second Order Delay Hamiltonian Systems

For a function $V \in C^2(\mathbb{R} \times \mathbb{R}^n, \mathbb{R})$ with $V(t+\tau, x) = V(t, x)$, we consider the $m\tau$-periodic solutions of following second order delay Hamiltonian system

$$\ddot{x}(t) = -[\nabla V(t, x(t-\tau)) + \nabla V(t, x(t-2\tau)) + \cdots + \nabla V(t, x(t-(m-1)\tau))]. \tag{8.45}$$

Let $z = (x, y) = (x_1, \cdots, x_m, y_1, \cdots, y_m)$, and $\mathcal{H}(t, z) = -\frac{1}{2}(\mathcal{A}_m y, y) - V(t, x_1) - \cdots - V(t, x_m)$. From Theorems 1.1 and 1.3, we obtain the following results.

Theorem 2.10 ([188]) *Suppose V satisfies the conditions (V1) and (V_∞) in Theorem 2.5. Then the system (8.45) possesses an $m\tau$-periodic solution x_0. Suppose $w_0(t)$ is the solution of (8.34) corresponding to x_0. Let $B_0(t) = \mathcal{H}''(t, w_0(t))$,*

$$B_\infty(t) = \begin{pmatrix} -R_\infty(t) & 0 \\ 0 & -\mathcal{A}_m \end{pmatrix}, \quad R_\infty(t) = \begin{pmatrix} C_\infty(t) & 0 & \cdots & 0 \\ 0 & C_\infty(t) & \cdots & 0 \\ \vdots & \vdots & \cdots & \vdots \\ 0 & 0 & \cdots & C_\infty(t) \end{pmatrix}_{mn\times mn}.$$

Set $(i_0, \nu_0) = (i^{\mathcal{J}}_{\mathcal{M}}(B_0), \nu^{\mathcal{J}}_{\mathcal{M}}(B_0))$, $(i_\infty, \nu_\infty) = (i^{\mathcal{J}}_{\mathcal{M}}(B_\infty), \nu^{\mathcal{J}}_{\mathcal{M}}(B_\infty))$ with $\mathcal{J} = J_{mn}$, $\mathcal{M} = \mathcal{P}^{-1}_{n,m}$ defined in Sect. 8.2 after (8.34). If (8.7) holds, the system (8.45) possesses at least two solutions. Furthermore, if w_0 is not pseudo-degenerated, and (8.40) holds, then the system (8.45) possesses at least three solutions.

Theorem 2.11 ([188]) *Suppose V satisfies the conditions (V2),(V3), (V4) and (V_∞) in Theorem 2.7. Then the system (8.45) possesses a nontrivial solution, provided the twisted condition (8.43) holds with $\mathcal{J} = J_{mn}$, $\mathcal{M} = \mathcal{P}^{-1}_{n,m}$,*

$$B_0(t) = \begin{pmatrix} -R_0(t) & 0 \\ 0 & -\mathcal{A}_m \end{pmatrix}, \quad R_0(t) = \begin{pmatrix} C_0(t) & 0 & \cdots & 0 \\ 0 & C_0(t) & \cdots & 0 \\ \vdots & \vdots & \cdots & \vdots \\ 0 & 0 & \cdots & C_0(t) \end{pmatrix}_{mn\times mn}$$

and $B_\infty(t)$ defined as in Theorem 2.10.

8.3 The Minimal Period Problem for P-Symmetric Solutions

In this section, we apply the P-index theory and its iteration theory to the P-boundary problem of the following autonomous Hamiltonian system

$$\begin{cases} \dot{x} = JH'(x), & x \in \mathbb{R}^{2n}, \\ x(\tau) = Px(0), \end{cases} \tag{8.46}$$

where $P \in Sp(2n)$ satisfying $P^k = I$, here k is assumed to be the smallest positive integer such that $P^k = I$(this condition for P is called $(P)_k$ condition in the sequel). And $H(x) \in C^2(\mathbb{R}^{2n}, \mathbb{R})$ satisfying $H(Px) = H(x)$, H' denote the gradient of H. We note that the matrix $P \in Sp(2n)$ satisfying $P^k = I$ is not necessary orthogonal symplectic, $P = \begin{pmatrix} a & b \\ -\frac{a^2+a+1}{b} & -a-1 \end{pmatrix}$ is an example with $k = 3$ and $n = 1$. A solution (τ, x) of the problem (8.46) is called *P-solution* of the Hamiltonian system (in [293] it was called a P-periodic orbit). Since $P^k = I$, so the P-*solution* (τ, x) can be extended as a $k\tau$-periodic solution $(k\tau, x^k)$. We say that a T-periodic solution (T, x) of a Hamiltonian system in (8.46) is P-symmetric if $x(\frac{T}{k}) = Px(0)$. T is the P-*symmetric* period of x. We say T to be the minimal P-*symmetric* period of x if $T = \min\{\tau > 0 \mid x(t + \frac{\tau}{k}) = Px(t), \forall t \in \mathbb{R}\}$.

For the minimal P-symmetric periodic problem, we have the following result.

Theorem 3.1 ([194]) *Suppose $P \in Sp(2n)$ satisfying the $(P)_k$ condition, and the Hamiltonian function H satisfies the following conditions:*

(H1) *$H \in C^2(\mathbb{R}^{2n}, \mathbb{R})$ satisfying $H(Px) = H(x)$, $\forall x \in \mathbb{R}^{2n}$.*

(H2) *There exist constants $\mu > 2$ and $R_0 > 0$ such that*

$$0 < \mu H(x) \leq H'(x) \cdot x, \quad \forall\, |x| \geq R_0.$$

(H3) *$H(x) = o(|x|^2)$ at $x = 0$.*

(H4) *$H(x) \geq 0$ for all $x \in \mathbb{R}^{2n}$.*

Then for every $\tau > 0$, the system (8.46) *possesses a non-constant P-solution (τ, x) satisfying*

$$\dim \ker_{\mathbb{R}}(P - I) + 2 - \nu^P(x) \leq i^P(x) \leq \dim \ker_{\mathbb{R}}(P - I) + 1, \tag{8.47}$$

Moreover, if this solution x further satisfies the following conditions:

(HC) $H''(x(t)) > 0$ for every $t \in \mathbb{R}$.

Then the minimal P-symmetric period of x is $k\tau$ or $\frac{k\tau}{k+1}$.

Suppose $\bar{\gamma}(t) \in \mathcal{P}_\tau(2n)$ is the fundamental solution of the Hamiltonian system $\dot{z}(t) = JB(t)z(t)$ with $B(t) = H''(x(t))$. If $\bar{\gamma} \notin {}_P\mathcal{P}^e_\tau(2n) = \{\bar{\gamma} \in \mathcal{P}_\tau(2n) | P^{-1}\bar{\gamma}(\tau) \in Sp^e(2n)\}$, then the minimal P-symmetric period of x is $k\tau$, i.e., the P-symmetric periodic solution $(k\tau, x^k)$ generated from x possesses minimal P-symmetric period.

In order to get the information about the Maslov-type indices of a critical point, we need the following result which was proved in [110, 163, 274].

Theorem 3.2 *Let E be a real Hilbert space with orthogonal decomposition $E = X \oplus Y$, where $\dim X < +\infty$. Suppose $f \in C^2(E, \mathbb{R})$ satisfies the (PS) condition, and the following conditions:*

(F1) *There exist ρ and $\alpha > 0$ such that*

$$f(w) \geq \alpha, \quad \forall w \in \partial B_\rho(0) \cap Y.$$

(F2) *There exist $e \in \partial B_1(0) \cap Y$ and $R > \rho$ such that*

$$f(w) < \alpha, \quad \forall w \in \partial Q.$$

where $Q = (\overline{B_R(0)} \cap X) \oplus \{re \mid 0 \le r \le R\}$.

Then

(1) *f possesses a critical value $c \ge \alpha$, which is given by*

$$c = \inf_{h \in \Lambda} \max_{w \in Q} f(h(w)),$$

where $\Lambda = \{h \in C(\overline{Q}, E) \mid h = id \text{ on } \partial Q\}$.

(2) *If $f''(w)$ is Fredholm for $w \in \mathcal{K}_c(f) \equiv \{w \in E : f'(w) = 0, f(w) = c\}$, then there exists an element $w_0 \in \mathcal{K}_c(f)$ such that the negative Morse index $m^-(w_0)$ and nullity $m^0(w_0)$ of f at w_0 satisfy*

$$m^-(w_0) \le \dim X + 1 \le m^-(w_0) + m^0(w_0). \tag{8.48}$$

(3) *Suppose that there is an S^1 action on E, f is S^1-invariant, and for w_0 defined in (2) the set $S^1 * w_0$ is not a single point. Then* (8.48) *can be further improved to*

$$m^-(w_0) \le \dim X + 1 \le m^-(w_0) + m^0(w_0) - 1. \tag{8.49}$$

Let $W_P = \gamma_P W^{1/2,2}(S_\tau, \mathbb{R}^{2n})$, it is the space of all $W^{1/2,2}$ functions z defined in $\mathbb{R}$ satisfying $z(t+\tau) = Pz(t)$. Its is an inner product space with inner product $\langle \cdot, \cdot \rangle$ and norm $\| \cdot \|$. We will denote the L^s norm by $\| \cdot \|_s$ for $s \ge 1$. The space W_P can be continuously embedded into $L^s([0, \tau], \mathbb{R}^{2n})$, i.e. there is $\alpha_s > 0$ such taht

$$\|z\|_s \le \alpha_s \|z\|, \quad \forall z \in W_P. \tag{8.50}$$

Let A and B be the self-adjoint operators defined on W_P by the following bilinear forms:

$$\langle Ax, y \rangle = \int_0^\tau (-J\dot{x}(t), y(t))dt, \quad \langle Bx, y \rangle = \int_0^\tau (B(t)x(t), y(t))dt. \tag{8.51}$$

Suppose that $\cdots \le \lambda_{-k} \le \cdots \le \lambda_{-1} < 0 < \lambda_1 \le \cdots \le \lambda_k \le \cdots$ are all nonzero eigenvalues of the operator A(count with multiplicity), and correspondingly, e_j is the eigenvector of λ_j satisfying $\langle e_j, e_l \rangle = \delta_{jl}$. We denote the kernel of the operator A by W_P^0 which is exactly the space $\ker_{\mathbb{R}}(P - I)$. We define the subspaces of W_P by

$$W_P^m = W_m^- \oplus W_P^0 \oplus W_m^+$$

with $W_m^- = \{z \in W_P \mid z(t) = \sum_{j=1}^m a_{-j} e_{-j}(t),\ a_{-j} \in \mathbb{R}\}$ and $W_m^+ = \{z \in W_P \mid z(t) = \sum_{j=1}^m a_j e_j(t),\ a_j \in \mathbb{R}\}$.

For $z \in W_P$, we define

$$f(z)=\frac{1}{2}\int_0^{k\tau}(-J\dot{z}(t),z(t))dt-\int_0^{k\tau}H(z)dt=k\left(\frac{1}{2}\langle Az,z\rangle-\int_0^{\tau}H(z)dt\right). \tag{8.52}$$

It is well known that $f\in C^2(W_P,\mathbb{R})$ whenever

$$H\in C^2(\mathbb{R}^{2n},\mathbb{R}) \ \text{ and } \ |H''(z)|\le a_1|z|^s+a_2 \tag{8.53}$$

for some $s\in(1,+\infty)$ and all $z\in\mathbb{R}^{2n}$. Looking for solutions of (8.46) is equivalent to looking for critical points of f on W_P.

Proof of Theorem 3.1. We carry out the proof in several steps.

Step 1. Since the growth condition (8.53) has not been assumed for H, we need to truncate the function H suitably to get a function H_K satisfying the condition (8.53).

We follow the method in Rabinowitz's pioneering work [254] (cf. also [96] and [255]). Let $K>R_0$ and select $\chi\in C^\infty(\mathbb{R},\mathbb{R})$ such that $\chi(y)\equiv 1$ if $y\le K$, $\chi(y)\equiv 0$ if $y\ge K+1$, and $\chi'(y)<0$ if $y\in(K,K+1)$, where K is free for now. Set

$$H_K(z)=\chi(|z|)H(z)+(1-\chi(|z|))R_K|z|^4, \tag{8.54}$$

where the constant R_K satisfies

$$R_K\ge\max_{K\le|z|\le K+1}\frac{H(z)}{|z|^4}.$$

Then $H_K\in C^2(\mathbb{R}^{2n},\mathbb{R})$, satisfies (H3), (H4) and (8.53) with $s=2$. Moreover a straightforward computation shows (H2) hold with μ replaced by $\nu=\min\{\mu,4\}$, i.e., there exist $R_0>0$ such that

$$0<\nu H_K(z)\le H_K'(z)\cdot z,\ \ \forall\,|z|\ge R_0. \tag{8.55}$$

Since $H_K\in C^2(\mathbb{R}^{2n},\mathbb{R})$, then $H_K(z)$ is bounded for $|z|\le R_0$. Thus for $K>R_0$ there exist positive constant K_1, K_2 independent of K such that

$$R_K\nu|z|^4-K_1\le\nu H_K(z)\le H_K'(z)\cdot z+K_2,\ \ \forall z\in\mathbb{R}^{2n} \tag{8.56}$$

via (8.54) the form of H_K and (8.55). Integrating (8.55) then yields

$$H_K(z)\ge a_3|z|^\nu-a_4 \tag{8.57}$$

for all $z\in\mathbb{R}^{2n}$, where $a_3,a_4>0$ are independent of K.

Define a functional f_K on W_P by

$$\begin{aligned}f_K(z)&=\tfrac{1}{2}\textstyle\int_0^{k\tau}(-J\dot{z}(t),z(t))dt-\int_0^{k\tau}H(z)dt\\&=\tfrac{k}{2}\langle Az,z\rangle-\textstyle\int_0^{k\tau}H_K(z)dt,\ \ \forall z\in W_P,\end{aligned} \tag{8.58}$$

then $f_K \in C^2(W_P, \mathbb{R})$.

Step 2. For $m > 0$, let $f_{K,m} = f_K|_{W_P^m}$. We will show that $f_{K,m}$ satisfies the hypotheses of Theorem 3.2. In fact, by (H3), for any $\epsilon > 0$, there is a $\delta > 0$ such that $H_K(z) \le \epsilon|z|^2$ for $|z| \le \delta$. Since $H_K(z)|z|^{-4}$ is uniformly bounded as $|z| \to +\infty$, there is an $M_1 = M_1(\epsilon, K)$ such that $H_K(z) \le M_1|z|^4$ for $|z| \ge \delta$. Hence

$$H_K(z) \le \epsilon|z|^2 + M_1|z|^4, \quad \forall z \in \mathbb{R}^{2n}. \tag{8.59}$$

Therefore by (8.59) and the Sobolev embedding theorem,

$$\int_0^{k\tau} H_K(z)dt \le C_K\left(\epsilon\|z\|_2^2 + M_1\|z\|_4^4\right) \le C_K(\epsilon\alpha_2 + M_1\alpha_4\|z\|^2)\|z\|^2, \tag{8.60}$$

where C_K is a constant depending on K. Let

$$X_m = W_m^- \oplus W_P^0, \quad Y_m = W_m^+. \tag{8.61}$$

Consequently for $z \in Y_m$, we have

$$\begin{aligned} f_{K,m}(z) &= \frac{k}{2}\langle Az, z\rangle - \int_0^{k\tau} H_K(z)dt \\ &\ge \frac{k\lambda_1}{2}\|z\|^2 - C_K(\epsilon\alpha_2 + M_1\alpha_4\|z\|^2)\|z\|^2. \end{aligned} \tag{8.62}$$

So there are constants $\rho = \rho(K) > 0$ and $\alpha = \alpha(K) > 0$, which are sufficiently small and independent of m, such that

$$f_{K,m}(z) \ge \alpha, \quad \forall z \in \partial B_\rho(0) \cap Y_m. \tag{8.63}$$

Let $e = e_1 \in \partial B_1(0) \cap Y_m$ and set

$$Q_m = \{re \mid 0 \le r \le r_1\} \oplus (B_{r_1} \cap X_m),$$

where r_1 is free for the moment. Let $z = z^- + z^0 \in W_m^- \oplus W_P^0$, then

$$\begin{aligned} & f_{K,m}(z + re) \\ =& \frac{k}{2}\langle Az^-, z^-\rangle + \frac{k}{2}r^2\langle Ae, e\rangle - \int_0^{k\tau} H_K(z + re)dt \\ \le& \frac{k\lambda_{-1}}{2}\|z^-\|^2 + \frac{k\lambda_1}{2}r^2 - \int_0^{k\tau} H_K(z + re)dt. \end{aligned} \tag{8.64}$$

If $r = 0$, from condition (H4), there holds

$$f_{K,m}(z+re) \leq \frac{k\lambda_{-1}}{2}\|z^-\|^2 \leq 0. \tag{8.65}$$

If $r = r_1$ or $\|z\| = r_1$, by (8.57), there holds

$$\begin{aligned}
\int_0^{k\tau} H_K(z+re)dt &\geq \int_0^{\tau} H_K(z+re)dt \geq a_3 \int_0^{\tau} |z+re|^\nu dt - \tau a_4 \\
&\geq a_5 \left(\int_0^{\tau} |z+re|^2 dt\right)^{\nu/2} - a_6 \\
&= a_5 (\int_0^{\tau} (|z^0|^2 + |z^-|^2 + r^2|e|^2)dt)^{\nu/2} - a_6 \\
&\geq a_7(|z^0|^\nu + r^\nu) - a_6.
\end{aligned} \tag{8.66}$$

Combining (8.66) with (8.64) yields

$$f_{K,m}(z+re) \leq \frac{k\lambda_1 r^2}{2} + \frac{k\lambda_{-1}}{2}\|z^-\|^2 - a_7(\|z^0\|^\nu + r^\nu) + a_6.$$

So we can choose r_1 large enough which is independent of K and m such that

$$f_{K,m}(z+re) \leq 0, \quad \forall z \in \partial Q_m. \tag{8.67}$$

Next we will show that $f_{K,m}$ satisfies (P.S) condition on W_P^m for $m > 0$, i.e., any sequence $\{z_j\} \subset W_P^m$ possesses a convergent subsequence in W_P^m, provided $f_{K,m}(z_j)$ is bounded and $f'_{K,m}(z_j) \to 0$ as $j \to \infty$. We suppose $\|f_{K,m}(z_j)\| \leq C$, then for large j:

$$\begin{aligned}
C + \|z_j\| &\geq f_{K,m}(z_j) - \frac{1}{2} f'_{K,m}(z_j) z_j \\
&= \int_0^{k\tau} [\frac{1}{2} H'_K(z_j) \cdot z_j - H_K(z_j)]dt \\
&\geq \nu(2^{-1} - \nu^{-1}) \int_0^{k\tau} H_K(z_j)dt - C_1 \\
&\geq \nu(2^{-1} - \nu^{-1}) \int_0^{\tau} H_K(z_j)dt - C_1 \\
&\geq C_2\|z_j\|_4^4 - C_3
\end{aligned} \tag{8.68}$$

via (8.56). In (8.68), C_1 is independent of K, but both C_2 and C_3 depend on K. So $\{z_j\}$ is bounded in W_P^m. Since W_P^m is finite dimensional, the sequence $\{z_j\}$ has a convergent subsequence.

We have already verified all the conditions of Theorem 3.2, then $f_{K,m}$ has a critical value $c_{K,m} \geq \alpha$, which is given by

$$c_{K,m} = \inf_{g\in\Lambda_m} \max_{w\in Q_m} f_{K,m}(g(w)), \tag{8.69}$$

where $\Lambda_m = \{g \in C(Q_m, W_P^m) \mid g = id \text{ on } \partial Q_m\}$. Note that there is a natural S^1-invariant on W_P and W_P^m defined by

$$\theta * x(t) = x(t+\theta), \quad \forall x \in W_P, \theta \in [0, k\tau]/\{0, k\tau\} = S^1. \tag{8.70}$$

Now since W_P^m is finite dimensional and then $f''_{K,m}(x)$ is Fredholm for any critical point x, and $f_{K,m}$ is S^1-invariant under the above S^1-action (8.70) on W_P^m. So there is a critical point $x_{K,m}$ of $f_{K,m}$ which satisfies

$$m^-(x_{K,m}) \leq \dim X_m + 1 = m + \dim \ker_{\mathbb{R}}(P-I) + 1 \leq m^-(x_{K,m}) + m^0(x_{K,m}) - 1. \tag{8.71}$$

Step 3. We prove that there exists a nonconstant *P-solution* (τ, x_K) of the following problem

$$\begin{cases} \dot{x} = JH'_K(x), \\ x(\tau) = Px(0). \end{cases} \tag{8.72}$$

On the one hand, since $id \in \Lambda_m$, by (8.64) and (H4) we have

$$c_{K,m} \leq \sup_{w\in Q_m} f_{K,m}(w) \leq \frac{k\lambda_1}{2} r_1^2. \tag{8.73}$$

Then in the sense of subsequence we have

$$c_{K,m} \to c_K, \quad \alpha \leq c_K \leq \frac{k\lambda_1}{2} r_1^2. \tag{8.74}$$

On the other hand, we need to prove that f_K satisfies the (P.S)* condition on W_P, i.e., any sequence $\{z_m\} \subset W_P$ satisfying $z_m \in W_P^m$, $f_{K,m}(z_m)$ is bounded and $f'_{K,m}(z_m) \to 0$ possesses a convergent subsequence in W_P. It is a well-known result in the case of general periodic solution. For the reader's convenience, we give the proof following the idea in the appendix of [14].

$f'_{K,m}(z_m) \to 0$ as $m \to +\infty$ implies

$$-J\dot{z}_m - P_m H'_K(z_m) = \epsilon_m, \tag{8.75}$$

with

$$\|\epsilon_m\|_{(W_P^m)'} \to 0, \quad \text{as } m \to +\infty.$$

We remind that the dual space of W is denoted by W'. Writing $z_m = z_m^0 + z_m^+ + z_m^-$. Using the same arguments as (8.68) and by some direct estimates, we see that $\{z_m\}$ is bounded in W_P. Thus by passing to a subsequence, we may assume that

$$\begin{aligned} &z_m \to z \quad \text{in } W_P \text{ weakly}, \\ &z_m \to z \quad \text{in } L^p \text{ strongly for } 1 \le p < +\infty, \\ &z_m^0 \to z^0 \text{ in } \mathbb{R}^{2n}. \end{aligned}$$

By (8.54) the form of H_K, there exists constant M_2 such that

$$|H_K'(z)| \le M_2|z|^3 + M_2, \quad \forall z \in \mathbb{R}^{2n}.$$

This implies $H_K'(z_m) \to H_K'(z)$ strongly in L^2. Thus $P_m H_K'(z_m) \to H_K'(z)$ strongly in L^2, and thus in W_P'. Therefore (8.75) implies that

$$\dot{z}_m = \varsigma_m + \epsilon_m \tag{8.76}$$

holds in W_P', where $\varsigma_m \to \varsigma = JH_K'(z)$ in L^2. This implies

$$\dot{z} = \varsigma \tag{8.77}$$

in W_P'. Since $\varsigma \in L^2$, then $z \in W^{1,2}$ and thus $z \in C^2$, i.e., (8.77) holds in the classical sense. Because W_P^m is a subspace of W_P, $P_m : W_P \to W_P^m$ is projection.

$$\|z_m - P_m z\|_{W_P^m}^2 = \|\dot{z}_m - P_m \dot{z}\|_{(W_P^m)'}^2 + |z_m^0 - z^0|^2.$$

Then

$$\|z_m - P_m z\|_{W_P^m}^2 \le (\|\varsigma_m - P_m\varsigma\|_{(W_P^m)'} + \|\epsilon_m\|_{(W_P^m)'})^2 + |z_m^0 - z^0|^2.$$

From

$$\|\varsigma_m - P_m\varsigma\|_{(W_P^m)'} \le M_3\|\varsigma_m - P_m\varsigma\|_{L^2} \to 0$$

for some $M_3 > 0$ independent of m, we then obtain

$$\|z_m - P_m z\|^2 = \|z_m - P_m z\|_{W_P^m}^2 \to 0.$$

This proves $z_m \to z$ in W strongly. We have thus proved that f_K satisfies (P.S)* condition. Hence in the sense of the subsequence we have

$$x_{K,m} \to x_K, \quad f_K(x_K) = c_K, \quad f'_K(x_K) = 0. \tag{8.78}$$

From above we conclude that f_K possesses a critical value $c_K \geq \alpha = \alpha(K) > 0$ with a corresponding critical point x_K. By the standard arguments as (6.35)–(6.37) in [255], x_K is a classical nonconstant *P-solution* of (8.72).

Indeed, if $x_K(t)$ is a constant solution of (8.72), then it should belong to $\ker_{\mathbb{R}}(P-I)$ and

$$f_K(x_K) = \frac{k}{2}\langle Ax_K, x_K\rangle - \int_0^{k\tau} H_K(x_K)dt \leq 0.$$

This contradicts to $f_K(x_K) = c_K \geq \alpha > 0$.

Step 4. We show that there is a $K_0 > 0$ such that for all $K \geq K_0$, $\|x_K\|_{L^\infty} < K$. Then $H'_K(x_K) = H'(x_K)$ and $x = x_K$ is a nonconstant *P-solution* of (8.46). By (8.74), $c_K \leq \frac{k\lambda_1}{2}r_1^2$ independently of K. By (8.56), we obtain

$$\begin{aligned}\frac{\lambda_1}{2}r_1^2 &\geq f_K(x_K) - \frac{1}{2}f'_K(x_K)x_K \\ &\geq (2^{-1} - \nu^{-1})\int_0^\tau H'_K(x_K)\cdot x_K dt - C.\end{aligned} \tag{8.79}$$

with $C = \nu^{-1}K_2\tau$ is independent of K. Therefore (8.79) provides a K independent upper bound for $\int_0^\tau H'_K(x_K)\cdot x_K dt$. By (8.56),

$$H_K(\zeta) \leq \nu^{-1}H'_K(\zeta)\cdot\zeta + C/\tau, \quad \forall\zeta \in \mathbb{R}^{2n}. \tag{8.80}$$

Recalling that $H_K(x_K) \equiv constant$ since x_K satisfies an autonomous Hamiltonian system, so replacing ζ by x_K, integrating (8.80) over $[0, \tau]$, (8.80) yields

$$k\tau H_K(x_K) \leq \nu^{-1}\int_0^{k\tau} H'_K(x_K)\cdot x_K + C. \tag{8.81}$$

The right hand side of (8.81) is bounded from above independently of K. Then (8.57) and (8.81) yield a K independent L^∞ bound for x_K. So choose K large such that $\|x_K\|_{L^\infty} < K$ thus x_K is a P-solution of the problem (8.46). We denote it simply by x.

Step 5. We prove that $\dim\ker_{\mathbb{R}}(P-I) + 2 - \nu^P(x) \leq i^P(x) \leq \dim\ker_{\mathbb{R}}(P-I) + 1$.

Let $B = H''(x(t))$ and B be the operator defined by (8.51) corresponding to $B(t)$. By direct computation, we get

$$\langle f_K''(z)w, w\rangle - k\langle (A-B)w, w\rangle$$
$$= \int_0^{k\tau} [(H_K''(x_K(t))w, w) - (H_K''(z(t))w, w)]dt, \quad \forall w \in W.$$

Then by the continuity of H_K'',

$$\| f_K''(z) - k(A-B)\| \to 0 \quad \text{as} \quad \| z - x_K \| \to 0. \tag{8.82}$$

Let $d = \frac{1}{4}\|(A-B)^\sharp\|^{-1}$. By (8.82), there exists $r_0 > 0$ such that

$$\| f_K''(z) - k(A-B)\| < \frac{1}{2}d, \quad \forall z \in V_{r_0} = \{z \in W : \|z - x_K\| \le r_0\}.$$

Hence for m large enough, there holds

$$\| f_{K,m}''(z) - kP_m(A-B)P_m\| < \frac{1}{2}d, \quad \forall z \in V_{r_0} \cap W_P^m. \tag{8.83}$$

For $x_{K,m} \in V_{r_0} \cap W_P^m$, $\forall w \in M_d^-(P_m(A-B)P_m) \setminus \{0\}$, from (8.83) we have

$$\begin{aligned}\langle f_{K,m}''(x_{K,m})w, w\rangle &\le k\langle P_m(A-B)P_m w, w\rangle + \| f_{K,m}''(x_{K,m}) \\ &\quad - kP_m(A-B)P_m\| \cdot \|w\|^2 \\ &\le -kd\|w\|^2 + \frac{1}{2}d\|w\|^2 = -\frac{1}{2}d\|w\|^2 < 0.\end{aligned}$$

Then there holds

$$\dim M^-(f_{K,m}''(x_{K,m})) \ge \dim M_d^-(P_m(A-B)P_m). \tag{8.84}$$

By (8.71), (8.78), (8.84) and Theorem IV.2.5, for large m we have

$$\begin{aligned}m + \dim \ker_{\mathbb{R}}(P-I) + 1 = \dim X_m + 1 &\ge m^-(x_{K,m}) \\ &\ge \dim M_d^-(P_m(A-B)P_m) = m + i^P(x_K).\end{aligned}$$

Then $i^P(x_K) \le \dim \ker_{\mathbb{R}}(P-I) + 1$.

Similarly, $\forall w \in M_d^+(P_m(A-B)P_m) \setminus \{0\}$, from (8.83) we have

$$\begin{aligned}\langle f_{K,m}''(x_{K,m})w, w\rangle &\ge \langle P_m(A-B)P_m w, w\rangle - \| f_{K,m}''(x_{K,m}) \\ &\quad - P_m(A-B)P_m\| \cdot \|w\|^2 \\ &\ge d\|w\|^2 - \frac{1}{2}d\|w\|^2 = \frac{1}{2}d\|w\|^2 > 0.\end{aligned}$$

Then

$$\dim M^+(f''_{K,m}(x_{K,m})) \geq \dim M_d^+(P_m(A-B)P_m). \tag{8.85}$$

By (8.71), (8.78), (8.85) and Theorem IV.2.5, for large m we have

$$\begin{aligned} m + \dim \ker_{\mathbb{R}}(P-I) + 1 &= \dim X_m + 1 \\ &\leq m^-(x_{K,m}) + m^0(x_{K,m}) - 1 \\ &= \dim W_P^m - m^+(x_{K,m}) - 1 \\ &\leq 2m + \dim \ker_{\mathbb{R}}(P-I) - \dim M_d^+(P_m(A-B)P_m) - 1 \\ &= m + i_P(x_K) + \nu^P(x_K) - 1. \end{aligned}$$

This implies $\dim \ker_{\mathbb{R}}(P-I) + 2 \leq i_P(x_K) + \nu^P(x_K)$.

Step 6. Estimate the minimal P-*symmetric* period of solution x.

If $k\tau$ is not the minimal P-*symmetric* period of x, i.e., $\tau > \min\{\lambda > 0 \mid x(t+\lambda) = Px(t), \forall t \in \mathbb{R}\}$, then there exists some l such that

$$T \equiv \frac{\tau}{l} = \min\{\lambda > 0 \mid x(t+\lambda) = Px(t), \forall t \in \mathbb{R}\}.$$

Thus $x(\tau - T) = x(0)$, both $(l-1)T$ and kT are the period of x. Since kT is the minimal P-*symmetric* period, we obtain $kT \leq (l-1)T$ and then $k \leq l-1$.

Note that $x|_{[0,kT]}$ is the k-th iteration of $x|_{[0,T]}$. Suppose $\gamma \in \mathcal{P}_T(2n)$ is the fundament solution of the following linear Hamiltonian system

$$\dot{z}(t) = JB(t)z(t)$$

with $B(t) = H''(x|_{[0,T]}(t))$. Suppose ξ be any symplectic path in $\mathcal{P}_T(2n)$ such that $\xi(T) = P^{-1}$, since $P^k = I$, then

$$\nu(\xi, 1) = \nu(\xi, k+1) = \nu(\xi, l). \tag{8.86}$$

All eigenvalues of P and P^{-1} are all on the unit circle, then the elliptic height

$$e(P^{-1}) = e(P) = 2n. \tag{8.87}$$

Since the system (8.46) is autonomous, we have

$$\nu_1(x|_{[0,kT]}) \geq 1 \ \text{ and } \ \nu^{P^{l-1}}(\gamma, l-1) = \nu_1(x|_{[0,(l-1)T]}) \geq 1. \tag{8.88}$$

By Corollary IV.6.15, $P^{l-1} = I$ and (8.86)–(8.87), we have

$$
\begin{aligned}
i^I(\gamma, l-1) &= i^{P^{l-1}}(\gamma, l-1)\\
&\le i^{P^l}(\gamma, l) - i^P(\gamma, 1) + \nu(\xi, 1) - \nu(\xi, l) + \frac{e(P^{-1}\gamma(T))}{2}\\
&\quad + \frac{e(P^{-1})}{2} - \nu^{P^{l-1}}(\gamma, l-1)\\
&\le i^{P^l}(\gamma, l) - i^P(\gamma, 1) + \frac{e(P^{-1}\gamma(T))}{2} + n - 1\\
&\le i^{P^l}(\gamma, l) - i^P(\gamma, 1) + 2n - 1.
\end{aligned}
\tag{8.89}
$$

Note that $i^{P^l}(\gamma, l) = i^P_{[0,\tau]}(x_K) \le \dim \ker_{\mathbb{R}}(P-I)+1$, here we write $i^P_{[0,\tau]}(x_K)$ for $i^P(x_K)$ to remind the solution x_K is defined in the interval $[0, \tau]$. By the definition of Maslov P-index,

$$
i^I(\gamma, l-1) = i_1(\gamma, l-1) + n.
$$

So we get

$$
i_1(\gamma, l-1) \le \dim \ker_{\mathbb{R}}(P-I) - i^P(\gamma, 1) + n. \tag{8.90}
$$

By the condition (HC) and (4.39) in Lemma IV.2.4, we have

$$
i^P(\gamma, 1) = i^P(B) = \sum_{s\in[0,1)} \nu^P(sB) = \sum_{s\in[0,1)} \dim \ker_{\mathbb{R}}(\gamma_B(sT) - P). \tag{8.91}
$$

Here we remind that $B(t) = H''(x|_{[0,T]}(t))$ and γ_B is the fundamental solution of the linear Hamiltonian system

$$
\dot{z}(t) = JB(t)z(t).
$$

Since $\gamma_B(0) = I$, so $\dim \ker_{\mathbb{R}}(\gamma_B(s\tau) - P) = \dim \ker_{\mathbb{R}}(P-I)$ when $s = 0$. Thus we have

$$
i^P(\gamma, 1) \ge \dim \ker_{\mathbb{R}}(P-I). \tag{8.92}
$$

From (8.90), it implies

$$
i_1(\gamma, l-1) \le n. \tag{8.93}
$$

By the convex condition (HC) We also have

$$
i_1(x|_{[0,kT]}) \ge n \quad \text{and} \quad i_1(x|_{[0,(l-1)T]}) \ge n. \tag{8.94}
$$

We set $m = \frac{l-1}{k}$. Note that $x|_{[0,(l-1)T]}$ is the m-th iteration of $x|_{[0,kT]}$. By (8.88), (8.94), (8.93) and Lemma 4.1 in [197], we obtain $m = 1$ and then $k = l-1$. From the process of the proof, we see that only if $e(P^{-1}\gamma(T)) = 2n$, we can obtain $k = l-1$. In this case, the minimal P-symmetric period of x is $\frac{k\tau}{k+1}$.

Note that

$$\bar{\gamma}(\tau) = P^{l-1}\gamma(T)(P^{-1}\gamma(T))^{l-1} = P(P^{-1}\gamma(T))^{l}.$$

So we have $e(P^{-1}\gamma(T)) = e((P^{-1}\gamma(T))^{l}) = e(P^{-1}\bar{\gamma}(\tau))$. If $\bar{\gamma} \notin {}_P\mathcal{P}_{\tau}^{e}(2n)$, then $e(P^{-1}\gamma(T)) \leq 2n-2$. We get $i_1(\gamma, l-1) < n$ by taking the same process as (8.89)–(8.90). It contradicts to the second inequality of (8.94). At the moment, the minimal P-symmetric period of x is $k\tau$.

We set

$$\mathrm{Sp}(2n)_k \equiv \{P \in Sp(2n) \mid P \ satisfies \ (P)_k \ condition\},$$

In the follows we consider for what $P \in \mathrm{Sp}(2n)_k$ the minimal P-symmetric period of x is $k\tau$. □

Lemma 3.3 ([216]) *If $P \in Sp(2n)_k$, then there exists a matrix $I_{2p} \diamond R(\frac{2\pi}{k})^{\diamond j_1} \diamond \cdots \diamond R(\frac{2r\pi}{k})^{\diamond j_r} \in \Omega_0(P^{-1})$, with $p + \sum_{m=1}^{r} j_m = n$.*

Proof For $P \in Sp(2n)_k$, we have

$$\sigma(P^{-1}) = \sigma(P) \subseteq \{1, e^{\frac{2\pi\sqrt{-1}}{k}}, e^{\frac{4\pi\sqrt{-1}}{k}}, \cdots, e^{\frac{2(k-1)\pi\sqrt{-1}}{k}}\} \subseteq \mathbf{U}.$$

By the Theorem 1.8.10 of [223], there exists $M_1(\omega_1) \diamond M_2(\omega_2) \diamond \cdots \diamond M_s(\omega_s) \in \Omega_0(P^{-1})$ where $M_i(\omega_i)$ is a basic normal form of a eigenvalue ω_i of P^{-1}, $1 \leq i \leq s$. And the following are the basic normal forms for eigenvalues in $\mathbf{U}$.

Case 1. $N_1(\lambda, b) = \begin{pmatrix} \lambda & b \\ 0 & \lambda \end{pmatrix}$, $\lambda = \pm 1, b = \pm 1, 0$.

Since $P \in Sp(2n)_k$, we have $b = 0$ and $\lambda \in \{-1, 1\} \cap \sigma(P^{-1})$.

Case 2. $R(\theta) = \begin{pmatrix} \cos\theta & -\sin\theta \\ \sin\theta & \cos\theta \end{pmatrix}$, $\theta \in (0, \pi) \cup (\pi, 2\pi)$.

Since $P \in Sp(2n)_k$, we have $\theta \in \{\frac{2\pi}{k}, \frac{4\pi}{k}, \cdots \frac{2(k-1)\pi}{k}\}$.

Case 3. $N_2(\omega, b) = \begin{pmatrix} R(\theta) & b \\ 0 & R(\theta) \end{pmatrix}$, $\theta \in (0, \pi) \cup (\pi, 2\pi)$, $b = \begin{pmatrix} b_1 & b_2 \\ b_3 & b_4 \end{pmatrix}$, $b_i \in \mathbb{R}$, $b_2 \neq b_3$.

From direct computation, it is easy to check that the matrix $S = \begin{pmatrix} \frac{1}{\sqrt{2}} & \frac{\sqrt{-1}}{\sqrt{2}} \\ \frac{\sqrt{-1}}{\sqrt{2}} & \frac{1}{\sqrt{2}} \end{pmatrix}$ satisfies that $SR(\theta)S^{-1} = \begin{pmatrix} e^{\sqrt{-1}\theta} & 0 \\ 0 & e^{-\sqrt{-1}\theta} \end{pmatrix}$. Then there holds

$$\begin{pmatrix} S & 0 \\ 0 & S \end{pmatrix} N_2(\omega, b) \begin{pmatrix} S^{-1} & 0 \\ 0 & S^{-1} \end{pmatrix} = \begin{pmatrix} SR(\theta)S^{-1} & SbS^{-1} \\ 0 & SR(\theta)S^{-1} \end{pmatrix},$$

where

$$SbS^{-1} = \begin{pmatrix} \frac{1}{2}(b_1 + b_4) - \frac{\sqrt{-1}}{2}(b_2 - b_3) & \frac{1}{2}(b_2 + b_3) - \frac{\sqrt{-1}}{2}(b_1 - b_4) \\ \frac{1}{2}(b_2 + b_3) + \frac{\sqrt{-1}}{2}(b_1 - b_4) & \frac{1}{2}(b_1 + b_4) + \frac{\sqrt{-1}}{2}(b_2 - b_3) \end{pmatrix}. \tag{8.95}$$

Denoted by

$$\begin{pmatrix} SR(i\theta)S^{-1} & X(i) \\ 0 & SR(k\theta)S^{-1} \end{pmatrix} = \begin{pmatrix} SR(\theta)S^{-1} & SbS^{-1} \\ 0 & SR(\theta)S^{-1} \end{pmatrix}^i, \quad i \in \mathbb{N}, \tag{8.96}$$

where $X(i) = \begin{pmatrix} x_1(i) & x_2(i) \\ x_3(i) & x_4(i) \end{pmatrix}$ and $X(1) = SbS^{-1} = \begin{pmatrix} x_1(1) & x_2(1) \\ x_3(1) & x_4(1) \end{pmatrix}$. By direct computation, we have

$$\begin{aligned} x_1(k) &= ke^{\sqrt{-1}(k-1)\theta} x_1(1), \\ x_4(k) &= ke^{-\sqrt{-1}(k-1)\theta} x_4(1). \end{aligned} \tag{8.97}$$

Thus from $P^k = I$ we have $X(k) = 0$, so $x_1(1) = x_4(1) = 0$, i.e.

$$\begin{aligned} \frac{1}{2}(b_1 + b_4) - \frac{\sqrt{-1}}{2}(b_2 - b_3) &= 0, \\ \frac{1}{2}(b_1 + b_4) + \frac{\sqrt{-1}}{2}(b_2 - b_3) &= 0. \end{aligned}$$

Therefore we have $b_2 = b_3$, which is contradict to the definition of the basic normal form $N_2(\omega, b)$.

From Case1 - Case3, we get $M_i(\omega_i) = R(\theta_i)$ where $\theta_i \in \{0, \frac{2\pi}{k}, \frac{4\pi}{k}, \cdots \frac{2(k-1)\pi}{k}\}$, $1 \le i \le s$. And the lemma is proved. □

For the notations in Lemma 3.3, we define the set of k-admissible symplectic matrices by

$$Sp(2n)_k^{ad} \equiv \left\{ P \in Sp(2n)_k \mid k - 2\sum_{m=1}^{r} m \cdot j_m > 1, r < \frac{k}{2} \right\}.$$

Theorem 3.4 ([216]) *Suppose $P \in Sp(2n)_k^{ad} \cup \{I\}$, and the Hamiltonian function H satisfies (H1)–(H4) and (HC), then for every $\tau > 0$, the system* (8.72) *possesses a non-constant P-solution (τ, x) such that the minimal P-symmetric period of the extended $k\tau$-periodic solution $(k\tau, x^k)$ is $k\tau$.*

We remind that when $P = I$, it implies $k = 1$. This is the case of periodic solution which was proved in [66] by D. Dong and Y. Long, and in [197] by C. Liu and Y. Long. So we only need to prove the case when $P \in Sp(2n)_k^{ad}$.

Lemma 3.5 ([216]) *For $P \in Sp(2n)_k^{ad}$ and $\xi \in \mathcal{P}_\tau(2n)$ with $\xi(\tau) = P^{-1}$, there holds*

$$(k+1)i(\xi) - i(\xi, k+1) = \sum_{m=1}^{r}(k-2m)j_m - kp. \tag{8.98}$$

Proof By Theorem 9.3.1 of [223], we have

$$\begin{aligned} i(\xi, k+1) &= (k+1)(i(\xi) + S^+_{P^{-1}}(1) - C(P^{-1})) \\ &\quad + 2\sum_{\theta\in(0,2\pi)} E(\frac{(k+1)\theta}{2\pi})S^-_{P^{-1}}(e^{\sqrt{-1}\theta}) - (S^+_{P^{-1}}(1) + C(P^{-1})) \\ &= (k+1)i(\xi) + kS^+_{P^{-1}}(1) - (k+2)C(P^{-1}) \\ &\quad + 2\sum_{\theta\in(0,2\pi)} E(\frac{(k+1)\theta}{2\pi})S^-_{P^{-1}}(e^{\sqrt{-1}\theta}), \end{aligned} \tag{8.99}$$

where $S^\pm_M(\omega)$ denote the splitting number of $M \in Sp(2n)$ at $\omega \in \mathbf{U}$, $C(M) = \sum_{\theta\in(0,2\pi)} S^-_M(e^{\sqrt{-1}\theta})$ and $E(a) = \min\{m \in \mathbb{Z} \mid m \geq a\}$.

For $P \in Sp(2n)_k^{ad}$, $I_{2p} \diamond R(\frac{2\pi}{k})^{\diamond j_1} \diamond \cdots \diamond R(\frac{2r\pi}{k})^{\diamond j_r} \in \Omega^0(P^{-1})$ with $2r < k$. By direct computation, for $\theta \in (0, \pi) \cup (\pi, 2\pi)$, we have

$$\begin{aligned} S^+_{P^{-1}}(1) &= p\,; \\ C(P^{-1}) &= \sum_{m=1}^{r} j_m\,; \end{aligned} \tag{8.100}$$

$$S^-_{P^{-1}}(e^{\sqrt{-1}\theta}) = \begin{cases} j_m, & if\ \theta = \frac{2m\pi}{k},\ 1 \leq m \leq r,\ e^{\frac{2m\pi\sqrt{-1}}{k}} \in \sigma(P^{-1})\,; \\ 0, & otherwise. \end{cases} \tag{8.101}$$

So there holds

$$\begin{aligned} &\sum_{\theta\in(0,2\pi)} E(\frac{(k+1)\theta}{2\pi})S^-_{P^{-1}}(e^{\sqrt{-1}\theta}) \\ &= E(\frac{k+1}{k})j_1 + E(\frac{2(k+1)}{k})j_2 + \cdots + E(\frac{r(k+1)}{k})j_r \\ &= 2j_1 + 3j_2 + \cdots + (r+1)j_r. \end{aligned} \tag{8.102}$$

Then we have

$$
\begin{aligned}
i(\xi, k+1) &= (k+1)i(\xi) + kp - (k+2)(j_1 + j_2 + \cdots + j_r) \\
&\quad + 2(2j_1 + 3j_2 + \cdots + (r+1)j_r) \\
&= (k+1)i(\xi) + kp - (k-2)j_1 - (k-4)j_2 - \cdots - (k-2r)j_r\,,
\end{aligned}
\tag{8.103}
$$

thus $(k+1)i(\xi) - i(\xi, k+1) = \sum_{m=1}^{r}(k-2m)j_m - kp.$ □

Proof of Theorem 3.4 Following the proof of Theorem 3.1. If the minimal P-symmetric period of x is $\frac{k\tau}{k+1}$, then there hold $e(P^{-1}\gamma(T)) = 2n$, and

$$
\begin{aligned}
i^P(\gamma, 1) &= \dim \ker_{\mathbb{R}}(P - I); \\
i^P(\gamma, l) &= i^P(\gamma, k+1) = \dim \ker_{\mathbb{R}}(P - I) + 1, \\
\nu^{P^k}(\gamma, k) &= \nu^P(\gamma, 1) = 1.
\end{aligned}
\tag{8.104}
$$

Here we remind that the left inequality in (4.62) of Proposition IV.6.13 holds independent of the choice of $\xi \in \mathcal{P}_\tau(2n)$, then for any $\xi \in \mathcal{P}_\tau(2n)$ we have

$$
i^P(\gamma, k+1) \geq (k+1)(i^P(\gamma, 1) + \nu^P(\gamma, 1) - n) + n - 1 + (k+1)i_1(\xi) - i(\xi, k+1).
\tag{8.105}
$$

By the condition $P \in Sp(2n)_k^{ad}$, we get

$$
\dim \ker_{\mathbb{R}}(P - I) = 2p,
\tag{8.106}
$$

$$
k - 2\sum_{m=1}^{r} m \cdot j_m > 1.
\tag{8.107}
$$

Applying (8.104), (8.106) and Lemma 3.5 to (8.105), we get

$$
k - 2\sum_{m=1}^{r} m \cdot j_m \leq 1.
\tag{8.108}
$$

It is contradict to the inequality (8.107). So the minimal P-*symmetric* period of $(k\tau, x^k)$ is $k\tau$. □

We note that the matrix $P = \begin{pmatrix} a & b \\ -\frac{a^2+a+1}{b} & -a-1 \end{pmatrix} \in \mathrm{Sp}(2)_3^{ad}$ for $b > 0$ with $(r, p; j_1) = (1, 0; 1)$, i.e., $R(\frac{2\pi}{3}) \in \Omega^0(P^{-1})$. Thus for the solution x found by variational method with the saddle point theorem as in Theorem 3.1, its three times iteration $(3\tau, x^3)$ is the minimal P-symmetric periodic solution.

Chapter 9
Applications of L-Index

In this chapter, we apply the L-index theory developed in Chap. 5 to study the existence and multiplicity of L-solutions of nonlinear Hamiltonian systems. In Sect. 9.1, we consider the existence of brake solution of asymptotically linear Hamiltonian system via the variational method and the L-index theory. In Sect. 9.2, by using the iteration theory of the L-index, we consider the minimal periodic problem for brake solutions of super-quadratic autonomous Hamiltonian systems. In Sect. 9.3, we obtain an infinitely many brake solutions of super-quadratic non-autonomous Hamiltonian systems.

9.1 The Existence of L-Solutions of Nonlinear Hamiltonian Systems

In this section, we consider the solutions of the following nonlinear Hamiltonian systems with Lagrangian boundary condition

$$\begin{cases} \dot{x}(t) = JH'(t, x(t)), & x(t) \in \mathbb{R}^{2n}, \\ x(0) \in L, & x(1) \in L, \end{cases} \tag{9.1}$$

where $L \in \Lambda(n)$, here $\Lambda(n)$ is the set of all Lagrangian subspaces of $(\mathbb{R}^{2n}, \omega_0)$. The Hamiltonian function $H \in C^2([0, 1] \times \mathbb{R}^{2n}, \mathbb{R})$ satisfying condition

(H_0) $H'(t, 0) \equiv 0, t \in [0, 1]$.

(H_∞) There exist continuous symmetric matrix functions $B_1(t),\ B_2(t)$ with $i_L(B_1) = i_L(B_2),\ \nu_L(B_2) = 0$ such that

$$B_1(t) \le H''(t, x) \le B_2(t),\ \forall (t, x) \text{ with } |x| \ge r \text{ for some large } r > 0, \forall\, t \in [0, 1].$$

For two symmetric matrices A and B, $A \leq B$ means that $A - B$ is a semi-positive definite matrix, and $A < B$ means that $A - B$ is a positive definite matrix similarly.

We have the following result.

Theorem 1.1 *Let H satisfy conditions (H_0) and (H_∞). Suppose $JB_1(t) = B_1(t)J$ and $B_0(t) = H''(t,0)$ satisfying one of the following twisted conditions*

$$B_1(t) + kI \leq B_0(t), \tag{9.2}$$

or

$$B_0(t) + kI \leq B_1(t), \tag{9.3}$$

with the constant $k \geq \pi$. Then (9.1) *possesses at least one nontrivial solution. Further more if $\nu_L(B_0) = 0$, the system* (9.1) *possesses at least two nontrivial solutions.*

We use the following result to prove Theorem 1.1.

Lemma 1.2 (Theorem 5.1 and Corollary II.5.2 of [33]) *Suppose $f \in C^2(\mathcal{L}, \mathbb{R})$ satisfies the (PS) condition, $f'(0) = 0$ and there is $r \notin [m^-(f''(0)), m^-(f''(0)) + m^0(f''(0))]$ with $H_q(\mathcal{L}, f_a; \mathbb{R}) \cong \delta_{q,r}\mathbb{R}$, then f has at leat one nontrivial critical point $u_1 \neq 0$. Moreover, if $m^0(f''(0)) = 0$ and $m^0(f''(u_1)) \leq |r - m^-(f''(0))|$, then f has one more nontrivial critical point $u_2 \neq u_1$.*

Without loss any generality we can suppose $H(t,0) = 0$ and $L = L_0$. By the condition (H_∞) and the remark after Corollary V.6.2, we get that $i_L(B_1)+\nu_L(B_1) \leq i_L(B_2) + \nu_L(B_2)$, so we have $\nu_L(B_1) = 0$. We shall first prove that under the above conditions (9.2) or (9.3), there holds

$$i_L(B_1) \notin [i_L(B_0), i_L(B_0) + \nu_L(B_0)]. \tag{9.4}$$

More clearly, under the condition (9.2), it is claimed

$$i_L(B_1) = i_L(B_1) + \nu_L(B_1) < i_L(B_0), \tag{9.5}$$

and under the condition (9.3), it is claimed

$$i_L(B_0) + \nu_L(B_0) < i_L(B_1). \tag{9.6}$$

We first prove (9.5). By Corollary 2.2 and condition (9.2), we have

$$i_L(B_1) \leq i_L(B_1 + kI) \leq i_L(B_0).$$

We shall prove

$$i_L(B_1) < i_L(B_1 + kI). \tag{9.7}$$

In fact, suppose $\gamma_1(t) = \begin{pmatrix} S_1(t) & V_1(t) \\ T_1(t) & U_1(t) \end{pmatrix} \in \mathcal{P}_\tau(2n)$ is a symplectic path which is the fundamental solution of the linear Hamiltonian system associate with the matrix function $B_1(t)$. Since $JB_1(t) = B_1(t)J$, one can show that $\exp(Jkt)\gamma_1(t)$ is the fundamental solution of the linear Hamiltonian system

$$\dot{z} = J(B_1(t) + kI)z. \tag{9.8}$$

One has

$$\exp(Jkt)\gamma_1(t) = \begin{pmatrix} S_1(t)\cos kt - T_1(t)\sin kt & V_1(t)\cos kt - U_1(t)\sin kt \\ S_1(t)\sin kt + T_1(t)\cos kt & V_1(t)\sin kt + U_1(t)\cos kt \end{pmatrix}.$$

The associate unitary $n \times n$ matrix $\mathcal{Q}(t)$ defined by (5.6) with respect to the above matrix is

$$\begin{aligned} \mathcal{Q}(t) &= [U_1(t) - \sqrt{-1}V_1(t)][U_1(t) + \sqrt{-1}V_1(t)]^{-1}\exp(2k\sqrt{-1}t) \\ &= \mathcal{Q}_1(t)\exp(2k\sqrt{-1}t). \end{aligned}$$

In (5.7) and Definition V.1.2, $\Delta_j = \theta_j(1) - \theta_j(0)$ and $\Delta_j^1 = \theta_j^1(1) - \theta_j^1(0)$ associate to $\mathcal{Q}(t)$ and $\mathcal{Q}_1(t)$ respectively satisfy

$$\Delta_j = \theta_j(1) - \theta_j(0) = \Delta_j^1 + 2k = \theta_j^1(1) - \theta_j^1(0) + 2k.$$

Since $k \geq \pi$, there holds

$$i_L(B_1) + n \leq i_L(B_1 + kI). \tag{9.9}$$

Thus we have proved (9.5). Equation (9.6) can be proved similarly.

By the condition (H_∞), $H''(t,x)$ is bounded and there exist $\mu_1,\ \mu > 0$ such that

$$I \leq H''(t,x) + \mu I \leq \mu_1 I,\ \forall (t,x). \tag{9.10}$$

We define a convex function $N(t,x) = H(t,x) + \frac{1}{2}\mu|x|^2$. Its Fenchel dual defined by $N^*(t,x) = \sup_{y\in\mathbb{R}^{2n}}\{(x,y) - N(t,y)\}$ satisfies $N^* \in C^2([0,1]\times\mathbb{R}^{2n},\mathbb{R})$ and (cf. [78])

$$N^{*\prime\prime}(t,y) = N''(t,x)^{-1}, \text{ for } y = N'(t,x). \tag{9.11}$$

From (9.10) we have

$$\mu_1^{-1} I \leq N^{*\prime\prime}(t, y) \leq I, \ \forall (t, y). \tag{9.12}$$

So we have $|x| \to \infty$ if and only if $|y| \to \infty$ with $y = N'(t, x)$. Thus there exists $r_1 > 0$ such that

$$(B_2(t) + \mu I)^{-1} \leq N^{*\prime\prime}(t, y) \leq (B_1(t) + \mu I)^{-1}, \ \forall (t, y) \text{ with } |y| \geq r_1. \tag{9.13}$$

We choose $\mu > 0$ satisfying (9.10) and $\mu \notin \sigma(A)$, and recall that $(\Lambda_\mu x)(t) = -J\dot{x}(t) + \mu x(t)$. We define the following functional

$$f(u) = -\int_0^1 [\frac{1}{2}(\Lambda_\mu^{-1} u(t), u(t)) - N^*(t, u(t))]\, dt, \ \forall u \in \mathcal{L}. \tag{9.14}$$

It is easy to see that $f \in C^2$ and satisfies (PS) condition (cf. [78]). There is a one to one corresponding from the critical points of f to the solutions of Hamiltonian systems (9.1). We note that 0 is a trivial critical point of f and $N^{*\prime}(t, 0) = 0$. At every critical point u_0, the second variation of f defines a quadratical form on $\mathcal{L}$ by

$$(f''(u_0)u, u) = -\int_0^1 [(\Lambda_\mu^{-1} u(t), u(t)) - (N^{*\prime\prime}(t, u_0(t))u(t), u(t))]\, dt, \ \forall u \in \mathcal{L}.$$

Its Morse index and nullity are both finite. We denote the index pair by $(i_\mu^*(u_0), \nu_\mu^*(u_0))$. The critical point u_0 corresponding to a solution $x_0 = \Lambda_\mu^{-1} u_0$ of (9.1), and $N^{*\prime\prime}(t, u_0(t)) = N''(t, x_0(t))^{-1}$. So by Theorem V.7.1, we have

$$i_\mu^*(u_0) = i_L(x_0) + n + n\left[\frac{\mu}{\pi}\right], \ \nu_\mu^*(u_0) = \nu_L(x_0).$$

The index pair $(i_L(x_0), \nu_L(x_0))$ is the L-index of the following linear Hamiltonian system

$$\dot{y}(t) = JH''(t, x_0(t))y(t).$$

By condition (9.2) and the result (9.9), we have

$$i_L(B_1) + \nu_L(B_1) + n \leq i_L(B_0). \tag{9.15}$$

By condition (9.3), similarly we have

$$i_L(B_0) + \nu_L(B_0) + n \leq i_L(B_1).$$

From (9.15) and the above inequality, we have that

$$|i_L(B_0) - i_L(B_1)| \geq n, \ \text{ and } |i_\mu^*(B_0) - i_\mu^*(B_1)| \geq n. \tag{9.16}$$

In the following, we need to prove that the Homological groups satisfy

$$H_q(\mathcal{L}, f_a; \mathbb{R}) \cong \delta_{qr}\mathbb{R}, \ q = 0, 1, \cdots, \tag{9.17}$$

for some $a \in \mathbb{R}$ and $r = i_\mu^*(B_1)$. $f_a = \{x \in \mathcal{L} | f(x) \le a\}$ is the level set below a. In the following we follow the ideas of the proof of Lemma II.5.1 in [33] to prove (9.17). See [67] and [187] for some similar arguments.

Step 1. Under the condition (H_∞), there holds

$$\mathcal{L} = \mathcal{L}_\mu^-(B_1) \oplus \mathcal{L}_\mu^+(B_2), \tag{9.18}$$

where $\mathcal{L}_\mu^*(B)$ for $* = \pm, 0$ is defined in Sect. 5.6. In fact, it is clear that $\mathcal{L}_\mu^-(B_1) \cap \mathcal{L}_\mu^+(B_2) = \{0\}$. By $\nu_\mu^*(B_2) = \nu_L(B_2) = 0$, we have $\mathcal{L} = \mathcal{L}_\mu^-(B_2) \oplus \mathcal{L}_\mu^+(B_2)$. By condition (H_∞), we have $i_\mu^*(B_1) = i_\mu^*(B_2) = r$. Suppose $\xi_1, \xi_2, \cdots, \xi_r$ is a base of $\mathcal{L}_\mu^-(B_1)$. Decompose ξ_j by $\xi_j = \xi_j^- + \xi_j^+$ with $\xi_j \in \mathcal{L}_\mu^\pm(B_2)$. It is clear that $\xi_1^-, \cdots, \xi_r^-$ is linear independent, so it is a base of $\mathcal{L}_\mu^-(B_2)$. For any $\xi \in \mathcal{L}$, there holds $\xi = \xi^- + \xi^+$ with $\xi^\pm \in \mathcal{L}_\mu^\pm(B_2)$. Suppose $\xi^- = a_1\xi_1^- + \cdots + a_r\xi_r^-$. Then $\xi = \sum_{j=1}^r a_j\xi_j + (\xi^+ - \sum_{j=1}^r a_j\xi_j^+) = \xi_1 + \xi_2$ with $\xi_1 \in \mathcal{L}_\mu^-(B_1)$ and $\xi_2 \in \mathcal{L}_\mu^+(B_2)$.

Step 2. For sufficiently small $s > 0$, from the structure of the symplectic group and the definition of the Maslov-type index, we know that $\nu_L(B_1 - sI) = \nu_L(B_1) = 0$, $\nu_L(B_2 + sI) = \nu_L(B_2) = 0$, and so $i_L(B_1 - sI) = i_L(B_1) = i_L(B_2) = i_L(B_2 + sI)$. Denote the so called deformation space by $D_R = \mathcal{L}_\mu^-(B_1 - sI) \oplus \{u \in \mathcal{L}_\mu^+(B_2 + sI) | \, \|u\| \le R\}$. For $R > 0$ and $-a > 0$ large, we have the following deformation result

$$H_q(\mathcal{L}, f_a; \mathbb{R}) = H_q(D_R, D_R \cap f_a; \mathbb{R}). \tag{9.19}$$

The proof of (9.19) is standard in the Morse theory [29]. We only need to use the negative gradient flow of f to deform $(\mathcal{L}, f_a)$ to $(D_R, D_R \cap f_a)$. For any $u = u_1 + u_2 \in \mathcal{L}$ with $u_1 \in \mathcal{L}_\mu^-(B_1 - sI)$ and $u_2 \in \mathcal{L}_\mu^+(B_2 + sI)$, by the self-adjointness, we have

$$\begin{aligned}
(f'(u), u_2 - u_1) &= -\textstyle\int_0^1 [(\Lambda^{-1}u, u_2 - u_1) - (N^{*\prime}(t, u), u_2 - u_1)]\, dt \\
&= \textstyle\int_0^1 (\Lambda^{-1}u_1, u_1)dt - \int_0^1 (\Lambda^{-1}u_2, u_2)dt + \\
&\quad + \textstyle\int_0^1 (\int_0^1 N^{*\prime\prime}(t, \tau u)d\tau(u_1 + u_2), u_2 - u_1)dt \\
&= \textstyle\int_0^1 (\Lambda^{-1}u_1, u_1)dt - \int_0^1 (\int_0^1 N^{*\prime\prime}(t, \tau u)d\tau u_1, u_1)dt \\
&\quad - \textstyle\int_0^1 (\Lambda^{-1}u_2, u_2)dt + \int_0^1 (\int_0^1 N^{*\prime\prime}(t, \tau u)d\tau u_2, u_2)dt.
\end{aligned} \tag{9.20}$$

By (9.12) and (9.13), we have

$$
\begin{aligned}
&\int_0^1 (\int_0^1 N^{*''}(t,\tau u)d\tau u_1, u_1)dt \\
&\quad = \int_0^1 \int_0^{h(t,u)} (N^{*''}(t,\tau u)d\tau u_1, u_1)dt + \int_0^1 \int_{h(t,u)}^1 (N^{*''}(t,\tau u)d\tau u_1, u_1)dt \\
&\quad \le c_0\|u\| + \int_0^1 ((B_1(t) + \mu I - sI)u_1, u_1)dt,
\end{aligned} \tag{9.21}
$$

where $h(t,u) = \frac{r_1}{|u(t)|}$. Similarly, We have

$$
\begin{aligned}
&\int_0^1 (\int_0^1 N^{*''}(t,\tau u)d\tau u_2, u_2)dt \ge \int_0^1 \int_{h(t,u)}^1 (N^{*''}(t,\tau u)d\tau u_2, u_2)dt \\
&\quad \ge \int_0^1 ((B_2(t) + \mu I + sI)u_2, u_2)dt - c\|u\|, \text{ for some } c > 0
\end{aligned} \tag{9.22}
$$

So by (9.20), (9.21), and (9.22), we have

$$
(f'(u), u_2 - u_1) \ge c_1\|u_1\|^2 + c_2\|u_2\|^2 - c_3(\|u_1\| + \|u_2\|). \tag{9.23}
$$

Thus for large R with $\|u_1\| \ge R$ or $\|u_2\| \ge R$, we have

$$
(-f'(u), u_2 - u_1) < -1. \tag{9.24}
$$

We know from (9.24) that f has no critical point outside D_R, and that $-f'(u)$ points inward to D_R on ∂D_R. So we can define the deformation by the negative gradient flow of f. In fact, for any $u = u_1 + u_2 \notin D_R$, let $\sigma(\theta, u) = e^\theta u_1 + e^{-\theta} u_2$, and $d_u = \log\|u_2\| - \log R$. We define the deformation map $\eta : [0,1] \times \mathcal{L} \to \mathcal{L}$ by

$$
\eta(\theta, u_1 + u_2) = \begin{cases} u_1 + u_2, & \|u_2\| \le R, \\ \sigma(d_u\theta, u), & \|u_2\| > R. \end{cases}
$$

η satisfies the following properties

$$
\begin{aligned}
&\eta(0,\cdot) = id, \quad \eta(1,\mathcal{L}) \subset D_R, \quad \eta(1, f_a) \subset D_R \cap f_a \\
&\eta(\theta, f_a) \subset f_a, \quad \eta(\theta,\cdot)|_{D_R} = id|_{D_R}.
\end{aligned}
$$

Thus the pair $(D_R, D_R \cap f_a)$ is a deformation retract of the pair $(\mathcal{L}, f_a)$.

Step 3. For large R, $-a > 0$, there holds

$$
H_q(D_R, D_R \cap f_a) \cong \delta_{q,r}\mathbb{R}. \tag{9.25}
$$

In fact, similarly to the above computation, for large number $m > 0$, we have

$$\begin{aligned}
\int_0^1 N^*(t,u(t))dt &= \int_0^1 dt \iint_{[0,1]\times[0,1]} \tau(N^{*\prime\prime}(t,\tau s u(t))u(t),u(t))d\tau ds \\
&\quad + \int_0^1 N^*(t,0)dt \\
&\leq \int_{|u(t)|\geq mr_1} dt \iint_{[0,1]\times[0,1]} \tau(N^{*\prime\prime}(t,\tau s u(t))u(t),u(t))d\tau ds + c_m \\
&\leq \int_{|u(t)|\geq mr_1} dt \iint_{|s\tau u(t)|\geq r_1,\tau,s\in[0,1]} \tau(N^{*\prime\prime}(t,\tau s u(t))u(t),u(t))d\tau ds + \\
&\quad + \int_{|u(t)|\geq mr_1} dt \iint_{|s\tau u(t)|\leq r_1,\tau,s\in[0,1]} \tau(N^{*\prime\prime}(t,\tau s u(t))u(t),u(t))d\tau ds + c_m \\
&\leq \frac{1}{2}\int_0^1((B_1(t)+\mu I)^{-1}u(t),u(t))dt + k_m\|u\| + c_m,
\end{aligned}$$

where c_m and k_m are constants depending only on m, and $k_m \to 0$ as $m \to +\infty$. So for the small s in the step 2 above, we can choose a large number m such that

$$\int_0^1 N^*(t,u(t))dt \leq \frac{1}{2}\int_0^1((B_1(t)+\mu I - sI)^{-1}u(t),u(t))dt + C,\ \forall u \in \mathcal{L} \tag{9.26}$$

for some constant $C > 0$. Thus for any $u = u_1 + u_2$ with $u_1 \in \mathcal{L}_\mu^-(B_1 - sI)$ and $u_2 \in \mathcal{L}_\mu^+(B_2 + sI)$ with $\|u_2\| \leq R$, there holds

$$f(u) \leq -C_1\|u_1\|^2 + C_2\|u_1\| + C_3, \tag{9.27}$$

where $C_j,\ j = 1,2,3$ are constants and $C_1 > 0$. It implies that $f(u) \to -\infty$ if and only if $\|u_1\| \to \infty$ uniformly for $u_2 \in \mathcal{L}_\mu^+(B_2+sI)$ with $\|u_2\| \leq R$. In the following we denote $B_r = \{x \in \mathcal{L} |\ \|x\| \leq r\}$ the ball with radius r in $\mathcal{L}$. Therefore for $-a_1 > -a_2$ sufficiently large, there exist three numbers with $R < R_1 < R_2 < R_3$ satisfying

$$\begin{aligned}
&(\mathcal{L}_\mu^+(B_2+sI)\cap B_{R_3}) \oplus (\mathcal{L}_\mu^-)(B_1 - sI)\setminus B_{R_2}) \subset f_{a_1}\cap D_{R_3} \\
&\subset (\mathcal{L}_\mu^+(B_2+sI)\cap B_{R_3}) \oplus (\mathcal{L}_\mu^-)(B_1 - sI)\setminus B_{R_1}) \subset f_{a_2}\cap D_{R_3}.
\end{aligned}$$

Recall that $\sigma(\theta,u) = e^\theta u_1 + e^{-\theta}u_2$. By definition, we have $f(\sigma(0,u)) = f(u) > a_1$ and $f(\sigma(\theta,u)) \to -\infty$ as $\theta \to \infty$ if $u = u_1+u_2 \in D_{R_3}\cap(f_{a_2}\setminus f_{a_1})$. It implies that there exists $\theta_0 = \theta_0(u) > 0$ such that $f(\sigma(\theta_0,u)) = a_1$. But by (9.24), there holds

$$\frac{d}{d\theta}f(\sigma(\theta,u)) \leq -1, \text{ at any point } \theta > 0.$$

By the implicit function theorem, $\theta_0(u)$ is continuous with respect to the variable u. We define another deformation map $\eta_0 : [0,1]\times f_{a_2}\cap D_{R_3} \to f_{a_2}\cap D_{R_3}$ by

$$\eta_0(\theta,u) = \begin{cases} u, & u \in f_{a_1}\cap D_{R_3}, \\ \sigma(\theta_0(u)\theta,u), & u \in D_{R_3}\cap(f_{a_2}\setminus f_{a_1}). \end{cases}$$

It is clearly that η_0 is a deformation from $f_{a_2}\cap D_{R_3}$ to $f_{a_1}\cap D_{R_3}$. We now define

$$\tilde{\eta}(u) = d(\eta_0(1, u)) \text{ with } d(u) = \begin{cases} u, \ \|u_1\| \geq R_1, \\ u_2 + \frac{u_1}{\|u_1\|} R_1, \ 0 < \|u_1\| < R_1. \end{cases}$$

This map defines a strong deformation retract:

$$\tilde{\eta} : D_{R_3} \cap d_{a_2} \to (\mathcal{L}_\mu^+(B_2 + sI) \cap B_{R_3}) \oplus (\mathcal{L}_\mu^-(B_1 - sI) \cap \{u \in \mathcal{L} | \ \|u\| \geq R_1\}).$$

Proof Now we can compute the Homological groups

$$\begin{aligned} & H_q(D_{R_3}, D_{R_3} \cap f_{a_2}; \mathbb{R}) \\ \cong\ & H_q(D_{R_3}, (\mathcal{L}_\mu^+(B_2 + sI) \cap B_{R_3}) \oplus (\mathcal{L}_\mu^-(B_1 - sI) \cap \{u \in \mathcal{L} | \ \|u\| \geq R_1\}); \mathbb{R}) \\ \cong\ & H_q(\mathcal{L}_\mu^-(B_1 - sI) \cap B_{R_3}, \partial(\mathcal{L}_\mu^-(B_1 - sI) \cap B_{R_3}); \mathbb{R}) \cong \delta_{qr}\mathbb{R}. \end{aligned}$$

From (9.16), (9.17), and by using Lemma 1.2, we complete the proof. □

Corollary 1.3 *Let H satisfy the conditions (H_0) and (H_∞), and suppose $B_0(t) = H''(t, 0)$ satisfying one of the following twisted conditions*

$$B_1(t) < B_0(t), \text{ there exists } \lambda \in (0,1) \text{ such that } \nu_L((1-\lambda)B_1 + \lambda B_0) \neq 0 \tag{9.28}$$

and

$$B_0(t) < B_1(t), \text{ there exists } \lambda \in (0,1) \text{ such that } \nu_L((1-\lambda)B_0 + \lambda B_1) \neq 0. \tag{9.29}$$

Then (9.1) *possesses at least one nontrivial solution. Furthermore, if $\nu_L(B_0) = 0$ and in* (9.28), *we replace the second condition by $\sum_{\lambda \in (0,1)} \nu((1-\lambda)B_1 + \lambda B_0) \geq n$, or in* (9.29), *we replace the second condition by $\sum_{\lambda \in (0,1)} \nu((1-\lambda)B_0 + \lambda B_1) \geq n$, the Hamiltonian systems* (9.1) *possesses at least two non-trivial solutions.*

Proof It follows from Theorem V.6.4, the above proof of Theorem 1.1 and Lemma 1.2. In the first case, we have $r = i_L(B_1) \notin [i_L(B_0), i_L(B_0) + \nu_L(B_0)]$. In the second case we have $|i_L(B_0) - i_L(B_1)| \geq n$. □

The proof of the Theorem 1.1 in fact proves the following result.

Theorem 1.4 *Let H satisfy conditions (H_0) and (H_∞). Suppose $B_0(t) = H''(t, 0)$ satisfying the following twisted conditions*

$$i_L(B_1) \notin [i_L(B_0), i_L(B_0) + \nu_L(B_0)]. \tag{9.30}$$

Then the problem (9.1) *possesses at least one nontrivial solution. Moreover, if $\nu_L(B_0) = 0$ and $|i_L(B_1) - i_L(B_0)| \geq n$, the problem* (9.1) *possesses at least two nontrivial solutions.*

Remark 1.5 The condition $B_1(t) < B_2(t)$ in Theorem V.6.4 can be replaced by $B_1(t) \leq B_2(t)$ for all t and $B_2 - B_1 \geq \delta > 0$ for some constant δ as an operator in $\mathcal{L}$. So the condition in (9.28) and (9.29) can be replaced by this kind of conditions. The condition $JB_1(t) = B_1(t)J$ in (H_∞) can be replaced by that $JB_0(t) = B_0(t)J$.

9.2 The Minimal Period Problem for Brake Solutions

We now apply Theorem V.10.3 to the brake orbit problem of autonomous Hamiltonian system

$$\begin{cases} -J\dot{x} = Bx + H'(x), \qquad x \in \mathbb{R}^{2n}, \\ x(\tau/2 + t) = Nx(\tau/2 - t), \\ x(\tau + t) = x(t), \ t \in \mathbb{N}, \end{cases} \tag{9.31}$$

where $H(Nx) = H(x)$ and $B = \begin{pmatrix} B_1 & 0 \\ 0 & B_2 \end{pmatrix}$ is a $2n \times 2n$ symmetric semi-positive definite matrix whose operator norm is denoted by $\|B\|$, B_1 and B_2 are $n \times n$ symmetric matrices. A solution (τ, x) of the problem (9.31) is a brake orbit of the Hamiltonian system, and τ is the brake period of x. To find a brake orbit of the Hamiltonian system in (9.31), it is sufficient to solve the following problem

$$\begin{cases} -J\dot{x}(t) = Bx + H'(x(t)), \ \ x \in \mathbb{R}^{2n}, \ t \in [0, \tau/2], \\ x(0) \in L_0, \ x(\tau/2) \in L_0. \end{cases} \tag{9.32}$$

Any solution x of problem (9.32) can be extended to a brake orbit (τ, x) via the mirror symmetry about L_0 by $x(\tau/2 + t) = Nx(\tau/2 - t), \ t \in [0, \tau/2]$ and $x(\tau + t) = x(t), \ t \in \mathbb{R}$.

Theorem 2.1 *Suppose the Hamiltonian function H satisfies the conditions:*

(H1) $H \in C^2(\mathbb{R}^{2n}, \mathbb{R})$ *satisfying* $H(Nx) = H(x), \ \forall x \in \mathbb{R}^{2n}$.
(H2) *there are constants* $\mu > 2$ *and* $r_0 > 0$ *such that*

$$0 < \mu H(x) \leq H'(x) \cdot x, \quad \forall |x| \geq r_0.$$

(H3) $H(x) = o(|x|^2)$ *at* $x = 0$.
(H4) $H(x) \geq 0 \quad \forall x \in \mathbb{R}^{2n}$.

Then for every $0 < \tau < \frac{2\pi}{\|B\|}$*, the system* (9.31) *possesses a non-constant brake orbit* (τ, x) *satisfying*

$$i_{L_0}(x, \tau/2) \leq 1. \tag{9.33}$$

Moreover, if x further satisfies the following condition:

(HX) $H''(x(t)) \geq 0\ \forall t \in \mathbb{R}$ *and* $\int_0^{\tau/2} H''(x(t))\,dt > 0$.

Then the minimal brake period of x *is* τ *or* $\tau/2$.

We remind that if $B = 0$, then $\frac{2\pi}{\|B\|} = +\infty$.

Proof We divide the proof into two steps.
Step 1. Show that there exists a brake orbit (τ, x) satisfying (9.33) for $0 < \tau < \frac{2\pi}{\|B\|}$.

Fix $\tau \in (0, \frac{2\pi}{\|B\|})$. Without loss generality, we suppose $\tau = 2$, then $\tau < \frac{2\pi}{\|B\|}$ implies $\|B\| < \pi$. By conditions (H1)–(H4), we can find a non-constant τ-periodic solution x of (9.32) via the saddle point theorem such that (9.33) holds. For reader's convenience, we sketch the proof here and refer the reader to Theorem 3.5 of [197] for the case of periodic solution. We note that the main ideas here are the same as that in the periodic case. We refer the paper [170] for some details. See the similar argument in Sect. 8.3 for the case of minimal P-symmetric periodic problem.

In fact, following Rabinowitz's pioneering work [253], let $K > 0$ and $\chi \in C^{\infty}(\mathbb{R}, \mathbb{R})$ such that $\chi(t) = 1$ if $t \leq K$, $\chi(t) = 0$ if $t \geq K + 1$, and $\chi'(t) < 0$ if $y \in (K, K + 1)$. The number K will be determined later. Set

$$\hat{H}_K(z) = \frac{1}{2}(Bz, z) + H_K(z),$$

with

$$H_K(z) = \chi(|z|)H(z) + (1 - \chi(|z|))R_K|z|^4,$$

where the constant R_K satisfies

$$R_K \geq \max_{K \leq |z| \leq K+1} \frac{H(z)}{|z|^4}.$$

We set $L^2 = L^2([0, 1], \mathbb{R}^{2n})$ and define a Hilbert space $E := \mathcal{W}_{L_0} = W_{L_0}^{1/2,2}([0, 1], \mathbb{R}^{2n})$ with L_0 boundary conditions by

$$\begin{aligned}\mathcal{W}_{L_0} &= \{z \in L^2 |\ z(t) \\ &= \sum_{k \in \mathbb{Z}} \exp(k\pi t J)a_k,\ a_k \in L_0,\ \|z\|^2 := \sum_{k \in \mathbb{Z}} (1 + |k|)|a_k|^2 < \infty\}.\end{aligned}$$

We denote its inner product by $\langle \cdot, \cdot \rangle$. By the well-known Sobolev embedding theorem, for any $s \in [1, +\infty)$, there is a constant $C_s > 0$ such that

$$\|z\|_{L^s} \leq C_s \|z\|, \quad \forall z \in \mathcal{W}_{L_0}.$$

Define a functional f_K on E by

$$f_K(z) = \int_0^1 (\frac{1}{2}\dot{z} \cdot Jz - \hat{H}_K(z))\, dt, \quad \forall z \in E. \tag{9.34}$$

For $m \in \mathbb{N}$, define $E^0 = L_0$,

$$E_m = \{z \in E \mid z(t) = \sum_{k=-m}^{m} \exp(k\pi t J)a_k,\ a_k \in L_0\},$$

$$E^{\pm} = \{z \in E \mid z(t) = \sum_{\pm k > 0} \exp(k\pi t J)a_k,\ a_k \in L_0\},$$

and $E_m^+ = E_m \cap E^+$, $E_m^- = E_m \cap E^-$. We have $E_m = E_m^- \oplus E^0 \oplus E_m^+$. Let P_m be the projection $P_m : E \to E_m$. Then $\{E_m, P_m\}_{m\in\mathbb{N}}$ form a Galerkin approximation scheme of the operator $-Jd/dt$ on E. Denote by $f_{K,m} = f_K|_{E_m}$. Set $Q_m = \{re : 0 \le r \le r_1\} \oplus \{B_{r_1}(0) \cap (E_m^- \oplus E^0)\}$ with some $e \in \partial B_1(0) \cap E_m^+$. Then for large $r_1 > 0$ and small $\rho > 0$, ∂Q_m and $B_\rho(0) \cap E_m^+$ form a topological (in fact homologically) link (cf. P84 of [34]). By the condition $\|B\| < \pi$, we obtain a constant $\beta = \beta(K) > 0$ such that

(I) $f_{K,m}(z) \ge \beta > 0, \quad \forall z \in \partial B_\rho(0) \cap E_m^+$,
(II) $f_{K,m}(z) \le 0, \quad \forall z \in \partial Q_m$.

In fact, by (H3), for any $\varepsilon > 0$, there is a $\delta > 0$ such that $H_K(z) \le \varepsilon|z|^2$ if $|z| \le \delta$. Since $\hat{H}_K(z)|z|^{-4}$ is uniformly bounded as $|z| \to +\infty$, there is an $M_1 = M_1(K)$ such that $\hat{H}_K(z) \le M_1|z|^4$ for $|z| \ge \delta$. Hence

$$\hat{H}_K(z) \le \varepsilon|z|^2 + M_1|z|^4, \quad \forall z \in \mathbb{R}^{2n}.$$

For $z \in \partial B_\rho(0) \cap E_m^+$, we have

$$\int_0^1 H_K(t,z)dt \le \varepsilon\|z\|_{L^2}^2 + M_1\|z\|_{L^4}^4 \le (\varepsilon C_2^2 + M_1 C_4^4\|z\|^2)\|z\|^2.$$

So we have

$$\begin{aligned} f_{K,m}(z) &= \frac{1}{2}\langle Az, z\rangle - \frac{1}{2}\langle Bz, z\rangle - \int_0^1 H_K(z(t))dt \\ &\ge \frac{\pi}{2}\|z\|^2 - \frac{\|B\|}{2}\|z\|^2 - (\varepsilon C_2^2 + M_1 C_4^4\|z\|^2)\|z\|^2 \\ &= \frac{\pi}{2}\rho^2 - \frac{\|B\|}{2}\rho^2 - (\varepsilon C_2^2 + M_1 C_4^4\rho^2)\rho^2. \end{aligned}$$

Since $\|B\| < \pi$, we can choose constants $\rho = \rho(K) > 0$ and $\beta = \beta(K) > 0$, which are sufficiently small and independent of m, such that for $z \in \partial B_\rho(0) \cap E_m^+$,

$$f_{K,m}(z) \geq \beta > 0.$$

Hence (I) holds.

Let $e \in E_m^+ \cap \partial B_1$ and $z = z^- + z^0 \in E_m^- \oplus E^0$. We have

$$\begin{aligned} f_{K,m}(z+re) &= \frac{1}{2}\langle Az^-, z^-\rangle + \frac{1}{2}r^2\langle Ae, e\rangle \\ &\quad - \frac{1}{2}\langle B(z+re), z+re\rangle - \int_0^1 \hat{H}_K(z+re)dt \\ &\leq -\frac{\pi}{2}\|z^-\|^2 + \frac{\pi}{2}r^2 - \int_0^1 \hat{H}_K(z+re)dt, \end{aligned}$$

If $r = 0$, from condition (H4), there holds

$$f_{K,m}(z+re) \leq -\frac{\pi}{2}\|z^-\|^2 \leq 0.$$

If $r = r_1$ or $\|z\| = r_1$, then from (H2), We have

$$H_K(z) \geq b_1|z|^\mu - b_2,$$

where $b_1 > 0,\ b_2$ are two constants independent of K and m. Then there holds

$$\begin{aligned} \int_0^1 \hat{H}_K(z+re)dt &\geq b_1 \int_0^1 |z+re|^\mu dt - b_2 \\ &\geq b_3 \left(\int_0^1 |z+re|^2 dt\right)^{\frac{\mu}{2}} - b_4 \\ &\geq b_5 \left(\|z^0\|^\mu + r^\mu\right) - b_4, \end{aligned}$$

where $b_3,\ b_4$ are constants and $b_5 > 0$ independent of K and m. Thus there holds

$$f_{K,m}(z+re) \leq -\frac{\pi}{2}\|z^-\|^2 + \frac{\pi}{2}r^2 - b_5\left(\|z^0\|^\mu + r^\mu\right) + b_4,$$

So we can choose large enough r_1 independent of K and m such that

$$\varphi_m(z+re) \leq 0, \quad on\ \partial Q_m.$$

Then (II) holds.

Now define $\Omega = \{\Phi \in C(Q_m, E_m) \mid \Phi(x) = x \ \text{ for } x \in \partial Q_m\}$, and set

$$c_{K,m} = \inf_{\Phi \in \Omega} \sup_{x \in \Phi(Q_m)} f_{K,m}(x).$$

It is well known that f_K satisfies the usual (P.S)* condition on E, i.e. a sequence $\{x_m\}$ with $x_m \in E_m$ possesses a convergent subsequence in E, provided $f'_{K,m}(x_m) \to 0$ as $m \to \infty$ and $|f_{K,m}(x_m)| \leq b$ for some $b > 0$ and all $m \in \mathbb{N}$ (see [170] for a proof). Thus by the saddle point theorem (cf. [255], or Theorem VIII.3.2), we see that $c_{K,m} \geq \beta > 0$ is a critical value of $f_{K,m}$, we denote the corresponding critical point by $x_{K,m}$. The Morse index of $x_{K,m}$ satisfies

$$m^-(x_{K,m}) \leq \dim Q_m = mn + n + 1.$$

By taking $m \to +\infty$, we obtain a critical point x_K such that $x_{K,m} \to x_K$, $m \to +\infty$ and $m_d^-(x_K) \leq \dim Q_m = 1 + n + mn$, $0 < c_K \equiv f_K(x_K) \leq M_1$, where the d-Morse index $m_d^-(x_K)$ is defined to the total number of the eigenvalues of f''_K belonging to $(-\infty, -d]$ for $d > 0$ small enough, and M_1 is a constant independent of K. Moreover, by the Galerkin approximation method, Theorem V.6.1, we have the d-Morse index satisfying

$$m_d^-(x_K) = mn + n + i_{L_0}(x_K, 1) \leq 1 + n + mn.$$

Thus we have

$$i_{L_0}(x_K, 1) \leq 1.$$

Now the similar arguments as in the section 6 of [255] yields a constant M_2 independent of K such that $\|x_K\|_\infty \leq M_2$. Choose $K > M_2$. Then $x \equiv x_K$ is a non-constant solution of the problem (9.32) satisfying (9.33). By extending the domain with mirror symmetry of L_0, we obtain a 2-periodic brake orbit $(2, x)$ of problem (9.31).

Step 2. Estimate the brake period of $(2, x)$.

Denote the minimal period of the brake orbit x by $2/k$ for some $k \in \mathbb{N}$, i.e., $(x, 1/k)$ is a solution of the problem (9.32). By the condition (HX) and B being semi-positive definite, using (9.17) of [66], we have that $i_1(x, 2/k) \geq n$ for every $2/k$-periodic solution $(x, 2/k)$, and by (5.110) we see that $i_{L_0}(x, 1/k) \geq 0$ for the L_0-solution $(x, 1/k)$. Together with (5.181), we obtain

$$i^{L_0}_{\sqrt{-1}}(x, 1/k) \geq i_{L_0}(x, 1/k) \geq 0, \quad i_1(x, 2/k) \geq n. \tag{9.35}$$

Since the system (9.31) is autonomous, we have

$$\nu_1(x, 2/k) \geq 1. \tag{9.36}$$

Therefore, by Theorem V.10.3, (9.33) and (9.35)–(9.36), we obtain $k = 1, 2, 3, 4$.

If $k=3$, by (9.33) and (9.35)–(9.36), and by using Theorem V.10.3 again we find the left equality of (5.218) holds for $k=3$ and $i_{L_0}(x,1/3)=0$, $i_1(x,2/3)=n$, and $\nu_1(x,2/3)=1$.

The left side hand equality in the inequality (5.218) holds if and only if $I_{2p}\diamond N_1(1,-1)^{\diamond q}\diamond K\in\Omega_0(\gamma(2/3))$ for some non-negative integers p and q satisfying $p+q\le n$ and some $K\in \mathrm{Sp}(2(n-p-q))$ satisfying $\sigma(K)\subset \mathbf{U}\setminus\mathbb{R}$. If $r=n-p-q>0$, then by List in Lemma II. 3.14, we have $R(\theta_1)\diamond\cdots\diamond R(\theta_r)\in\Omega_0(K)$ for some $\theta_j\in(0,\pi)$. In this case, all eigenvalues of K on $\mathbf{U}^+$ (on $\mathbf{U}^-$) are located on the arc between 1 and $\exp(2\pi\sqrt{-1}/k)$ (and $\exp(-2\pi\sqrt{-1}/k)$) on $\mathbf{U}^+$ (in $\mathbf{U}^-$) and are all Krein negative (positive) definite. We remind that $\gamma(t)$ is the fundamental solution of the linearized system at $(2/3,x)$. By the condition $\nu_1(x,2/3)=1$, we have $p=0$, $q=1$. By Lemma II.2.7, there are paths $\alpha\in\mathcal{P}_{2/3}(2)$, $\beta\in\mathcal{P}_{2/3}(2n-2)$ such that $\gamma\sim\alpha\diamond\beta$, $\alpha(2/3)=N_1(1,-1)$, $\beta(\tau)=K$. By the locations of the end point matrix $\alpha(2/3)$ and $\beta(2/3)$, there are two integers $k_1,\ k_2$ such that (see the proof of Theorem 4.3 in [197], specially (4.18) and (4.19) there).

$$i_1(\alpha,2/3)=2k_1,\ i_1(\beta,2/3)=2k_2+n-1.$$

From this result, we see that if $n=1$, then $N_1(1,-1)\in\Omega_0(\gamma(2/3))$, and $i_1(x,2/3)$ must be even, so $i_1(x,2/3)=n=1$ is impossible. If $n>1$, we have $n-1>0$ and

$$i_1(x,2/3)=2(k_1+k_2)+n-1.$$

But $i_1(x,2/3)=n$, so $k_1+k_2=\frac{1}{2}$. It is also impossible.

If $k=4$, the solution $(1/2,x)$ itself is a brake orbit. Thus $i_1(x,1/2)$ and $i_1(x,1)$ are well defined and by Theorem V.10.3, we have that the left hand side equality in (5.219) holds for $k=4$ and

$$i_1(x,1/2)=n,\ \nu_1(x,1/2)=1, i_{L_0}(x,1/4)=i^{L_0}_{\sqrt{-1}}(x,1/4)=0.$$

By the same arguments as above, we still get $i_1(x,1/2)=2(k_1+k_2)+n-1$. This is also impossible. □

Remark 2.2 If $B=0$, the results of Theorem 2.1 hold for every $\tau>0$. The following condition is more accessible than (HX) but it implies the condition (HX).

(H6) $H''(x)\ge 0$ for all $x\in\mathbb{R}^{2n}$, the set $D=\{x\in\mathbb{R}^{2n}|H'(x)\ne 0,\ 0\in\sigma(H''(x))\}$ is hereditarily disconnected, i.e. every connected component of D contains only one point.

For the brake orbit (τ,x) in (9.33), writing $H''(x(t))=\begin{pmatrix}U_{11}(t) & U_{12}(t)\\ U_{21}(t) & U_{22}(t)\end{pmatrix}$, it was proved in [307] that the minimal brake period of x is τ or $\frac{\tau}{2}$ provided $U_{22}(t)>0,\ t\in[0,\tau]$.

Similarly, we consider the brake orbit minimal periodic problem for the following autonomous second order Hamiltonian system

$$\begin{cases} \ddot{x} + V'(x) = 0, & x \in \mathbb{N}^n, \\ x(0) = x(\tau/2) = 0 \\ x(\tau/2 + t) = -x(\tau/2 - t), & x(\tau + t) = x(t). \end{cases} \tag{9.37}$$

A solution (τ, x) of (9.37) is a kind of brake orbit for the second order Hamiltonian system.

In this paper, we consider the following conditions on V:

(V1) $V \in C^2(\mathbb{R}^n, \mathbb{R})$.

(V2) There exist constants $\mu > 2$ and $r_0 > 0$ such that

$$0 < \mu V(x) \leq V'(x) \cdot x, \qquad \forall |x| \geq r_0.$$

(V3) $V(x) \geq V(0) = 0 \ \forall x \in \mathbb{R}^n$.

(V4) $V(x) = o(|x|^2)$, at $x = 0$.

(V5) $V(-x) = V(x)$, $\forall x \in \mathbb{R}^n$.

(V6) $V''(x) > 0$, $\forall x \in \mathbb{R}$.

Theorem 2.3 *Suppose V satisfies the conditions (V1)–(V6). Then for every $\tau > 0$, the problem* (9.37) *possesses a non-constant solution (τ, x) such that the minimal period of x is τ or $\tau/2$.*

Proof Without loss generality, we suppose $\tau = 2$. We define a Hilbert space W which is a subspace of $W^{1,2}([0, 1], \mathbb{R}^n)$ by

$$W = \{x \in W^{1,2}([0, 1], \mathbb{R}^n) \,|\, x(t) = \sum_{k=1}^{\infty} \sin k\pi t \cdot a_k, \ a_k \in \mathbb{R}^n\}.$$

The inner product of W is still the $W^{1,2}$ inner product.

We consider the following functional

$$\psi(x) = \int_0^1 (\frac{1}{2}|\dot{x}|^2 - V(x))\, dt, \ \ \forall x \in W. \tag{9.38}$$

A critical point x of ψ is a solution of the problem (9.37) by extending the domain to $\mathbb{R}$ via $x(1+t) = -x(1-t)$ and $x(2+t) = x(t)$. The condition (V3) implies $\psi(0) = 0$. The condition (V4) implies $\psi(\partial B_\rho(0)) \geq \alpha_0$ with $\partial B_\rho(0) = \{x \in W \mid \|x\| = \rho\}$ for some small $\rho > 0$ and $\alpha_0 > 0$. In fact, there exists a constant $c_1 > 0$ such that

$$\int_0^1 |\dot{x}|^2\, dt \geq c_1 \|x\|_W^2. \tag{9.39}$$

If $\|x\|_W \to 0$, then $\|x\|_\infty \to 0$. So by condition (V4), for any $0 < \varepsilon < \frac{c_1}{2}$, there exists small $\rho > 0$ such that

$$\int_0^1 V(x(t))dt \le \varepsilon\|x\|_2^2 \le \varepsilon\|x\|_W^2, \quad \|x\|_W = \rho.$$

Thus we have

$$\psi(x) = \int_0^1 (\frac{1}{2}|\dot{x}|^2 - V(x))\,dt \ge (\frac{c_1}{2} - \varepsilon)\rho^2 := \alpha_0 > 0.$$

The condition (V2) implies that there exists an element $x_0 \in W$ with $\|x_0\| > \rho$, such that $\psi(x_0) < 0$. In fact, we take an element $e \in W$ with $\|e\| = 1$ and by (V3) we assume $\int_0^1 V(e(t))dt > 0$. Consider $x = \lambda e$ for $\lambda > 0$. Condition (V2) implies that there is a constant $c_2 > 0$ such that $V(\lambda e) \ge \lambda^\mu V(e) - c_2$ for λ large enough, and there holds

$$\psi(\lambda e) \le \lambda^2 \int_0^1 \frac{1}{2}|\dot{e}|^2 dt - \lambda^\mu \int_0^1 V(e(t))dt + c_2 < 0.$$

Then we take $x_0 = \lambda e$ for large λ such that the above inequalities holds.

We define

$$\Gamma = \{h \in C([0,1], W) \mid h(0) = 0, \quad h(1) = x_0\}$$

and

$$c = \inf_{h\in\Gamma} \sup_{s\in[0,1]} \psi(h(s)).$$

By using the Mountain pass theorem (cf. Theorem 2.2 of [255]), from the conditions (V2)–(V4) it is well known that there exists a critical point $x \in W$ of ψ with critical value $c > 0$ which is a Mountain pass point such that its Morse index satisfying $m^-(x,1) \le 1$. If we set $y = \dot{x}$ and $z = (x, y) \in \mathbb{R}^{2n}$, the problem (9.37) can be transformed into the following problem

$$\begin{cases} \dot{z} = -JH'(z), \\ z(0) \in L_0, \quad z(1) \in L_0 \end{cases}$$

with $H(z) = H(x,y) = \frac{1}{2}|y|^2 + V(x)$. We note that (V5) implies $H(Nz) = H(z)$, so $(2, z)$ is a brake orbit with brake period 2. We remind that in this case the complex structure is $-J$, but it does not cause any difficult to apply the index theory. By Theorem V.4.1, the Morse index $m^-(x,1)$ of x is just the L_0-index $i_{L_0}(z,1)$ of $(1, z)$. I.e., there holds

$$m^-(x,1) = i_{L_0}(z,1), \quad m^0(x,1) = \nu_{L_0}(z,1).$$

We can suppose the minimal period of x is $2/k$ for $k \in \mathbb{N}$. But $i_{L_0}(z, 1/k) = m^-(x, 1/k) \geq 0$, and from the convexity condition (V6), we have $i_1(z, 2/k) \geq n$. With the same arguments as in the proof of Theorem 2.1, we get $k \in \{1, 2\}$. □

We note that the functional ψ is even, there may be infinite many solutions (τ, x) satisfying Theorem 2.3. We also note that Theorem 2.3 is not a special case of Theorem 2.1, since the Hamiltonian function $H(x, y) = \frac{1}{2}|y|^2 + V(x)$ is quadratic in the variables y, in this case $B = \begin{pmatrix} 0 & 0 \\ 0 & I_n \end{pmatrix}$ with $\|B\| = 1$. Thus when applying Theorem 2.1 to this case, we can only get the result of Theorem 2.3 for $0 < \tau < 2\pi$.

We now consider the following problem

$$\begin{cases} \ddot{x} + V'(x) = 0, & x \in \mathbb{N}^n, \\ \dot{x}(0) = \dot{x}(\tau/2) = 0, \\ x(\tau/2 + t) = x(\tau/2 - t), & x(\tau + t) = x(t). \end{cases} \tag{9.40}$$

A solution of (9.40) is also a kind of brake orbit for the second order Hamiltonian system.

By set $y = \dot{x}$, $z = (y, x)$ and $H(z) = H(y, x) = \frac{1}{2}|y|^2 + V(x)$, the problem (9.40) can be transformed into the following L_0-boundary value problem

$$\begin{cases} \dot{z} = JH'(z) \\ z(0) \in L_0, \quad z(\tau/2) \in L_0. \end{cases}$$

In this case the condition $H(Nz) = H(z)$ is satisfied automatically. Set $B = \begin{pmatrix} I_n & 0 \\ 0 & 0 \end{pmatrix}$, then $\|B\| = 1$. The following result is a direct consequence of Theorem 2.1.

Corollary 2.4 *Suppose V satisfies the conditions (V1)–(V4) and (V6). Then for every $0 < \tau < 2\pi$, the problem* (9.40) *possesses a non-constant solution (τ, x) such that x has minimal period τ or $\tau/2$.*

We note that if we directly solve the problem (9.40) by the same way as in the proof of Theorem 2.3, the formation of the functional is still ψ as defined in (9.38), but the domain should be

$$W_1 = \{x \in W^{1,2}([0, 1], \mathbb{R}^n) | \, x(t) = \sum_{k=0}^{\infty} \cos k\pi t \cdot a_k, \; a_k \in \mathbb{R}^n\}.$$

In this time, it is not able to apply the Mountain pass theorem to get a critical point directly due to the fact $\mathbb{R}^n \subset W_1$, so the inequality (9.39) is not true.

9.3 Brake Subharmonic Solutions of First Order Hamiltonian Systems

We consider the first order non-autonomous Hamiltonian systems

$$\dot{z}(t) = J\nabla H(t, z(t)), \ \forall z \in \mathbb{R}^{2n}, \ \forall t \in \mathbb{R}, \tag{9.41}$$

where $H \in C^2(\mathbb{R} \times \mathbb{R}^{2n}, \mathbb{R})$.

Suppose that $H(t, z) = \frac{1}{2}(\hat{B}(t)z, z) + \hat{H}(t, z)$ and $H \in C^2(\mathbb{R} \times \mathbb{R}^{2n}, \mathbb{R})$ satisfies the following conditions:

(H1) $\hat{H}(T + t, z) = \hat{H}(t, z)$, for all $z \in \mathbb{R}^{2n}$, $t \in \mathbb{R}$,

(H2) $\hat{H}(t, z) = \hat{H}(-t, Nz)$, for all $z \in \mathbb{R}^{2n}$, $t \in \mathbb{R}$, $N = \begin{pmatrix} -I_n & 0 \\ 0 & I_n \end{pmatrix}$,

(H3) $\hat{H}''(t, z) > 0$, for all $z \in \mathbb{R}^{2n} \backslash \{0\}$, $t \in \mathbb{R}$,

(H4) $\hat{H}(t, z) \geq 0$, for all $z \in \mathbb{R}^{2n}$, $t \in \mathbb{R}$,

(H5) $\hat{H}(t, z) = o(|z|^2)$ at $z = 0$,

(H6) There is a $\theta \in (0, 1/2)$ and $\bar{r} > 0$ such that

$$0 < \frac{1}{\theta}\hat{H}(t, z) \leq (z, \nabla\hat{H}(t, z)), \ \text{for all } z \in \mathbb{R}^{2n}, |z| \geq \bar{r}, t \in \mathbb{R},$$

(H7) $\hat{B}(t)$ is a symmetrical continuous matrix, $|\hat{B}|_{C^0} \leq \beta_0$ for some $\beta_0 > 0$, and $\hat{B}(t)$ is a semi-positively definite for all $t \in \mathbb{R}$,

(H8) $\hat{B}(T + t) = \hat{B}(t) = \hat{B}(-t)$, $\hat{B}(t)N = N\hat{B}(t)$, for all $t \in \mathbb{R}$.

Recall that a T-periodic solution (z, T) of (9.41) is called *brake solution* if $z(t + T) = z(t)$ and $z(t) = Nz(-t)$, the later is equivalent to $z(T/2+t) = Nz(T/2-t)$, in this time T is called the *brake period* of z.

Theorem 3.1 ([172]) *Suppose that $H \in C^2(\mathbb{R} \times \mathbb{R}^{2n}, \mathbb{R})$ satisfies (H1)–(H8), then for each integer $1 \leq j < 2\pi/\beta_0 T$, there is a jT-periodic nonconstant brake solution z_j of* (9.41) *such that z_j and z_{kj} are distinct for $k \geq 5$ and $kj < 2\pi/\beta_0 T$. Furthermore, $\{z_{k^p} | p \in \mathbb{N}\}$ is a pairwise distinct brake solution sequence of* (9.41) *for $k \geq 5$ and $1 \leq k^p < 2\pi/\beta_0 T$.*

Especially, if $\hat{B}(t) \equiv 0$, then $2\pi/\beta_0 T = +\infty$. Therefore, one can state the following theorem.

Theorem 3.2 ([172]) *Suppose that $H \in C^2(\mathbb{R} \times \mathbb{R}^{2n}, \mathbb{R})$ with $\hat{B}(t) \equiv 0$ satisfies (H1)–(H6), then for each integer $j \geq 1$, there is a jT-periodic nonconstant brake solution z_j of* (9.41). *Furthermore, given any integers $j \geq 1$ and $k \geq 5$, z_j and z_{kj} are distinct brake solutions of* (9.41), *in particularly, $\{z_{k^p} | p \in \mathbb{N}\}$ is a pairwise distinct brake solution sequence of* (9.41).

The first result on subharmonic periodic solutions for the Hamiltonian systems $\dot{z}(t) = J\nabla H(t, z(t))$, where $z \in \mathbb{R}^{2n}$ and $H(t, z)$ is T-periodic in t, was obtained by P. Rabinowitz in his pioneer work [258]. Since then, many mathematician made their contributions. See for example [78, 81, 184, 223, 269] and the references therein. Especially, in [81], I. Ekeland and H. Hofer proved that under a strict convex condition and a superquadratic condition, the Hamiltonian system $\dot{z}(t) = J\nabla H(t, z(t))$ possesses subharmonic solution z_k for each integer $k \geq 1$ and all of these solutions are pairwise geometrically distinct. In [184], C. Liu obtained a result of subharmonic solutions for the non-convex case by using the Maslov-type index iteration theory.

We first consider the following Hamiltonian systems

$$\begin{cases} \dot{z}(t) = J\nabla H(t, z(t)), \ \forall z \in \mathbb{R}^{2n}, \ \forall t \in [0, jT/2], \\ z(0) \in L_0, \ z(jT/2) \in L_0, \end{cases} \tag{9.42}$$

where $j \in \mathbb{N}$.

In order to prove Theorem 3.1, as in Theorem 2.1 and Theorem VIII.3.1, by using saddle point theorem (Theorem VIII.3.2), we have the following result. The proof is omitted.

Lemma 3.3 *Suppose $H(t, z) \in C^2(\mathbb{R} \times \mathbb{R}^{2n}, \mathbb{R})$ satisfies (H4)–(H7), then for $1 \leq j < 2\pi/\beta_0 T$, (9.42) possesses at least one nontrivial solution z_j whose L_0-index pair $(i_{L_0}(z_j), \nu_{L_0}(z_j))$ satisfies*

$$i_{L_0}(z_j) \leq 1 \leq i_{L_0}(z_j) + \nu_{L_0}(z_j).$$

For a solution (z_k, k) of the problem (9.42), we define $(\tilde{z}_k, 2k)$ as

$$\tilde{z}(t) = \begin{cases} z(t), & t \in [0, \frac{kT}{2}], \\ Nz(T - t), & t \in (\frac{kT}{2}, kT]. \end{cases}$$

It is clear that $(\tilde{z}_k, 2k)$ is a brake solution of (9.41) with period kT.

We are ready to give a proof of Theorem 3.1.

Proof of Theorem 3.1. For $1 \leq k < \pi/\beta_0$, by Lemma 3.3, we obtain that there is a nontrivial solution (z_k, k) of the problem (9.42) and its L_0-index pair satisfies

$$i_{L_0}(z_k, k) \leq 1 \leq i_{L_0}(z_k, k) + \nu_{L_0}(z_k, k). \tag{9.43}$$

Then $(\tilde{z}_k, 2k)$ is a nonconstant brake solution of (9.41).

For $k \in 2\mathbb{N} - 1$, we suppose that $(\tilde{z}_1, 2)$ and $(\tilde{z}_k, 2k)$ are not distinct. By (9.43), Proposition V.9.8 and Theorem V.10.3, we have

$$
\begin{aligned}
1 &\geq i_{L_0}(z_k, k) \geq i_{L_0}(z_1, 1) + \frac{k-1}{2}\,(i_1(\tilde{z}_1, 2) + \nu_1(\tilde{z}_1, 2) - n) \\
&\geq i_{L_0}(z_1, 1) + \frac{k-1}{2}\left(i_{L_0}(z_1, 1) + i_{L_1}(z_1, 1) + n + \nu_{L_0}(z_1, 1) + \nu_{L_1}(z_1, 1) - n\right) \\
&= i_{L_0}(z_1, 1) + \frac{k-1}{2}\left(i_{L_0}(z_1, 1) + i_{L_1}(z_1, 1) + \nu_{L_0}(z_1, 1) + \nu_{L_1}(z_1, 1)\right), \qquad (9.44)
\end{aligned}
$$

where $L_1 = \mathbb{R}^n \oplus \{0\} \in \Lambda(n)$. By (H3), (H7) and Corollary V.4.5, we have $i_{L_1}(z_1, 1) \geq 0$. We also know that $\nu_{L_1}(z_1, 1) \geq 0$ and $i_{L_0}(z_1, 1) + \nu_{L_0}(z_1, 1) \geq 1$. Then from (9.44) we deduce that

$$
1 \geq i_{L_0}(z_1, 1) + \frac{k-1}{2}. \qquad (9.45)
$$

By $0 \leq i_{L_0}(z_1, 1) \leq 1$, from (9.45) we have $\frac{k-1}{2} \leq 1$, i.e., $k \leq 3$. Since we have the condition $k \geq 5$, $(\tilde{z}_1, 2)$ and $(\tilde{z}_k, 2k)$ must be distinct. Similarly, we have that for each $k \in 2\mathbb{N}-1, k \geq 5$ and $kj < \frac{\pi}{\beta_0}, 1 \leq j < \frac{\pi}{\beta_0}$, $(\tilde{z}_j, 2j)$ and $(\tilde{z}_{kj}, 2kj)$ are distinct brake solutions of (9.41). Furthermore, $(\tilde{z}_1, 2)$, $(\tilde{z}_k, 2k)$, $(\tilde{z}_{k^2}, 2k^2)$, $(\tilde{z}_{k^3}, 2k^3)$, $\cdots$, $(\tilde{z}_{k^p}, 2k^p)$ are pairwise distinct brake solutions of (9.41), where $k \in 2\mathbb{N} - 1, k \geq 5$ and $1 \leq k^p < \frac{\pi}{\beta_0}$ with $p \in \mathbb{N}$.

For $k \in 2\mathbb{N}$, as above, we suppose that $(\tilde{z}_1, 2)$ and $(\tilde{z}_k, 2k)$ are not distinct. By (9.43), Proposition V.9.8 and Theorem V.10.3, we have

$$
\begin{aligned}
1 &\geq i_{L_0}(z_k, k) \geq i_{L_0}(z_1, 1) + i^{L_0}_{\sqrt{-1}}(z_1, 1) + \left(\frac{k}{2} - 1\right)(i_1(\tilde{z}_1, 2) + \nu_1(\tilde{z}_1, 2) - n) \\
&\geq i_{L_0}(z_1, 1) + i^{L_0}_{\sqrt{-1}}(z_1, 1) \\
&\quad + \left(\frac{k}{2} - 1\right)\left(i_{L_0}(z_1, 1) + i_{L_1}(z_1, 1) + n + \nu_{L_0}(z_1, 1) + \nu_{L_1}(z_1, 1) - n\right) \\
&= i_{L_0}(z_1, 1) + i^{L_0}_{\sqrt{-1}}(z_1, 1) \\
&\quad + \left(\frac{k}{2} - 1\right)\left(i_{L_0}(z_1, 1) + i_{L_1}(z_1, 1) + \nu_{L_0}(z_1, 1) + \nu_{L_1}(z_1, 1)\right). \qquad (9.46)
\end{aligned}
$$

Similarly, we also know that $i_{L_1}(z_1, 1) \geq 0$, $\nu_{L_1}(z_1, 1) \geq 0$, $i_{L_0}(z_1, 1) + \nu_{L_0}(z_1, 1) \geq 1$. By Lemma V.8.2 and (5.181), we have $i^{L_0}_{\sqrt{-1}}(z_1, 1) \geq i_{L_0}(z_1, 1) \geq 0$. Then from (9.46) we have

$$
1 \geq i_{L_0}(z_1, 1) + \left(\frac{k}{2} - 1\right). \qquad (9.47)
$$

By $0 \leq i_{L_0}(z_1, 1) \leq 1$, from (9.47) we have $\frac{k}{2} - 1 \leq 1$, i.e., $k \leq 4$. It contradicts to the condition $k \geq 5$. Similarly we have that for each $k \in 2\mathbb{N}$, $k \geq 6$ and $kj < \frac{\pi}{\beta_0}$, $1 \leq j < \frac{\pi}{\beta_0}$, $(\tilde{z}_j, 2j)$ and $(\tilde{z}_{kj}, 2kj)$ are distinct brake solutions of (9.41). Furthermore, $(\tilde{z}_1, 2)$, $(\tilde{z}_k, 2k)$, $(\tilde{z}_{k^2}, 2k^2)$, $(\tilde{z}_{k^3}, 2k^3)$, $\cdots$, $(\tilde{z}_{k^p}, 2k^p)$ are pairwise distinct brake solutions of (9.41), where $k \in 2\mathbb{N}$, $k \geq 6$ and $1 \leq k^p < \frac{\pi}{\beta_0}$ with $p \in \mathbb{N}$.

In all, for any integer $1 \leq j < \frac{\pi}{\beta_0}$, $\tilde{z}_j$ and $\tilde{z}_{kj}$ are distinct brake solutions of (9.41) for $k \geq 5$ and $kj < \frac{\pi}{\beta_0}$. Furthermore, $\{\tilde{z}_{k^p} | p \in \mathbb{N}\}$ is a pairwise distinct brake solution sequence of (9.41) for $k \geq 5$ and $1 \leq k^p < \frac{\pi}{\beta_0}$. The proof of Theorem 3.1 is complete. □

Chapter 10
Multiplicity of Brake Orbits on a Fixed Energy Surface

10.1 Brake Orbits of Nonlinear Hamiltonian Systems

For the standard symplectic space $(\mathbf{R}^{2n}, \omega_0)$ with $\omega_0(x, y) = \langle Jx, y\rangle$, where $J = \begin{pmatrix} 0 & -I \\ I & 0 \end{pmatrix}$ is the standard symplectic matrix and I is the $n \times n$ identity matrix, an involution matrix defined by $N = \begin{pmatrix} -I & 0 \\ 0 & I \end{pmatrix}$ is clearly anti-symplectic, i.e., $NJ = -JN$. The fixed point set of N and $-N$ are the Lagrangian subspaces $L_0 = \{0\}\times\mathbf{R}^n$ and $L_1 = \mathbf{R}^n \times \{0\}$ of $(\mathbf{R}^{2n}, \omega_0)$ respectively.

Suppose $H \in C^2(\mathbb{R}^{2n} \setminus \{0\}, \mathbb{R}) \cap C^1(\mathbb{R}^{2n}, \mathbb{R})$ satisfies the following reversible condition

$$H(Nx) = H(x), \qquad \forall x \in \mathbb{R}^{2n}. \tag{10.1}$$

We consider the following fixed energy problem of nonlinear Hamiltonian system with Lagrangian boundary conditions

$$\dot{x}(t) = JH'(x(t)), \tag{10.2}$$

$$H(x(t)) = h, \tag{10.3}$$

$$x(0) \in L_0, \; x(\tau/2) \in L_0. \tag{10.4}$$

Here we require that $h \in \mathbf{R}$ is a regular value of H. It is clear that a solution (τ, x) of (10.2), (10.3), and (10.4) is a characteristic chord on the contact submanifold $\Sigma := H^{-1}(h) = \{y \in \mathbb{R}^{2n} \mid H(y) = h\}$ of $(\mathbf{R}^{2n}, \omega_0)$ and it can be extended to the whole real number set satisfying

$$x(-t) = Nx(t), \tag{10.5}$$

$$x(\tau + t) = x(t). \tag{10.6}$$

In this paper this kind of τ-periodic characteristic (τ, x) is called a *brake orbit* on the hypersurface Σ. We denote by $\mathcal{J}_b(\Sigma, H)$ the set of all brake orbits on Σ. Two brake orbits $(\tau_i, x_i) \in \mathcal{J}_b(\Sigma, H)$, $i = 1, 2$ are equivalent if the two brake orbits are geometrically the same, i.e., $x_1(\mathbb{R}) = x_2(\mathbb{R})$. We denote by $[(\tau, x)]$ the equivalence class of $(\tau, x) \in \mathcal{J}_b(\Sigma, H)$ in this equivalence relation and by $\tilde{\mathcal{J}}_b(\Sigma, H)$ the set of $[(\tau, x)]$ for all $(\tau, x) \in \mathcal{J}_b(\Sigma, H)$. In fact $\tilde{\mathcal{J}}_b(\Sigma, H)$ is the set of geometrically distinct brake orbits on Σ, which is independent on the choice of H. So from now on we simply denote it by $\tilde{\mathcal{J}}_b(\Sigma)$ and in the notation $[(\tau, x)]$ we always assume x has minimal period τ. We also denote by $\tilde{\mathcal{J}}(\Sigma)$ the set of all geometrically distinct closed characteristics on Σ. The number of elements in a set S is denoted by ${}^{\#}S$. It is well known that ${}^{\#}\tilde{\mathcal{J}}_b(\Sigma)$ (and also ${}^{\#}\tilde{\mathcal{J}}(\Sigma)$) is only depending on Σ, that is to say, for simplicity we take $h = 1$, if H and G are two C^2 functions satisfying (10.1) and $\Sigma_H := H^{-1}(1) = \Sigma_G := G^{-1}(1)$, then ${}^{\#}\mathcal{J}_b(\Sigma_H) ={}^{\#} \mathcal{J}_b(\Sigma_G)$. So we can consider the brake orbit problem in a more general setting. Let Σ be a C^2 compact hypersurface in $\mathbb{R}^{2n}$ bounding a compact set C with nonempty interior. Suppose Σ has non-vanishing Gaussian curvature and satisfies the reversible condition $N(\Sigma - x_0) = \Sigma - x_0 := \{x - x_0 | x \in \Sigma\}$ for some $x_0 \in C$. Without loss of generality, we may assume $x_0 = 0$ the origin. We denote the set of all such hypersurfaces in $\mathbb{R}^{2n}$ by $\mathcal{H}_b(2n)$. For $x \in \Sigma$, let $n_\Sigma(x)$ be the unit outward normal vector at $x \in \Sigma$. Note that here by the reversible condition there holds $n_\Sigma(Nx) = Nn_\Sigma(x)$. We consider the dynamics problem of finding $\tau > 0$ and a C^1 smooth curve $x : [0, \tau] \to \mathbb{R}^{2n}$ such that

$$\dot{x}(t) = Jn_\Sigma(x(t)), \qquad x(t) \in \Sigma, \tag{10.7}$$

$$x(-t) = Nx(t), \qquad x(\tau + t) = x(t), \qquad \text{for all } t \in \mathbb{R}. \tag{10.8}$$

A solution (τ, x) of the problem (10.7) and (10.8) determines a brake orbit on Σ.

Definition 1.1 We denote by

$$\mathcal{H}_b^c(2n) = \{\Sigma \in \mathcal{H}_b(2n) |\ \Sigma \text{ is strictly convex }\},$$
$$\mathcal{H}_b^{s,c}(2n) = \{\Sigma \in \mathcal{H}_b^c(2n) |\ -\Sigma = \Sigma\}.$$

We will prove the following multiplicity result for brake orbits on a hypersurface $\Sigma \in \mathcal{H}_b^{s,c}(2n)$.

Theorem 1.2 ([210]) *For any $\Sigma \in \mathcal{H}_b^{s,c}(2n)$, there holds*

$${}^{\#}\tilde{\mathcal{J}}_b(\Sigma) \geq n.$$

Remark 1.3 Theorem 1.2 is a kind of multiplicity result related to the Arnold chord conjecture. The Arnold chord conjecture is an existence result which was prove by K. Mohnke in [241]. Another kind of multiplicity result related to the Arnold chord conjecture was proved in [128].

10.1.1 Seifert Conjecture

Let us introduce the famous conjecture proposed by H. Seifert in his pioneer work [268] concerning the multiplicity of brake orbits of certain Hamiltonian systems in $\mathbb{R}^{2n}$.

As a special case of (10.1), we assume that $H \in C^2(\mathbb{R}^{2n}, \mathbb{R})$ possesses the following form

$$H(p,q) = \frac{1}{2}A(q)p \cdot p + V(q), \tag{10.9}$$

where $p, q \in \mathbb{R}^n$, $A(q)$ is a positive definite $n \times n$ matrix for any $q \in \mathbb{R}^n$ and A is C^2 on q, $V \in C^2(\mathbb{R}^n, \mathbb{R})$ is the potential energy. It is clear that a solution of the following Hamiltonian system

$$\dot{x} = JH'(x), \quad x = (p,q), \tag{10.10}$$

$$p(0) = p(\frac{\tau}{2}) = 0. \tag{10.11}$$

is a brake orbit. Moreover, if h is the total energy of a brake orbit (q, p), i.e., $H(p(t), q(t)) = h$, then $V(q(0)) = V(q(\tau)) = h$ and $q(t) \in \bar{\Omega} \equiv \{q \in \mathbb{R}^n | V(q) \leq h\}$ for all $t \in \mathbb{R}$.

In 1948, H. Seifert in [268] studied the existence of brake orbit for system (10.10) and (10.11) with the Hamiltonian function H in the form of (10.9) and proved that $\mathcal{J}_b(\Sigma) \neq \emptyset$ provided $V' \neq 0$ on $\partial\Omega$, V is analytic and $\bar{\Omega}$ is bounded and homeomorphic to the unit ball $B_1^n(0)$ in $\mathbb{R}^n$. Then in the same paper he proposed the following conjecture which is still open for $n > 2$ now (that it is true for the case $n = 2$ was proved recently in [114] by R. Giambò, F. Giannoni and P. Piccione):

$$^{\#}\tilde{\mathcal{J}}_b(\Sigma) \geq n \textit{ under the same conditions.}$$

We note that for the Hamiltonian function

$$H(p,q) = \frac{1}{2}|p|^2 + \sum_{j=1}^{n} a_j^2 q_j^2, \qquad q, p \in \mathbb{R}^n,$$

where $a_i/a_j \notin \mathbb{Q}$ for all $i \neq j$ and $q = (q_1, q_2, \ldots, q_n)$. There are exactly n geometrically distinct brake orbits on the energy hypersurface $\Sigma = H^{-1}(h)$.

10.1.2 Some Related Results Since 1948

As a special case, letting $A(q) = I$ in (10.9), the problem corresponds to the following classical fixed energy problem of the second order autonomous Hamiltonian system

$$\ddot{q}(t) + V'(q(t)) = 0, \quad \text{for } q(t) \in \Omega, \tag{10.12}$$

$$\frac{1}{2}|\dot{q}(t)|^2 + V(q(t)) = h, \qquad \forall t \in \mathbb{R}, \tag{10.13}$$

$$\dot{q}(0) = \dot{q}(\frac{\tau}{2}) = 0, \tag{10.14}$$

where $V \in C^2(\mathbb{R}^n, \mathbb{R})$ and h is constant such that $\Omega \equiv \{q \in \mathbb{R}^n | V(q) < h\}$ is nonempty, bounded and connected.

A solution (τ, q) of (10.12), (10.13), and (10.14) is still called a *brake orbit* in $\bar{\Omega}$. Two brake orbits q_1 and $q_2 : \mathbb{R} \to \mathbb{R}^n$ are *geometrically distinct* if $q_1(\mathbb{R}) \neq q_2(\mathbb{R})$. We denote by $\mathcal{O}(\Omega, V)$ and $\tilde{\mathcal{O}}(\Omega)$ the sets of all brake orbits and geometrically distinct brake orbits in $\bar{\Omega}$ respectively.

Remark 1.4 It is well known that via

$$H(p, q) = \frac{1}{2}|p|^2 + V(q),$$

$x = (p, q)$ and $p = \dot{q}$, the elements in $\mathcal{O}(\Omega, V)$ and the solutions of (10.2), (10.3), and (10.4) are one to one correspondent.

Definition 1.5 For $\Sigma \in \mathcal{H}_b^{s,c}(2n)$, a brake orbit (τ, x) on Σ is called symmetric if $x(\mathbb{R}) = -x(\mathbb{R})$. Similarly, for a C^2 convex symmetric bounded domain $\Omega \subset \mathbb{R}^n$, a brake orbit $(\tau, q) \in \mathcal{O}(\Omega, V)$ is called symmetric if $q(\mathbb{R}) = -q(\mathbb{R})$.

Note that a brake orbit $(\tau, x) \in \mathcal{J}_b(\Sigma, H)$ with minimal period τ is symmetric if $x(t + \tau/2) = -x(t)$ for $t \in \mathbb{R}$, a brake orbit $(\tau, q) \in \mathcal{O}(\Omega, V)$ with minimal period τ is symmetric if $q(t + \tau/2) = -q(t)$ for $t \in \mathbb{R}$.

After 1948, many studies have been carried out for the brake orbit problem. In 1978, S. Bolotin proved in [23] the existence of brake orbits in general setting. K. Hayashi in [136], H. Gluck and W. Ziller in [123], and V. Benci in [19] proved $^{\#}\tilde{\mathcal{O}}(\Omega) \geq 1$ if V is C^1, $\bar{\Omega} = \{V \leq h\}$ is compact, and $V'(q) \neq 0$ for all $q \in \partial\Omega$. P. Rabinowitz in [253] proved that if H satisfies (10.1), $\Sigma \equiv H^{-1}(h)$ is star-shaped, and $x \cdot H'(x) \neq 0$ for all $x \in \Sigma$, then $^{\#}\tilde{\mathcal{J}}_b(\Sigma) \geq 1$. V. Benci and F. Giannoni gave a different proof of the existence of one brake orbit in [21]. It has been pointed out in [112] that the problem of finding brake orbits is equivalent to find orthogonal geodesic chords on manifold with concave boundary. R. Giambó, F. Giannoni and P. Piccione in [113] proved the existence of an orthogonal geodesic chord on a Riemannian manifold homeomorphic to a closed disk and with concave boundary. For multiplicity of the brake problems, A. Weinstein in [293] proved a localized result: *Assume H satisfies* (10.1). *For any h sufficiently close to* $H(z_0)$ *with* z_0 *is a nondegenerate local minimum of* H, *there are n geometrically distinct brake orbits on the energy surface* $H^{-1}(h)$. In [24] and in [123], under assumptions of Seifert in [268], it was proved the existence of at least n brake orbits while a very strong assumption on the energy integral was used to ensure that different minimax critical levels correspond to geometrically distinct brake orbits. A. Szulkin in [278] proved that $^{\#}\tilde{\mathcal{J}}_b(H^{-1}(h)) \geq n$, if H satisfies conditions in [253] of Rabinowitz and the energy hypersurface $H^{-1}(h)$ is $\sqrt{2}$-pinched. E. van Groesen in [126] and A. Ambrosetti, V. Benci, Y. Long in [6] also proved $^{\#}\tilde{\mathcal{O}}(\Omega) \geq n$ under different pinching conditions. Without pinching condition, in [232] Y. Long, C. Zhu and D. Zhang proved that: *For any* $\Sigma \in \mathcal{H}_b^{s,c}(2n)$ *with* $n \geq 2$, $^{\#}\tilde{\mathcal{J}}_b(\Sigma) \geq 2$. C. Liu and D. Zhang in [209] proved that $^{\#}\tilde{\mathcal{J}}_b(\Sigma) \geq \left[\frac{n}{2}\right] + 1$ for $\Sigma \in \mathcal{H}_b^{s,c}(2n)$. Moreover it was proved that if all brake orbits on Σ are nondegenerate, then $^{\#}\tilde{\mathcal{J}}_b(\Sigma) \geq n + \mathfrak{A}(\Sigma)$, where $2\mathfrak{A}(\Sigma)$ is the number of geometrically distinct asymmetric brake orbits on Σ. Recently, in [308] the authors improved the results of [209] to that $^{\#}\tilde{\mathcal{J}}_b(\Sigma) \geq \left[\frac{n+1}{2}\right] + 1$ for $\Sigma \in \mathcal{H}_b^{s,c}(2n)$, $n \geq 3$. In [309] the authors proved that $^{\#}\tilde{\mathcal{J}}_b(\Sigma) \geq \left[\frac{n+1}{2}\right] + 2$ for $\Sigma \in \mathcal{H}_b^{s,c}(2n)$, $n \geq 4$.

10.1.3 Some Consequences of Theorem 1.2 and Further Arguments

As direct consequences of Theorem 1.2 we have the following two important Corollaries.

Corollary 1.6 ([210]) *If* $H(p, q)$ *defined by* (10.9) *is even and convex, then Seifert conjecture holds.*

Remark 1.7 If the function H in Remark 1.1 is convex and even, then V is convex and even, and Ω is convex and central symmetric. Hence Ω is homeomorphic to the unit open ball in $\mathbb{R}^n$.

Corollary 1.8 ([210]) *Suppose* $V(0) = 0$, $V(q) \geq 0$, $V(-q) = V(q)$ *and* $V''(q)$ *is positive definite for all* $q \in \mathbb{R}^n \setminus \{0\}$. *Then for any given* $h > 0$ *and* $\Omega \equiv \{q \in \mathbb{R}^n | V(q) < h\}$, *there holds*

$$^{\#}\tilde{\mathcal{O}}(\Omega) \geq n.$$

It is interesting to ask the following question: *whether all closed characteristics on any hypersurfaces* $\Sigma \in \mathcal{H}_b^{s,c}(2n)$ *are symmetric brake orbits after suitable time translation provided that* $^{\#}\tilde{\mathcal{J}}(\Sigma) < +\infty$? In this direction, we have the following result.

Theorem 1.9 ([210]) *For any* $\Sigma \in \mathcal{H}_b^{s,c}(2n)$*, suppose*

$$^{\#}\tilde{\mathcal{J}}(\Sigma) = n.$$

Then all of the n closed characteristics on Σ *are symmetric brake orbits after suitable time translation.*

For $n = 2$, it was proved in [138] that $^{\#}\tilde{\mathcal{J}}(\Sigma)$ is either 2 or $+\infty$ for any C^2 compact convex hypersurface Σ in $\mathbb{R}^4$. Hence Theorem 1.9 gives a positive answer to the above question in the case $n = 2$. We also note that for the hypersurface $\Sigma = \{(x_1, x_2, y_1, y_2) \in \mathbf{R}^4 | \ x_1^2 + y_1^2 + \frac{x_2^2+y_2^2}{4} = 1\}$ we have $^{\#}\tilde{\mathcal{J}}_b(\Sigma) = +\infty$ and $^{\#}\tilde{\mathcal{J}}_b^s(\Sigma) = 2$, where we have denoted by $\tilde{\mathcal{J}}_b^s(\Sigma)$ the set of all symmetric brake orbits on Σ. We also note that on the hypersurface $\Sigma = \{x \in \mathbf{R}^{2n} | \ |x| = 1\}$ there are some non-brake closed characteristics.

10.2 Proofs of Theorems 1.2 and 1.9

In this section we prove Theorems 1.2 and 1.9.

For $\Sigma \in \mathcal{H}_b^{s,c}(2n)$, let $j_\Sigma : \Sigma \to [0, +\infty)$ be the gauge function of Σ defined by

$$j_\Sigma(0) = 0, \quad \text{and} \quad j_\Sigma(x) = \inf\{\lambda > 0 \mid \frac{x}{\lambda} \in C\}, \quad \forall x \in \mathbb{R}^{2n} \setminus \{0\},$$

where C is the domain enclosed by Σ.

Define

$$H_\alpha(x) = (j_\Sigma(x))^\alpha, \ \alpha > 1, \quad H_\Sigma(x) = H_2(x), \ \forall x \in \mathbb{R}^{2n}. \tag{10.15}$$

Then $H_\Sigma \in C^2(\mathbb{R}^{2n}\backslash\{0\}, \mathbb{R}) \cap C^{1,1}(\mathbb{R}^{2n}, \mathbb{R})$.

We consider the following fixed energy problem

$$\dot{x}(t) = JH_\Sigma'(x(t)), \tag{10.16}$$

$$H_\Sigma(x(t)) = 1, \tag{10.17}$$

$$x(-t) = Nx(t), \tag{10.18}$$

$$x(\tau + t) = x(t), \quad \forall t \in \mathbb{R}. \tag{10.19}$$

Denote by $\mathcal{J}_b(\Sigma, 2)$ ($\mathcal{J}_b(\Sigma, \alpha)$ for $\alpha = 2$ in (10.15)) the set of all solutions (τ, x) of problem (10.16), (10.17), (10.18), and (10.19) and by $\tilde{\mathcal{J}}_b(\Sigma, 2)$ the set of all geometrically distinct solutions of (10.16), (10.17), (10.18), and (10.19). By

Remark 1.2 of [209] or discussion in [232], elements in $\mathcal{J}_b(\Sigma)$ and $\mathcal{J}_b(\Sigma,2)$ are in one to one correspondence. So we have ${}^{\#}\tilde{\mathcal{J}}_b(\Sigma)={}^{\#}\tilde{\mathcal{J}}_b(\Sigma,2)$.

For readers' convenience in the following we introduce some results which will be used in the proof of Theorem 1.2. In the following we write $(i_{L_0}(\gamma,k),\nu_{L_0}(\gamma,k))=(i_{L_0}(\gamma^k),\nu_{L_0}(\gamma^k))$ for any symplectic path $\gamma\in\mathcal{P}_\tau(2n)$ and $k\in\mathbb{N}$, where γ^k is defined by (5.185) and (5.186). We have the following jumping formulas for the L_0-index.

Lemma 2.1 ([209]) *Let $\gamma_j\in\mathcal{P}_{\tau_j}(2n)$ for $j=1,\cdots,q$. Let $M_j=\gamma_j^2(2\tau_j)=N\gamma_j(\tau_j)^{-1}N\gamma_j(\tau_j)$, for $j=1,\cdots,q$. Suppose*

$$\bar{i}_{L_0}(\gamma_j)>0,\quad j=1,\cdots,q.$$

Then there exist infinitely many $(R,m_1,m_2,\cdots,m_q)\in\mathbb{N}^{q+1}$ such that

(i) $\nu_{L_0}(\gamma_j,2m_j\pm1)=\nu_{L_0}(\gamma_j)$,
(ii) $i_{L_0}(\gamma_j,2m_j-1)+\nu_{L_0}(\gamma_j,2m_j-1)=R-(i_{L_1}(\gamma_j)+n+S_{M_j}^+(1)-\nu_{L_0}(\gamma_j))$,
(iii) $i_{L_0}(\gamma_j,2m_j+1)=R+i_{L_0}(\gamma_j)$,
(iv) $\nu(\gamma_j^2,2m_j\pm1)=\nu(\gamma_j^2)$,
(v) $i(\gamma_j^2,2m_j-1)+\nu(\gamma_j^2,2m_j-1)=2R-(i(\gamma_j^2)+2S_{M_j}^+(1)-\nu(\gamma_j^2))$,
(vi) $i(\gamma_j^2,2m_j+1)=2R+i(\gamma_j^2)$,

where we have set $i(\gamma_j^2,n_j)=i(\gamma_j^{2n_j})$, $\nu(\gamma_j^2,n_j)=\nu(\gamma_j^{2n_j})$ for $n_j\in\mathbb{N}$.

In order to prove Lemma 2.1, we need the following jumping formulas for the Maslov-type index.

Proposition 2.2 (Theorem 4.3 in [233]) *Let $\gamma_j\in\mathcal{P}_{\tau_j}(2n)$ for $j=1,\cdots,q$ be a finite collection of symplectic paths. Extend γ_j to $[0,+\infty)$ by $\gamma_j(t+\tau_j)=\gamma_j(t)\gamma_j(\tau_j)$ and let $M_j=\gamma(\tau_j)$, for $j=1,\cdots,q$ and $t>0$. Suppose*

$$\bar{i}(\gamma_j)>0,\quad j=1,\cdots,q.$$

Then there exist infinitely many $(R,m_1,m_2,\cdots,m_q)\in\mathbb{N}^{q+1}$ such that

(i) $\nu(\gamma_j,2m_j\pm1)=\nu(\gamma_j)$,
(ii) $i(\gamma_j,2m_j-1)+\nu(\gamma_j,2m_j-1)=2R-(i(\gamma_j)+2S_{M_j}^+(1)-\nu(\gamma_j))$,
(iii) $i(\gamma_j,2m_j+1)=2R+i(\gamma_j)$,

where we have set $i(\gamma_j,n_j)=i(\gamma_j,[0,n_j\tau_j])$, $\nu(\gamma_j,n_j)=\nu(\gamma_j,[0,n_j\tau_j])$ for $n_j\in\mathbb{N}$.

We divide our proof in three steps.

Step 1. Application of Proposition 2.2.

By Corollary V.10.2, we have

$$\bar{i}(\gamma_j^2)=2\bar{i}_{L_0}(\gamma_j)>0.\tag{10.20}$$

So we have

$$\bar{i}(\gamma_j^2) > 0, \quad j = 1, \cdots, q, \tag{10.21}$$

where γ_j^2 is the 2-times iteration of γ_j defined by (5.186). Hence the symplectic paths γ_j^2, $j = 1, 2, \cdots, q$ satisfy the condition in Proposition 2.2, so there exist infinitely $(R, m_1, m_2, \cdots, m_q) \in \mathbb{N}^{q+1}$ such that

$$\nu(\gamma_j^2, 2m_j \pm 1) = \nu(\gamma_j^2), \tag{10.22}$$

$$i(\gamma_j^2, 2m_j - 1) + \nu(\gamma_j^2, 2m_j - 1) = 2R - (i(\gamma_j^2) + 2S_{M_j}^+(1) - \nu(\gamma_j^2)), \tag{10.23}$$

$$i(\gamma_j^2, 2m_j + 1) = 2R + i(\gamma_j^2). \tag{10.24}$$

Step 2. Verification of (i).
By Theorems V.9.1 and V.9.7, we have

$$\nu_{L_0}(\gamma_j, 2m_j \pm 1) = \nu_{L_0}(\gamma_j) + \frac{\nu(\gamma_j^2, 2m_j \pm 1) - \nu(\gamma_j^2)}{2}, \tag{10.25}$$

$$\nu_{L_1}(\gamma_j, 2m_j \pm 1) = \nu_{L_1}(\gamma_j) + \frac{\nu(\gamma_j^2, 2m_j \pm 1) - \nu(\gamma_j^2)}{2}. \tag{10.26}$$

Hence (i) follows from (10.22) and (10.25).
Step 3. Verifications of (ii) and (iii).

Proof By Theorems V.9.1 and V.9.7, we have

$$i_{L_0}(\gamma^m) - i_{L_1}(\gamma^m) = i_{L_0}(\gamma) - i_{L_1}(\gamma), \quad \forall m \in 2\mathbb{N} - 1, \tag{10.27}$$

$$i_{L_0}(\gamma^m) - i_{L_1}(\gamma^m) = i_{L_0}(\gamma^2) - i_{L_1}(\gamma^2), \quad \forall m \in 2\mathbb{N}. \tag{10.28}$$

By Proposition V.9.8 and (10.27) we have

$$2i_{L_0}(\gamma_j, 2m_j \pm 1) = i(\gamma_j^2, 2m_j \pm 1) - n + i_{L_0}(\gamma_j) - i_{L_1}(\gamma_j). \tag{10.29}$$

By (10.22), (10.23), and (10.29) we have

$$2i_{L_0}(\gamma_j, 2m_j - 1) = 2R - (i(\gamma_j^2) - 2S_{M_j}^+(1) + n - i_{L_0}(\gamma_j) + i_{L_1}(\gamma_j)). \tag{10.30}$$

So by Proposition V.9.8 we have

$$i_{L_0}(\gamma_j, 2m_j - 1) = R - (i_{L_1}(\gamma_j) + n + S_{M_j}^+(1)). \tag{10.31}$$

Together with (i), this yields (ii).

By (10.24) and (10.29) we have

$$2i_{L_0}(\gamma_j, 2m_j+1) = 2R + i(\gamma_j^2) - n + i_{L_0}(\gamma_j) - i_{L_1}(\gamma_j). \tag{10.32}$$

By Proposition V.9.8 and (10.32) we have

$$i_{L_0}(\gamma_j, 2m_j+1) = R + i_{L_0}(\gamma_j). \tag{10.33}$$

Hence (iii) holds and the proof of Lemma 2.1 is complete. □

For any $(\tau, x) \in \mathcal{J}_b(\Sigma, 2)$, there is a corresponding path $\gamma_x \in \mathcal{P}_\tau(2n)$. For $m \in \mathbb{N}$, we denote by $i_{L_j}(x, m) = i_{L_j}(\gamma_x^m)$ and $\nu_{L_j}(x, m) = \nu_{L_j}(\gamma_x^m)$ for $j = 0, 1$. Also we denote by $i(x, m) = i(\gamma_x^{2m})$ and $\nu(x, m) = \nu(\gamma_x^{2m})$. We remind that the symplectic path γ_x^m is defined in the interval $[0, \frac{m\tau}{2}]$ and the symplectic path γ_x^{2m} is defined in the interval $[0, m\tau]$. If $m = 1$, we denote by $i(x) = i(x, 1)$ and $\nu(x) = \nu(x, 1)$.

For $S^1 = \mathbb{R}/\mathbb{Z}$, as in [232] we define the Hilbert space E by

$$E = \left\{ x \in W^{1,2}(S^1, \mathbb{R}^{2n}) \,\middle|\, x(-t) = Nx(t), \quad \text{for all } t \in \mathbb{R} \text{ and } \int_0^1 x(t)dt = 0 \right\}.$$

The inner product on E is given by

$$(x, y) = \int_0^1 \langle \dot{x}(t), \dot{y}(t) \rangle dt. \tag{10.34}$$

The $C^{1,1}$ Hilbert manifold $M_\Sigma \subset E$ associated to Σ is defined by

$$M_\Sigma = \left\{ x \in E \,\middle|\, \int_0^1 H_\Sigma^*(-J\dot{x}(t))dt = 1 \text{ and } \int_0^1 \langle J\dot{x}(t), x(t) \rangle dt < 0 \right\}, \tag{10.35}$$

where H_Σ^* is the Fenchel conjugate function of the function H_Σ defined by

$$H_\Sigma^*(y) = \max\{(x \cdot y - H_\Sigma(x)) \mid x \in \mathbb{R}^{2n}\}. \tag{10.36}$$

Let $\mathbb{Z}_2 = \{-id, id\}$ be the usual $\mathbb{Z}_2$ group. We define the $\mathbb{Z}_2$-action on E by

$$-id(x) = -x, \quad id(x) = x, \qquad \forall x \in E.$$

Since H_Σ^* is even, M_Σ is symmetric to 0, i.e., $\mathbb{Z}_2$ invariant. M_Σ is a paracompact $\mathbb{Z}_2$-space. We define

$$\Phi(x) = \frac{1}{2}\int_0^1 \langle J\dot{x}(t), x(t)\rangle dt, \tag{10.37}$$

then Φ is a $\mathbb{Z}_2$ invariant function and $\Phi \in C^\infty(E, \mathbb{R})$. We denote by Φ_Σ the restriction of Φ to M_Σ, we remind that Φ and Φ_Σ here are the functionals A and A_Σ in [232] respectively.

Lemma 2.3 ([209, 232]) *If* $^{\#}\tilde{\mathcal{J}}_b(\Sigma) < +\infty$, *there is a sequence* $\{c_k\}_{k\in\mathbb{N}}$, *such that*

$$-\infty < c_1 < c_2 < \cdots < c_k < c_{k+1} < \cdots < 0, \tag{10.38}$$

$$c_k \to 0 \quad \text{as } k \to +\infty. \tag{10.39}$$

For any $k \in \mathbb{N}$, *there exists a brake orbit* $(\tau, x) \in \mathcal{J}_b(\Sigma, 2)$ *with* τ *being the minimal period of* x *and* $m \in \mathbb{N}$ *satisfying* $m\tau = (-c_k)^{-1}$ *such that for*

$$z(x)(t) = (m\tau)^{-1}x(m\tau t) - \frac{1}{(m\tau)^2}\int_0^{m\tau} x(s)ds, \quad t \in S^1, \tag{10.40}$$

$z(x) \in M_\Sigma$ *is a critical point of* Φ_Σ *with* $\Phi_\Sigma(z(x)) = c_k$ *and*

$$i_{L_0}(x, m) \le k - 1 \le i_{L_0}(x, m) + \nu_{L_0}(x, m) - 1, \tag{10.41}$$

where we denote by $(i_{L_0}(x, m), \nu_{L_0}(x, m)) = (i_{L_0}(\gamma_x, m), \nu_{L_0}(\gamma_x, m))$ *and* γ_x *the associated symplectic path of* (τ, x).

We refer [232] for a complete proof of Lemma 2.3.

Definition 2.4 $(\tau, x) \in \mathcal{J}_b(\Sigma, 2)$ with minimal period τ is called (m, k)-*variational visible*, if there is some m and $k \in \mathbb{N}$ such that (10.40) and (10.41) hold. We call that $(\tau, x) \in \mathcal{J}_b(\Sigma, 2)$ with minimal period τ *infinite variationally visible*, if there exist infinitely many (m, k) such that (τ, x) is (m, k)-variationally visible. We denote by $\mathcal{V}_{\infty,b}(\Sigma, 2)$ the subset of $\tilde{\mathcal{J}}_b(\Sigma)$ in which a representative $(\tau, x) \in \mathcal{J}_b(\Sigma, 2)$ of each $[(\tau, x)]$ is infinite variationally visible.

It is clear that if $^{\#}\tilde{\mathcal{J}}_b(\Sigma) < +\infty$, then $\mathcal{V}_{\infty,b}(\Sigma, 2) \ne \emptyset$.

Lemma 2.5 ([209]) *Suppose* $^{\#}\tilde{\mathcal{J}}_b(\Sigma) < +\infty$. *Then there exist an integer* $K \ge 0$ *and an injection map* $\phi : \mathbb{N} + K \mapsto \mathcal{V}_{\infty,b}(\Sigma, 2) \times \mathbb{N}$ *such that*

(i) *For any* $k \in \mathbb{N} + K$, $[(\tau, x)] \in \mathcal{V}_{\infty,b}(\Sigma, 2)$ *and* $m \in \mathbb{N}$ *satisfying* $\phi(k) = ([(\tau\ , x)], m)$, *there holds*

$$i_{L_0}(x, m) \le k - 1 \le i_{L_0}(x, m) + \nu_{L_0}(x, m) - 1,$$

where x *has minimal period* τ.

(ii) *For any* $k_j \in \mathbb{N} + K$, $k_1 < k_2$, $(\tau_j, x_j) \in \mathcal{J}_b(\Sigma, 2)$ *satisfying* $\phi(k_j) = ([(\tau_j, x_j)], m_j)$ *with* $j = 1, 2$ *and* $[(\tau_1, x_1)] = [(\tau_2, x_2)]$, *there holds*

$$m_1 < m_2.$$

Proof Since ${}^{\#}\tilde{\mathcal{J}}_b(\Sigma) < +\infty$, there is an integer $K \geq 0$ such that all critical values c_{k+K} with $k \in \mathbb{N}$ come from iterations of elements in $\mathcal{V}_{\infty,b}(\Sigma, 2)$. Together with Lemma 2.3, for each $k \in \mathbb{N}$, there is a $(\tau, x) \in \mathcal{J}_b(\Sigma, 2)$ with minimal period τ and $m \in \mathbb{N}$ such that (10.40) and (10.41) hold for $k + K$ instead of k. So we define a map $\phi : \mathbb{N} + K \mapsto \mathcal{V}_{\infty,b}(\Sigma, 2) \times \mathbb{N}$ by $\phi(k + K) = ([(\tau, x)], m)$.

For any $k_1 < k_2 \in \mathbb{N}$, if $\phi(k_j) = ([(\tau_j, x_j)], m_j)$ for $j = 1, 2$. Write $[(\tau_1, x_1)] = [(\tau_2, x_2)] = [(\tau, x)]$ with τ being the minimal period of x, then by Lemma 2.3 we have

$$m_j\tau = (-c_{k_j+K})^{-1}, \quad j = 1, 2. \tag{10.42}$$

Since $k_1 < k_2$ and c_k increases strictly to 0 as $k \to +\infty$, we have

$$m_1 < m_2. \tag{10.43}$$

So the map ϕ is injective, also (ii) is proved. The proof of this Lemma 2.5 is complete. □

Lemma 2.6 ([209]) *Let* $\gamma \in \mathcal{P}_\tau(2n)$ *be extended to* $[0, +\infty)$ *by* $\gamma(\tau + t) = \gamma(t)\gamma(\tau)$ *for all* $t > 0$. *Suppose* $\gamma(\tau) = M = P^{-1}(I_2 \diamond \tilde{M})P$ *with* $\tilde{M} \in \mathrm{Sp}(2n-2)$ *and* $i(\gamma) \geq n$. *Then we have*

$$i(\gamma, 2) + 2S^+_{M^2}(1) - \nu(\gamma, 2) \geq n + 2.$$

Proof By the Bott formula (2.13), (2.21), and (2.23), we have

$$\begin{aligned}
& i(\gamma, 2) + 2S^+_{M^2}(1) - \nu(\gamma, 2) \\
&= 2i(\gamma) + 2S^+_M(1) + \sum_{\theta\in(0,\pi)} S^+_M(e^{\sqrt{-1}\theta}) \\
&\quad -(\sum_{\theta\in(0,\pi)} S^-_M(e^{\sqrt{-1}\theta}) + (\nu(M) - S^-_M(1)) + (\nu_{-1}(M) - S^-_M(-1))) \\
&\geq 2n + 2S^+_M(1) - n \\
&= n + 2S^+_M(1) \\
&\geq n + 2,
\end{aligned} \tag{10.44}$$

where in the last inequality we have used $\gamma(\tau) = M = P^{-1}(I_2 \diamond \tilde{M})P$ and the fact $S^+_{I_2}(1) = 1$. □

Lemma 2.7 ([209]) *For any* $(\tau, x) \in \mathcal{J}_b(\Sigma, 2)$ *and* $m \in \mathbb{N}$, *there hold*

$$i_{L_0}(x, m+1) - i_{L_0}(x, m) \geq 1, \tag{10.45}$$

$$i_{L_0}(x, m+1) + \nu_{L_0}(x, m+1) - 1 \geq i_{L_0}(x, m+1) > i_{L_0}(x, m) + \nu_{L_0}(x, m) - 1. \tag{10.46}$$

Proof Let γ be the associated symplectic path of (τ, x) and we extend γ to $[0, +\infty)$ by $\gamma|_{[0, \frac{k\tau}{2}]} = \gamma^k$ with γ^k defined in (5.187) for any $k \in \mathbb{N}$. Due to the autonomous Hamiltonian systems, for any $m \in \mathbb{N}$, we have

$$\nu_{L_0}(x, m) \geq 1, \qquad \forall m \in \mathbb{N}. \tag{10.47}$$

Since H_Σ is strictly convex, $H''_\Sigma(x(t))$ is positive for all $t \in \mathbb{R}$. So by Theorem V.4.1 and Lemma V.4.2, we have

$$\begin{aligned} i_{L_0}(x, m+1) &= \sum_{0<t<\frac{(m+1)\tau}{2}} \nu_{L_0}(\gamma(t)) \\ &\geq \sum_{0<t\leq\frac{m\tau}{2}} \nu_{L_0}(\gamma(t)) \\ &= \sum_{0<t<\frac{m\tau}{2}} \nu_{L_0}(\gamma(t)) + \nu_{L_0}(\gamma(\frac{m\tau}{2})) \\ &= i_{L_0}(x, m) + \nu_{L_0}(x, m) \\ &> i_{L_0}(x, m) + \nu_{L_0}(x, m) - 1. \end{aligned} \tag{10.48}$$

Thus we get (10.45) and (10.46) from (10.47) and (10.48). This proves Lemma 2.7. □

Proof of Theorem 1.2 It is suffices to consider the case ${}^\#\tilde{\mathcal{J}}_b(\Sigma) < +\infty$. Since $-\Sigma = \Sigma$, for $(\tau, x) \in \mathcal{J}_b(\Sigma, 2)$ we have

$$\begin{aligned} H_\Sigma(x) &= H_\Sigma(-x), \\ H'_\Sigma(x) &= -H'_\Sigma(-x), \\ H''_\Sigma(x) &= H''_\Sigma(-x). \end{aligned} \tag{10.49}$$

It follows that $(\tau, -x) \in \mathcal{J}_b(\Sigma, 2)$ and, in view of the definition of γ_x, we obtain that

$$\gamma_x = \gamma_{-x}.$$

Hence

$$(i_{L_0}(x,m), \nu_{L_0}(x,m)) = (i_{L_0}(-x,m), \nu_{L_0}(-x,m)),$$
$$(i_{L_1}(x,m), \nu_{L_1}(x,m)) = (i_{L_1}(-x,m), \nu_{L_1}(-x,m)), \quad \forall m \in \mathbb{N}. \quad (10.50)$$

We can write

$$\tilde{\mathcal{J}}_b(\Sigma,2) = \{[(\tau_j,x_j)] | j=1,\cdots,p\} \cup \{[(\tau_k,x_k)], [(\tau_k,-x_k)] | k=p+1,\cdots,p+q\}. \quad (10.51)$$

with $x_j(\mathbb{R}) = -x_j(\mathbb{R})$ for $j=1,\cdots,p$ and $x_k(\mathbb{R}) \neq -x_k(\mathbb{R})$ for $k=p+1,\cdots,p+q$. Here we recall that (τ_j,x_j) has minimal period τ_j for $j=1,\cdots,p+q$ and $x_j(\frac{\tau_j}{2}+t) = -x_j(t),\ t\in\mathbb{R}$ for $j=1,\cdots,p$.

In view of Lemma 2.5 there exists an integer $K \geq 0$ and an injective map ϕ : $\mathbb{N}+K \to \mathcal{J}_b(\Sigma,2)\times\mathbb{N}$. By (10.50), (τ_k,x_k) and $(\tau_k,-x_k)$ have the same (i_{L_0},ν_{L_0})-indices. So by Lemma 2.5, without loss of generality, we can further require that

$$\mathrm{Im}(\phi) \subseteq \{[(\tau_k,x_k)] | k=1,2,\cdots,p+q\} \times \mathbb{N}. \quad (10.52)$$

By the strict convexity of H_Σ and (5.210), we have

$$\bar{i}_{L_0}(x_k) > 0, \quad k=1,2,\cdots,p+q.$$

Applying Lemma 2.1 to symplectic paths

$$\gamma_1,\ \cdots,\ \gamma_{p+q},\ \gamma_{p+q+1},\ \cdots,\ \gamma_{p+2q}$$

associated with $(\tau_1,x_1),\ \cdots,\ (\tau_{p+q},x_{p+q}),\ (2\tau_{p+1},x^2_{p+1}),\ \cdots,\ (2\tau_{p+q},x^2_{p+q})$ respectively, there exists a vector $(R,m_1,\cdots,m_{p+2q}) \in \mathbb{N}^{p+2q+1}$ such that $R > K+n$ and

$$i_{L_0}(x_k,2m_k+1) = R + i_{L_0}(x_k), \quad (10.53)$$
$$i_{L_0}(x_k,2m_k-1) + \nu_{L_0}(x_k,2m_k-1)$$
$$= R - (i_{L_1}(x_k) + n + S^+_{M_k}(1) - \nu_{L_0}(x_k)), \quad (10.54)$$

for $k=1,\cdots,p+q$, $M_k = \gamma_k^2(\tau_k)$, and

$$i_{L_0}(x_k,4m_k+2) = R + i_{L_0}(x_k,2), \quad (10.55)$$
$$i_{L_0}(x_k,4m_k-2) + \nu_{L_0}(x_k,4m_k-2)$$
$$= R - (i_{L_1}(x_k,2) + n + S^+_{M_k}(1) - \nu_{L_0}(x_k,2)), \quad (10.56)$$

for $k=p+q+1,\cdots,p+2q$ and $M_k = \gamma_k^4(2\tau_k) = \gamma_k^2(\tau_k)^2$.

By Lemma 2.1, we also have

$$i(x_k, 2m_k + 1) = 2R + i(x_k), \tag{10.57}$$

$$\begin{aligned} &i(x_k, 2m_k - 1) + \nu(x_k, 2m_k - 1) \\ &\quad = 2R - (i(x_k) + 2S^+_{M_k}(1) - \nu(x_k)), \end{aligned} \tag{10.58}$$

for $k = 1, \cdots, p+q$, $M_k = \gamma_k^2(\tau_k)$, and

$$i(x_k, 4m_k + 2) = 2R + i(x_k, 2), \tag{10.59}$$

$$\begin{aligned} &i(x_k, 4m_k - 2) + \nu(x_k, 4m_k - 2) \\ &\quad = 2R - (i(x_k, 2) + 2S^+_{M_k}(1) - \nu(x_k, 2)), \end{aligned} \tag{10.60}$$

for $k = p+q+1, \cdots, p+2q$ and $M_k = \gamma_k^4(2\tau_k) = \gamma_k^2(\tau_k)^2$.

From (10.52), we can set

$$\phi(R - (s-1)) = ([(\tau_{k(s)}, x_{k(s)})], m(s)), \qquad \forall s \in S := \{1, 2, \cdots, n\},$$

where $k(s) \in \{1, 2, \cdots, p+q\}$ and $m(s) \in \mathbb{N}$.

We continue our proof to study the symmetric and asymmetric orbits separately. Let

$$S_1 = \{s \in S | k(s) \leq p\}, \qquad S_2 = S \setminus S_1.$$

We shall prove that $^{\#}S_1 \leq p$ and $^{\#}S_2 \leq 2q$. These estimates together with the definitions of S_1 and S_2 yield Theorem 1.2. □

Claim 3.1 $^{\#}S_1 \leq p$.

Proof of Claim 3.1. By the definition of S_1, $([(\tau_{k(s)}, x_{k(s)})], m(s))$ is symmetric when $k(s) \leq p$. We further prove that $m(s) = 2m_{k(s)}$ for $s \in S_1$.

In fact, by the definition of ϕ and Lemma 2.5, for all $s = 1, 2, \cdots, n$ we have

$$\begin{aligned} i_{L_0}(x_{k(s)}, m(s)) &\leq (R - (s-1)) - 1 = R - s \\ &\leq i_{L_0}(x_{k(s)}, m(s)) + \nu_{L_0}(x_{k(s)}, m(s)) - 1. \end{aligned} \tag{10.61}$$

By the strict convexity of H_Σ and (5.110) or (5.111), we have $i_{L_0}(x_{k(s)}) \geq 0$, so that

$$\begin{aligned} i_{L_0}(x_{k(s)}, m(s)) &\leq R - s < R \leq R + i_{L_0}(x_{k(s)}) \\ &= i_{L_0}(x_{k(s)}, 2m_{k(s)} + 1), \end{aligned} \tag{10.62}$$

for every $s = 1, 2, \cdots, n$, where we have used (10.53) in the last equality. Note that the proofs of (10.61) and (10.62) do not depend on the condition $s \in S_1$.

It is easy to see that γ_{x_k} satisfies conditions of Theorem VII.4.3 with $\tau = \frac{\tau_k}{2}$. Note that by definition $i_{L_1}(x_k) = i_{L_1}(\gamma_{x_k})$ and $\nu_{L_0}(x_k) = \nu_{L_0}(\gamma_{x_k})$. So by Theorem VII.4.3 we have

$$i_{L_1}(x_k) + S^+_{M_k}(1) - \nu_{L_0}(x_k) \geq 0, \quad \forall k = 1, \cdots, p. \tag{10.63}$$

Hence by (10.61) and (10.63), if $k(s) \leq p$, it follows that

$$\begin{aligned}
& i_{L_0}(x_{k(s)}, 2m_{k(s)} - 1) + \nu_{L_0}(x_{k(s)}, 2m_{k(s)} - 1) - 1 \\
&= R - (i_{L_1}(x_{k(s)}) + n + S^+_{M_{k(s)}}(1) - \nu_{L_0}(x_{k(s)})) - 1 \\
&\leq R - \frac{1-n}{2} - 1 - n \\
&< R - s \\
&\leq i_{L_0}(x_{k(s)}, m(s)) + \nu_{L_0}(x_{k(s)}, m(s)) - 1.
\end{aligned} \tag{10.64}$$

Thus by (10.62) and (10.64) and Lemma 2.7 we obtain

$$2m_{k(s)} - 1 < m(s) < 2m_{k(s)} + 1.$$

Hence

$$m(s) = 2m_{k(s)}$$

and

$$\phi(R - s + 1) = ([(\tau_{k(s)}, x_{k(s)})], 2m_{k(s)}), \qquad \forall s \in S_1.$$

Then the injectivity of the map ϕ induces an injective map

$$\phi_1 : S_1 \to \{1, \cdots, p\}, \ s \mapsto k(s).$$

Therefore, ${}^\# S_1 \leq p$ and Claim 3.1 is proved. □

Claim 3.2 ${}^\# S_2 \leq 2q$.

Proof of Claim 3.2. By the formulas (10.57), (10.58), (10.59), and (10.60), and (59) of [198] (also Claim 4 on p. 352 of [223]), we have

$$m_k = 2m_{k+q} \quad \text{for } k = p+1, p+2, \cdots, p+q. \tag{10.65}$$

By Theorem VII.4.3 there holds

$$i_{L_1}(x_k, 2) + S^+_{M_k}(1) - \nu_{L_0}(x_k, 2) \geq 0, \quad p+1 \leq k \leq p+q. \tag{10.66}$$

By (10.56), (10.61), (10.65), and (10.66), for $p+1 \le k(s) \le p+q$ we have

$$\begin{aligned}
& i_{L_0}(x_{k(s)}, 2m_{k(s)}-2) + \nu_{L_0}(x_{k(s)}, 2m_{k(s)}-2) - 1 \\
&= i_{L_0}(x_{k(s)}, 4m_{k(s)+q}-2) + \nu_{L_0}(x_{k(s)}, 4m_{k(s)+q}-2) - 1 \\
&= R - (i_{L_1}(x_{k(s)}, 2) + n + S^+_{M_{k(s)}}(1) - \nu_{L_0}(x_{k(s)}, 2)) - 1 \\
&= R - (i_{L_1}(x_k, 2) + S^+_{M_k}(1) - \nu_{L_0}(x_k, 2)) - 1 - n \\
&\le R - 1 - n \\
&< R - s \\
&\le i_{L_0}(x_{k(s)}, m(s)) + \nu_{L_0}(x_{k(s)}, m(s)) - 1.
\end{aligned} \tag{10.67}$$

Thus (10.62), (10.67) and Lemma 2.7 imply

$$2m_{k(s)} - 2 < m(s) < 2m_{k(s)} + 1, \qquad p < k(s) \le p+q.$$

So

$$m(s) \in \{2m_{k(s)} - 1, 2m_{k(s)}\}, \qquad \text{for } \ p < k(s) \le p+q.$$

Especially this yields that for any s_0 and $s \in S_2$, if $k(s) = k(s_0)$, then

$$m(s) \in \{2m_{k(s)} - 1, 2m_{k(s)}\} = \{2m_{k(s_0)} - 1, 2m_{k(s_0)}\}.$$

Then, in view of the injectivity of the map ϕ from Lemma 2.5, we have

$$^{\#}\{s \in S_2 | k(s) = k(s_0)\} \le 2.$$

This proves Claim 3.2. □

By Claims 3.1 and 3.2, we obtain

$$^{\#}\tilde{\mathcal{J}}_b(\Sigma) =^{\#} \tilde{\mathcal{J}}_b(\Sigma, 2) = p + 2q \ge^{\#} S_1 +^{\#} S_2 = n.$$

The proof of Theorem 1.2 is completed. ∎

Definition 2.8 We call a closed characteristic x on Σ a *dual brake orbit* on Σ if $x(-t) = -Nx(t)$.

Lemma 2.9 *For $\Sigma \in \mathcal{H}_b(2n)$, and $(\tau, x) \in \mathcal{J}(\Sigma, 2)$, we have the following statements:*

(i) *if $x(\mathbb{R}) = Nx(\mathbb{R})$, then x is a brake orbit after suitable time translation;*
(ii) *if $x(\mathbb{R}) = -Nx(\mathbb{R})$, then x is a dual brake orbit after suitable time translation.*

Proof Since $x \in \mathcal{J}(\Sigma, 2)$, we have

$$\dot{x}(t) = JH_2'(x(t)).$$

So let $y(t) = Nx(-t)$ for $t \in \mathbb{R}$, we have

$$\dot{y}(t) = -NJH_2'(x(-t)) = JNH_2'(x(-t)) = JH_2'(Nx(-t)) = JH_2'(y(t)).$$

Hence $(\tau, y) \in \mathcal{J}(\Sigma, 2)$. If $x(\mathbb{R}) = Nx(\mathbb{R})$, (τ, x) and (τ, y) are geometrically the same closed characteristics on Σ. By the existence and uniqueness of first order ordinary differential equations, there exists a unique $t_0 \in [0, \tau)$ such that $x(t+t_0) = y(t)$ for all $t \in \mathbb{R}$. i.e.

$$x(t + t_0) = Nx(-t).$$

Let $z(t) = x(t + \frac{t_0}{2})$ for all $t \in \mathbb{R}$, then we have

$$z(-t) = x(-t - \frac{t_0}{2} + t_0) = y(-t - \frac{t_0}{2}) = Nx(t + \frac{t_0}{2}) = Nz(t).$$

Hence (τ, z) is a brake orbit. This complete the proof of statement (i). The proof of statement (ii) is similar, so we omit it. □

Proof of Theorem 1.9 By Lemma 2.9, if a closed characteristic x on Σ can both become brake orbits and dual brake orbits after suitable translation, then $x(\mathbb{R}) = Nx(\mathbb{R}) = -Nx(\mathbb{R})$, Thus $x(\mathbb{R}) = -x(\mathbb{R})$.

Since we also have $-N\Sigma = \Sigma$, $(-N)^2 = I_{2n}$ and $(-N)J = -J(-N)$, dually by the same proof of Theorem 1.2 (with the estimate (7.145) in Theorem VII.4.3), there are at least n geometrically distinct dual brake orbits on Σ.

If there are exactly n closed characteristics on Σ, then Theorem 1.2 implies that all of them are brake orbits on Σ after suitable time translation. By the same argument all the n closed characteristics must be dual brake orbits on Σ. Then by the argument in the first paragraph of the proof of this theorem, all these n closed characteristics on Σ must be symmetric. Hence all of them are symmetric brake orbits after suitable time translation. The proof of Theorem 1.9 is completed. □

Chapter 11
The Existence and Multiplicity of Solutions of Wave Equations

In this chapter, we apply the index theories defined in Chap. 3 to study the existence and multiplicity of solutions of wave equations. We will use the same concepts and notations as in Sect. 3.3.

11.1 Variational Setting and Critical Point Theories

11.1.1 Critical Point Theorems in Case 1 and Case 2

In this subsection, we will consider the operator equation

$$Au = F'(u),\ u \in D(A) \subset H, \tag{$O.E.$}$$

where H is an infinite-dimensional separable Hilbert space, A is a self-adjoint operator on H with its domain $D(A)$, F is a nonlinear functional on H with the self-adjoint operator $A \in \mathcal{O}_e^{\mp}(\lambda^{\mp})$ which are defined in Sect. 3.3. Firstly, assume $F \in C^2(H, \mathbb{R})$. In order to use the Morse theory to find the solutions of $(O.E.)$, in addition to the index theory developed in Sect. 3.3, we need some further preparations. Assume F satisfying the following conditions according to $A \in \mathcal{O}_e^{\mp}(\lambda^{\mp})$.

$(F_0^{\mp})$ $F'(0) = 0$, and there exists $B_0 \in \mathcal{L}_s^{\mp}(H, \lambda^{\mp})$ satisfying

$$F'(z) = B_0 z + o(\|z\|_H),\ \ \|z\|_H \to 0.$$

$(F_1^{\mp})$ There exist $c \in \mathbb{R}$ and $\delta > 0$ satisfying

$$\pm c \cdot I < \pm F''(z) < \pm(\lambda^{\mp} \mp \delta) \cdot I,\ \ \forall z \in H.$$

$(F^{\mp}_{\infty,1})$ There exists $B_\infty \in \mathcal{L}_s^{\mp}(H, \lambda^{\mp})$ with $\nu_A(B_\infty) = 0$, such that

$$F'(z) = B_\infty z + o(\|z\|_H),\ \ \|z\|_H \to \infty.$$

$(F^{\mp}_{\infty,2})$ There exist $B_{\infty,\alpha}, B_{\infty,\beta} \in \mathcal{L}_s^{\mp}(H, \lambda^{\mp})$ with $\pm B_{\infty,\alpha} \leq \pm B_{\infty,\beta}$, $i_A^{\mp}(B_{\infty,\alpha}) = i_A^{\mp}(B_{\infty,\beta})$ and $\nu_A(B_{\infty,\beta}) = 0$, such that

$$\pm B_{\infty,\alpha} \leq \pm F''(z) \leq \pm B_{\infty,\beta},\ \ \|z\|_H > R \text{ for a constant R} > 0.$$

In the following, we assume $A \in \mathcal{O}_e^{\mp}(\lambda^{\mp})$, then $(O.E.)$ is equivalent to

$$A_k u = F_k'(u),\ \ u \in D(A) \subset H, \tag{11.1}$$

where $A_k = \pm(A - k \cdot I)$ and $k \notin \sigma(A)$ as defined in Sect. 3.3 with $\pm k < \pm c$,

$$F_k(u) := \pm(F(u) - \frac{k}{2}(u, u)_H).$$

Since $F \in C^2(H, \mathbb{R})$ satisfying condition $(F_1^{\mp})$, $F_k \in C^2(H, \mathbb{R})$ and its Hessian is positive definite everywhere, that is

$$(F_k''(u)v, v)_H > 0,\ \ \forall u,\ v \in H \text{ and } v \neq 0.$$

Let F_k^* be the Fenchel conjugate or Legendre transform of F_k, defined by

$$F_k^*(z) := \sup_{u \in H}\{(z, u)_H - F_k(u)\},\ \ \forall z \in H.$$

The readers can refer [78] for more information about the Fenchel conjugate of a convex functional on Banach space. From the property of F_k^*, we have that $u \in D(A)$ is a solution of (11.1) if and only if

$$z = A_k u \tag{11.2}$$

is a solution of the following equation

$$A_k^{-1} z = F_k^{*\prime}(z),\ \ z \in H. \tag{11.3}$$

So there exists a one-to-one corresponding between the solutions of (11.1) and solutions of (11.3). Define the functional Ψ on H by

$$\Psi(z) = F_k^*(z) - \frac{1}{2}(A_k^{-1} z, z)_H,\ \ \forall z \in H. \tag{11.4}$$

The critical points of Ψ are the solutions of (11.3). In order to use the Morse theory to find the critical points of Ψ we need the following lemmas.

Lemma 1.1 ([284]) *Assume $A \in \mathcal{O}_e^{\mp}(\lambda^{\mp})$ and $F \in C^2(H,\mathbb{R})$, satisfying $F_1^{\mp}$ and $F_{\infty,1}^{\mp}$, then the functional Ψ satisfies the (PS) condition. That is to say for any sequence $\{z_n\} \subset H$ satisfying that $\{\Psi(z_n)\}$ is bounded and $\Psi'(z_n) \to 0$ in H, there is a convergent subsequence.*

Proof We only prove the case $A \in \mathcal{O}_e^-(\lambda^-)$. Assume $\{z_n\} \subset H$ is a (PS) sequence of Ψ, i.e., $\{\Psi(z_n)\}$ is bounded and $\Psi'(z_n) \to 0$ in H. From the definition of Ψ, $\Psi'(z_n) = F_k^{*'}(z_n) - A_k^{-1}z_n$, by setting $y_n = -\Psi'(z_n)$, we have

$$A_k^{-1}z_n - F_k^{*'}(z_n) = y_n \to 0.$$

Let $A_k^{-1}z_n - y_n = u_n$, that is $F_k^{*'}(z_n) = u_n$ and

$$A_k(u_n + y_n) = F_k'(u_n). \tag{11.5}$$

Since $\nu_A(B_\infty) = 0$, $A_k - B_{\infty,k}$ is invertible on H, and we have the following decomposition

$$H = H^+ \oplus H^-,$$

where $A_k - B_{\infty,k}$ is positive definite on H^+ and negative define on H^-, for any $u \in H, u = u^+ + u^-$ with $u^+ \in H^+$ and $u^- \in H^-$. Then we have

$$\begin{aligned} &((A_k - B_{\infty,k})(u_n + y_n), (u_n + y_n)^+ - (u_n + y_n)^-)_H \\ &= (F_k'(u_n) - B_{\infty,k}u_n, (u_n + y_n)^+ - (u_n + y_n)^-)_H - (B_{\infty,k}y_n, (u_n + y_n)^+ \\ &\quad -(u_n + y_n)^-)_H. \end{aligned}$$

It is easy to see

$$((A_k - B_{\infty,k})(u_n + y_n), (u_n + y_n)^+ - (u_n + y_n)^-)_H \geq c_1\|u_n + y_n\|_H^2,$$

for some $c_1 > 0$. On the other hand, from condition (F_∞^-) and the fact $y_n \to 0$, we have

$$\begin{aligned} &|(F_k'(u_n) - B_{\infty,k}u_n, (u_n + y_n)^+ - (u_n + y_n)^-)_H - (B_{\infty,k}y_n, (u_n + y_n)^+ \\ &-(u_n + y_n)^-)_H| < (o(\|u_n\|_H) + c_2)\|u_n + y_n\|_H, \end{aligned}$$

for $c_2 > 0$. That is to say $\{u_n\}$ are bounded in H. Further more, since F satisfies conditions (F_0^-) and (F_1^-), consider the Eq. (11.5), with the similar idea used in Lemma 3.1, concretely speaking, by the implicit function theorem, there exists u $\in C^1(H_0 \oplus H_1, H_1)$, such that

$$u_{n,1} = \mathrm{u}(u_{n,0}, y_{n,1}),$$

where $u_n = u_{n,0} + u_{n,1}$, $y_n = y_{n,0} + y_{n,1}$ with $u_{n,*}, y_{n,*} \in H_*(* = 0, 1)$ and H_* is defined in Sect. 3.3. Since $\{u_n\}$ are bounded in H, $\{u_{n,0}\}$ are bounded in H_0. Recall that H_0 is a finite dimensional space, so $\{u_{n,0}\}$ has a convergent subsequence, and from the fact that u is C^1 continuous and $y_n \to 0$, we have $\{u_{n,1}\}$ has a convergent subsequence, so $\{u_n\}$ has a convergent subsequence. From equation $z_n = F_k'(u_n)$ and the C^1 smooth of F_k', we have $\{z_n\}$ has a convergent subsequence and the proof is complete. □

Similarly, we have the following lemma.

Lemma 1.2 ([284]) *Assume $A \in \mathcal{O}_e^{\mp}(\lambda^{\mp})$ and $F \in C^2(H, \mathbb{R})$, satisfying $(F_1^{\mp})$ and $(F_{\infty,2}^{\mp})$, then the functional Ψ satisfies the (PS) condition.*

Proof We only prove the case $A \in \mathcal{O}_e^-(\lambda^-)$. The proof is similar to the proof of Lemma 1.1. Assume $\{z_n\} \subset H$ is a (PS) sequence of Ψ, that is $\{\Psi(z_n)\}$ are bounded and $\Psi'(z_n) \to 0$ in H. Recall that $y_n = -\Psi'(z_n) \to 0$, $F_k^{*'}(z_n) = u_n$ and

$$A_k(u_n + y_n) = F_k'(u_n). \tag{11.6}$$

As in the proof of Lemma 1.1, what we need to do is to prove the boundedness of $\{u_n\}$. Arguing indirectly, assume $\|u_n\|_H \to \infty$, as $n \to \infty$, thus for any $\varepsilon > 0$, there exists $N(\varepsilon) > 0$ such that

$$\|u_n\|_H > \frac{R}{\varepsilon}, \ \forall n > N(\varepsilon), \tag{11.7}$$

with R defined in condition $(F_{\infty,2}^-)$. Let

$$C_n = \int_0^1 F_k''(su_n)ds.$$

Consider $C_n(1) = \int_0^\varepsilon F_k''(su_n)ds$ and $C_n(2) = \int_\varepsilon^1 F_k''(su_n)ds$ respectively. From (F_1^-) we have

$$\varepsilon c \cdot I \le C_n(1) \le \varepsilon\lambda^- \cdot I,$$

where the constant c is defined in condition (F_1^-). From $(F_{\infty,2}^-)$ and (11.7), we have

$$(1-\varepsilon)B_{\infty,\alpha} \le C_n(2) \le (1-\varepsilon)B_{\infty,\beta},$$

thus from $C_n = C_n(1) + C_n(2)$, we have

$$B_{\infty,\alpha} - \varepsilon(B_{\infty,\alpha} - c \cdot I) \le C_n \le B_{\infty,\beta} - \varepsilon(B_{\infty,\beta} - \lambda^- \cdot I), \ \forall n > N(\varepsilon).$$

Since $B_{\infty,\alpha} < B_{\infty,\beta}$ and $i_A^-(B_{\infty,\alpha}) = i_A^-(B_{\infty,\beta})$, by the property of our index stated in Lemma 3.3, we have $\nu_A(B_{\infty,\alpha}) = 0$, and since $\nu_A(B_{\infty,\beta}) = 0$, we can choose ε small enough such that

$$i_A^-(B_{\infty,\alpha} - \varepsilon(B_{\infty,\alpha} - c \cdot I)) = i_A^-(B_{\infty,\beta} - \varepsilon(B_{\infty,\beta} - \lambda^- \cdot I)) = i_A^-(B_{\infty,\beta}),$$

and

$$\nu_A(B_{\infty,\alpha} - \varepsilon(B_{\infty,\alpha} - c \cdot I)) = \nu_A(B_{\infty,\beta} - \varepsilon(B_{\infty,\beta} - \lambda^- \cdot I)) = 0.$$

So from the properties of our index pair, we have

$$i_A^-(C_n) = i_A^-(B_{\infty,\beta}), \text{ and } \nu_A(C_n) = 0.$$

On the other hand, from condition (F_0^-) and the definition of C_n, we have

$$F_k'(u_n) = C_n u_n, \ \forall n = 1, 2, \cdots.$$

Now, as done in Lemma 1.1, if we replace B_∞ by C_n, the rest part of the proof will be the same. We omit the details here. For the case $A \in \mathcal{O}_e^+(\lambda^+)$, the proof is similar. □

Now we are ready to state the following abstract critical point theorem. One of the key points in this theorem is the index twist conditions about origin and infinity, i.e. (11.8) and (11.9) below. It is well known that these twist conditions are related to the famous Poincaré-Birkhoff Theorem on the existence of fixed points of area preserving homeomorphisms on an annulus under twist conditions on the opposite sides of the boundary. The readers can refer [222] for more information about this topic.

Theorem 1.3 ([284]) *Assume $A \in \mathcal{O}_e^\mp(\lambda^\mp)$ and $F \in C^2(H, \mathbb{R})$, satisfying $(F_0^\mp)$, $(F_1^\mp)$ and $(F_{\infty,1}^\mp)$ (or $(F_{\infty,2}^\mp)$). We divided the result into the following two parts.*

Part 1. *If the indices satisfying*

$$i_A^\mp(B_\infty) \notin [i_A^\mp(B_0), i_A^\mp(B_0)+\nu_A(B_0)] (or\ i_A^\mp(B_{\infty,a}) \notin [i_A^\mp(B_0), i_A^\mp(B_0)+\nu_A(B_0)]), \tag{11.8}$$

then the problem $(O.E.)$ possesses at least one nontrivial solution. Moreover, denote this nontrivial solution by u_1, and denote $\nu_A(u_1) := \nu_A(B_1)$, $B_1 = F''(u_1)$. If $\nu_A(B_0) = 0$, and

$$\nu_A(u_1) < |i_A^\mp(B_1) - i_A^\mp(B_0)|,$$

then the problem $(O.E.)$ possesses one more nontrivial solution.

Part 2. *If F is even in u and*

$$i_A^{\mp}(B_0) \neq i_A^{\mp}(B_\infty), \tag{11.9}$$

then the problem $(O.E.)$ possesses at least $|i_A^{\mp}(B_0) - i_A^{\mp}(B_\infty)|$ pairs of nontrivial solutions.

We only consider the case $A \in \mathcal{O}_e^-(\lambda^-)$, and from the proof of Lemma 1.2, condition $(F_{\infty,2}^-)$ will be similar to $(F_{\infty,1}^-)$. So we assume F satisfies conditions (F_0^-), (F_1^-) and $(F_{\infty,1}^-)$. Without loss of generality, we can assume $F(0) = 0$. Since F satisfies condition (F_1^-), choose $k \in \mathbb{R}$ satisfying $k < c$ and $k \notin \sigma(A)$. Then $F_k(u) := F(u) - \frac{k}{2}(u,u)_H$ is a convex functional on H satisfying $F_k(0) = 0$ and $F_k'(0) = 0$, so it's dual F_k^* is well defined and also a convex functional on H satisfying

$$F_k^*(0) = 0, \ F_k^{*'}(0) = 0, \tag{11.10}$$

and

$$F_k^{*''}(0) = [F_k''(0)]^{-1} = B_{0,k}^{-1}. \tag{11.11}$$

Proof of Part 1. Recall that the solutions of $(O.E.)$ correspond to the critical points of the functional Ψ defined in (11.4). From Lemma 3.2 and condition (F_1^-), for any critical point z of Ψ, its Morse index $m_\Psi^-(z)$ and nullity $\nu_\Psi(z)$ are finite. From (3.32), (11.10) and (11.11), 0 is a critical point of Ψ with

$$m_\Psi^-(0) = i_{A,k}^-(B_0), \text{ and } \nu_\Psi(0) = \nu_A(B_0). \tag{11.12}$$

For any $z \in H$, let $u = F_k^{*'}(z)$, that is $F_k'(u) = z$, and $F_k^{*''}(z) = [F_k''(u)]^{-1}$. From condition (F_1^-) the boundedness of F'', we have $\|u\|_H \to \infty \Leftrightarrow \|z\|_H \to \infty$. And from $(F_{\infty,1}^-)$, we have

$$F_k^{*'}(z) = B_{\infty,k}^{-1} z + o(\|z\|_H). \tag{11.13}$$

As in [34], for any pair of topological spaces (X, Y) with $Y \subset X$, let $H_q(X, Y; \mathbb{R})$ denote the singular q-relative homology group. Since Ψ satisfies (PS) condition, from (3.32) and (11.13), with similar arguments as in Lemma II 5.1 of [34], we have

$$H_q(H, \Psi_c; \mathbb{R}) \cong \delta_{q\gamma}\mathbb{R} \tag{11.14}$$

for $-c$ large enough and $\gamma = i_{A,k}^-(B_\infty)$. Now, by Definition 3.6, (11.12) and (11.14), with Morse inequality, as done in Theorem II 5.1 and Corollary II 5.2 in [34], we will get a proof of the part 1. We omit the details here.

In order to prove part 2 of Theorem 1.3, instead of Morse theory we make use of the minimax arguments for multiplicity of critical points for even functional, actually the following lemmas.

Assume $\phi \in C^2(H, \mathbb{R})$ is an even functional, satisfying the (PS) condition and $\phi(0) = 0$. Denote $S_a = \{u \in H \mid \|u\| = a\}$.

Lemma 1.4 (See [111, Corollary 10.19] and [222, Lemma 3.4]) *Assume Y and Z are subspaces of H satisfying* $\dim Y = j > k = \mathrm{codim} Z$. *If there exist $R > r > 0$ and $\alpha > 0$ such that*

$$\inf \phi(S_r \cap Z) \geq \alpha, \ \sup \phi(S_R \cap Y) \leq 0,$$

then ϕ has $j - k$ pairs of nontrivial critical points.

Lemma 1.5 (See [34, Theorem II 4.1] and [222, Lemma 3.5]) *Assume Y and Z are subspaces of H satisfying* $\dim Y = j > k = \mathrm{codim} Z$. *If there exist $r > 0$ and $\alpha > 0$ such that*

$$\inf \phi(Z) > -\infty, \ \sup \phi(S_r \cap Y) \leq -\alpha,$$

then ϕ has $j - k$ pairs of nontrivial critical points.

Proof of Part 2 of Theorem 1.3. Let $\phi = \Psi$, ϕ is an even functional satisfying $\phi(0) = 0$. From Lemma 1.1, ϕ satisfies the (PS) condition. Since $i_A^-(B_0) \neq i_A^-(B_\infty)$, firstly, if $i_A^-(B_0) > i_A^-(B_\infty)$, define

$$Y = H^-_{T^-_{B_0,k}} \text{ and } Z := H^+_{T^-_{B_\infty,k}},$$

where the operator $T^-_{B,k}$ is defined in (3.31). Since our dual functional Ψ has the following two forms

$$\Psi(z) = \frac{1}{2}(B^{-1}_{0,k}z, z)_H - \frac{1}{2}(A^{-1}_k z, z)_H + (F^*_k(z) - \frac{1}{2}(B^{-1}_{0,k}z, z)_H), \ \forall z \in H,$$

and

$$\Psi(z) = \frac{1}{2}(B^{-1}_{\infty,k}z, z)_H - \frac{1}{2}(A^{-1}_k z, z)_H + (F^*_k(z) - \frac{1}{2}(B^{-1}_{\infty,k}z, z)_H), \ \forall z \in H,$$

where $B^{-1}_{*,k} = (B_* - k)^{-1}$ with $* = 0, \infty$. That is to say ϕ has the following two forms

$$\phi(z) = \frac{1}{2}(T^-_{B_0,k}z, z)_H + (F^*_k(z) - \frac{1}{2}(B^{-1}_{0,k}z, z)_H), \ \forall z \in H,$$

and

$$\phi(z) = \frac{1}{2}(T^-_{B_\infty,k}z, z)_H + (F^*_k(z) - \frac{1}{2}(B^{-1}_{\infty,k}z, z)_H), \ \forall z \in H.$$

From Lemma 3.2, we have

$$\frac{1}{2}(T^-_{B_0,k}z, z)_H \le -c\|z\|^2_H, \ \forall z \in Y,$$

and

$$\frac{1}{2}(T^-_{B_\infty,k}z, z)_H \ge c\|z\|^2_H, \ \forall z \in Z,$$

for some constant $c > 0$. So we only need to prove the following two equalities

$$F^*_k(z) - \frac{1}{2}(B^{-1}_{0,k}z, z)_H = o(\|z\|^2_H), \ z \in Y \text{ and } \|z\|_H \to 0, \tag{11.15}$$

and

$$F^*_k(z) - \frac{1}{2}(B^{-1}_{\infty,k}z, z)_H = o(\|z\|^2_H), \ z \in Z \text{ and } \|z\|_H \to \infty. \tag{11.16}$$

From (11.10) and (11.11), we have (11.15). From (11.13), we have (11.16). Thus by Lemma 1.4, $(O.E.)$ has $i^-_A(B_0) - i^-_A(B_\infty)$ pairs of nontrivial solutions.

If $i^-_A(B_0) < i^-_A(B_\infty)$, define

$$Y = H^-_{T^-_{B_\infty,k}}, \text{ and } Z = H^+_{T^-_{B_0,k}}.$$

By the same discussion and Lemma 1.5, $(O.E.)$ has $i^-_A(B_\infty) - i^-_A(B_0)$ pairs of nontrivial solutions. Thus we have proved Theorem 1.3. □

If $F \in C^1(H, \mathbb{R})$, then the above theorem will not work. To deal with this situation, we first make the following assumptions.

$(F^\mp_3)$ There exist $B_1, B_2 \in \mathcal{L}^\mp_s(H, \lambda^\mp)$ with $\pm B_1 < \pm B_2$, $i^\mp_A(B_1) = i^\mp_A(B_2)$, $\nu_A(B_2) = 0$, such that $\pm(F(z) - \frac{1}{2}(B_1 z, z)_H)$ is convex and

$$\pm(F(z) - \frac{1}{2}(B_2 z, z)_H) \le c, \ \forall\, z \in H,$$

where $c \in \mathbb{R}$ is a constant.

$(F^\mp_4)$ $F(0) = 0$, $F'(0) = 0$ and there exists $B_3 \in \mathcal{L}^\mp_s(H, \lambda^\mp)$ with $\pm B_3 > \pm B_1$, such that

$$\pm(F(z) - \frac{1}{2}(B_3 z, z)_H) \ge 0, \text{ for } \|z\|_H < r,$$

where $r > 0$ is a small number.

We now consider the case $A \in \mathcal{O}_e^-(\lambda^-)$, the case $A \in \mathcal{O}_e^+(\lambda^+)$ is similar. As claimed in Remark 3.7, we can replace the number k by some suitable operator B with $B \in \mathcal{L}_s^-(H, \lambda^-)$, denote $B_\varepsilon := B_1 - \varepsilon \cdot I$ for some $\varepsilon > 0$, from condition $B_1 < B_2$, $i_A^-(B_1) = i_A^-(B_2)$ and $\nu_A(B_2) = 0$, we have $\nu_A(B_1) = 0$, so we can choose ε small enough, such that $\nu_A(B_\varepsilon) = 0$ and

$$(B_1 - B_\varepsilon)^{-1} - (A - B_\varepsilon)^{-1} > 0. \tag{11.17}$$

Now, define $F_\varepsilon(z) := F(z) - \frac{1}{2}(B_\varepsilon z, z)_H \in C^1(H, \mathbb{R})$ and $A_\varepsilon := A - B_\varepsilon$. It is easy to see F_ε is convex and A_ε is invariable on H, define the functional Ψ by

$$\Psi(z) = F_\varepsilon^*(z) - \frac{1}{2}(A_\varepsilon^{-1} z, z), \ \forall z \in H.$$

Then the critical points of Ψ correspond to the solutions of $(O.E.)$. With the idea of Theorem 3.1.7 in [69], we have the following result.

Theorem 1.6 ([284]) *Suppose $A \in \mathcal{O}_e^\mp(\lambda^\mp)$, and $F \in C^1(H, \mathbb{R})$ satisfying $(F_3^\mp)$. If Ψ satisfies (PS) condition, then the problem $(O.E.)$ has a solution. Further more, if F satisfying $(F_4^\mp)$ and*

$$i_A^\mp(B_3) > i_A^\mp(B_2), \tag{11.18}$$

then the problem $(O.E.)$ has a nontrivial solution.

Proof Only consider the case $A \in \mathcal{O}_e^-(\lambda^-)$. Firstly, since $F(z) \le \frac{1}{2}(B_2 z, z)_H + c$, we have

$$\Psi(z) \ge \frac{1}{2}((B_2 - B_\varepsilon)^{-1} z, z) - \frac{1}{2}(A_\varepsilon^{-1} z, z) + c.$$

From the definition of the index, condition $i_A^-(B_1) = i_A^-(B_2)$ and (11.17), we have $(B_2 - B_\varepsilon)^{-1} - A_\varepsilon^{-1} > 0$, so

$$\Psi(z) \to +\infty, \text{ as } \|z\|_H \to \infty$$

and Ψ is bounded from below, then by Ekeland's variational principle and the (PS) condition, we have Ψ gets its minimal value at some point z_0. Of cause, z_0 is a critical point of Ψ. Thus we have proved the first part of the theorem. Secondly, if $F'(0) = 0$, then 0 is a trivial solution of $(O.E.)$. We will prove $z_0 \ne 0$. In fact, from (11.18), we have

$$\frac{1}{2}((B_3 - B_\varepsilon)^{-1} z, z) - \frac{1}{2}(A_\varepsilon^{-1} z, z)$$

has an $i_A^-(B_3) - i_A^-(B_2)$- dimensional negative define space, denote it by Z. From condition $(F_4^{\mp})$, we have

$$\Psi(z) \leq \frac{1}{2}((B_3 - B_\varepsilon)^{-1}z, z) - \frac{1}{2}(A_\varepsilon^{-1}z, z), \ z \in Z \text{ and } \|z\|_H \text{ small enough.}$$

Thus 0 is not a minimal value point of Ψ, and $z_0 \neq 0$. The proof is complete. □

11.1.2 Critical Point Theorems in Case 3

In this subsection, we will consider the operator equation $(O.E.)$ on H with the self-adjoint operator $A \in \mathcal{O}_e^0(\lambda_a, \lambda_b)$.

Assume $F \in C^2(H, \mathbb{R})$ satisfying the following conditions

(F_0) $F'(0) = 0$, and there exists $B_0 \in \mathcal{L}_s^0(H, \lambda_a, \lambda_b)$ satisfying

$$F'(z) = B_0 z + o(\|z\|_H), \ \|z\|_H \to 0.$$

(F_1) There exists $\delta > 0$ satisfying

$$(\lambda_a + \delta) \cdot I < F''(z) < (\lambda_b - \delta) \cdot I, \ \forall z \in H.$$

$(F_{\infty,1})$ There exists $B_\infty \in \mathcal{L}_s^0(H, \lambda_a, \lambda_b)$ with $\nu_A(B_\infty) = 0$, such that

$$F'(z) = B_\infty z + o(\|z\|_H), \ z \to \infty.$$

$(F_{\infty,2})$ There exist $B_{\infty,\alpha}, B_{\infty,\beta} \in \mathcal{L}_s^0(H, \lambda_a, \lambda_b)$ with $B_{\infty,\alpha} \leq B_{\infty,\beta}$, $i_A^0(B_{\infty,\alpha}) = i_A^0(B_{\infty,\beta})$ and $\nu_A(B_{\infty,\beta}) = 0$, such that

$$B_{\infty,\alpha} \leq F''(z) \leq B_{\infty,\beta}, \ \|z\|_H > R,$$

where $R > 0$ is a constant.

Let $k = \frac{\lambda_a + \lambda_b}{2}$ in this case and P_k the projection map on the eigenspace of k if $k \in \sigma(A)$. Recall that $(O.E.)$ is equivalent to (11.1) the following equation

$$A_k u = F_k'(u), \ u \in D(A) \subset H,$$

with A_k, F_k defined by

$$A_k := \begin{cases} A - k \cdot I, & k \notin \sigma(A), \\ A - k \cdot I + P_k, & k \in \sigma(A), \end{cases}$$

and

$$F_k(u) := \begin{cases} F(u) - \frac{k}{2}(u,u)_H, & k \notin \sigma(A), \\ F(u) - \frac{k}{2}(u,u)_H + (P_k u, u)_H, & k \in \sigma(A). \end{cases}$$

Instead of using dual variational method to find the critical points, we consider the saddle point reduction here. As done in Sect. 3.3, let

$$P_0 = \int_{-(\lambda_b-\lambda_a)/2}^{(\lambda_b-\lambda_a)/2} 1dE(z),$$

with $E(z)$ the spectrum measure of A_k and

$$P_1 = I - P_0.$$

Let

$$H_* = P_* H, \ * = 0, 1.$$

Decompose the space $E = D(|A|^{1/2})$ as follows

$$E = E_0 \oplus E_1, \tag{11.19}$$

where $E_* = E \cap H_*$, $* = 0, 1$. Then for any $z \in E$, we have the decomposition $z = x + y$ with $x \in E_0$ and $y \in E_1$. Define a functional Φ on E as follows:

$$\Phi(z) = \frac{1}{2}(A_k z, z)_H - F_k(z), \ \forall z \in E.$$

The critical points of Φ are the solutions of $(O.E.)$, since F is C^2 continuous and satisfies condition (F_1), by standard argument of saddle point reduction, there exists $\xi \in C^1(E_0, E_1)$, denote $z(x) = x + \xi(x)$, then we have the following functional a on finite dimensional space E_0

$$a(x) = \Phi(z(x)), \ \forall x \in E_0,$$

and the following theorem duo to Amann and Zehnder [4], Chang [34] and Long [223].

Theorem 1.7 ([284]) *Suppose $A \in \mathcal{O}_e^0(\lambda_a, \lambda_b)$. If F is C^2 continuous and satisfies condition (F_1), then there is a one-to-one correspondence between the critical points of the C^2-function $a \in C^2(E_0, \mathbb{R})$ and the solutions of the operator equation $(O.E.)$. Moreover, the functional a satisfies*

$$a'(x) = Az(x) - F'(z(x)) = Ax - P_0 F'(z(x)),$$

$$a''(x) = [A - F''(z(x))]z'(x) = AP_0 - P_0 F''(z(x))z'(x).$$

Since E_0 is a finite dimensional space, for every critical point x of a in E_0, the Morse index $m_a^-(x)$ and nullity $\nu_a(x)$ are finite, and by the definition of the index, it is easy to see $m_a^-(x) = i_A^0(F''(z(x)))$ and $\nu_a(x) = \nu_A(F''(z(x)))$. Similar to Lemmas 1.1 and 1.2, we have the following lemma.

Lemma 1.8 *Assume $F \in C^2(H, \mathbb{R})$, satisfying (F_1) and $(F_{\infty,1})$ (or $F_{\infty,2}$), then the functional a satisfies the (PS) condition.*

The proof is similar to Lemmas 1.1 and 1.2, so we omit it here. Similar to Theorem 1.3, we have the following theorem.

Theorem 1.9 ([284]) *Assume (F_0), (F_1) and $(F_{\infty,1})$ (or $(F_{\infty,2})$) hold. We state the result as two parts.*

Part 1. *If*

$$i_A^0(B_\infty) \notin [i_A^0(B_0), i_A^0(B_0)+\nu_A(B_0)] (or\ i_A^0(B_{\infty,\alpha}) \notin [i_A^0(B_0), i_A^0(B_0)+\nu_A(B_0)]), \tag{11.20}$$

then the problem $(O.E.)$ has at least one nontrivial solution. Moreover, denote this nontrivial solution by u_1, and denote $\nu_A(u_1) := \nu_A(F''(u_1))$. If $\nu_A(B_0) = 0$, and

$$\nu_A(u_1) < |i_A^0(B_1) - i_A^0(B_0)|,$$

then the problem $(O.E.)$ has one more nontrivial solution.

Part 2. *If F is even in u and*

$$i_A^0(B_0) \neq i_A^0(B_\infty), \tag{11.21}$$

then the problem $(O.E.)$ has at least $|i_A^0(B_0) - i_A^0(B_\infty)|$ pairs of nontrivial solutions.

11.2 Applications: The Existence and Multiplicity of Solutions for Wave Equations

11.2.1 One Dimensional Wave Equations

In this subsection, we will consider the following one dimensional wave equation

$$\begin{cases} \Box u \equiv u_{tt} - u_{xx} = f(x, t, u), \\ u(0, t) = u(\pi, t) = 0, \qquad \forall (x, t) \in [0, \pi] \times \mathbb{R}, \\ u(x, t + T) = u(x, t), \end{cases} \tag{W.E.}$$

where $T > 0$, $f \in C^1([0,\pi]\times\mathbb{R}^2,\mathbb{R})$ and T-periodic in variable t. In what follows we assume systematically that T is a rational multiple of π. So, there exist coprime integers $a,\ b$, such that $T = \frac{2\pi b}{a}$. Let

$$L^2 := \left\{ u,\, u = \sum_{j>0,k\in\mathbb{Z}} u_{j,k}\sin jx \exp ik\frac{a}{b}t \right\},$$

where $i = \sqrt{-1}$ and $u_{j,k}\in\mathbb{C}$ with $u_{j,k} = \bar{u}_{j,-k}$, its inner product is

$$(u,v)_2 = \sum_{j>0,k\in\mathbb{Z}} (u_{j,k},\bar{v}_{j,k}),\ \ u,v\in L^2,$$

the corresponding norm is

$$\|u\|_2^2 = \sum_{j>0,k\in\mathbb{Z}} |u_{j,k}|^2\ u,v\in L^2.$$

Consider $\Box$ as an unbounded self-adjoint operator on L^2. It's spectrum set is

$$\sigma(\Box) = \{(a^2k^2 - b^2j^2)/b^2 | j>0, k\in\mathbb{Z}\}.$$

It is easy to see $\Box$ has only one essential spectrum $\lambda_0 = 0$. Take the working space $H = L^2$ and the operator $A = \Box$. As defined in Sect. 3.3, we can define the index pair $(i_A^{\mp}(B), \nu_A(B))$ for any $B\in\mathcal{L}_s^{\mp}(L^2,\lambda_0)$. Denote $L^\infty = L^\infty([0,\pi]\times S^1,\mathbb{R})$ the set of all essentially bounded functions, where $S^1 = \mathbb{R}/T$. For any $g\in L^\infty$, it is easy to see g determines a bounded self-adjoint operator on L^2, by

$$u(x,t)\mapsto g(x,t)u(x,t),\ \forall u\in L^2. \tag{11.22}$$

Without confusion, we still denote this operator by g, that is to say we have the continuous embedding $L^\infty\hookrightarrow\mathcal{L}_s(L^2)$. Thus for any $g\in L^\infty\cap\mathcal{L}_s^{\mp}(L^2,\lambda_0)$ we have the index pair $(i_A^{\mp}(g),\nu_A(g))$. Besides, for any $g_1, g_2\in L^\infty$, $g_1\le g_2$ means that

$$g_2(x,t) - g_1(x,t)\ge 0,\ \ a.e.(x,t)\in[0,\pi]\times S^1.$$

Denote $\Omega = [0,\pi]\times S^1$ for simplicity, and assume $f\in C^1(\Omega\times\mathbb{R},\mathbb{R})$ satisfying the following conditions.

($f_0^{\mp}$) $f(x,t,0)\equiv 0$, and denote $g_0{:=}f_u'(x,t,0)\in L^\infty\cap\mathcal{L}_s^{\mp}(L^2,\lambda_0)$.

($f_1^{\mp}$) There exists a constant $\delta > 0$ such that

$$\pm f_u'(x,t,u) < -\delta,\ \forall (x,t,u)\in\Omega\times\mathbb{R}.$$

$(f_2^{\mp})$ There exist $g_1, g_2 \in L^\infty \cap \mathcal{L}_s^{\mp}(L^2, \lambda_0)$ with $\pm g_1 < \pm g_2, i_A^{\mp}(g_1)+\nu_A(g_1) = i_A^{\mp}(g_2)$ and $\nu_A(g_2) = 0$, such that

$$\pm f_u'(x,t,u) \geq \pm g_1(x,t), \ \forall (x,t,u) \in \Omega \times \mathbb{R}$$

and

$$\pm f_u'(x,t,u) \leq \pm g_2(x,t), \ \forall (x,t) \in \Omega \text{ and } |u| > R \text{ for some constant } R > 0.$$

Define

$$\mathcal{F}(x,t,u) := \int_0^u f(x,t,s)ds, \ \forall u \in \mathbb{R},$$

and

$$F(u) := \int_\Omega \mathcal{F}(x,t,u)dxdt, \ \forall u \in H.$$

If $f \in C^1(\Omega \times \mathbb{R}, \mathbb{R})$ satisfies $(f_1^{\mp})$, we have $F \in C^1(H, \mathbb{R})$, the solutions of the operator equation

$$Au = F'(u), \ u \in D(A)$$

are the solutions of $(W.E.)$. And we have the following result.

Theorem 2.1 *For* $A = \Box \in \mathcal{O}_e^{\mp}(\lambda^{\mp})$, $\lambda^{\mp} = 0$, *assume* $f \in C^1(\Omega \times \mathbb{R}, \mathbb{R})$ *satisfies conditions* $(f_1^{\mp})$ *and* $(f_2^{\mp})$, *then* $(W.E.)$ *has a weak solution. Further more, if* f *satisfies condition* $(f_0^{\mp})$ *and*

$$i_A^{\mp}(g_0) > i_A^{\mp}(g_2),$$

then $(W.E.)$ *has a nontrivial weak solution.*

Proof Only consider $A = \Box \in \mathcal{O}_e^-(\lambda^-)$, by virtue of the similar reason as stated before Theorem 1.6, we can choose $\varepsilon > 0$ small enough, such that

$$(g_1 - g_\varepsilon)^{-1} - A_\varepsilon^{-1} > 0,$$

where $g_{1,\varepsilon} := g_1 - \varepsilon \cdot I$ and $A_\varepsilon := A - g_{1,\varepsilon}$. Denote $F_\varepsilon(z) := F(z) - \frac{1}{2}(g_{1,\varepsilon}z, z)_H$. In order to prove this theorem, we only need to check the conditions in Theorem 1.6 step by step.

Firstly, by condition (f_2^-), we have $F(z) - \frac{1}{2}(g_1 z, z)_H$ is convex. Since $\nu_A(g_2) = 0$, we can choose $\eta > 0$ small enough, such that $i_A^-(g_2 + \eta \cdot I) = i_A^-(g_2)$ and $\nu_A(g_2 + \eta \cdot I) = 0$, denote $g_{2,\eta} := g_2 + \eta \cdot I$, then we have $F(z) \leq \frac{1}{2}(g_{2,\eta}z, z)_H + c$, that is to say F satisfies (F_3^-).

Secondly, from (f_0^-) and condition $i_A^-(g_0) > i_A^-(g_2)$, we can choose $\zeta > 0$ small enough, such that

$$\mathcal{F}(x,t,u) \geq \frac{1}{2} g_{0,\zeta} u^2, \text{ for } |u| \text{ small enough}, \tag{11.23}$$

and

$$i_A^-(g_{0,\zeta}) = i_A^-(g_0) > i_A^-(g_2),$$

where $g_{0,\zeta} := g_0 - \zeta$. Consider the proof of Theorem 1.6, in stead of verifying (F_4^-), we only need to prove the following inequality

$$F_\varepsilon^*(z) \leq \frac{1}{2}((g_{0,\zeta} - g_{1,\varepsilon})^{-1} z, z)_H, \forall z \in Z, \text{ with } \|z\| \text{ small enough}, \tag{11.24}$$

where Z is an arbitrary finite dimensional subspace of H. By the definition of F_ε^*, it is easy to verify that

$$F_\varepsilon^*(z) = \int_\Omega \mathcal{F}_\varepsilon^*(x,t,z)dxdt,$$

where $\mathcal{F}_\varepsilon(x,t,u) := \mathcal{F}(x,t,u) - \frac{1}{2} g_{1,\varepsilon} u^2$ and $\mathcal{F}_\varepsilon^*$ is its Fenchel conjugate corresponding to u. From (11.23), we have $\mathcal{F}_\varepsilon^*(x,t,z) \leq \frac{1}{2}(g_{0,\zeta} - g_{1,\varepsilon})^{-1} z^2$ for $|z|$ small enough, and since Z is a finite dimensional space, we have proved (11.24).

Now, what we need to prove is the (PS) condition. Recall the functional Ψ on H defined by

$$\Psi(z) = F_\varepsilon^*(z) - \frac{1}{2}(A_\varepsilon^{-1} z, z)_H, \ \forall z \in H.$$

Recall the constant R in condition (f_2^-), define $\eta \in C([0,\infty),[0,1])$ by

$$\eta(r) := \begin{cases} 1, & r > R+1, \\ 0, & 0 \leq r \leq R. \end{cases}$$

Define $b \in C(\Omega \times \mathbb{R}, \mathbb{R})$ by

$$b(x,t,u) := \eta(|u|) f_u'(x,t,u) + (1 - \eta(|u|)) g_2(x,t), \ \forall (x,t,u) \in \Omega \times \mathbb{R}.$$

From conditon (f_2^-), we have $g_1 \leq b \leq g_2$ and

$$b(x,t,u) \equiv f_u'(x,t,u), \ \forall |u| > R+1. \tag{11.25}$$

Define $B \in C(\Omega \times \mathbb{R}, \mathbb{R})$ by

$$B(x,t,u) := \int_0^1 b(x,t,su)ds.$$

Of course, we have

$$g_1 \leq B \leq g_2. \tag{11.26}$$

For any $\epsilon > 0$, if $|u| > (R+1)/\epsilon$, then

$$\begin{aligned}
|f(x,t,u) - B(x,t,u)u| = & \left| f(x,t,0) + \int_0^1 [f_u'(x,t,su) - b(x,t,su)]dsu \right| \\
\leq & |f(x,t,0)| + \left| \int_0^\epsilon [f_u'(x,t,su) - b(x,t,su)]ds \right| |u| \\
& + \left| \int_\epsilon^1 [f_u'(x,t,su) - b(x,t,su)]dsu \right|.
\end{aligned}$$

Since Ω is compact set, from the continuity of f, the first term $|f(x,t,0)|$ of the above inequality is bounded by some constant c_1. From boundedness condition in $(f_1^{\mp})$ and the boundedness of function b, the second term satisfies

$$\left| \int_0^\epsilon [f_u'(x,t,su) - b(x,t,su)]ds \right| |u| \leq \epsilon c_2 |u|$$

for some constant c_2. From (11.25), the third term satisfies

$$\int_\epsilon^1 [f_u'(x,t,su) - b(x,t,su)]dsu \equiv 0$$

for $|u| > (R+1)/\epsilon$. So we have the following estimate

$$|f(x,t,u) - B(x,t,u)u| \leq c_1 + \epsilon c_2 |u|, \ \forall \epsilon > 0, \ |u| > (R+1)/\epsilon.$$

That is to say

$$f(x,t,u) - B(x,t,u)u = o(|u|), \text{ as } |u| \to \infty. \tag{11.27}$$

Therefore we have

$$\begin{aligned}
\|F'(u) - B(u)u\|_H^2 = & \int_\Omega |f(x,t,u) - B(x,t,u)u|^2 dxdt \\
= & o(\|u\|_H^2), \ \|u\|_H \to \infty.
\end{aligned} \tag{11.28}$$

Let $\{z_n\}$ be a (PS) sequence of Ψ, and denote $u_n = F_\varepsilon^{*'}(z_n)$ (or equivalent $F_\varepsilon'(u_n) = z_n$), we have $y_n := -\Psi'(z_n) \to 0$ in H and

$$A_\varepsilon(u_n + y_n) = F_\varepsilon'(u_n).$$

or equivalent

$$A(u_n + y_n) = F'(u_n) + g_{1,\varepsilon} y_n. \tag{11.29}$$

With the similar discussion as in Lemmas 1.1 and 1.2, we can prove that $\{u_n + y_n\}$ is bounded in $E = D(|A^{1/2}|)$. For details, firstly, from (f_2^-) and (11.26), for any $u \in H$, $A - B(u)$ is invertible and there exists a constant $c > 0$, such that

$$c\|u_n + y_n\|_E^2 \le ((A - B(u_n))(u_n + y_n), (u_n + y_n)^+ - (u_n + y_n)^-)_H,$$

where the decomposition $u = u^+ + u^-$ corresponds to the positive and negative space of $A - B(u)$. Secondly, from (11.28) and $y_n \to 0$, we have

$$\begin{aligned}
&((A - B(u_n))(u_n + y_n), (u_n + y_n)^+ - (u_n + y_n)^-)_H \\
=&((F'(u_n) - B(u_n))u_n, (u_n + y_n)^+ - (u_n + y_n)^-)_H \\
&+ ((g_{1,\varepsilon} + B(u_n))y_n, (u_n + y_n)^+ - (u_n + y_n)^-)_H \\
\le&(o(\|u_n\|_H) + o(1))\|u_n + y_n\|_H.
\end{aligned}$$

thus $\{u_n + y_n\}$ is bounded in E. Now, we can not use saddle point reduction method to receive the convergent subsequence of $\{u_n\}$, because F is not C^2 continuous on H. But we can use the property of the spectrum of A in this specific situation. Since 0 is the unique isolate essential spectrum of A, we have the following orthogonal decomposition

$$E = \ker A \oplus E_1$$

with the compact embedding $E_1 \hookrightarrow H$. Assume $u_n + y_n = v_n + w_n$ with $v_n \in \ker A$ and $w_n \in E_1$, from the boundedness of $\{u_n + y_n\}$, without loss of generality, we have $v_n \rightharpoonup v \in \ker A$ and $w_n \rightharpoonup w \in E_1$, thus $w_n \to w$ in H. Let $\delta > 0$ be the constant in (f_1^-), then we have

$$\begin{aligned}
\lim_{n\to\infty} \frac{\delta}{2}\|v_n - v\|_H^2 &= \lim_{n\to\infty} \frac{\delta}{2}(v_n - v, v_n - v)_H \\
&= \lim_{n\to\infty} \frac{\delta}{2}(v_n, v_n - v)_H - \lim_{n\to\infty} \frac{\delta}{2}(v, v_n - v)_H.
\end{aligned}$$

Since $v_n \rightharpoonup v$ in H, $\lim_{n\to\infty} \frac{\delta}{2}(v, v_n - v)_H = 0$. And since $\ker(A) \perp E_1$, we have

$$\begin{aligned}\lim_{n\to\infty}\frac{\delta}{2}\|v_n-v\|_H^2&=\lim_{n\to\infty}\frac{\delta}{2}(v_n,v_n-v)_H\\&=\lim_{n\to\infty}((A+\frac{\delta}{2}I)(v_n+w_n),v_n-v)_H.\end{aligned}$$

Next, by (11.29), the property $y_n\to 0$, $w_n\to w$, $v_n\rightharpoonup v$ in H and boundedness of f'_u, we have the following equalities

$$\begin{aligned}\lim_{n\to\infty}\frac{\delta}{2}\|v_n-v\|_H^2&=\lim_{n\to\infty}(F'(v_n+w_n-y_n)+g_{1,\varepsilon}y_n+\frac{\delta}{2}(v_n+w_n),v_n-v)_H\\&=\lim_{n\to\infty}(F'(v_n+w_n-y_n)\\&\quad+\frac{\delta}{2}(v_n+w_n),v_n-v)_H\\&=\lim_{n\to\infty}\int_\Omega(f(x,t,v_n+w_n-y_n))\\&\quad+\frac{\delta}{2}(v_n+w_n-y_n),v_n-v)dxdt\\&=\lim_{n\to\infty}\int_\Omega(f_{\delta/2}(x,t,v_n+w_n-y_n)\\&\quad-f_{\delta/2}(x,t,v+w-y),v_n-v)dxdt\\&=\lim_{n\to\infty}\int_\Omega(f'_{\delta/2}(x,t,\xi_n)\\&\quad(v_n-v+w_n-w-y_n+y),v_n-v)dxdt\\&=\lim_{n\to\infty}\int_\Omega(f'_{\delta/2}(x,t,\xi_n)(v_n-v),v_n-v)dxdt,\end{aligned}$$

where $f_{\delta/2}(x,t,u):=f(x,t,u)+\frac{\delta}{2}u$, $f'_{\delta/2}:=f'_u+\frac{\delta}{2}$ and ξ_n depends on v_n, w_n and y_n. From $(f_1^{\mp})$, we have $f'_{\delta/2}<-\frac{\delta}{2}$, so we have

$$\lim_{n\to\infty}\frac{\delta}{2}\|v_n-v\|_H^2\le 0.$$

That is to say $v_n\to v$ in H, $u_n\to v+w$ in H and $z_n\to F'_\varepsilon(v+w)$. So, we have proved that Ψ satisfied (PS) condition, and the proof of this theorem is complete. □

Further more, in order to use Theorem 1.3 to receive the solutions of $(W.E.)$, we can use the idea of [58], corresponding to the parity of b, we restrict the wave operator □ on an invariant subspace L_b^2 of L^2, such that □ is reversible on L_b^2, thus □ can be regard as an unbounded self-adjoint operator on L_b^2 with compact resolvent. For details, recall the space

$$L^2 := \left\{ u, u = \sum_{j>0,k\in\mathbb{Z}} u_{j,k} \sin jx \exp ik\frac{a}{b}t \right\},$$

defined above, we consider the following two situations.

Situation 1. b is odd.

In this case, we assume the nonlinear term f satisfying the following condition

(f_{odd}) f is $\frac{T}{2}$-periodic in t and $f(x,t,u) = f(\pi - x, t, u)$, $\forall (x,t,u) \in [0,\pi] \times \mathbb{R}^2$.

Let L_1^2 be the closed subspace of L^2 defined by

$$L_1^2 = \{u \in L^2, u(\pi - x, t) = u(x,t), u(x, t+\frac{T}{2}) = u(x,t),\ a.e. x \in [0,\pi], t \in \mathbb{R}\}$$

Then L_1^2 is invariant by f, that is to say that $u \in L_1^2 \Rightarrow f(x,t,u) \in L_1^2$ and we have

$$u \in L_1^2 \Leftrightarrow u_{j,k} = 0 \text{ if } j \text{ is even or } k \text{ is odd.} \tag{11.30}$$

Let $\Box_1 \equiv \Box|_{L_1^2}$, similarly we have $\Box_1$ is self-adjoint in L_1^2 and $L_1^2 \cap N(\Box) = \{0\}$.

In this case, we can consider the problem $(W.E.)$ for $u \in L_1^2$, that is

$$\begin{cases} \Box_2 u \equiv u_{tt} - u_{xx} = f(x,t,u), \\ u(0,t) = u(\pi,t) = 0, \\ u(x, t+\frac{T}{2}) = u(x,t), \\ u(\pi - x, t) = u(x,t) \end{cases} \tag{W.E.1}$$

Situation 2. b is even.

In this case, we need the nonlinear term f satisfying the following condition

(f_{even}) f is $\frac{T}{2}$-periodic in t and odd in u.

Let L_2^2 be the closed subspace of L^2 defined by

$$L_2^2 = \left\{ u \in L^2, u(x,t) = -u(x, t+\frac{T}{2}),\ a.e. x \in [0,\pi], t \in \mathbb{R} \right\}.$$

Then L_2^2 is invariant by f, and we have

$$u \in L_2^2 \Leftrightarrow u_{j,k} = 0 \text{ for any even } k. \tag{11.31}$$

Let $\Box_2 \equiv \Box|_{L_2^2}$, then we have $\Box_2$ is self-adjoint in L_2^2 and $L_2^2 \cap N(\Box) = \{0\}$, where $N(\Box)$ is the kernel of $\Box$.

In this case, we can consider the problem $(W.E.)$ for $u \in L_2^2$, that is

$$\begin{cases} \Box_1 u \equiv u_{tt} - u_{xx} = f(x, t, u), \\ u(0, t) = u(\pi, t) = 0, \\ u(x, t + \frac{T}{2}) = -u(x, t), \end{cases} \tag{W.E.2}$$

In these two cases, the key point is that $\Box_i$ has compact resolvent on L_i^2 for $i = 1, 2$.

For the sake of simplicity we shall write L instead of $L_i^2 (i = 1, 2)$, correspondingly we write $\Box$ instead of $\Box_i (i = 1, 2)$, and we shall renumber that $\Box$ has compact resolvent on L. In order to use Theorem 1.3, we can not regard L as the working space, since the functional F defined above will not be C^2 continuous on L. So, we need to find a suitable working space to protect the smoothness of F. Assume $\sigma(\Box) = \{\lambda_n\}$, satisfying

$$-\infty \leftarrow \lambda_{-l} \leq \cdots \leq \lambda_l \to +\infty, \ l \to +\infty,$$

and for any $n \in \mathbb{Z}$, λ_n is a point spectrum of $\Box$ with its eigenvector e_n. That is to say $\Box e_n = \lambda_n e_n$. Thus for any $u \in L$, we can rewrite it as

$$u = \sum_{n=-\infty}^{+\infty} u_n e_n, \ u_n \in \mathbb{R},$$

and it's norm $\|u\|_{L^2}^2 = \sum_{n=-\infty}^{+\infty} |u_n|^2$. Since $\lambda_n \neq 0$, we define a new Hilbert space H by

$$H = \left\{ u \in L \,\middle|\, \sum_{n=-\infty}^{+\infty} |\lambda_n|^{1/2} |u_n|^2 < \infty \right\}, \tag{11.32}$$

with its inner product defined by

$$(u, v)_H := \sum_{n=-\infty}^{+\infty} |\lambda_n|^{1/2} u_n v_n, \ \forall u, v \in H,$$

where $u = \sum_{n=-\infty}^{+\infty} u_n e_n$ and $v = \sum_{n=-\infty}^{+\infty} v_n e_n$. It is easy to see the embedding $\tau : H \hookrightarrow L^2$ is compact. Now, define a self-adjoint operator A on H by

$$(Au, v)_H := (\Box u, v)_{L^2}, \tag{11.33}$$

it is easy to see A is unbounded, its spectrum $\sigma(A) = \{\lambda_n / |\lambda_n|^{1/2}\}$ satisfying

$$Ae_n = \frac{\lambda_n}{|\lambda_n|^{1/2}} e_n.$$

Thus A has compact resolvent. By Remark 3.7, $\mathcal{L}_s^-(H, \lambda^-) = \mathcal{L}_s(H)$, for any $g \in L^\infty$, by (11.22), g will define a bounded self-adjoint operator $\tau^* g$ on H by

$$(\tau^* gu, v)_H := (gu, v)_{L^2}, \forall u, v \in H,$$

where τ^* is the dual operator of τ. Thus we have the index pair $(i_A^-(\tau^* g), \nu_A(\tau^* g))$ for any $g \in L^\infty$. For simplicity, we denote it by $(i_A(g), \nu_A(g))$. In addition to Theorem 2.1, assume $f \in C^1(\Omega \times \mathbb{R}, \mathbb{R})$ satisfying the following conditions.

(f_0) $f(x, t, 0) \equiv 0$, and denote $g_0 := f_u'(x, t, 0) \in L^\infty$.

(f_1) There exists $c_f > 0$ such that

$$|f_u'(x, t, u)| \leq c_f, \ \forall (x, t, u) \in \Omega \times \mathbb{R}.$$

$(f_{\infty,1})$ There exists $g_\infty \in L^\infty$ with $\nu_A(g_\infty) = 0$ such that

$$f(x, t, u) = g_\infty(x, t)u + o(|u|), \text{ as } |u| \to \infty, \ \forall (x, t) \in \Omega.$$

$(f_{\infty,2})$ There exists a number $R > 0$, and $g_{\infty,a}, g_{\infty,b} \in L^\infty$ with $i_A(g_{\infty,a}) = i_A(g_{\infty,b})$ and $\nu_A(g_{\infty,b}) = 0$ satisfying

$$g_{\infty,a}(x, t) \leq f_u'(x, t, u) \leq g_{\infty,b}(x, t), \ \forall (x, t) \in \Omega, \ |u| > R,$$

The condition (f_0) means that $u \equiv 0$ is a trivial solution of $(W.E.)$. Corresponding to Theorem 1.3, we have the following result.

Theorem 2.2 ([284]) *Assume $f \in C^1(\Omega \times \mathbb{R}, \mathbb{R})$ satisfying (f_0), (f_1) and $(f_{\infty,1})$ (or $(f_{\infty,2})$). Further more, corresponding to the parity of b, assume f satisfying (f_{odd}) or (f_{even}). For the results, we have the following two parts.*

Part 1. *If*

$$i_A(g_\infty) \notin [i_A(g_0), i_A(g_0) + \nu_A(g_0)] (or \ i_A(g_{\infty,a}) \notin [i_A(g_0), i_A(g_0) + \nu_A(g_0)]), \quad (11.34)$$

then the problem $(W.E.)$ has at least one nontrivial weak solution. More over, denote this nontrivial solution by u_1, and denote $\nu_A(u_1) := \nu_A(f_u'(x, t, u_1))$. If $\nu_A(g_0) = 0$, and

$$\nu_A(u_1) < |i_A(g_1) - i_A(g_0)|,$$

then the problem $(W.E.)$ has one more nontrivial weak solution.

Part 2. *If f is odd in u and*

$$i_A(g_0) > i_A(g_\infty), \tag{11.35}$$

then the problem $(W.E.)$ has at least $i_A(g_0) - i_A(g_\infty)$ pairs of nontrivial weak solutions.

Proof Since the proof of this theorem is similar Theorem 2.1, we only give a brief proof here. Corresponding to the parity of b, let the working space H defined in (11.32) and operator A defined in (11.33). Recall the function $\mathcal{F}$ and functional F defined as above with f satisfying (f_{odd}) or (f_{even}), then the solutions of $(W.E.1)$ and $(W.E.2)$ are the solutions of $(O.E.)$. Since the embedding $\tau : H \hookrightarrow L^2$ is compact, $f \in C^1([0,\pi] \times \mathbb{R}^2, \mathbb{R})$ and f'_u is bounded, we have $F \in C^2(H, \mathbb{R})$ satisfying

$$(F'(u), v)_H = (f(x,t,u), v)_{L^2}, \ \forall u, v \in H,$$

and

$$(F''(u)v, w)_H = (f'_u(x,t,u)v, w)_{L^2}, \ \forall u, v, w \in H.$$

That is to say $F'(u) = \tau^* f(x,t,u)$ and $F''(u) = \tau^* f'_u(x,t,u)$. Now we only need to check the conditions of Theorem 1.3 step by step. Firstly, from condition (f_0), we have $F'(0) = 0$ and $F''(0) = \tau^* g_0$. Secondly, from condition (f_1), we have F'' is bounded on H. Thirdly, from condition $(f_{\infty,1})$, we have $F'(u) = \tau^* g_\infty u + o(\|u\|_H)$ as $\|u\|_H \to \infty$. Lastly, instead of verifying condition $(F^{\mp}_{\infty,2})$, with the same method in Theorem 2.1, from condition $(f_{\infty,2})$, we can get a similar estimate as (11.28), so Lemma 1.2 and then Theorem 1.3 will keep valid. □

11.2.2 n-Dimensional Wave Equations

In this subsection, we consider the existence and multiplicity of radially symmetric solutions for the n-dimensional wave equation:

$$\begin{cases} \Box u \equiv u_{tt} - \Delta_x u = h(x,t,u), \ t \in \mathbb{R}, & x \in B_R, \\ u(x,t) = 0, t \in \mathbb{R}, & t \in \mathbb{R}, \ x \in \partial B_R, \\ u(x,t+T) = u(x,t), & t \in \mathbb{R}, \ x \in B_R, \end{cases} \tag{n–$W.E.$}$$

where $B_R = \{x \in \mathbb{R}^n, |x| < R\}$, $\partial B_R = \{x \in \mathbb{R}^n, |x| = R\}$, $n > 1$ and the nonlinear term h is T-periodic in variable t. Restriction of the radially symmetry allows us to know the nature of spectrum of the wave operator. Let $r = |x|$ and $S^1 := \mathbb{R}/T$, if $h(x,t,u) = h(r,t,u)$ then the n-dimensional wave equation (n–W.E.) can be transformed into:

$$\begin{cases} A_0 u := u_{tt} - u_{rr} - \frac{n-1}{r} u_r = h(r,t,u), \\ u(R,t) = 0, & (r,t) \in \Omega := [0,R] \times S^1. \\ u(r,0) = u(r,T),\ u_t(r,0) = u_t(r,T), \end{cases} \tag{RS–W.E.}$$

A_0 is symmetric on $L^2(\Omega, \rho)$, where $\rho = r^{n-1}$ and

$$L^2(\Omega,\rho) := \left\{ u \,|\, \|u\|^2_{L^2(\Omega,\rho)} := \int_\Omega |u(t,r)|^2 r^{n-1} dt dr < \infty \right\}.$$

By the method of separation of variables, we obtain the eigenvalues of the operator A_0 are

$$\lambda_{jk} = \left(\frac{\gamma_j}{R}\right)^2 - \left(\frac{2k\pi}{T}\right)^2, \quad j \in \mathbb{Z}_+,\ k \in \mathbb{Z},$$

where γ_j is the j-th positive zero of $J_\nu(x)$, $\nu = (n-2)/2$, and J_ν is the Bessel function of the first kind of order ν. Denote the corresponding eigenfunctions by $\{\psi_{jk}(t,r)\}$. We can check that $\{\psi_{jk}\}$ form a complete orthonormal sequence in $L^2(\Omega,\rho)$. By the asymptotic properties of the Bessel functions (see [290]), the spectrum of the wave operator can be characterized as follows.

Lemma 2.3 (See [266, Theorem 2.1]) *Assume that* $8R/T = a/b$, $(a,b) = 1$, *then* A_0 *has a self-adjoint extension* A *having no essential spectrum other than the point* $\lambda_0 = -(n-3)(n-1)/4R^2$. *If* $n \not\equiv 3(mod(4,a))$, *then* A *has no essential spectrum. If* $n = 3(mod(4,a))$, *the essential spectrum of* A *is precisely the point* $\lambda_0 = -(n-3)(n-1)/4R^2$.

Thus for $n = 3(mod(4,a))$, the operator A has only one essential spectrum $\lambda_0 = -(n-3)(n-1)/4R^2$, that is to say $\lambda^\mp = \lambda_0$, we can define the index pair $(i_A^\mp(B), \nu_A(B))$ for any $B \in \mathcal{L}_s^\mp(L^2(\Omega,\rho),\lambda_0) \subset \mathcal{L}_s(L^2(\Omega,\rho))$ and with the same discussion in above subsection, we have the index pair $(i_A^\mp(g), \nu_A(g))$ for any $g \in L^\infty(\Omega,\mathbb{R}) \cap \mathcal{L}_s^\mp(L^2(\Omega,\rho),\lambda_0)$. Assume $h \in C^1(\Omega \times \mathbb{R}, \mathbb{R})$ satisfying the following conditions.

$(h_0^\mp)$ $h(r,t,0) \equiv 0$, and denote $g_0 := h'_u(r,t,0) \in L^\infty(\Omega,\mathbb{R}) \cap \mathcal{L}_s^\mp(L^2(\Omega,\rho),\lambda_0)$.

$(h_1^\mp)$ There exists a constant $\delta > 0$ such that

$$\pm h'_u(r,t,u) < \pm(\lambda_0 \mp \delta), \ \forall (r,t,u) \in \Omega \times \mathbb{R}.$$

$(h_2^\mp)$ There exist $g_1, g_2 \in L^\infty(\Omega,\mathbb{R}) \cap \mathcal{L}_s^\mp(L^2(\Omega,\rho),\lambda_0)$ with $\pm g_1 < \pm g_2$, $i_A^\mp(g_1) + \nu_A(g_1) = i_A^\mp(g_2)$ and $\nu_A(g_2) = 0$, such that

$$\pm h'_u(r,t,u) \geq \pm g_1(r,t), \ \forall (r,t,u) \in \Omega \times \mathbb{R}$$

and

$$\pm h'_u(r,t,u) \le \pm g_2(r,t),\ \forall (r,t) \in \Omega \text{ and } |u| > K$$

for some constant $K > 0$.

We have the similar result as Theorem 2.1 for (n–W.E.).

Theorem 2.4 ([284]) *For $n = 3(mod(4,a))$ and the operator A defined above, assume $h \in C^1(\Omega \times \mathbb{R}, \mathbb{R})$ satisfying conditions $(h_1^{\mp})$ and $(h_2^{\mp})$, then (n–W.E.) has a radially symmetrical weak solution. Further more, if h satisfies condition $(h_0^{\mp})$ and*

$$i_A^{\mp}(g_0) > i_A^{\mp}(g_2),$$

then (n–W.E.) has a nontrivial radially symmetrical weak solution.

For $n \neq 3(mod(4,a))$, since A has no essential spectrum, we have $\lambda^- = +\infty$ and $\mathcal{L}_s^-(L^2(\Omega,\rho),\lambda^-) = \mathcal{L}_s(L^2(\Omega,\rho))$ as claimed in Remark 3.7. In addition to Theorem 2.4, we have the similar result as Theorem 2.2. Assume $h \in C^1(\Omega \times \mathbb{R}, \mathbb{R})$ satisfying the following conditions

(h_0) $h(r,t,0) \equiv 0$, and denote $g_0 := h'_u(r,t,0) \in L^\infty(\Omega,\mathbb{R})$.
(h_1) There exists $c_h > 0$ such that

$$|h'_u(r,t,u)| \le c_h,\ \forall (r,t,u) \in \Omega \times \mathbb{R}.$$

$(h_{\infty,1})$ There exists $g_\infty \in L^\infty$ with $\nu_A(g_\infty) = 0$ such that

$$h(r,t,u) = g_\infty(r,t)u + o(|u|),\ \text{as } |u| \to \infty,\ \forall (r,t) \in \Omega.$$

$(h_{\infty,2})$ There exists a number $K > 0$, and $g_{\infty,a}, g_{\infty,b} \in L^\infty(\Omega,\mathbb{R})$ with $i_A(g_{\infty,a}) = i_A(g_{\infty,b})$ and $\nu_A(g_{\infty,b}) = 0$ satisfying

$$g_{\infty,a}(r,t) \le h'_u(r,t,u) \le g_{\infty,b}(r,t),\ \forall (r,t) \in \Omega,\ |u| > K.$$

Theorem 2.5 ([284]) *For $n \neq 3(mod(4,a))$ and the operator A defined above, assume $h \in C^1(\Omega \times \mathbb{R}, \mathbb{R})$ satisfying (h_0), (h_1) and $(h_{\infty,1})$ (or $(h_{\infty,2})$). We have the following two parts of results.*

Part 1. *If*

$$i_A(g_\infty) \notin [i_A(g_0), i_A(g_0) + \nu_A(g_0)] (or\ i_A(g_{\infty,a}) \notin [i_A(g_0), i_A(g_0) + \nu_A(g_0)]), \tag{11.36}$$

then the problem (n–W.E.) has at least one nontrivial radially symmetrical weak solution. More over, denote this nontrivial solution by u_1, and denote $\nu_A(u_1) := \nu_A(h'_u(r,t,u_1))$. If $\nu_A(g_0) = 0$, and

$$\nu_A(u_1) < |i_A(g_1) - i_A(g_0)|,$$

then the problem (n–W.E.) has one more nontrivial radially symmetrical weak solution.

Part 2. *If h is odd in u and*

$$i_A(g_0) > i_A(g_\infty), \tag{11.37}$$

then the problem (n–W.E.) has at least $i_A(g_0) - i_A(g_\infty)$ pairs of nontrivial radially symmetrical weak solutions.

Remark 2.6

A. Generally, it is hard to compute the index of a bounded self-adjoint operator though we have given its definition, but in some special conditions we can do it easily. For example, from Lemma 2.3, let $\sigma(A) = \{\lambda_n^{\mp}\} \cup \{\lambda_0\}$ satisfying

$$-\infty \leftarrow \lambda_{-l}^- < \cdots < \lambda_l^- \to \lambda_0, \ l \to +\infty,$$

and

$$\lambda_0 \leftarrow \lambda_{-l}^+ < \cdots < \lambda_l^+ \to +\infty, \ l \to +\infty,$$

where λ_0 is the unique essential spectrum of A (if A has essential spectrum). If the functions g_0, g_∞ in conditions $(f_0^{\mp})$ and $(f_{\infty,1}^{\mp})$, for example, satisfying the following inequality

$$\lambda_{k-1}^- < g_\infty(x,t) < \lambda_k^- \cdots \le \lambda_l^- < g_0(x,t) < \lambda_0, \ \forall (x,t) \in \Omega,$$

or

$$\lambda_0 < g_0(x,t) < \lambda_k^+ \cdots \le \lambda_l^+ < g_\infty(x,t) < \lambda_{l+1}^+, \ \forall (x,t) \in \Omega,$$

for some $k, l \in \mathbb{Z}$, it is easy to see $\nu_A(g_\infty) = 0$ and $i_A^-(g_0) - i_A^-(g_\infty) > l - k + 1$ (or $i_A^+(g_0) - i_A^+(g_\infty) > l - k + 1$) satisfying the twisted conditions.

B. The discussion in above two subsections can also be used in the existence and multiplicity of nontrivial solutions of the nonlinear beam equation

$$\begin{cases} u_{tt} + u_{xxxx} = f(x,t,u), \\ u(0,t) = u(\pi,t) = 0, \\ u_{xx}(0,t) = u_{xx}(\pi,t) = 0, \\ u(x,t+T) = u(x,t), \end{cases} \forall (x,t) \in [0,\pi] \times \mathbb{R}, \tag{B.E.}$$

where $T > 0$, $f \in C^1([0,\pi] \times \mathbb{R}^2, \mathbb{R})$ and T-periodic in variable t. If we also assume systematically that T is a rational multiple of π, then the spectrum of operator $\partial_{tt}^2 + \partial_{xxxx}^4$ will be similar to the spectrum of one dimensional wave operator, so all of the results will keep valid in this problem.

Bibliography

1. Abbondandolo, A.: Index estimates for strongly indefinite functionals, periodic orbits and homoclinic solutions of first order Hamiltonian systems. Calc. Var. Partial Differ. Equ. **11**(4), 395–430 (2000)
2. Abbondandolo, A.: Morse Theory for Hamiltonian Systems. Chapman Hall CRC Research Notes in Mathematics, vol. 425. Chapman Hall CRC, Boca Raton (2001)
3. Abbondandolo, A.: On the Morse index of Lagrangian systems. Nonlinear Anal. **53**(3–4), 551–566 (2003)
4. Amann, H.: Saddle points and multiple solutions of differential equations. Math. Z. **169**, 122–166 (1979)
5. Amann, H., Zehnder, E.: Nontrivial solutions for a class of non–resonance problems and applications to nonlinear differential equations. Ann. Scoula Norm. Sup. Pisa. Cl. Sci. Ser. **4.7**, 539–603 (1980)
6. Ambrosetti, A., Benci, V., Long, Y.: A note on the existence of multiple brake orbits. Nonlinear Anal. T. M. A. **21**, 643–649 (1993)
7. An, T., Long, Y.: On the index theories for second order Hamiltonian systems. Nonlinear Anal. **34**, 585–592 (1998)
8. Arnold, V. I.: Sur une proprétés des application globalement canoniques de la méchanique classique. COMPTES Rendus de I'Académic. Paris **261**, 3719–3722 (1965)
9. Arnold, V.: On the characteristic class entering in the quantization condition. Funct. Anal. Appl. **1**, 1–13 (1967)
10. Arnold, V.: The sturm theorems and symplectic geometry (AMS Trans.) Funkts. Anal. Prilozh. **19**, 11–12 (1985)
11. Atiyah, M., Patodi, V., Singer, I.: Spectral asymmetry and Riemannian geometry III. Proc. Camb. Phic. Soc. **79**, 71–99 (1976)
12. Babyaga, A., Houenou, D.: A Brief Introduction to Symplectic and Contact Manifolds. Nankai Tracts in Mathematics, vol. 15. World Scientific, Singapore (2016)
13. Bahri, A.: Critical Points at Infinity in Some Variational Problems. Longman Scientific and Technical, New York (1989)
14. Bahri, A., Berestycki, H.: Forced vibrations of superquadratic Hamiltonian systems. Acta Math. **152**, 143–197 (1984)
15. Bartsch, T.: Topological Methods for Variational Problems with Symmetries. Lecture Notes in Mathematics, vol. 1560. Springer, Berlin (1993)
16. Bartsch, T.: A generalization of the Weinstein-Moser theorems on periodic orbits of a Hamiltonian system near an equilibrium. Ann. Inst. H. Poincaré Anal. Non Linéaire **14**(6), 691–718 (1997)

17. Bartsch, T., Ding, Y.: Solutions of nonlinear Dirac equations. J. Differ. Equ. **226**(1), 210–249 (2006)
18. Benci, V.: A geometric index for the group S^1 and some applications to the periodic solutions of ordinary differential equations. Commun. Pure Appl. Math. **34**, 393–432 (1981)
19. Benci, V.: Closed geodesics for the Jacobi metric and periodic solutions of prescribed energy of natural Hamiltonian systems. Ann. I. H. P. Analyse Nonl. **1**, 401–412 (1984)
20. Benci, V.: A new approach to the Morse-Conley theory and some applications. Annali Mat. Pura Appl. **158**, 231–305 (1991)
21. Benci, V., Giannoni, F.: A new proof of the existence of a brake orbit. In: Advanced Topics in the Theory of Dynamical Systems. Notes and Reports in Mathematics in Science and Engineering Series, vol. 6, pp. 37–49. Academic, Boston (1989)
22. Benci, V., Rabinowitz, P.: Critical point theorem for indefinite functionals. Inv. Math. **52**, 241–273 (1979)
23. Bolotin, S.: Libration motions of natural dynamical systems. Vestnik Moskov Univ. Ser. I. Mat. Mekh. **6**, 72–77 (1978) (in Russian)
24. Bolotin, S., Kozlov, V.: Librations with many degrees of freedom. J. Appl. Math. Mech. **42**, 245–250 (1978) (in Russian)
25. Booβ-Bavnbek, B., Zhu, C.: The Maslov index in weak symplectic functional analysis. Ann. Global Anal. Geom. **44**(3), 283–318 (2013)
26. Booβ-Bavnbek, B., Chen, G., Lesch, M., Zhu, C.: Perturbation of sectorial projections of elliptic pseudo-differential operators. J. Pseudo-Differ. Oper. Appl. **3**(1), 49–79 (2012)
27. Booβ-Bavnbek, B., Lesch, M., Zhu, C.: The Calderón projection: new definition and applications. J. Geom. Phys. **59**(7), 784–826 (2009)
28. Bott, R.: On the iteration of closed geodesics and the Sturm intersection theory. Commun. Pure Appl. Math. **9**, 171–206 (1956)
29. Bott, R.: Lectures on Morse theory, old and new. Bull. Am. math. Soc. **7**(2), 331–358 (1982)
30. Brezis, H.: Functional Analysis. Sobolev Spaces and Partial Differential Equations. Springer, New York (2011)
31. Broöcker, T., tom Dieck, T.: Representations of Compact Lie Groups. Springer, New York (1985)
32. Cappell, S., Lee, R., Miller, E.: On the Maslov index. Commun. Pure Appl. Math. **47**, 121–186 (1994)
33. Chang, K.: Solutions of asymptotically linear operator equations via Morse theory. Commun. Pure Appl. Math. **34**, 693–712 (1981)
34. Chang, K.: Infinite Dimensional Morse Theory and Multiple Solution Problems. Birkhäuser, Basel (1993)
35. Chang, K.: Methods in Nonlinear Analysis. Springer, Berlin/Heidelberg (2005)
36. Chang, K.: Critical Point Theory and Its Applications. Shanghai Sci. Tech. Press, Shanghai (1986) (in Chinese)
37. Chang, K., Jiang, M.: The Lagrangian intersections for (CP^n, RP^n). Manuscripta Math. **68**, 89–100 (1990)
38. Chang, K., Liu, J., Liu, M.: Nontrivial periodic solutions for strong resonance Hamiltonian systems. Ann. Inst. H. Poincaré Anal. Nonlinéaire **14**(1), 103–117 (1997)
39. Chen, K.: Existence and minimizing properties of retrograd orbits in three-body problem with various choice of mass. Ann. Math. **167**(2), 325–348 (2008)
40. Chen, C., Hu, X.: Maslov index for homoclinic orbits of Hamiltonian systems. Ann. Inst. H. Poincaré, Anal. Non linéaire **24**, 589–603 (2007)
41. Chen, Y., Liu, C.: Ground state solutions for non-autonomous fractional Choquard equations. Nonlinearity **29**, 1827–1842 (2016)
42. Chen, Y., Liu, C., Zheng, Y.: Existence results for the fractional Nirenberg problem. J. Funct. Anal. **270**, 4043–4086 (2016)
43. Chen, Z., Zou, W.: Existence and symmetry of positive ground states for a doubly critical Schrödinger system. Trans. Am. Math. Soc. **367**(5), 3599–3646 (2015)

44. Cheng, C., Sun, Y.: Regular and Stochastic Motions in Hamiltonian Systems. The Publishing House, Shanghai (1996) (in Chinese)
45. Chenciner, A., Montgomery, R.: A remarkable periodic solution of the three body problem in the case of equal masses. Ann. Math. **152**(3), 881–901 (2000)
46. Chern, S., Shen, Z.: Riermann-Finsler Geometry. Nankai Tracts in Mathematics. World Scientific, Singapore (2005)
47. Cieliebak, K., Eliashberg, Y.: The topology of rationally and polynomially convex domains. Invent. Math. **199**(1), 215–238 (2015)
48. Cieliebak, K., Floer, A., Hofer, H.: Symplectic homology. II. A general construction. Math. Z. **218**(1), 103–122 (1995)
49. Clarke, F.: Solutions périodiques des équations hamiltoniennes. Note CRAS Paris **287**, 951–952 (1978)
50. Clarke, F.: A classical variational principle for periodic Hamiltonian trajectories. Proc. Am. Math. Soc. **76**, 186–189 (1979)
51. Clarke, F.: Periodic solutions of Hamiltonian inclusions. J. Differ. Equ. **40**, 1–6 (1980)
52. Clarke, F., Ekeland, I.: Solutions périodiques, de période donnée, des é quations hamiltoniennes. Note CRAS Paris **287**, 1013–1015 (1978)
53. Clarke, F., Ekeland, I.: Hamiltonian trajectories having prescribed minimal period. Commun. Pure Appl. Math. **33**, 103–116 (1980)
54. Clarke, F., Ekeland, I.: Nonlinear oscillations and boundary value problems for Hamiltonian systems. Arch. Rational Mech. Anal. **78**(4), 315–333 (1982)
55. Conley, C.: An Oscillation Theorem for Linear System with More than One Degree of Freedom. IBM Technical report:18004, IBM Watson Research Center, Yorktown Heights, Now York (1972)
56. Conley, C.: Isolated Invariant Sets and the Morse Index. CBMS, vol. 38. American Mathematical Society, Providence (1978)
57. Conley, C., Zehnder, E.: Morse-type index theory for flows and periodic solutions for Hamiltonian equations. Commun. Pure. Appl. Math. **37**, 207–253 (1984)
58. Coron, J.: Periodic solutions of a nonlinear wave equation without assumption of monotonicity. Math. Ann. **262**, 273–285 (1983)
59. Deimling, K.: Nonlinear Functional Analysis. Springer, Berlin (1985)
60. del Pino, M., Kowalczyk, M., Wei, J.: On De Giorgi's conjecture in dimension $N = 9$. Ann. Math. (2) **174**(3), 1485–3569 (2011)
61. Ding, W.: A generalization of the Poincaré-Birkhoff theorem. Proc. Am. Math. Soc. **88**, 341–346 (1983)
62. Ding, W., Tang, H., Zeng, C.: Self-similar solutions of Schrödinger flows. Calc. Var. Partial Differ. Equ. **34**(2), 267–277 (2009)
63. Ding, W., Li, J., Liu, Q.: Evolution of minimal torus in Riemannian manifolds. Invent. Math. **165**(2), 225–242 (2006)
64. Ding, Y.: Variational Methods for Strongly Indefinite Problems. Interdisciplinary Mathematical Sciences, vol. 7. World Scientific Publishing Co. Pte. Ltd., Hackensack (2007)
65. Ding, Y., Xu, T.: Concentrating patterns of reaction-diffusion systems: a variational approach. Trans. Am. Math. Soc. **369**(1), 97–138 (2017)
66. Dong, D., Long, Y.: The iteration formula of the Maslov-type index theory with applications to nonlinear Hamiltonian systems. Trans. Am. Math. Soc. **349**, 2619–2661 (1997)
67. Dong, Y.: Maslov type index theory for linear Hamiltonian systems with Bolza boundary value conditions and multiple solutions for nonlinear Hamiltonian systems. Pac. J. Math. **221**(2), 253–280 (2005)
68. Dong, Y.: P-index theory for linear Hamiltonian systems and multiple solutions for nonlinear Hamiltonian systems. Nonlinearity **19**(6), 1275–1294 (2006)
69. Dong, Y.: Index theory for linear selfadjoint operator equations and nontrivial solutions for asymptotically linear operator equations. Calc. Var. **38**, 75–109 (2010)
70. Dong, Y.: Index Theory for Hamiltonian Systems and Multiple Solution Problems. Science Press, Beijing (2015)

71. Dong, Y., Long, Y.: Closed characteristics on partially symmetric convex hypersurfaces in $\mathbb{R}^{2n}$. J. Differ. Equ. **196**, 226–248 (2004)
72. Duan, H.: Non-hyperbolic closed geodesics on positively curved Finsler spheres. J. Funct. Anal. **269**(11), 3645–3662 (2015)
73. Duan, H.: Two elliptic closed geodesics on positively curved Finsler spheres. J. Differ. Equ. **260**(12), 8388–8402 (2016)
74. Duan, H., Long, Y.: The index growth and multiplicity of closed geodesics. J. Funct. Anal. **259**(7), 1850–1913 (2010)
75. Duan, H., Long, Y., Wang, W.: Two closed geodesics on compact simply connected bumpy Finsler manifolds. J. Differ. Geom. **104**(2), 275–289 (2016)
76. Duistermaat, J.: Fourier Integral Operators. Birkhäuser, Basel (1996)
77. Duistermaat, J.: On the Morse index in variational calculus. Adv. Math. **21**, 173–195 (1976)
78. Ekeland, I.: Convexity Methods in Hamiltonian Mechanics. Springer, Berlin (1990)
79. Ekeland, I.: An index theory for periodic solutions of convex Hamiltonian systems. Proc. Symp. Pure Math. **45**, 395–423 (1986)
80. Ekeland, I., Hofer, H.: Periodic solutions with prescribed period for convex autonomous hamiltonian systems. Invent. Math. **81**, 155–188 (1985)
81. Ekeland, I., Hofer, H.: Subharmonics of convex Hamiltonian systems. Commun. Pure Appl. Math. **40**, 1–37 (1987)
82. Ekeland, I., Hofer, H.: Symplectic topology and Hamiltonian dynamics. Math. Z. **200**, 355–378 (1990)
83. Ekeland, I., Lasry, J.: On the number of periodic trajectories for a Hamiltonian flow on a convex energy surface. Ann. Math. **112**, 283–319 (1980)
84. Ekeland, I., Lassoued, L.: Multiplicité des trajectoires fermées d'un systéme hamiltonien sur une hypersurface d'energie convexe. Ann. IHP. Anal. Nonlinéaire. **4**, 1–37 (1987)
85. Ekeland, I., Long, Y., Zhou, Q.: A new class of problems in the calculus of variations. Regul. Chaotic Dyn. **18**(6), 553–584 (2013)
86. Eliashberg, Y.: Recent advances in symplectic flexibility. Bull. Am. Math. Soc. **52**(1), 1–26 (2015)
87. Everitt, W., Markus, L.: Complex symplectic spaces and boundary value problems. Bull. AMS. **42**(4), 461C500 (2005)
88. Fadell, E., Husseini, S., Rabinowitz, P., Borsuk-Ulam theorems for arbitrary S^2 actions and applications. Trans. Am. Math. Soc. **274**, 345–360 (1982)
89. Fadell, E., Rabinowitz, P.: Generalized cohomological theories for Lie group actions with an application to bifurcation questions for Hamiltonian systems. Invent. Math. **45**, 139–174 (1978)
90. Fan, H., Jarvis, T., Ruan, Y.: The Witten equation, mirror symmetry, and quantum singularity theory. Ann. Math. (2) **178**(1), 1–106 (2013)
91. Fei, G.: Relative morse index and its application to Hamiltonian systems in the presence of symmetries. J. Differ. Equ. **122**, 302–315 (1995)
92. Fei, G.: Multiple periodic solutions of differential delay equations via Hamiltonian systems(I). Nonlinear Anal. **65**, 25–39 (2006)
93. Fei, G.: Multiple periodic solutions of differential delay equations via Hamiltonian systems(II). Nonlinear Anal. **65**, 40–58 (2006)
94. Fei, G., Qiu, Q.: Periodic solutions of asymptotically linear Hamiltonian systems. Chin. Ann. Math. **18B**(3), 359–372 (1997)
95. Fei, G., Qiu, Q.: Minimal period solutions of nonlinear Hamiltonian systems. Nonlinear Anal. Theory Methods Appl. **27**, 821–839 (1996)
96. Fei, G., Kim, S., Wang, T.: Minimal period estimates of period solutions for superquadratic Hamiltonian systems. J. Math. Anal. Appl. **238**, 216–233 (1999)
97. Ferrario, D., Terracini, S.: On the existence of collisionless equivariant minimizers for the classical n-body problem. Invent. Math. **155**(2), 305–362 (2004)
98. Floer, A.: A relative Morse index for the symplectic action. Commun. Pure Appl. Math. **41**, 393–407 (1988)

99. Floer, A.: Morse theory for Lagrangian intersections. J. Differ. Geom. **28**(3), 513–547 (1988)
100. Floer, A.: Cuplength estimates on Lagrangian intersections. Commun. Pure Appl. Math. **42**(4), 335–356 (1989)
101. Floer, A.: Symplectic fixed points and holomorphic spheres. Commun. Math. Phys. **120**(4), 575–611 (1989)
102. Floer, A., Hofer, H.: Symplectic homology. I. Open sets in C^n. Math. Z. **215**(1), 37–88 (1994)
103. Franks, J.: Generalizations of the Poincaré-Birkhoff theorem. Ann. Math. **128**, 131–151 (1988)
104. Fukaya, K., Ono, K.: Arnold conjecture and Gromov-Witten invariant for general symplectic manifolds. Topology **38**, 933–1048 (1999)
105. Fukaya, K., Oh, Y., Ohta, H., Ono, K.: Lagrangian Floer theory and mirror symmetry on compact toric manifolds. Astérisque **376**, vi+340 (2016)
106. Fukaya, K., Oh, Y., Ohta, H., Ono, K.: Lagrangian intersection Floer theory: anomaly and obstruction. Part I. AMS/IP Studies in Advanced Mathematics, vol. 46.1. American Mathematical Society/International Press, Providence/Somerville (2009)
107. Fukaya, K., Oh, Y., Ohta, H., Ono, K.: Lagrangian intersection Floer theory: anomaly and obstruction. Part II. AMS/IP Studies in Advanced Mathematics, vol. 46.2. American Mathematical Society/International Press, Providence/Somerville (2009)
108. Gan, S., Wen, L.: Nonsingular star flows satisfy Axiom A and the no-cycle condition. Invent. Math. **164**(2), 279–315 (2006)
109. Ge, W.: Number of simple periodic solutions of differential difference equation $x'(t) = f(x(t-1))$. Chin. Ann. Math. **14A**, 472–479 (1993)
110. Ghoussoub, N.: Location, multiplicity and Morse indices of min-max critical points. J. Reine Angew Math. **417**, 27–76 (1991)
111. Ghoussoub, N.: Duality and Perturbation Methods in Critical Point Theory. Cambridge University Press, Cambridge (1993)
112. Giambò, R., Giannoni, F., Piccione, P.: Orthogonal geodesic chords, brake orbits and homoclinic orbits in Riemannian manifolds. Adv. Differ. Equ. **10**, 931–960 (2005)
113. Giambò, R., Giannoni, F., Piccione, P.: Existence of orthogonal geodesic chords on Riemannian manifolds with concave boundary and homeomorphic to the N-dimensional disk. Nonlinear Anal. T. M. A. **73**, 290–337 (2010)
114. Giambò, R., Giannoni, F., Piccione, P.: Multiple brake orbits in m-dimensional disks. Calc. Var. Partial Differ. Equ. **54**(3), 2553–2580 (2015)
115. Gilkey, P.B.: Invariance Theory, the Heat Equation, and the Atiyah-Singer Index Theorem. CRC Press, Boca Raton (1994)
116. Ginzburg, V.: The Conley conjecture. Ann. Math. (2) **172**(2), 1127–1180 (2010)
117. Ginzburg, V., Gürel, B.: Hyperbolic fixed points and periodic orbits of Hamiltonian diffeomorphisms. Duke Math. J. **163**(3), 565–390 (2014)
118. Ginzburg, V., Gürel, B.: The Conley conjecture and beyond. Arnold Math. J. **1**(3), 299–337 (2015)
119. Girardi, M., Matzeu, M.: Some results on solutions of minimal period to superquadratic Hamiltonian equations. Nonlinear Anal. T. M. A. **7**, 475–482 (1983)
120. Girardi, M., Matzeu, M.: Solutios of minimal period for a class of nonconvex Hamiltonian systems and applications to the fixed energy problem. Nonlinear Anal. T. M. A. **10**, 371–382 (1986)
121. Girardi, M., Matzeu, M.: Dual Morse index estimates for periodic solutions of Hamiltonian systems in some nonconvex superquadratic case. Nonlinear Anal. T. M. A. **17**, 481–497 (1991)
122. Gohberg, I., Goldberg, S., Krupnik, N.: Thaces and Determinants of Linear Operators. Birkhäuser, Berlin (2000)
123. Gluck, H., Ziller, W.: Existence of periodic solutions of conservative systems. In: Seminar on Minimal Submanifolds, pp. 65–98. Princeton University Press, Princeton (1983)
124. Gopalsamy, K., Li, J., He, X.: On the construction of periodic solutions of Kaplan Yorke type for some differential delay equations. Appl. Anal. **59**, 65–80 (1995)

125. Greenberg, L.: A Prüfer method for calculating eigenvalues of self–adjoint systems of ordinary differential equations. Part 1 (Preprint)
126. van Groesen, E.: Analytical mini-max methods for Hamiltonian brake orbits of prescribed energy. J. Math. Anal. Appl. **132**, 1–12 (1988)
127. Guillemin, V., Sternberg, S.: Geometric Asymptotics. Mathematical Surveys, vol. 14. AMS, Providence (1977)
128. Guo, F., Liu, C.: Multiplicity of Lagrangian orbits on symmetric star-shaped hypersurfaces. Nonlinear Anal. **69**(4), 1425–1436 (2008)
129. Guo, F., Liu, C.: Multiplicity of characteristics with Lagrangian boundary values on symmetric star-shaped hypersurfaces. J. Math. Anal. Appl. **353**(1), 88–98 (2009)
130. Guo, Z., Yu, J.: Multiplicity results for periodic solutions to delay differential equations via critical point theory. J. Differ. Equ. **218**, 15–35 (2005)
131. Guo, Z., Yu, J.: Multiplicity results on period solutions to higher dimensional differential equations with multiple delays. J. Dyn. Differ. Equ. **23**, 1029–1052 (2011)
132. Hale, J.: Theory of Functional Differential Equations. Springer, New York (1977)
133. Hale, J., Lunel, S.: Introduction to Functional Differential Equations. Springer, New York (1993)
134. Han, Z., Li, Y., Teixeira, E.: Asymptotic behavior of solutions to the sk-Yamabe equation near isolated singularities. Invent. Math. **182**(3), 635–684 (2010)
135. Hasselblatt, B., Katok, A.: A First Course in Dynamics. Cambridge University Press, Cambridge (2003)
136. Hayashi, K.: Periodic solution of classical Hamiltonian systems. Tokyo J. Math. **6**, 473–486 (1983)
137. Hofer, H.: Lusternik-Schnirelman theory for Lagrangian intersections. Ann I. H. P. Analyse Nonl **5**(5), 465–499 (1988)
138. Hofer, H., Wysocki, K., Zehnder, E.: The dynamics on three-dimensional strictly convex energy surfaces. Ann. Math. **148**(2), 197–289 (1998)
139. Hofer, H., Wysocki, K., Zehnder, E.: Pseudoholomorphic curves and dynamics in three dimensions. In: Handbook of Dynamical Systems, vol. 1A, pp. 1129–1188. North-Holland, Amsterdam (2002)
140. Hofer, H., Wysocki, K., Zehnder, E.: A characterization of the tight 3-sphere. II. Commun. Pure Appl. Math. **52**(9), 1139–1177 (1999)
141. Hofer, H., Wysocki, K., Zehnder, E.: Properties of pseudo-holomorphic curves in symplectisations. II. Embedding controls and algebraic invariants. Geom. Funct. Anal. **5**(2), 270–328 (1995)
142. Hofer, H., Zehnder, E.: Symplectic Invariants and Hamiltonian Dynamics. Birkhäuser, Basel (1994)
143. Hörmander, L.: Fourier integral operators I. Acta math. **127**, 79–183 (1971)
144. Hörmander, L.: The Analysis of Linear Partial Differential Operators, I–IV. Springer, Berlin (1983)
145. Hörmander, L.: Symplectic classification of quadratic forms and general Mahler formulas. Math. Z. **219**, 413–449 (1995)
146. Hu, J., Li, T., Ruan, Y.: Birational cobordism invariance of uniruled symplectic manifolds. Invent. Math. **172**(2), 231–275 (2008)
147. Hu, X., Long, Y., Sun, S.: Linear stability of elliptic Lagrangian solutions of the planar three-body problem via index theory. Arch. Ration. Mech. Anal. **213**(3), 993–1045 (2014)
148. Hu, X., Ou, Y., Wang, P.: Trace formula for linear Hamiltonian systems with its applications to elliptic Lagrangian solutions. Arch. Ration. Mech. Anal. **216**(1), 313–357 (2015)
149. Hu, X., Sun, S.: Morse index and stability of elliptic Lagrangian solutions in the planar three-body problem. Adv. Math. **223**(1), 98–119 (2010)
150. Hu, X., Sun, S.: Index and stability of symmetric periodic orbits in Hamiltonian systems with application to figure- eight orbit. Commun. Math. Phys. **290**(2), 737–777 (2009)
151. Hu, X., Wang, P.: Conditional Fredholm determinant for the S-periodic orbits in Hamiltonian systems. J. Funct. Anal. **261**(11), 3247–3278 (2011)

152. Huang, W., Song, S., Ye, X.: Nil Bohr-sets and almost automorphy of higher order. Mem. Am. Math. Soc. **241**(1143), v+83pp (2016)
153. Huang, W., Ye, X., Zhang, G.: Lowering topological entropy over subsets revisited. Trans. Am. Math. Soc. **366**(8), 4423–4442 (2014)
154. Hou, X., You, J.: Almost reducibility and non-perturbative reducibility of quasi-periodic linear systems. Invent. Math. **190**(1), 209–260 (2012)
155. Jabri, Y.: The Mountain Pass Theorem. Cambridge University Press, Cambridge (2003)
156. Jiang, M.: Hofer-Zehnder syplectic capacity for two dimensional manifolds. Proc. R. Soc. Edinb. **123A**, 945–950 (1993)
157. Jiang, M.: Periodic solutions of Hamiltonian systems on hypersurfaces in torus. Manuscript Math. **85**, 307–321 (1994)
158. Kaplan, J., Yorke, J.: Ordinary differential equations which yield periodic solutions of differential delay equations. J. Math. Anal. Appl. **48**, 317–324 (1974)
159. Kato, T.: Perturbation Theory for Linear Operators. Springer, Berlin (1995)
160. Katok, A., Nitica, V.: Rigidity in Higher Rank Abelian Group Actions. Volume I. Introduction and Cocycle Problem. Cambridge Tracts in Mathematics, vol. 185. Cambridge University Press, Cambridge (2011)
161. Klingenberg, W.: Riemannian Geometry. Walter de Gruyter, Berlin/New York (1982)
162. Kriegl, A., Michor, P.: The Convenient Setting of Global Analysis. American Mathematical Society, Providence (1997)
163. Lazer, A., Solomini, S.: Nontrivial solution of operator equations and Morse indices of critical points of min-max type. Nonlinear Anal. **12**, 761–775 (1988)
164. Lefschetz, S.: Algebraic Topology. American Mathematical Society, New York (1942)
165. Leray, J.: Lagrangian Analysis and Quantum Mechanics. The MIT Press, Cambridge (1982)
166. Li, A., Ruan, Y.: Symplectic surgery and Gromov-Witten invariants of Calabi-Yau 3-folds. Invent. Math. **145**(1), 151–218 (2001)
167. Li, C.: Generalized Poincaré-Hopf theorem and application to nonlinear elliptic problem. J. Funct. Anal. **267**(10), 3783–3814 (2014)
168. Li, C., Li, S.: Gaps of consecutive eigenvalues of Laplace operator and the existence of multiple solutions for superlinear elliptic problem. J. Funct. Anal. **271**(2), 245–263 (2016)
169. Li, C., Li, S., Liu, J.: Splitting theorem, Poincaré-Hopf theorem and jumping nonlinear problems. J. Funct. Anal. **221**, 439–455 (2005)
170. Li, C., Liu, C.: Nontrivial solutions of superquadratic Hamiltonian systems with Lagrangian boundary conditions and the L-index theory. Chin. Ann. Math. **29**(6), 597–610 (2008)
171. Li, C., Liu, C.: The existence of nontrivial solutions of Hamiltonian systems with Lagrangian boundary conditions. Acta Mathematica Scientia **29B**(2), 313–326 (2009)
172. Li, C., Liu, C.: Brake subharmonic solutions of first order Hamiltonian systems. Sci. China (Math.) **53**(10), 2719–2732 (2010)
173. Li, J., He, X.: Multiple periodic solutions of differential delay equations created by asymptotically linear Hamiltonian systems. Nonlinear Anal. T. M. A. **31**, 45–54 (1998)
174. Li, J., He, X.: Proof and generalization of Kaplan-Yorke's conjecture under the condition $f'(0) > 0$ on periodic solution of differential delay equations. Sci. China (Ser. A) **42**(9), 957–964 (1999)
175. Li, J., He, X., Liu, Z.: Hamiltonian symmetric groups and multiple periodic solutions of differential delay equations. Nonlinear Anal. **35**, 457–474 (1999)
176. Li, S.: A new Morse theory and strong resonance problems. Topol. Methods Nonlinear Anal. **21**(1), 81–100 (2003)
177. Li, S., Liu, J.: Morse theory and asymptotically linear Hamiltonian systems. J. Differ. Equ. **78**, 53–73 (1989)
178. Li, S., Liu, Z.: Multiplicity of solutions for some elliptic equation involving critical and supercritical Sobolev exponents. Topol. Methods Nonlinear Anal. **28**(2), 235–261 (2006)
179. Li, Y., Lin, Z.: A constructive proof of the Poincaré-Birkhoff theorem. Trans. Am. Math. Soc. **347**(6), 2111–2126 (1995)

180. Li, Y., Yi, Y.: Persistence of hyperbolic tori in Hamiltonian systems. J. Differ. Equ. **208**(2), 344–387 (2005)
181. Lieb, E., Loss, M.: Analysis. Graduate Studies in Mathematics, vol. 14. American Mathematical Society, Providence (2001)
182. Lin, C., Wei, J., Ye, D.: Classification and nondegeneracy of SU(n+1) Toda system with singular sources. Invent. Math. **190**(1), 169–207 (2012)
183. Lin, L., Liu, Z.: Multi-bump solutions and multi-tower solutions for equations on RN. J. Funct. Anal. **257**(2), 485–505 (2009)
184. Liu, C.: Subharmonic solutions of Hamiltonian systems. Nonlinear Anal. T. M. A. **42**, 185–198 (2000)
185. Liu, C.: Dual Morse index estimates for subquadratic functional and the stability of the closed characteristics of Hamiltonian systems. Acta Math. Sin. (Chinese ed.) **44**(6), 1073–1088 (2001)
186. Liu, C.: Maslov P-index theory for a symplectic path with applications. Chin. Ann. Math. **4**, 441–458 (2006)
187. Liu, C.: A note on the monotonicity of Maslov-type index of linear Hamiltonian systems with applications. Proc. R. Soc. Edinb. **135A**, 1263–1277 (2005)
188. Liu, C.: Periodic solutions of asymptotically linear delay differential systems via Hamiltonian systems. J. Differ. Equ. **252**, 5712–5734 (2012)
189. Liu, C.: Maslov-type index theory for symplectic paths with Lagrangian boundary conditions. Adv. Nonlinear Stud. **7**(1), 131–161 (2007)
190. Liu, C.: The relation of the Morse index of closed geodesics with the Maslov-type index of symplectic paths. Acta Math. Sin. (English Series) **21**(2), 237–248 (2005)
191. Liu, C.: Cup-length estimate for Lagrangian intersections. J. Differ. Equ. **209**, 57–76 (2005)
192. Liu, C.: Asymptotically linear Hamiltonian systems with Lagrangian boundary conditions. Pacific J. Math. **232**(1), 233–255 (2007)
193. Liu, C.: Minimal period estimates for brake orbits of nonlinear symmetric Hamiltonian systems. Discret. Continuous Dyn. Syst. **27**(1), 337–355 (2010)
194. Liu, C.: Relative index theories and applications. Topol. Methods Nonlinear Anal. **49**(2), 587–614 (2017)
195. Liu, C.: A note on the relations between the various index theories. J. Fixed Point Theory Appl. **19**(1), 617–648 (2017)
196. Liu, C., Long, Y.: An Optimal estimate of Maslov-type index. Chin. Sci. Bull. **43**, 1063–1066 (1998)
197. Liu, C., Long, Y.: Iteration inequalities of the Maslov-type index theory with applications. J. Differ. Equ. **165**, 355–376 (2000)
198. Liu, C., Long, Y., Zhu, C.: Multiplicity of closed characteristics on symmetric convex hypersurfaces in $\mathbb{R}^{2n}$. Math. Ann. **323**(2), 201–215 (2002)
199. Liu, C., Ren, Q.: On the steady-state solutions of a nonlinear photonic lattice model. J. Math. Phys. **56**, 031501, 1–12 (2015)
200. Liu, C., Ren, Q.: Infinitely many non-radial solutions for fractional Nirenberg problem. Calc. Var. Partial Differ. Equ. **56**, 52–91 (2017)
201. Liu, C., Ren, Q.: Multi-bump solutions for fractional Nirenberg problem. Nonlinear Anal. **171**, 177–207 (2018)
202. Liu, C., Tang, S.: Maslov (P, ω)-index theory for symplectic paths. Adv. Nonlinear Stud. **15**, 963–990 (2015)
203. Liu, C., Tang, S.: Iteration inequalities of the Maslov P-index theory with applications. Nonlinear Anal. **127**, 215–234 (2015)
204. Liu, C., Tang, S.: Subharmonic P-solutions of first order Hamiltonian systems. J. Math. Anal. Appl. **453**(1), 338–359 (2017)
205. Liu, C., Wang, Q.: Some abstract critical point theorem for self-adjoint operator equations. Chin. Ann. Math. **32B**(1), 1–14 (2011)
206. Liu, C., Wang, Q.: Symmetrical symplectic capacity with applications. Discret. Continuous Dyn. Syst. **32**(6), 2253–2270 (2012)

207. Liu, C., Wang, Q., Lin, X.: An index theory for symplectic paths associated with two Lagrangian subspaces with applications. Nonlinearity **24**(1), 43–70 (2011)
208. Liu, C., Wang, Y.: Existence results for the fractional Q-curvature problem on three dimensional CR sphere. Commun. Pure Appl. Anal. **17**(3), 849–885 (2018)
209. Liu, C., Zhang, D.: Iteration theory of L-index and multiplicity of brake orbits. J. Differ. Equ. **257**(4), 1194C1245 (2014)
210. Liu, C., Zhang, D.: Seifert conjecture in the even convex case. Commun. Pure Appl. Math. **67**(10), 1563–1604 (2014)
211. Liu, C., Zhang, Q.: Nontrivial solutions for asymptotically linear Hamiltonian systems with Lagrangian boundary conditions. Acta Matematica Scientia **32B**(4), 1545–1558 (2012)
212. Liu, C., Zhang, X.: Subharmonic solutions and minimal periodic solutions of first-order Hamiltonian systems with anisotropic growth. Discret. Continuous Dyn. Syst. **37**(3), 1559–1574 (2017)
213. Liu, C., Zheng, Y.: Linking solutions for p-Laplace equations with nonlinear boundary conditions and indefinite weight. Calc. Var. Partial Differ. Equ. **41**(1–2), 261–284 (2011)
214. Liu, C., Zheng, Y.: Existence of nontrivial solutions for p-Laplacian equations in $\mathbb{R}^N$. J. Math. Anal. Appl. **380**(2), 669–679 (2011)
215. Liu, C., Zheng, Y.: On soliton solutions of a class of Schrodinger-KdV system. Proc. Am. Math. Soc. **141**(10), 3477–3484 (2013)
216. Liu, C., Zhou, B.: Minimal P-symmetric period problem of first-order autonomous Hamiltonian systems. Front. Math. China **12**(3), 641–654 (2017)
217. Liu, H.: Multiple P-invariant closed characteristics on partially symmetric compact convex hypersurfaces in $\mathbb{R}^{2n}$. Calc. Var. Partial Differ. Equ. **49**(3–4), 1121–1147 (2014)
218. Liu, H., Long, Y.: Resonance identities and stability of symmetric closed characteristics on symmetric compact star-shaped hypersurfaces. Calc. Var. Partial Differ. Equ. **54**(4), 3753–3787 (2015)
219. Liu, H., Long, Y.: Resonance identity for symmetric closed characteristics on symmetric convex Hamiltonian energy hypersurfaces and its applications. J. Differ. Equ. **255**(9), 2952–2980 (2013)
220. Liu, H., Long, Y., Wang, W.: Resonance identities for closed characteristics on compact star-shaped hypersurfaces in $\mathbb{R}^{2n}$. J. Funct. Anal. **266**(9), 5598–5638 (2014)
221. Liu, Z., Wang, Z.: On Clark's theorem and its applications to partially sublinear problems. Ann. Inst. H. Poincaré Anal. Non Linéaire **32**(5), 1015–1037 (2015)
222. Liu, Z., Su, J., Wang, Z.: A twist condition and periodic solutions of Hamiltonian system. Adv. Math. **218**, 1895–1913 (2008)
223. Long, Y.: Index Theory for Symplectic Path with Applications. Progress in Mathematics, vol. 207. Birkhäuser, Basel (2002)
224. Long, Y.: Maslov-type index, degenerate critical points, and asymptotically linear Hamiltonian systems. Sci. China **33**, 1409–1419 (1990)
225. Long, Y.: A Maslov-type index theory for symplectic paths. Top. Methods Nonlinear Anal. **10**, 47–78 (1997)
226. Long, Y.: Bott formula of the Maslov-type index theory. Pacific J. Math. **187**, 113–149 (1999)
227. Long, Y.: Index Theory of Hamiltonian Systems with Applications. Science Press, Beijing (1993) (in Chinese)
228. Long, Y., An, T.: Indexing domains of instability for Hamiltonian systems. NoDEA **5**, 461–478 (1998)
229. Long, Y., Duan, H.: Multiple closed geodesics on 3-spheres. Adv. Math. **221**(6), 1757–1803 (2009)
230. Long, Y., Sun, S.: Collinear central configurations and singular surfaces in the mass space. Arch. Ration. Mech. Anal. **173**(2), 151–167 (2004)
231. Long, Y., Zehnder, E.: Morse theory for forced oscillations of asymptotically linear Hamiltonian systems. In: Albeverio, S., et al. (ed.) Stochastic Processes in Physics and Geometry, pp. 528–563. World Scientific, Singapore (1990)

232. Long, Y., Zhang, D., Zhu, C.: Multiple brake orbits in bounded convex symmetric domains. Adv. Math. **203**, 568–635 (2006)
233. Long, Y., Zhu, C.: Closed characteristics on compact convex hypersurfaces in $\mathbb{R}^{2n}$. Ann. Math. **155**, 317–368 (2002)
234. Long, Y., Zhu, C.: Maslov-type index theory for symplectic paths and spectral flow (II). Chin. Ann. Math. **21B**(1), 89–108 (2000)
235. Marchioro, C., Pulvirenti, M.: Mathematical Theory of Incompressible Nonviscous Fluids. Springer (1994)
236. Mawhin, J., Willem, M.: Critical Point Theory and Hamiltonian Systems. Springer, New York (1989)
237. Mather, J., Forni, G.: Action Minimizing Orbits in Hamiltonian Systems. Lecture Notes in Mathematics, vol. 1589. Springer, Berlin (1994)
238. McDuff, D., Salamon, D.: Introduction to Symplectic Topology. Clarendon Press, New York (1998)
239. McDuff, D., Salamon, D.: J-Holomorphic Curves and Symplectic Topology. Colloquium Publications, vol. 52. American Mathematical Society, Providence (2004)
240. Milnor, J.: Morse Theory. Princeton University Press, Princeton (1963)
241. Mohnke, K.: Holomorphic disks and the chord conjecture. Ann. Math. **154**, 219–222 (2001)
242. Morse, M.: The Calculus of Variations in the Large. AMS Colloquium Publications, vol. 18. American Mathematical Society, Providence (1934)
243. Moser, J.: Stable and Random Motions in Dynamical Systems. Princeton University Press, Princeton (1973)
244. Musso, M., Wei, J., Yan, S.: Infinitely many positive solutions for a nonlinear field equation with super-critical growth. Proc. Lond. Math. Soc. (3) **112**(1), 1–26 (2016)
245. Niconlaescu, L.: The Maslov index, the spectral flow, and decompositions of manifolds. Duke Math. J. **80**(2), 485–533 (1995)
246. Nussbaum, R.: Periodic solutions of special differential-delay equation: an example in nonlinear functional analysis. Proc. R. Soc. Edinb. (Sect. A) **81**, 131–151 (1978)
247. Olver, P.: Applications of Lie Groups to Differential Equations, 2nd edn. Springer, New York/Berlin (1993)
248. Palais, R.: The principle of symmetric criticality. Commun. Math. Phys. **69**, 19–30 (1979)
249. Palais, R.: Morse theory to Hilbert manifolds. Topology **2**, 299–340 (1963)
250. Palais, R., Smale, S.: A generalized Morse theory. Bull. Am. Math. Soc. **70**, 165–171 (1964)
251. Paternain, G.: Geodesic Flows. Birkhäuser, Berlin (1999)
252. Perera, K., Agarwal, R., O'Regan, D.: Morse Theoretic Aspects of p-Laplacian Type Operators. American Mathematical Society, Providence (2010)
253. Rabinowitz, P.: On the existence of periodic solutions for a class of symmetric Hamiltonian systems. Nonlinear Anal. T. M. A. **11**, 599–611 (1987)
254. Rabinowitz, P.: Periodic solutions of Hamiltonian systems. Commun. Pure Appl. Math. **31**, 157–184 (1978)
255. Rabinowitz, P.: Minimax Method in Critical Point Theory with Applications to Differential Equations. CBMS Regional Conference Series in Applied Mathematics, vol. 65. American Mathematical Society, Providence (1986)
256. Rabinowitz, P.: On a theorem of Hofer and Zednder. In: Periodic solutions of Hamiltonian systems and related topics (Il Ciocco, 1986), pp. 245–253, NATO Adv. Sci. Inst. Ser. C Math. Phys. Sci., vol. 209. Reidel, Dordrecht (1987)
257. Rabinowitz, P.: Periodic solutions of Hamiltonian systems on prescribed energy surfaces. J. Differ. Equ. **33**, 336–352 (1979)
258. Rabinowitz, P.: On subharmonic solutions of Hamiltonian systems. Commun. Pure Appl. Math. **33**, 609–633 (1980)
259. Robbin, J., Salamon, D.: Then Maslov index for paths. Topology **32**(4), 827–844 (1993)
260. Robbin, J., Salamon, D.: The spectral flow and the Morse index. Bull. Lond. Math. Soc. **27**(1), 1–33 (1995)

261. Rothe, E.: Critical points and the gradient fields of scalars in Hilbert space. Acta Math. **85**, 74–98 (1951)
262. Rothe, E.: Morse theory in Hilbert space. Rocky Mt. J. Math. **3**, 251–274 (1973)
263. Ruan, Y.: Symplectic topology on the algebraic 3-folds. J. Differ. Geom. **39**, 215–227 (1994)
264. Ruan, Y.: Virtual neighborhood and pseudo-holomorphic curves. Turkish J. Math. **23**(1), 161–231 (1999)
265. Ruan, Y., Tian, G.: A mathematical theory of quantum cohomology. J. Differ. Geom. **42**(2), 259–367 (1995)
266. Schechter, M.: Rotationally invariant periodic solutions of semilinear wave equations. Abstr. Appl. Anal. **3**, 171–180 (1998)
267. Schlenk, F.: Embedding Problems in Symplectic Geometry. Walter de Gruyter, Berlin (2005)
268. Seifert, H.: Periodische Bewegungen mechanischer Systeme. Math. Z. **51**, 197–216 (1948)
269. Silva, E.: Subharmonic solutions for subquadratic Hamiltonian systems. J. Differ. Equ. **115**, 120–145 (1995)
270. da Silva, A.: Lectures on Symplectic Geometry. Springer, Berlin (2008)
271. Simon, B.: Trace Ideals and Their Applications. Am. Math. Soc. (2005)
272. Smale, S.: Generalized Poincaré's conjecture in dimension greater than four. Ann. Math. **74**, 391–406 (1961)
273. Smale, S.: Morse theory and nonlinear generalization of Dirichlet problem. Ann. Math. **17**, 307–315 (1964)
274. Solimini, S.: Morse index estimates in min-max theorems. Manuscription Math. **63**, 421–453 (1989)
275. Struwe, M.: Variational Methods. Springer, Berlin/Heidelberg (1996)
276. Su, J.: Quasilinear elliptic equations on RN with singular potentials and bounded nonlinearity. Z. Angew. Math. Phys. **63**(1), 51–62 (2012)
277. Su, J., Wang, Z.: Sobolev type embedding and quasilinear elliptic equations with radial potentials. J. Differ. Equ. **250**(1), 223–242 (2011)
278. Szulkin, A.: An index theory and existence of multiple brake orbits for star-shaped Hamiltonian systems. Math. Ann. **283**, 241–255 (1989)
279. Tian, G.: On the mountain pass theorem. Kexue Tobao. (Chin. Sci. Bull.) **20**, 1150–1154 (1984)
280. Viterbo, C.: A new obstruction to embedding Lagrangian tori. Invent. Math. **100**, 301–320 (1990)
281. Viterbo, C.: Equivariant Morse theory for starshaped Hamiltonian systems. Trans. Am. Math. Soc. **311**, 621–655 (1989)
282. Wang, Q., Liu, C.: Periodic solutions of delay differential systems via Hamiltonian systems. Nonlinear Anal. **102**, 159C167 (2014)
283. Wang, Q., Liu, C.: The relative Morse index theory for infinite dimensional Hamiltonian systems with applications. J. Math. Anal. Appl. **427**(1), 17C30 (2015)
284. Wang, Q., Liu, C.: A new index theory for linear self-adjoint operator equations and its applications. J. Differ. Equ. **260**, 3749–3784 (2016)
285. Wang, W.: Closed characteristics on compact convex hypersurfaces in $\mathbb{R}^8$. Adv. Math. **297**, 93–148 (2016)
286. Wang, W.: Non-hyperbolic closed geodesics on Finsler spheres. J. Differ. Geom. **99**(3), 473–496 (2015)
287. Wang, W.: Irrationally elliptic closed characteristics on compact convex hypersurfaces in $\mathbb{R}^6$. J. Funct. Anal. **267**(3), 799–841 (2014)
288. Wang, X.: Convex solutions to the mean curvature flow. Ann. Math. (2), **173**(3), 1185–1239 (2011)
289. Wang, Z.: On a superlinear elliptic equation. Ann. Inst. H. poncaré Anal. non Linèaire **8**, 43–57 (1991)
290. Watson, G.: A Treatise on the Theory of Bessel Functions, 2nd edn. Cambridge University Press, Cambridge (1952)

291. Weinberger, H.: Variational Methods for Eigenvalue Approximation. CBMS-NSF Regional Conference Series in Applied Mathematics, vol. 15. SIAM, Philadelphia (1974)
292. Weinstein, A.: Perturbation of periodic manifolds of Hamiltonian systems. Bull. Am. Math. Soc. **77**(5), 814–818 (1971)
293. Weinstein, A.: Normal modes for nonlinear Hamiltonian systems. Inv. Math. **20**, 47–57 (1973)
294. Weinstein, A.: Lagrangian submanifolds and Hamiltonian systems. Ann. Math.(2) **98**(3), 377–410 (1973)
295. Weinstein, A.: Periodic orbits for convex Hamiltonian systems. Ann. Math. **108**, 507–518 (1978)
296. Weinstein, A.: On the hypothesis of Rabinowitz's periodic orbit theorems. J. Differ. Equ. **33**, 353–358 (1979)
297. Weinstein, A.: Lectures on the Symplectic Manifolds. CBMS regional Reports, vol. 29. American Mathematical Society, Providence (1977)
298. Willem, M.: Minimax Theorems. Birkhäuse, Berlin (1996)
299. Xia, Z.: The existence of noncollission singularities in Newtonian systems. Ann. Math. **135**, 411–468 (1992)
300. Xia, Z.: Existence of invariant tori in volume preserving diffeomorphisms. Ergod. Th. Dyn. Syst. **12**, 621–631 (1992)
301. You, J., Zhou, Q.: Phase transition and semi-global reducibility. Commun. Math. Phys. **330**(3), 1095–1113 (2014)
302. Zehnder, E.: Homoclinic points near elliptic fixed points. Commun. Pure Appl. Math. **26**, 131–182 (1973)
303. Zehnder, E.: A Poincaré-Birkhoff type result in higher dimensions. Travaux eu Cour, vol. 25, pp. 119–146. Hermann, Paris (1987)
304. Zeidler, E.: Nonlinear Functional Analysis and its Applications. III. Variational Methods and Optimization. Springer, Berlin (1985)
305. Zhang, D.: P-cyclic symmetric closed characteristics on compact convex P-cyclic symmetric hypersurface in $\mathbb{R}^{2n}$. Discret. Continuous Dyn. Syst. **33**(2), 947–964 (2013)
306. Zhang, D.: Brake type closed characteristics on reversible compact convex hypersurfaces in $\mathbb{R}^{2n}$. Nonlinear Anal. T. M. A. **74**, 3149–3158 (2011)
307. Zhang, D.: Minimal period problems for brake orbits of nonlinear autonomous reversible semipositive Hamiltonian systems. Discret. Continuous Dyn. Syst. **35**(5), 2227–2272 (2015)
308. Zhang, D., Liu, C.: Multiple brake orbits on compact convex symmetric reversible hypersurfaces in $\mathbb{R}^{2n}$. Annales de l'Institut Henri Poincaré: Analyse Non Linéaire **31**(3), 531–554 (2014)
309. Zhang, D., Liu, C.: Multiplicity of brake orbits on compact convex symmetric reversible hypersurfaces in $\mathbb{R}^{2n}$ for $n \geq 4$. Proc. Lond. Math. Soc. **107**(3), 1–38 (2013)
310. Zhang, Q., Liu, C.: Infinitely many homoclinic solutions for second order Hamiltonian systems. Nonlinear Anal. **72**(2), 894–903 (2010)
311. Zhang, Q., Liu, C.: Infinitely many periodic solutions for second order Hamiltonian systems. J. Differ. Equ. **251**(4–5), 816–833 (2011)
312. Zhang, Q., Liu, C.: Multiple solutions for a class of semilinear elliptic equations with general potentials. Nonlinear Anal. **75**(14), 5473–5481 (2012)
313. Zhang, S.: Multiple closed orbits of fixed energy for N-body type problems with gravitational potentials. J. Math. Anal. Appl. **208**, 462–475 (1997)
314. Zhang, S., Zhou, Q.: Symmetric periodic noncollision solutions for N-body type problems. Acta Math. Sin. New Ser. **11**, 37–43 (1995)
315. Zhou, Q., Long, Y.: Equivalence of linear stabilities of elliptic triangle solutions of the planar charged and classical three-body problems. J. Differ. Equ. **258**(11), 3851–3879 (2015)
316. Zhu, C., Long, Y.: Maslov index theory for symplectic paths and spectral flow(I). Chin. Ann. Math. **208**, 413–424 (1999)
317. Zou, W.: Sign-Changing Critical Point Theory. Springer, New York (2008)
318. Zou, W., Schechter, M.: Critical Point Theory and Its Applications. Springer, New York (2006)

Index

Symbols

Mathematics Monograph Series

1. Finite Element Methods:Accuracy and Improvement (有限元方法：精度及其改善) 2006.4 Qun Lin, Jiafu Lin
2. Spectral Analysis of Large Dimensional Random Matrices (大维随机矩阵的谱分析) 2006.9 Zhidong Bai
3. Spectral and High-Order Methods with Applications (谱方法和高精度算法及其应用) 2006.12 Jie Shen, Tao Tang
4. Functional Inequalities, Markov Semigroups and Spectral Theory (泛函不等式，马尔可夫半群和谱理论) 2005.3 Fengyu Wang
5. Kac-Moody Algebras and Their Representations (卡茨–穆迪代数及其表示) 2007.3 Xiaoping Xu
6. Adaptive Computations：Theory and Algorithms (自适应计算：理论与算法) 2007.3 Tao Tang, Jinchao Xu
7. On the Study of Singular Nonlinear Traveling Wave Equations: Dynamical System Approach (奇非线性行波方程研究的动力系统方法) 2007.5 Jibin Li, Huihui Dai
8. Growth Curve Models and Statistical Diagnostics (生长曲线模型及其统计诊断) 2007.8 Jianxin Pan, Kaitai Fang
9. Theory of Polyhedra (多面形理论) 2008.2 Yanpei Liu
10. Herz Type Spaces and Their Applications (赫兹型空间及其应用) 2008.2 Shangzhen Lu, Dachun Yang, Guoen Hu
11. Some Topics on Value Distribution and Differentiability in Complex and P-adic Analysis (复与P进位分析中有关值分布及微分性的一些论题) 2008. 5 Escassut, W.Tutschke, C.C.Yang
12. Nonlinear Complex Analysis and Its Applications (非线性复分析及其应用) 2008.5 Guochun Wen, Dechang Chen, Zuoliang Xu
13. Introduction to Mathematical Logic and Resolution Principle (数理逻辑引论与归结原理) 2009.4 Guojun Wang, Hongjun Zhou
14. General Theory of Map Census (地图计数通论) 2009.5 Yanpei Liu
15. Spectral Analysis of Large Dimensional Random Matrices(Second Edition) (大维随机矩阵的谱分析) (二版) 2010.2 Zhidong Bai
16. Association Schemes of Matrices (矩阵结合方案) 2010.4 Yangxian Wang, Yuanji Huo and Changli
17. Empirical Likelihood in Nonparametric and Semiparametric Models (非参数和半参数模型中的经验似然) 2010. 6 Liu Xue and Lixing Zhu
18. Mathematical Theory of Elasticity of Quasicrystals and Its Applications (准晶数学的弹性理论及应用) 2010.9 Tianyou Fan
19. Economic Operation of Electricity Market and Its Mathematical Methods (电力市场的经济运

行及其数学方法) 2011.12 Xiaojiao Tong Hongming Yang

20. Intuitionistic Fuzzy Information Aggregation Theory and Applications (直觉模糊信息集成理论与应用) 2012.5 Zeshui Xu and Xiaoqiang Cai
21. Linguistic Decision Making Theory and Methods (基于语言信息的决策理论和方法) 2012.5 Zeshui Xu
22. Modular Forms with Integral and Half-Integral Weights (整权与半整权模形式) 2012.5 Xueli Wang Dingyi Pei
23. Finsler Geometry-An Approach via Randers Spaces (Finsler 几何——从 Randers 空间进入) 2012.5 Xinyue Cheng and Zhongmin Shen
24. Global Superconvergence of Finite Elements for Ellipitic Equations and Its Applications (椭圆方程有限元的整体超收敛及其应用) 2012.5 Zi-Cai Li Hung-Tsai Huang Ningning Yan
25. Bifurcation Theory of Limit Cycles (极限环分支理论) 2013.1 Han Maoan
26. Combinatorial Theory in Networks (组合网络理论) 2013.1 Xu Junming
27. Singular Nonlinear Travelling Wave Equations: bifurcations and Exact Solutions (奇非线性波方程：分支和精确解) 2013.6 Li Jibin
28. Planar Dynamical Systems:Selected Classical Problems (平面动力系统的若干经典问题) 2014.6 Liu Yirong, Li Jibin, Huang Wentao
29. Complex Differences and Difference Equations (复域差分与差分方程) 2014.9 Chen Zongxuan
30. The Elements of Financial Econometrics (计量金融精要) 2015.2 Fan Jianqing, Yao Qiwei
31. Mathematical Foundations of Public Key Cryptography (公钥密码学的数学基础) 2016.3 Xiaoyun Wang Guangwu Xu Mingqiang Wang Xianmeng Meng
32. Mathematical Theory of Elasticity of Quasicrystals and Its Applications (准晶数学弹性理论及应用) 2017.1 Tian-You Fan
33. Harmonic Analysis and Fractal Analysis over Local Fields and Applications (局部域上的调和分析与分形分析及其应用) 2017.3 Su Weiyi
34. Generalized Metric Spaces and Mappings (广义度量空间与映射) 2017.3 Shou Lin Ziqiu Yun
35. Introduction to Stochastic Finance (金融数学引论) 2018.12 Jiaan Yan
36. Generalized Inverses: Theory and Computations (Second Edition) (广义逆：理论与计算) 2018.12 Guorong Wang Yimin Wei Sanzheng Qiao
37. Theory and Application of Uniform Experimental Designs (均匀试验设计的理论和应用) 2018.12 Kaitai Fang, Minqian Liu, Hong Qin, Yongdao Zhou
38. Index Theory in Nonlinear Analysis (非线性分析的指标理论) 2019.6 Chungen Liu